Spatial Cognition, Spatial Perception

Mapping the Self and Space

Spatial cognition is discussed in relation to the internal mapping of external stimuli (e.g. landmarks and sensory perception of environmental information), the internal mapping of internally perceived stimuli (e.g. kinesthetic and visual imagery), and their subsequent effects on behavior. The diverse ways in which spatial information is encapsulated in perceptual and cognitive processes, allowing the self to move in space, are then examined. Major points and controversies in human and non-human animal spatial cognition, spatial perception, and landmark recognition are discussed comparatively within an evoluntionary framework. Written for postgraduate students and researchers, the authors present theoretical and experimental accounts at multiple levels of analysis – perceptual, behavioral, developmental, and cognitive – providing a thorough review of the processes of spatial cognition.

FRANCINE L. DOLINS is a Comparative Psychologist focusing on the spatial cognitive abilities of non-human and human primates in the field and laboratory, examining use of landmarks in large- and small-scale space and in simple and complex environments. Francine Dolins has related interests and publications in animal welfare, captive environmental enrichment, and conservation education, including an edited volume on societal attitudes to animals, and is currently guest editing a special issue of *The American Journal of Primatology* on conservation education. Her education was at the Universities of Sussex and Stirling in the United Kingdom, and she is currently employed at the University of Michigan-Dearborn.

ROBERT W. MITCHELL has engaged in laboratory studies of cognition in primates, cetaceans, and canids, including human interactions with these animals, and is currently studying play and other social behavior in Galápagos sea lions. His graduate education was at the University of

Hawai'i and Clark University, and he is currently Foundation Professor of Psychology at Eastern Kentucky Unversity. He has edited books on various forms of animal and human cognition, including deception, pretense, self-awareness and anthropomorphism, and is on the boards of editors of the *Journal of Comparative Psychology* and *Society and Animals*.

Spatial Cognition, Spatial Perception

Mapping the Self and Space

Edited by
FRANCINE L. DOLINS
ROBERT W. MITCHELL

CAMBRIDGE
UNIVERSITY PRESS

University Printing House, Cambridge CB2 8BS, United Kingdom

Cambridge University Press is part of the University of Cambridge.

It furthers the University's mission by disseminating knowledge in the pursuit of education, learning and research at the highest international levels of excellence.

www.cambridge.org
Information on this title: www.cambridge.org/9781107646230

First published 2010
First paperback edition 2014

A catalogue record for this publication is available from the British Library

ISBN 978-0-521-84505-2 Hardback
ISBN 978-1-107-64623-0 Paperback

Francine Dolins dedicates this book to her husband, Christopher Klimowicz, whose patience, creativity and love made it possible.

Robert Mitchell dedicates this book to his partner, Randy Huff, whose frequent wonderment ("Aren't you done with that *yet?*") provided a helpful prod toward the book's completion.

Contents

Colour plates appear between pages 320 and 321.

Contributors

Angelo Arleo
Laboratory of Adaptive NeuroComputation, ANC
Unit of Neurobiology of Adaptive Processes, UMR 7102
Centre National de la Recherche Scientifique
University Pierre & Marie Curie
University of Paris 6
75005 Paris
France

Kim A. Bard
University of Portsmouth
Department of Psychology
King Henry Building
King Henry I Street
Portsmouth, PO1 2DY
UK

Megan Bloom Pickard
Child Study Center
102 Gilmer Hall
University of Virginia
P.O. Box 400400
Charlottesville, VA 22904
USA

Sarah T. Boysen
Department of Psychology
The Ohio State University
1885 Neil Avenue Mall
Columbus, OH 43210
USA

Victoria A. Braithwaite
Institute of Cell, Animal and Population Biology
The University of Edinburgh
King's Buildings
Edinburgh, EH9 3JT
UK

Claudio Cantalupo
Department of Psychology
Clemson University
Brackett Hall
Clemson, SC 29634
USA

Vanessa Chabanne
Lab. de Neurophysiologie & Neuropsychologie (EMI-U INSERM 9926)
Université de la Méditerranée
Faculté de Médecine Timone
27 Boulevard Jean Moulin1
3385 Marseille Cdx 5
France

Ken Cheng
Department of Psychology
Macquarie University
Sydney NSW 2109
Australia

Thomas Collett
Department of Biology and Environmental Science
School of Life Sciences
University of Sussex
Brighton, BN1 9QG
UK

Sarah H. Creem-Regehr
Department of Psychology
University of Utah
380 South 1530 East, Room 502
Salt Lake City, UT 84112
USA

Judy S. DeLoache
Department of Psychology
University of Virginia
P. O. Box 400400
Charlottesville, VA 22904
USA

Elizabeth Disbrow
Center for Neuroscience
Department of Psychology
University of California
1544 Newton Court
Davis, CA 95618
USA

Francine L. Dolins
Department of Behavioral Sciences
Psychology Program
University of Michigan-Dearborn
Dearborn, MI 48128
USA

Paul A. Garber
Department of Anthropology
Program in Ecology and Evolutionary Biology
University of Illinois Urbana-Champaign
Urbana, IL 61801
USA

Paul Graham
Department of Biology and Environmental Science
School of Life Sciences
University of Sussex
Brighton, BN1 9QG
UK

Susan D. Healy
School of Psychology
University of St Andrews
St Andrews
Fife, KY16 9JU
UK

Esmé Hoban
43 Indian Hill Rd.
Wilton, CT 06897
USA

William D. Hopkins
Psychology Department
Agnes Scott College
141 E. College Ave.
Decatur, GA 30030
USA

and

Division of Psychobiology
Yerkes National Primate Research Center
Emory University
Atlanta, Georgia 30322
USA

Atsushi Iriki
Brain Science Institute, RIKEN
BSI West Building 2F
2–1 Hirosawa Wako City
Saitama 351–0198
Japan

Ronald Killiany
Department of Anatomy and Neurobiology
Boston University School of Medicine
715 Albany St.
Boston MA 02118
USA

Leah Krubitzer
Center for Neuroscience
Department of Psychology
University of California
134 Young Hall
One Shields Avenue
Davis, CA 95616
USA

Amy E. Learmonth
Psychology Department
Science Hall, Room 262
William Paterson University
300 Pompton Road
Wayne, NJ 07470
USA

Charles Menzel
Language Research Center
Department of Psychology
Georgia State University
University Plaza
Atlanta, GA 30303
USA

Emil W. Menzel, Jr., Professor Emeritus
Department of Psychology
Stony Brook University
Stony Brook, NY 11794, USA

Robert W. Mitchell
Department of Psychology
Eastern Kentucky University
Richmond, KY 40475
USA

Mark B. Moss
Department of Anatomy and Neurobiology
Boston University School of Medicine
715 Albany St.
Boston MA 02118
USA

Nora S. Newcombe
Department of Psychology
Weiss Hall
Temple University
Philadelphia, PA 19122
USA

Patrick Péruch
INSERM U 751 Epilepsy and Cognition
University of Meditterranea
Faculty of Medecine La Timone
27, bd. Jean Moulin
F-13385 Marseille Cedex 13
France

Lucio Rehbein
Departamento de Psicología
Universidad de La Frontera
Av. Francisco, Salazar 01145, Temuco
Chile

Laure Rondi-Reig
Equipe Navigation Mémoire et Vieillissement (ENMVI)
Université Pierre et Marie Curie
UMR CNRS 7102
Bât B, 5ème étage
9 quai Saint Bernard
F-75005 Paris
France

Richard Sambrook
Department of Geography and Geology
Eastern Michigan University
205 Strong Hall
Ypsilanti, MI 48197
USA

Steve Schettler
Department of Anatomy and Neurobiology
Boston University School of Medicine
715 Albany St.
Boston, MA 02118
USA

Maxine Sheets-Johnstone
Department of Philosophy
University of Oregon
Eugene, OR 97403
USA

Catherine Thinus-Blanc
Laboratory of Cognitive Sciences UMR 6146
CNRS–Université de of Provence
Marseilles
France

Luca Tommasi
University of Chieti
Via dei Vestini, 29
I-66131 Chieti
Italy

Jacques Vauclair
Department of Psychology
Université de Provence
Aix-en-Provence
France

Thomas Wynn
Department of Anthropology
Dwire Hall
University of Colorado
1420 Austin Bluffs Pkwy
Colorado Springs, CO 80918
USA

David Zurick
Department of Geography and Geology
Eastern Kentucky University
Richmond, KY 40475
USA

Acknowledgments

We appreciate the assistance of several people who helped us produce this book. Dr Martin Griffiths, our commissioning editor at Cambridge University Press, has been a constant support and dealt effectively with all the quirky and frustrating details of copyright. His assistants, Alison Evans and Rachel Eley, also helped to put this volume together. The copy-editor, Gail Welsh, ensured the high quality of this book. We are grateful for the smooth and succesful final production of this book ensured by the extraordinary work of Joseph Bottrill from Out of House Publishing Solutions Ltd. Two representatives of Eastern Kentucky University – the Dean of the College of Arts and Sciences, Dr John Wade, and the Psychology Department Chair, Dr Robert Brubaker – provided much appreciated financial support. Francine Dolins would like to acknowledge the wonderful discussions with Dr Merelyn Dolins and Dr Gabriella Day on many aspects of this volume during its creation. Francine would also like to thank Christopher Klimowicz for his patient and detailed editing of her writing.

Finally, our authors generously contributed their work and have waited patiently for the volume's completion. We gratefully acknowledge the support of all.

FRANCINE L. DOLINS AND ROBERT W. MITCHELL

1

Linking spatial cognition and spatial perception

A boundary is that which is an extremity of anything.
Euclid's *Elements*: Book 1

Space is not a "final frontier" despite the familiar phrase heard on television. In fact, a frontier depicts a barrier of sorts, a wall, the opposite of space. Without borders, space, as in Gertrude Stein's (1937, p. 289) statement, can be defined as "*there is no there there*." That is, space is defined only by what encloses it, the boundaries imposed on space. Thus, the study of organisms that move through space is the study of their perception of boundaries and landmarks, and what lies within those distances.

Boundaries can be physically continuous or a few discrete points (landmarks). However, even discrete points can be perceived in gestalt-like fashion as continuous, as encompassing or imposing a shape on space; the perception is often relative to visual overview, near or far (Poucet, 1993). Knowledge of boundaries and landmarks is not trivial in survival terms. Localizing route and place are essential for fulfilling adaptive behaviors, such as goal-directed navigation to sleeping and foraging sites, locating mates, and avoiding predators. The capacity for spatial perception and cognition underlies the necessity for spatial updating and referencing of past and present locations during movement.

Navigating organisms respond to actual space through behavioral strategies based on what they perceive as space. There may not exist a one-to-one correspondence between their internal representation and the physical world. Representations encoding the distance, angle and direction between perceived landmarks may be skewed subjectively, where the absolute symmetry and metric organization required

Spatial Cognition, Spatial Perception: Mapping the Self and Space, ed. Francine L. Dolins and Robert W. Mitchell. Published by Cambridge University Press.

for Euclidean spatial understanding is not necessarily maintained. The question is, how do these intermediary internal representations or even external, two-dimensional symbolic representations of space (e.g., map-form) function to generate spatial behavior (Golledge, 1999)? Perception and use of boundary and landmark information reflect the experience, knowledge, goals and motivation of the organism. Thus, we can examine spatial perception and cognition according to how spatial knowledge is accrued, accounting for: (1) the attention and perception of salient spatial information from which subsequent strategies emerge; (2) changes during lifespan development and from experience affecting the processes of attention and perception of spatial information and subsequent spatial strategies; and (3) the emergence of spatial strategies and behaviors in response to environmental information derived from understanding the body's location and movement in space.

Spatial Cognition, Spatial Perception: Mapping the Self and Space examines the topic of spatial cognition from two different yet closely related perspectives: the encoding in representations of both external and internally perceived stimuli, and subsequently their effects on behavior. In this volume we integrate the study of spatial perception and spatial cognition to more fully understand how visual and other types of perception incorporate spatial knowledge at multiple levels, enabling complex spatial strategies and behaviors to emerge. The emergence of these behaviors involves the perception and encoding of spatial boundaries and location of objects relationally, as well as the body's relationship to those boundaries and landmarks in space, that is, shifts between egocentric and non-egocentric perspectives of the environment. Methodologically, we will examine the study of spatial perception and cognition from behavioral, cognitive, developmental, evolutionary, neuroanatomical, and neurophysiological perspectives.

What are the origins of spatial behavior? We address this question by examining the processes of deriving externally perceived stimuli (e.g., landmarks and sensory perception of environmental stimuli) in relation to internally generated perceptions (e.g., kinesthesis, visual imagery, etc.), translated into internal representations. Our goal is to elucidate what spatial-perceptual information derived from the physical movement and placement of the self within the environment informs and relates to cognitively understanding the surrounding environment itself. A broad question arising from this idea is how spatial cognition, and perhaps all cognition, might be derived from and/or influenced by perceptual processes. To this end, there is an in-depth examination in this book of representations of the self and of cross-modal spatial information (e.g., between vision and kinesthesis).

In this volume we examine the components of spatial cognition, exploring the bases of internal spatial representations, what kinds of information populate different kinds of representations, and how these representations determine behavior and cognitive abilities and strategies. We also explore the foundations of the self and the body underlying the development of spatial understanding and skills through how these are mapped onto representations of the body and the body in space.

We have organized this book into conceptual divisions to present a scope of the current body of research and theoretical debates in how spatial knowledge is accrued, as well as to identify questions for future research. The divisions of this volume address the five following topic areas:

1. What do animals know and how do they represent external space?
2. Perception and memory of landmarks: implications for spatial cognition and behavior.
3. Evolutionary perspectives of cognitive capacities in spatial perception and object recognition.
4. Does mapping of the body generate understanding of external space?
5. Comparisons of human and non-human primate spatial cognitive abilities.

Contributing authors discuss within their chapters spatial cognition and perception from multiple perspectives, for example, the perception of extant boundaries and landmarks and how these influence behavior and cognitive processes; how encoded spatial information is organized and applied to navigate toward specific goals; and the organism's understanding of external space as defined by understanding of the body's movement and position in space.

Part I examines the kinds of information required for animals when navigating or solving spatial problems. Chapter 2 by Mitchell and Dolins discusses the historical trajectory of spatial cognitive and perceptual research. Chapter 3 by Cheng elucidates diverse models of spatial cognition and details underlying commonalities between spatial and other forms of cognition, between spatial cognition and perception. It addresses the question of where spatial cognition falls within cognition as a learning and encoding process overall and what is unique about spatial learning and cognition. A discussion of the geometry of space and the use of discrete or grouped landmarks forms the basis of the implications for spatial cognitive abilities in animals. Chapter 4 by Emil Menzel introduces some of the major questions of the book and identifies important topics related to spatial cognition and perception, discussing major philosophical questions raised by attempting to understand other species' spatial knowledge and subsequent differences in spatial strategies and navigation.

The translation from what is "actual" in the physical world to what is "perceived" is fundamental to cognitive processing and encoding in representations. From a strict Euclidean framework, perceptual and spatial information from the external world would necessarily be encoded veridically. However, it is clear that our perceptions, and therefore our representations of space, are skewed toward subjectivity (Golledge, 1999). However, a system coding for spatial relations of the environmental features must by necessity have evolved for relative spatial accuracy, as a system consistently coding spatial misperceptions would have been selected against (Rozin, 1976). The question is, how accurately and how much of the veridical world do we actually perceive and encode? How is this difference between accuracy and subjectivity negotiated in perceptual, cognitive, and behavioral outcomes?

In Part II, the initial chapter discusses the use of landmarks, geometry of space and vision in navigation (see Chapter 5). Specifically, an argument is presented for understanding spatial learning and representations through differentiating the geometric encoding of spatial information compared to that of discrete elements in the environment. In Chapter 5, Thinus-Blanc *et al.* present distinctions in performance of species for the capacity to be able to abstractly process spatial information, i.e., processing geometric and local cues, to compute novel information about their environment.

Environments afford animals perceptual information dependent on their species-specific sensory and learning capabilities and ecological and dietary requirements (see Chapters 7 and 9, this volume; also Gibson, 1986; Greeno, 1994; Wells, 2002). It is essential for most animals to forage for food, find mates, and locate secure places for rest and sleep. The environment can provide numerous potential cues to achieve these goals. Small South American tamarin monkeys, for example, rely on "visual, olfactory, and auditory cues (i.e., landmarks, smell of fruits or flowers, calls of birds or other primates) [which] undoubtedly play a role in spatial orientation and the exploitation of nearby feeding trees" (Garber, 1989, p. 212). The role of experience permits "learning associated with food acquisition and sensitivity to changes in the location, orientation, and presence of objects in the environment" (Garber, 1989, p. 212).

In examining the navigational strategies of an animal very different from primates, Collett and Graham (see Chapter 6) present experimental and ecological evidence of ant species' recognition and knowledge of routes and local environmental information. What is fascinating, as illustrated in this chapter, is that the spatial learning mechanisms animals rely upon are based on some fundamentally shared principles, such as route following using path integration and based on a sequence of perceptual cues. In the case of some ant species, these are often pheromone/olfactory-based, but can also be visual (e.g., in the case of the giant tropical ant). By distinguishing between innate mechanisms and learning processes, Collett and Graham address questions about how the ants learn routes, what information these routes are based on, and how that information is used in navigation.

In Chapter 7, differences in landmark use and memory in small-scale and large-scale space across multiple species (e.g., birds, deer, bees, etc.) are compared, providing a comprehensive overview of the types of spatial strategies used by animals under free-ranging conditions (also see Chapter 4). Spatial learning in small- and large-scale space involves distinguishing cues that are available short- and long-term, near and far. Short-term cues such as changes in the color of ripening fruit or placement of rocks or sticks may provide more precise spatial information in small-scale, local space (Poucet, 1993). However, as these are not stable over time, global, distal cues such as trees, streams and rocky outcroppings that are relatively permanent may provide more dependable although less precise spatial information in large-scale space (see Chapters 3, 6, 8 and 9, this volume; also, Spetch and Edwards, 1988; Garber and Dolins, 1996). In a prescient (1949) statement, D. O. Hebb wrote, "Now it happens that the visual activity of lower species is dominated by the perception of place. This turns out experimentally to mean a dominance of cues from remote objects instead of near ones; and remote objects provide the most stable and constant stimulation of the animal's environment" (p. 47).

Chapters 8 and 9 address questions about (1) whether animals' spatial cognitive and behavioral strategies differ when they are afforded a simultaneous overview of the entire array of salient spatial-perceptual information as compared to a subset of landmarks; (2) how spatial memory functions in differing spatial contexts (e.g., small or large, simple, complex, etc); and (3) whether these strategies are parsed for emerging spatial behaviors from the same or different types of spatial representations.

Part III explores spatial perception, object recognition and emergent spatial strategies from evolutionary and neurophysiological perspectives. With objects as the basis of landmark arrays and spatial representations, these chapters examine how landmark information is organized and the functional and evolutionary importance of optimizing navigational and spatial behavior in relation to environmental context.

The question of how perceptual and cognitive representations of space are generated leads us to examine the way in which information is perceived and encoded, and what motivates an organism to explore its environment. One interpretation posits that as the physical world is a dynamic place, an organism's perceptual and cognitive processes have to adapt accordingly on an ongoing basis (Krechevsky, 1932; Tolman, 1948; Gibson, 1986). This presupposes that organisms are either active seekers or passive recipients of information, both consequences for developing animal's spatial behavior (see Chapters 21 and 23). The degree to which learning is active also has implications for understanding an organism's attention and response to internally-dervied input from kinesthesia and proprioception in combination with that from external environmentally-based information (see Chapters 3, 4, and 5).

Each of the four chapters in Part III reflects contributions from the authors' diverse disciplines. In Chapter 10, Wynn examines the evolution of modern human's visuo-spatial cognitive capacities necessary for recognizing constructed objects such as tools, and the combined cognitive and eye-hand coordination process required for creating these objects. Constructing objects such as tools requires use of egocentric and allocentric frames of reference by the individual creating the tool. Wynn discusses how object-centered spatial information about tools (and other objects), such as details of their size and shape, are linked to a perceptual frame of reference which includes the self and external stimuli. This information is typically perceived visually (Paillard, 1991) but can be also perceived by echolocation (see Chapter 12, this volume; Altringham, 1996; Price *et al.*, 2004) or haptic (tactile) perception (see Chapters 2 and 14). Haptic, auditory, as well as olfactory perceptions can translate into visual expectations that can be spatial, providing information of distance, angle and direction for localizing cues. These perceptual modalities provide bases that can be translated into behavioral strategies and offer animals means for understanding and taking action in the external world (see Chapter 12, this volume; also Garber and Dolins, 1996).

Species' perceptual modalities are constructed within the constraints of morphological and species-specific requirements and capabilities, and are often hierarchically organized in relation to the saliency of the perceptual cues (see Chapters 4 and 8, this volume; also Garber and Dolins, 1996). For instance, microchiropteran bat species as well as some birds (e.g., swiflets) navigate via echolocation and are therefore highly dependent upon their sense of hearing, while others such as the Megachiroptera rely principally on sight and smell (for bats, see Altringham, 1996; for birds, see Price *et al.*, 2004; see Chapter 12 on cetacean echolocation). Some sensory perceptual systems will be more relevant than others in allowing an animal to respond behaviorally to environmental contexts. For example, while rats use vestibular, auditory, olfactory and tactile sensory information to navigate, rats are highly reliant upon visual information even with their limited visual acuity (see Chapter 23, this volume; Zoladek and Roberts, 1978; Leonard and McNaughton, 1990). Visual information permits distance estimations whereas tactile and olfactory information enable only non-localized gradient estimations to be performed (see Chapter 6). There is also an inherent hierarchy and interplay of information derived internally and externally about space. Visual information and kinesthesis usually work in tandem, but can also be conflicting. The result leads to a prioritization in attentional capacities (see Chapter 15).

Both the integration of sensory systems for localization of cues and navigation in a species relying on perception of a different kind than that used by humans (i.e., cetaceans) is discussed in Chapter 12. For terrestrial animals, a cue may be a single distal landmark, such as the sun, or multiple landmarks, as in a constellation of stars or a mountain range. Across species and environments, cues can be obscured or hidden from direct perception. Learning about obvious or hidden cues involves generating a landmark-to-desired-object representation. Tasks evaluating understanding of object permanence in relation to cue/object localization can be used to show development of different forms of spatial learning. In this chapter, object permanence and cue localization are also discussed in relation to the hierarchy of some sensory systems over others for different species.

Areas of neural processing of spatial information are distributed corresponding to the type of spatial responses employed by an animal. Evidence shows that the hippocampus and related structures are selectively activated when animals are engaged in different types of spatial strategies (Holdstock *et al.*, 1999; Maguire *et al.*, 2000). In rats the CA1 of the hippocampus is more active for a response strategy, whereas CA3 and dentate gyrus are activated for a place strategy (Miranda *et al.*, 2005). Spatial experience also alters the neural areas responsible for processing spatial challenges encountered in the environment. Maguire *et al.* (2000) found that London taxi drivers with extensive navigational experience had a significantly larger posterior hippocampus than non-taxi driver controls. Moreover, the navigational performance of experienced taxi drivers was more efficient and optimal around a virtual London, indicating that experience modifies the hippocampal region responsible for place learning, resulting in greater behavioral flexibility and enhanced navigational skills and spatial memory. The authors write that, "differential changes in posterior and anterior hippocampus may represent two separate processes … our findings reflect an overall internal reorganization of hippocampal circuitry in response to a need to store an increasingly detailed spatial representation, where changes in one hippocampal region are very likely to affect others" (p. 4402).

The role of the hippocampus in relational as well as associative learning observed in the spatial domain extends more generally (Hartley *et al.*, 2003). Animals and humans with hippocampal damage show selective deficits in both the spatial and nonspatial domains in the ability to form associations between previously unrelated items or between items and context, even though the ability to learn single items in isolation is relatively spared (Gaffan *et al.*, 2002; Henke *et al.*, 2003). In addition, the processing in the hippocampus also incorporates temporal information into memory, thus contributing to the formation of memories of the sequence of events, a foundation for relational learning (Fortin *et al.*, 2002). Evidence from a radial maze task suggests that damage to the hippocampus impairs an animal's ability for sequential spatial learning (DeCotaeu and Kesner, 2000). For example, rats with hippocampal damage were selectively impaired on learning sequential orderings of odors despite their intact ability to recognize recently encountered odors (Dusek and Eichenbaum, 1997; see also Alvarez *et al.*, 2002).

Forming associations of objects and events in spatial and temporal contexts are two component abilities of relational learning. While lesions to the hippocampus are known to impair relational learning in spatial tasks (e.g., Aggleton and Brown, 2005), evidence also indicates that the hippocampus plays an important role in the formation of episodic memory. Episodic memory can be considered a foundation for relational memory in that it invokes associations of objects and events in spatial and temporal contexts, linking them into a large-scale framework that allows for inferences and generalizations from organized relations to new situations (Eichenbaum, 2004). Consistent with this idea, damage to the hippocampus has been shown to impair function of episodic memory despite largely spared semantic memory and simple recognition memory, which does not involve associations (Vargha-Khadem *et al.*, 1997; Holdstock *et al.*, 2004). In a study of sequential learning, hippocampal-damaged rats failed to make transitive inferences applicable to normal novel stimuli, which indicates that episodic relational memory is also impaired due to hippocampal lesions (Dusek and Eichenbaum, 1997).

In rats trained to perform identical olfactory recognition tasks in different spatial locations, the neural activity of most of the hippocampal cells was selective to particular combinations of spatial locations and nonspatial events (e.g., perceptual, behavioral, cognitive) (Wood *et al.*, 1999). The evidence suggests that the hippocampus is capable of encoding discrete episodes at the same time as representing features common/overlapping in different episodes and environments.

Thus, the hippocampus' role in associative, sequential and relational memory in the spatial domain seems to generalize to memory in general (Eichenbaum, 2000a). In producing more complex spatial strategies, the hippocampus is crucial for encoding relational information and for place learning. The hippocampus plays a role in forming associations between spatially present stimuli and the incorporation of temporal information in relational episodic memory. However, although accumulating evidence suggests the involvement of the hippocampus in general episodic memory, the precise degree to which it makes contributions is still not clear (Burgess *et al.*, 2002; Eichenbaum, 2004). For instance, Kumaran and Maguire (2005) showed that the hippocampus was selectively activated by a spatial navigation task but not by an equivalent navigation task in the nonspatial, social domain (see also Kumaran and Maguire, 2006). Nevertheless, the hippocampus does provide links between discrete objects and events in their spatial and temporal contexts. These links aid in the unification of spatial–temporal information into a relational representation in the spatial domain, fundamental for coherent and goal-oriented complex spatial behaviors to be performed.

In their discussion of the significant role of the hippocampus in spatial learning, Rehbein *et al.* (Chapter 11) present evidence based on investigations of nonhuman primates' spatial processing from a comparative perspective. Recent studies have added layers of understanding about the specificity of neural processing pathways for spatial learning, spatial cognition, and memory in various cortical regions and, in particular, in the hippocampus (e.g., Eichenbaum, 2000b; Ekstrom *et al.*, 2003; Maguire *et al.*, 2003). Processing of spatial information and generation of spatial representations in neuroanatomical structures appears to be specific to the type of spatial learning, with each area processing and contributing input for different types of spatial behaviors, such as in route, place or response learning. For example, in response to dynamic changes in an environment, separate neural pathways are activated when an individual plans a new route than when following one previously known (Spiers and Maguire, 2006).

Using neural processing as the basis for robotic spatial cognition, Arleo and Rondi-Reig (Chapter 13) present an excellent review and highlight the integration of sensory processes comparing real and artificial organisms in generating navigational strategies. These authors provide evidence of spatial learning in neuro-mimetic robots in a series of experiments in which visual cues were constant while the route to the goal was randomized within the familiar environment. They demonstrate that neuro-mimetic robots learned to distinguish between idiothetic and allothetic signals, thereby computing novel allocentric routes to known locations. Additional experiments with variations on origination points and landmarks are also discussed, with the implication of disorientation emerging from the de-coupling of the idiothetic and allothetic signals, interrupting the mapped representation of these landmark relations. The results from the neuro-mimetic robots are compared with biological models and ethological and neurophysiological evidence.

Part IV presents research and discussion about how humans and animals use their own interpretation of bodily space to understand their position in space and body movements (internal stimulus mapping). A relevant question is whether or not an individual's internal experience of their own body movements provides a developmental basis for their ability to learn about external space. This question is explored in developmental contexts in the next and final section of the book.

Gibson (1986) proposed that perception is not created through an individual's static picture of the environment and translated through the construct of an abstract notion of space; instead the environment is where the individual dynamically samples points in its world to create action, experiencing a "visual array," "olfactory array," or other perceptual arrangement, thereby discovering an "invariant structure" (a frame of reference) of the environment. Gibson states, "The environment of animals ... is what they perceive ... [it's] not the same as the physical world, if one means by that the world described by physics" (1986, p. 15). Similarly, Creem-Regehr (Chapter 17) sums up an individual's interaction of perception and action in the environment through the concept of "embodied cognition." Gibson (1986) writes,

> we understand ... an entirely new way of thinking about perception and behavior ... in which animals can move about (and in which objects can *be* moved about) is at the same time the medium for light, sound, and odor coming from sources in the environment ... [which function as] possible point[s] of observation ... [and] are continuously connected to one another by paths of possible locomotion. Instead of geometrical points and lines, then, we have points of observation and lines of locomotion. As the observer moves from point to point, the optical information, acoustic information, and the chemical information change accordingly. Each potential point of observation in the medium is unique in this respect ... therefore, [this] is not the same as the concept of space inasmuch as the points in space are not unique but equivalent to one another. (p. 17)

In this way, an individual's discovery of invariant structures or elements in its environment creates possibilities for the generation of novel spatial information based on perceived parameters, thus permitting action to occur (i.e., affordances).

The investigation of the interactive phenomenon of knowledge of the self in relation to external space is explored via the role of neural processing in the parietal cortex. Multiple regions of the parietal cortex process various aspects of navigational behavior. Regions are differentiated by when the action can or does take place. The medial parietal regions appear to play an active role when an individual is actually moving through space, while the lateral parietal regions are important for intentional planning of movement through a part of the environment that is out of sight (see Chapter 16, this volume; also Spiers and Maguire, 2006). Levels of information are necessary to respond to different types of cues in the environment, for example, a beacon or a landmark array are integrated from multiple sources of encoded information. Spatial information about the self and others is also processed in the parietal cortex (see Chapters 13, 15, 16, 17 and 18).

Creem-Regehr (Chapter 17) discusses the development of body mapping and spatial representations for understanding of the body's location in external space, specifically linking the representation of the body schema to that of embodied cognition. Iriki (Chapter 18) presents experimental evidence of neural correlates in the parietal cortex processing maps of the body in space. Specifically, he describes mapping within the parietal cortex not only of arm movements but also of arm movements that extend beyond the body's own spatial sphere with the use of tools and objects.

Krubitzer and Disbrow (Chapter 16) outline evidence in support of cortical representations and fields processing digit and hand movement in space and discuss this evidence within an evolutionary framework. Recognizing whether images are of left or right hands or feet apparently requires an imagined egocentric placement of oneself in the image, whereas recognizing whether non-body objects are the same or inverted requires imaginary movement of the object (see Chapters 17 and 19). Krubitzer and Disbrow's chapter, in conjunction with Chapters 17 and 18 on the movement of body and body part representations in space, incorporates elements of imagining one's body moving in space. Action, or imagining action, according to these authors, is deemed necessary as an organizing factor in facilitating cognitive thought (see also Wilson, 2002). While imagining an action differs from physical actions in the environment, there are some fundamental neurophysiological processes in common. Although this volume does not have a focus on the integration of motor control and spatial cognition, it is recognized that this is an integral component in spatial behavior as it is enacted in the environment (Schmahmann and Caplan, 2006). Locomotor behavior is an inherent part of planning navigation and goal-oriented behavior, while motorically driven behavior helps to provide both feedforward and feedback information from the environment. The role of the cerebellum has more recently been identified as crucial to spatial and other forms of cognition:

> There is increasing recognition that the cerebellum contributes to cognitive processing and emotional control in addition to its role in motor coordination ... cerebral association areas that subserve higher order behaviour are linked preferentially with the lateral hemispheres of the cerebellar posterior lobe – in feedforward loops via the nuclei of the basis pontis, and in feedback loops from deep cerebellar nuclei via the thalamus. There are also reciprocal connections between the cerebellum and hypothalamus. These pathways facilitate cerebellar incorporation into the distributed neural circuits governing intellect, emotion and autonomic function in addition to sensorimotor control ... [I]n the description of the cerebellar cognitive affective

> syndrome (CCAS) . . . in patients with lesions confined to the cerebellum . . . [are] characterized by impairments in executive function (planning, set shifting, verbal fluency, abstract reasoning, working memory), spatial cognition (visual spatial organization and memory) and linguistic processing (agrammatism and dysprosodia) when the lesions involve the hemispheric regions of the cerebellar posterior lobes. (Schmahmann and Caplan, 2006, p. 290)

Symmetry of most species' body plans is evident by a clear midline and a mirror-image distribution of limbs, eyes and other body parts located in the left/right center of the face/body (e.g., the mouth and nose). Despite this apparent symmetry, most humans are considered to be right-handed for many tasks. Handedness is associated with functional and processing differences in the cerebral hemispheres. Hemispheric specialization in humans has often been cited as uniquely associated not only with handedness but also with tool use and production, intentional gesture, language and other complex cognitive abilities (Corballis, 2002; Leavens *et al.*, 1996; Hopkins and Leavens, 1998; Hopkins *et al.*, 2007). From the subject's perspective, left–right discrimination can either be judged egocentrically or allocentrically, with egocentric left–right knowledge foundational for an allocentric discrimination as in left–right mirror-image discriminations. Given no extrinsic environmental cues to identify left from right, an animal's ability to discriminate left–right is of particular interest to researchers. Links have been proposed for human functional hemispheric specialization as the basis for the ability for left–right discrimination (e.g., Corballis and Beale, 1970). It is hypothesized that together with binocular vision, hemispheric asymmetries characterize the ability to make left–right discriminations, as is evident with humans (Corballis and Beale, 1970). However, functional and neuroanatomical evidence suggests that many vertebrate species do possess hemispheric asymmetries (e.g., Hopkins *et al.*, 2003), and should also be able to make left–right discriminations. This topic is examined in Chapter 19 particularly with reference to whether nonhuman primates possess the capacity for left–right discrimination.

It is often overlooked that proprioception and kinesthesis can function within spatial perceptual processing, and organisms' understanding of external spatial dimensions through body movement is often omitted in discussions of spatial cognition, an omission that this book aims to rectify. In an exploration of how bodily movement is generated, Sheets-Johnstone discusses the relationship between movement and kinesthesis (Chapter 14). The understanding of kinesthesis and perception via various sensory modalities provides insight into the relationship between body and space, body in space, and body moving in space (see Chapters 15, 16, 17 and 18). The tactile–kinesthetic spatiality described by Sheets-Johnstone during human development begins with (in her view) an infant's implicit sense of its body as moving in an expanse. The body image is a representation of the body, its parts, and the relations among these, and the body schema. From Sheets-Johnstone's description, the body schema is usually viewed as combining elements of body percept and body image (also see Chapters 15, 17 and 18, this volume). At the same time, the body schema can be extended to include tools, or restricted to remaining limbs following amputation (though phantom limbs can retain the somasthetic feeling of limbs) (see Chapters 10, 15 and 18).

Spatial representations based on one sensory modality can be usefully translated into other modalities, such as proprioception and kinesthesis. Functional equivalence, where encoded information is equally distributed within different areas of the brain (Squire, 1987), allows for behavioral flexibility in response to environmental stimuli. Evidence from studies with blindfolded but sighted humans (Landau *et al.*, 1981) and animals blinded after being trained (Watson and Carr, 1908) show that spatially determined behavior, even though learned visually, can be competently executed without visual processing. If one type of modality-specific information is lacking or unavailable the representation can still be a "useful" reference to guide spatial behavior, as other modalities such as proprioception can reinterpret or add information to make up for the deficit.

Expanding on concepts about movement, Mitchell (see Chapter 15) examines the relationship between kinesthesis, proprioception (somasthesis) and visual matching of movement for understanding of body location in relation to external markers. Other modalities allow animals to experience and comprehend their internal state. Kinesthesis, a combination of somasthesis, the nonvisual perception or feel of the body's form, and proprioception, the perception of muscle position and movement, provide a spatial map of exterior bodily movements and location in space. While this map may be inaccurate (i.e. subjective), as occurs with phantom limbs or kinesthetic–visual misalignments, this does not refute its usual mapping qualities just as visual illusions do not reflect the general accuracy of vision.

Experimental work concerning mental imagery and other kinds of representations (including symbolic) highlights the complexity, establishment and utilization of perceptual information as incorporated into spatial cognition and behavioral strategies (see Chapters 20 and 21). In the final section of the book, Part V, authors address issues from comparative, anthropological, and developmental perspectives about the type of information encoded in representations, how these representations are formed given specific environmental information, and the efficiency with which they are subsequently applied for spatial strategies. Maps created by people are external representations of place learning, displaying a useful allocentric alignment between the world and a scaled-down representation (see Chapters 20, 21, 22 and 23). As described in Chapter 20 by Sambrook and Zurick, these representations include real, two-dimensional maps as imagined and created by humans (for contrasting opinions on adaptationist views of sex differences in human spatial cognitive abilities see, for example, Wynn *et al.*, 1996). Maps, real and also imagined (represented), embody not only physical and external markers but also social and political confounds of space. These issues are discussed with relevance to the use of maps, history and human spatial cognitive abilities.

Spatial information present in most perceptual processes is used in cognition about the self and the physical environment. We represent space utilizing our perception of it, which is incorporated into our imagination. We can mentally transform space by restructuring perceptual components encoded in memory (see Chapter 9). The transformation is then employed in spatial cognition to solve real world problems (see Chapter 20). Perception and representation are integrally connected:

> attention to the motor image is itself a step towards bringing the movement about ... Every image has, in fact, for its physiological basis a very faint form of the same kind of excitation as the corresponding percept has. If I imagine a touch on the tip of my nose, the same kind of nervous change takes place there as if I had actually been touched there, only in a much weaker degree ... So when I imagine a movement, the same nerves and muscles are affected as when the actual movement takes place. (Ryland, 1909, pp. 168–169)

Ryland's 100-year-old summary posits a mutually influential and dependent relationship between perception and mental imagery. For example, visual-spatial memory, when experienced, is dimensional, much as is visual experience of the external world. Imagery is a kind of "perceptual priming" (Pearson *et al.*, 2008):

> imagery can have a pronounced facilitatory influence on subsequent conscious perception ... [and] in the absence of any incoming visual

> signals, leads to the formation of a short-term sensory trace that can bias future perception, suggesting a means by which high-level processes that support imagination and memory retrieval may shape low-level sensory representations. (pp. 982, 986)

Mental imagery may function as an essential component for generating spatial memory, spatial planning and navigation, tool production and use, as well as some nonspatial cognitive functions, for example, language and social comprehension (see Chapter 10, this volume; also, Pearson *et al.*, 2008). Spatial cognition, and perhaps all cognition, seem to derive from, and rely on perceptual processes.

The active process of imagery or visualization is often used to enhance physical performance. Of the various ways visualization can be used, the "cognitive specific" method involves imagining oneself successfully performing a specific set of motor actions, equivalent to a skill. This requires a realistic self-image of going through the motions and activating the necessary muscle groups for the skilled action to take place, imagining oneself in the active manner albeit in a relaxed, controlled state and with a positive outcome. Sensory stimuli from tactile, visual and kinesthetic experiences, emotional pleasure at the successful conclusion of an action as well as the timing of the action, create the basis of visualized imagery. The more accurate the sensory stimuli the individual can visualize during the imagined motor skill, the more successful that individual will be when physically carrying out an action such as serving the tennis ball or turning multiple pirouettes. Skilled sports players, for example, describe success in using imagery as similar to watching a very detailed and realistic film of themselves being played out, where they see their own perfect performance take shape before they take any action (see Chapter 10, this volume; also Morris *et al.*, 2005). Thus, matching between sensory modalities, especially between kinesthesis and vision, allows for a complex cognitive underpinning of the body's spatiality (see Chapters 14, 15, 16, 17 and 18).

Elaborate exploration, particularly during early developmental stages, should result in efficient spatial learning. In many species the ability to recognize spatial configurations from many orientations should afford an individual flexible navigation within a familiar environment. An animal's early experiences modify the neural structures involved in perception and learning dependent upon that function, thereby altering the strategies available by which to respond. Locomotor experience contributes to the development of increasingly complex spatial skills in normal infants (see Chapters 18, 21 and 23, this volume; also Acredolo, 1990). By contrast, a handicapped infant without self-mobility exhibited a delay in development of spatial cognitive abilities (Bertenthal *et al.*, 1984). It appears that limitations in self-directed locomotor activity lead to limitations in spatial relational representations. Exploration and motoric experience may be equally important in the development of body representations, spatial representations, sophisticated spatial strategies, and computation for taking action in the environment (see Chapters 5, 10 and 17). These early experiences and potential sensory-perceptual limitations have implications for the type of spatial representations that can be generated as well as spatial strategies employed. For example, while survey/place information includes an overview of veridical spatial relationships among locations, route information includes within it aspects of dead reckoning, which predictably would be more salient for those relying on visual or tactile information than other sensory modalities (see Chapters 3 and 20, this volume; see also Vanlierde and Wanet-Defalque, 2004; Noordzij *et al.*, 2006). A study using a visuo-spatial memory task to generate mental representations from verbally presented two-dimensional patterns in a grid (i.e., auditory information) tested subjects with early or late onset of blindness as well as visually competent subjects; they all showed equivalent performance (Vanlierde and Wanet-Defalque, 2004). Early blind subjects, however, described their problem-solving strategy as using an X, Y coordinate system to encode locations within the mental representation of the pattern grid (comparable to route learning). In comparison, late blind and sighted subjects relied on visuo-spatial imagery that included only the relevant information, comparable to place learning (see also Noordzij *et al.*, 2006).

Learmonth and Newcombe (Chapter 23) focus on developmental aspects of place learning within the context of four types of spatial learning they describe – place, cue, response learning and dead reckoning/path integration – where the development of allocentric and egocentric frames of reference are differentiated in the developmental/learning process (Gallistel, 1990). These authors compare human and rat spatial skills examining at what ages and cognitive abilities specific types of spatial strategies emerge in the development of each species.

Chapters 21 and 22 present evidence based on similar paradigms testing children of varying ages and adult chimpanzees. They examine perception of two-dimensional and three-dimensional space and use of a small-scale spatial model of objects in a room in relation to a real-sized room. DeLoache and Bloom modified the room-scale model to signify symbolic information, thus not only testing children's small-scale spatial knowledge but also two-dimensional symbolic representations of large-scale space where map and real-world spatial relations are maintained in both contexts. These chapters compare evidence for each species and children of different ages. The success and performance on models of space in relation to physical space has implications for how internal mental representations of real space are generated in each species and age group.

Scale is intriguing in all frames of reference, in that recognition of "identity" of objects and locations in differently scaled contexts, as well as translation between frames can be problematic. Scale, small and large, is determined by the relationship and distance among objects/landmarks, creating proportion to the representation it represents (see Chapters 4, 5, 6 and 7 *et al.*). Animals using landmarks as guides must be able to recognize the identity between smaller-appearing objects from a distance with those same objects close-up. Perception provides the algorithm mapping the distant version of a landmark with the closer version. Adult humans report ease in the ability to imagine a distant object as a small variant of a normally sized, closer object (e.g., people appearing tiny when viewed from a distance). Parallels occur in viewing a scale model, a variant of the normal-sized object it reflects. Under some testing conditions, small children and chimpanzees show some difficulty solving real-world scale problems to the same degree of performance as adult humans. However, with changes in conditions, some of these performance factors decreased (see Chapters 21 and 22, this volume).

Our perceptual system takes the distance of ourselves from an object into account in determining its likely size. Similarly, the typical human adult imagination about perceptual stimuli allows us to recognize scale models with ease. Adult humans use scale models for a variety of purposes including architectural creation, imaginary games and aesthetic enjoyment (Akre, 1983; Kurrent, 1996; Jodidio, 2007). They also create and use three-dimensional spatial models and two-dimensional symbolic representations of space (maps) for goal-directed behaviors and navigation (see Chapter 20). Young children perform successfully when presented with a scale model's relation to a larger "real" space as a problem of connecting a perceptually smaller version of the same space to a larger version, as in a photograph, videotape, or following "miniaturization" (see Chapter 21). Similarly, in the studies described by Boysen and Bard, chimpanzees perform well once the nature of the task is made clear to them (see Chapter 22, this volume). Findings that very young children do not see the relationship between models and modeled, and even older children need the identity between scale models to be specified for comprehension, lead DeLoache and Bloom Pickard (see Chapter 21, this volume) to posit that understanding of models requires understanding of symbols where the model is a symbol of the modeled. To date, as with young human children, young chimpanzees have not been tested for a developmental progression from an initial lack of performance of understanding scale models in relation to real space to that of understanding the models' significance. The question is whether chimpanzees understand models as symbols or use some other means to infer relations between the models and physical rooms/objects. We fully expect the developmental changes similar to those observed in children will be found once the scale-model studies are tested with very young chimpanzees, but this type of study has not yet been done.

How symbol-competent chimpanzees generate internal spatial representations and optimize routes or localize hidden, preferred food items is the topic of Charles Menzel's chapter (Chapter 24). Menzel focuses on chimpanzee memory recall of the types of information in small-, medium- and large-scale spaces for flexible navigation and spatial decision-making. The author describes use and discrimination of referential symbols, referred to as lexigrams, to assist in efficient navigation in relation to bonobo and chimpanzee ecology and memory. A fascinating study is also described with the female chimpanzee, Panzee, recruiting naïve human helpers to assist in obtaining hidden food and non-food items out of her reach and at times visually occluded by trees and leaves. She did so unprompted by any researcher or helper present. Panzee's gaze, intentional pointing and gestures provided precise direction and location of hidden food items not visible to the naïve helper. Moreover, in trials with presentation of hidden item locations from video representations, some of which were from angles not normally accessible to Panzee, her performance did not decrease or alter significantly in efficiency and success. Labeling items, some quite rarely seen by Panzee, using lexigrams was also extremely accurate. In conjunction with the earlier chapters presented in this volume, an intriguing picture of nonhuman primate, and particularly great ape, spatial abilities emerges.

POSTSCRIPT

Some terms and concepts used in the discussion of spatial cognition and perception

As editors, we would like to raise a few points about concepts presented and terminology used throughout this volume. References to concepts such as spatial strategies, representations, and navigational and spatial frameworks are used with certain assumptions attached. Authors have used often terms in a specific and distinctive manner stemming from their disciplinary viewpoint, although it is important to point out that authors' approaches to issues in spatial cognition and perception from different disciplines adds to the richness of knowledge on these subjects. In this postscript we briefly clarify types of spatial strategies, frames of reference, internal spatial representations, and some terminology as used in discussions of spatial cognition and spatial perception.

FOUR SPATIAL STRATEGIES BRIEFLY OUTLINED

There are four systems in which spatial information relevant to navigation is represented. These systems are associated with corresponding emergent behavioral/navigational strategies, which are: (1) dead reckoning; (2) use of a single cue or beacon; (3) use of sequential cues or route following; and (4) place learning (see Chapters 3, 4, 5, 6, 7, 8, 9, 13 and 23). These strategies fall along a continuum from the self as a cue (as in dead reckoning), to the self in response to a discrete cue (a beacon), to the self in response to individual cues within a sequence, and finally to navigational strategies based on the relational and simultaneous use of multiple cues perceived in an array or configuration (place learning).

Dead reckoning is the ability to compute on an ongoing basis the direction, speed and physical effort of movement from one point to another. This computation is measured internally by 'distance as effort' without correspondence to the external environment, that is, by the animal mapping spatial information relative to its own movement (Nadel, 1991; Dolins, 1994; Garber and Dolins, 1996; Waller *et al.*, 2008). Even though with dead reckoning the frame of reference is the individual's own body moving through space with respect to itself, it is typically used in conjunction with at least one external cue to so that the individual can halt at a destination. Corresponding to the second type of spatially represented information is the use of one landmark or cue (at a time) in the environment as a beacon, which the individual has learned to use in order to orient its own position in space with respect to that particular object. Both dead reckoning and beacon strategies (often used together) are categorized as egocentric frameworks.

Route-following based on a strict sequence of discrete cues is slightly more complex to categorize. It can be divided into an egocentric and non-egocentric (allocentric) type of framework, depending on how the information is encoded and then employed in a navigational strategy. Egocentrically framed route-following is where each consecutive cue within the sequence elicits an individual spatial behavioral response (e.g., at the tree stump turn right). The organism relies rigidly on these cues for directional information, but must be able to discriminate one cue from another to respond appropriately within the sequence. These cues may be visual, tactile or olfactory (e.g., pheromonal) (see Chapter 6) and encoded in what is referred to as a strip map (Collett *et al.*, 1992; Nicholson *et al.*, 1999; Collett and Collett, 2000 and 2002; Collett *et al.*, 2001).

Route-following based on an allocentric framework represents a more complex set of information for the organism but also provides greater behavioral flexibility. This kind of representation is referred to as a topological map, and is where pre-set, intersecting routes (like the London Underground map) are available to the individual. Novel routes can be generated along these pre-set routes by using familiar landmarks common to more than one pre-set route (i.e., nodes). These nodes provide the opportunity to transfer from one route to another; however, the individual cannot stray from the pre-set routes. Although more restricted than a geometric relational representation (cognitive map), in this regard a topological representation affords certain transformational rules; that is, the organism may start and end a navigational bout from many locations within a represented space, allowing for computation of novel routes. Thus the topological map is to some degree also a relational representation (for further discussion of topological maps, see Chapter 8, this volume; Dolins, 1994; Garber and Dolins, 1996; Dolins, 2009).

Finally, there is a veridically represented allocentric framework, as Tolman (1932, 1948) coined the term, a "cognitive map," or a geometric representation (Gallistel, 1990; Nadel, 1991). This spatial system involves more complex use of information in which the individual relies on the spatial relationships amongst several landmarks simultaneously. Such a map can be defined as an internal representation in which an organism is able to encode and recall for use the veridical and non-egocentric spatial relationship between two or more cues in order to localize additional points in space to generate novel spatial information/routes (O'Keefe and Nadel, 1979). This requires imposing a configural shape composed of discrete landmarks to demarcate a particular space.

An array of landmarks can either be perceived as multiple individual landmarks or as a geometric shape imposed on a configuration of landmarks. The perceived shape is based on the spatial relationships between the landmarks (Cheng, 1994; Spetch *et al.*, 1997). The type and placement of landmarks play a role in the type of internal representations formed and subsequent spatial behaviors exhibited.

The geometric shape of an array under a rotation can be a guide to the location of a hidden goal as long as the goal remains in a constant position with regard to the landmarks. For example, if the goal is located in the center of the equilateral triangle, then the "middle-rule" can be applied, that is, the rule to search in the computational center of the configuration regardless of whether the array has been rotated or from which direction a traveling animal has approached that configuration of landmarks.

If an array is asymmetrical (e.g., an isosceles triangle), then localization of the goal can be derived using the shape of the configuration regardless of perspective or rotation of the landmarks comprising the array. However, if a configuration is composed entirely of unique landmarks in a symmetrical or asymmetrical pattern, a relational strategy is possible although not the only alternative (Cheng, 1994; Spetch *et al.*, 1997); the animal could make use of unique landmarks as beacons in a vector strategy to localize the goal (see Chapters 3, 5 and 6, this volume *et al.*; Cheng and Newcombe, 2005; Dolins, 2009). To note, the terms *local* and *global* are contrasted by Collett and Graham (Chapter 6) in discussing vectors, where local vectors specify direction of movement from present location, and global vectors specify movement toward a goal.

FRAMES OF REFERENCE

The frames of reference organizing spatial information are integral to how we understand an organism's representation of, and response to, its own internal information and that in the external environment (see Chapters 4 and 6). Conceptually, frames of reference can be thought of as reference coordinate systems. A frame of reference is an active process: how information is perceived or framed affects how it will be used in spatial strategies, in updating present actions (navigational decisions), and in creating an organizational basis for future spatial behavior (Wickens *et al.*, 2005).

Frames of reference are divided into egocentric and allocentric. An egocentric (or intrinsic) frame of reference is based on the perspective of the perceiver in strict spatial relation to internal perceptions or to external objects or locations. An allocentric (or extracentric or extrinsic) frame of reference represents external objects and locations as defined by relationships amongst these discrete points, and is independent of the perceiver's spatial perspective (Klatzky, 1998; also see Chapters 13, 16 and 18, this volume).

An egocentric frame of reference can focus on the entire body, on body parts such as a shoulder or hand or on objects relative to the body, for example to the left or right (see Chapters 13, 14, 15, 16, 17, 18 and 19, this volume; also Gurfinkel and Levick, 1991). In this way, an egocentric framework uses self-movement in relation to the body (dead reckoning) or to static external reference points such as a tree or boulder as a beacon. Similarly, multiple sequential discrete points can integrated within an egocentric framework in route following/path integration (see Chapter 6).

In contrast, an allocentric framework is relational, not dependent on the self to provide distance, angle and direction to an object or multiple objects within a space, or to define a route (see Chapter 8). Allocentric frameworks (geometric, relational representations) provide the potential for generating novel routes during navigation. This process requires using cues in an array in related coordinate map-like points, allowing for precise computation of direction, angle and distance in localizating a goal and generation of novel spatial information (see Chapters 6 and 8, this volume; also Gallistel, 1990; Dolins, 2009).

INTERNAL AND EXTERNAL MAPPING

How do egocentric and allocentric frameworks relate to an understanding of the body in space? An egocentric framework changes with the body or body parts' movement, but is relatively fixed for any given bodily position (Gurfinkel and Levick, 1991; Paillard, 1991), whereas an allocentric frame of reference varies with regard to changes in position and location of the individual perceiver.

Some authors in this volume refer to an egocentric frame of reference when referring to the whole body as a "body schema." Looking across chapters for comparison, different usages of the terms "body schema," "kinesthesis," "body percept" and "body image" can cause confusion in how they are applied. Kinesthesis includes the nonvisual perception of the body's spatial extent (somasthesis) and the position of its parts in relation to each other (proprioception). The body percept incorporates sensory information about the body, including kinesthesis and vestibular sensations of body position. The body image is a representation of the body, its parts, and the relations among these. The body schema combines elements of body percept and body image. Similarly, idiothetic stimuli include "self-motion related signals," as well as representations of actions and "sensory flow information" such as visual kinesthesis and optical signals about one's own movements (for specific details on idiothetic stimuli, see Chapter 13).

In an allocentric framework points or objects are localized in relation to each other and create the framework for knowledge about that space. In this way, the individual can move freely about in familiar space from any starting place to any destination. The location of the self in an allocentric framework is not encoded in any direct spatial relationship to static (discrete) objects or landmarks, thus ensuring greater behavioral flexibility. In an allocentric framework, the "primitive" parameters as defined by Klatzky (1998) are the basis by which "derived" parameters are generated. Derived parameters are those that enable novel computations by relying on encoded details of distance, angle and direction between these static, primitive points in space regardless of the position of self (see also Chapters 3, 5 and 8, this volume; also Dolins, 2009).

An allocentric frame of reference remains fixed independent of the subject's movements, though the subject's position within the frame can change. The frame may remain fixed in relation to other moving objects, for example in relation to a running predator or a moving car, or the frame may incorporate moving elements themselves. Allothetic stimuli provide information about objects outside the organism (for specific details on allothetic stimuli, see Chapter 13).

INTERNAL SPATIAL REPRESENTATIONS

The development of an internal spatial representation is dependent on learning about cues in the environment, which is in turn dependent upon detailed exploration of that environment. Exploration is "an information-gathering behavior which is intended first to build and then to update cognitive maps" (O'Keefe and Nadel, 1979, p. 490). During exploration an individual attends to salient cues in the environment, and for relational (allocentric) representations, will encode spatial locations relative to each other. Exploration of space can be rule-based or guided by prior knowledge from a secondary source (e.g., two-dimensional map, verbal description, or even a photograph). Rule-based exploration can be based on simple but very effective algorithms for localizing cues/objects from either one static point (by positioning the self in that location) or even from a number of static points to produce route-based navigation (see Chapter 13). Alternatively rule-based learning can incorporate more elaborate cognitive processes whereby localized landmark cues are encoded relationally, and with transformational rules, provide a foundation for generating novel route information (see Chapters 8 and 23, this volume; also Dolins, 2009).

Route-based navigation relies on sequentially organized spatial cues, a form of response learning, that is, a behavioral response to each individual cue in the sequence (see Chapter 6). In contrast, survey or place navigation requires having an overview of spatial relationships among multiple cues (see Chapters 8 and 23, this volume; see also Poucet, 1993). Knowledge of a configuration or array of landmarks allows computation of optimal or novel route information from one directional perspective or from a rotation of the configuration in any orientation along the X and Y axes.

Generation of internal spatial representations also depend upon transformation of the perception of physical space into a neurally based representation. Veridical space is inherently three-dimensional, necessitating the incorporation of the three-dimensional information into neurally-based spatial representations and into any subsequent computations generated for action in the environment. Euclidean geometric relations are generally considered an appropriate model for adapting neurally encoded spatial relations to a physical space (O'Keefe and Nadel, 1978; Gallistel, 1990; Nadel, 1991; Golledge, 1999). Gallistel (1990) writes,

> If the formal structure of a coding system is rich enough to enable that system to capture the uniquely metric relations, which are invariant under displacement transformation ... then the structure is also rich enough to capture every other type of geometric relation. It follows that if one wants to know what categories of geometric relations the neural code is capable of capturing, one should begin by testing whether the code can capture the uniquely metric relations. (pp. 181–182)

Internal spatial representations can be either veridically (geometrically accurate in accordance with points in real space) or subjectively encoded to some degree, where some details of accuracy in accordance with the physical world are maintained but other aspects have been modified or are proportionally inaccurate (much like real maps – see Chapter 20). A veridical-type map can be defined as an internal representation in which an organism is able to encode and recall for use the geometric and non-egocentric spatial relationships between two or more cues in order to localize additional points in space (O'Keefe and Nadel, 1979). The same can be true of a less than veridical representation, where the encoded non-veridical (more subjective) spatial relationships between points in space can be used to localize areas or regions, and what encoded veridical information is available supplies the basis to re-adjust for accuracy. Both of these types of computations impose shape on space through a perceptual-cognitive process and require spatial relational learning, a most abstract cognitive process (see also Chapter 5).

An allocentric representation can be complete in the sense that all the geometrical relationships between landmarks in an environment are either recorded explicitly or computable from the encoded information. A Euclidean representation of this kind provides the knowledge required to plan routes within that environment. If it can be demonstrated that an individual plans detailed and novel routes, we can infer that the underlying representation is a rich repository of geometrical information about the environment (see Chapters 4, 6, 8 and 9, this volume; also Collett *et al.*, 1986, p. 836; Golledge, 1999).

The internal geometric or relational map encodes the metric relationship between salient landmarks referenced in a common coordinate system so that the coordinates correspond with those in real space (O'Keefe and Nadel, 1978; Gallistel, 1990). Through this representation an individual may compute novel and accurate routes derived from relations amongst landmarks represented in its map. In place learning, an internal relational map can be considered as "the representation of the metric spatial relations between three or more points on a plane [that] requires a system of coordinates or coordinate framework, by reference to which the positions of points are specified" (Gallistel, 1990, p. 42).

Generation of novel routes, as defined by Klatzky (1998), uses primitive parameters to compute derived parameters. For instance, an individual may have knowledge of the route that leads between points (or landmarks) X and Y; they may also have knowledge of the route between points X and Z. If they want to get to point Y when at Z, without an internal spatial relational map they would have to return to point X, the common point between the two routes, and then follow the route to Y. However, with an internal relational representation they have the ability to compute a novel route that leads directly between points Z and Y (Dolins, 1994). Thus, the individual has used the geometric relationship amongst known points to generate a novel and efficient route, a process that is parallel to transitive inference. With a series of these relational bridges or links between multiple known points or landmarks, flexibility in behavior and in spatial decision-making is significantly increased, such that a set of transformational rules can generate optimal and efficient novel routes to known destinations (Etienne *et al.*, 1999). These rules may be operations based on understanding rotations of a landmark configuration or translations of a landmark array in various directions (up, down, right or left). It could also include both a rotation and a translation. This means that the individual is not restricted to specific start and goal locations, as in route following, but can flexibly begin and end a navigational bout almost anywhere within the represented space. Detour situations along known routes can also function similarly, requiring planning an alternate route, which may not be as efficient and may even lead away from the goal location.

REFERENCES

Acredolo, L. P. (1990). Behavioral approaches to spatial orientation in infancy. In A. Diamond (Ed.), *The development and neural bases of higher cognitive functions* (pp. 596–607). New York: New York Academy of Sciences.

Aggleton, J. P. and Brown, M. W. (2005). Contrasting hippocampal and perirhinalcortex function using immediate early gene imaging. *The Quarterly Journal of Experimental Psychology*, **58** (3&4), 218–233.

Akre, N. (1983). *Miniatures*. Washington, DC: Smithsonian Institution.

Altringham, J. D. (1996). *Bats: biology and behaviour*. Oxford: Oxford University Press.

Alvarez, P., Wendelken, L. and Eichenbaum, H. (2002). Hippocampal formation lesions impair performance in an odor-odor association task independently of spatial context. *Neurobiology of Learning and Memory*, **78** (2), 470–476.

Bertenthal, B. I., Campos, J. J. and Barrett, K. C. (1984). Self-produced locomotion: an organizer of emotional, cognitive, and social development in infancy. In R. Emde and R. Harmon (Eds.), *Continuities and discontinuities in development* (pp. 175–210). New York: Plenum.

Burgess, N., Maguire, E. A. and O'Keefe, J. (2002). The human hippocampus and spatial and episodic memory. *Neuron*, **35**, 625–641.

Cheng, K. (1994). The determination of direction in landmark-based spatial search in pigeons: a further test of the vector sum model. *Animal Learning and Behavior*, **22**, 291–301.

Cheng, K. and Newcombe, N. S. (2005). Is there a geometric module for spatial orientation? Squaring theory and evidence. *Psychonomic Bulletin & Review*, **12**, 1–23.

Collett, T. S. and Collett, M. (2000). Path integration in insects. *Current Opinion in Neurobiology*, **10** (6), 757–762.

Collett, T. S. and Collett, M. (2002). Memory use in insect visual navigation. *Nature Reviews Neuroscience*, **3**, 542–552.

Collett, T. S., Cartwright, B. A. and Smith, B. A. (1986). Landmark learning and visuo-spatial memories in gerbils. *Journal of Comparative Physiology A: Sensory, Neural, and Behavioral Physiology*, **158**, 835–851.

Collett, T. S., Collett, M. and Wehner, R. (2001). The guidance of desert ants by extended landmarks. *Journal of Experimental Biology*, **240**, 1635–1639.

Collett, T. S., Dillmann, E., Giger, A. and Wehner, R. (1992). Visual landmarks and route following in desert ants. *Journal of Comparative Physiology A: Sensory, Neural and Behavioural Physiology*, **170**, 435–442.

Corballis, M. C. (2002). *From hand to mouth: the origins of language*. Princeton: Princeton University Press.

Corballis, M. C. and Beale, I. L. (1970). Bilateral symmetry and behavior. *Psychological Review*, **77**, 451–464.

DeCoteau, W. E. and Kesner, R. P. (2000). A double dissociation between the rat hippocampus and medial caudoputamen in processing two forms of knowledge. *Behavioral Neuroscience*, **114**, 1096–1108.

Dolins, F. L. (1994). *Spatial relational learning and foraging in cotton-top tamarins (Saguinus oedipus oedipus)*, University of Stirling, Scotland, unpublished PhD thesis.

Dolins, F. L. (2009). Captive cotton-top tamarins' (*Saguinus oedipus oedipus*) use of landmarks to localize hidden food items. *American Journal of Primatology*, **71**, 316–323.

Dusek, J. A. and Eichenbaum, H. (1997). The hippocampus and memory for orderly stimulus relations. *Proceedings of the National Academy of Sciences*, **94** (13), 7109–7114.

Eichenbaum, H. (2000a). Hippocampus: mapping or memory? *Current Biology*, **10** (21), R785–R787.

Eichenbaum, H. (2000b). A cortical-hippocampal system for declarative memory. *Nature Reviews Neuroscience*, **1** (1), 41–50.

Eichenbaum, H. (2004). Hippocampus: cognitive processes and neural representations that underlie declarative memory. *Neuron*, **44**, 109–120.

Ekstrom, A. D., Kahana, M. J., Caplan, J. B., Fields, T. A., Isham, E. A., Newman, E. L. and Fried, I. (2003). Cellular networks underlying human spatial navigation. *Nature*, **425**, 184–187.

Etienne, A. S., Maurer, R., Georgakopoulos, J. and Griffin, A. (1999). Dead reckoning (path integration), landmarks, and representation of space in a comparative perspective. In R. Golledge (Ed.), *Wayfinding behavior: cognitive mapping and other spatial processes* (pp. 197–208). Baltimore: Johns Hopkins University Press.

Fortin, N. J., Agster, K. L. and Eichenbaum, H. B. (2002). Critical role of the hippocampus in memory for sequences of events. *Nature Neuroscience*, **5**, 458–462.

Gaffan, D., Easton, A. and Parker, A. (2002). Interaction of inferior temporal cortex with frontal cortex and basal forebrain: double dissociation in strategy implementation and associative learning. *Journal of Neuroscience*, **22** (16), 7288–7296.

Gallistel, C. R. (1990). *The organization of learning*. London: MIT Press.

Garber, P. A. (1989). Role of spatial memory in primate foraging patterns: *Saguinus mystax* and *Saguinus fuscicollis*. *American Journal of Primatology*, **19**, 203–216.

Garber, P. A. and Dolins, F. L. (1996). Testing learning paradigms in the field: evidence for use of spatial and perceptual information and rule-based foraging in wild moustached tamarins. In M. A. Norconk, A. L. Rosenberger and P. A. Garber (Eds.), *Adaptive radiations of neotropica primates* (pp. 201–216). New York: Plenum.

Gibson, J. J. (1986). *The ecological approach to visual perception*. Hillsdale: Lawrence Erlbaum.

Golledge, R. G. (1999). *Wayfinding behavior: cognitive mapping and other spatial processes*. Baltimore: Johns Hopkins University Press.

Greeno, J. G. (1994). Gibson's affordances. *Psychological Review*, **101**, 336–342.

Gurfinkel, V. S. and Levick, Y. S. (1991). Perceptual and automatic aspects of the postural body scheme. In J. Paillard (Ed.), *Brain and space* (pp. 147–162). New York: Oxford University Press.

Hartley, T., Maguire, E. A., Spiers, H. J. and Burgess, N. (2003). The well-worn route and the path less traveled: distinct neural bases of route following and wayfinding in humans. *Neuron*, **37**, 877–888.

Heath, T. L. (1925/1956). *The thirteen books of Euclid's 'Elements'* (2nd edn). Cambridge: Cambridge University Press.

Hebb, D. O. (1949). *The organization of behavior*. New York: Wiley.

Henke, K., Treyer, V., Weber, B., Nitsch, R. M., Hock, C., Wieser, H. G. and Buck, A. (2003). Functional neuroimaging predicts individual memory outcome after amygdalohippocampectomy. *NeuroReport*, **14**, 1197–1202.

Holdstock, J. S., Mayes, A. R., Cezayirli, E., Aggleton, J. P. and Roberts, N. (1999). A comparison of egocentric and allocentric spatial memory in medial temporal lobe and Korsakoff anesics. *Cortex*, **35**, 479–501.

Holdstock, J. S., Mayes, A. R., Gong, Q. Y., Roberts, N. and Kapur, N. (2004). Item recognition is less impaired than recall and associative recognition in a patient with selective hippocampal damage. *Hippocampus*, **15** (2), 203–215.

Hopkins, W. D. and Leavens, D. A. (1998). Hand use and gestural communication in chimpanzees (*Pan troglodytes*). *Journal of Comparative Psychology*, **112**, 95–99.

Hopkins, W. D., Pilcher, D. and Cantalupo, C. (2003). Brain substrates for communication, cognition, and handedness. In D. Maestripieri (Ed.), *Primate psychology* (pp. 424–450). Cambridge, MA: Harvard University Press.

Hopkins, W. D., Dunham, L., Cantalupo, C. and Taglialatela, J. (2007). The association between handedness, brain asymmetries, and corpus callosum size in chimpanzees (*Pan troglodytes*). *Cerebral Cortex*, **17**, 1757–1765.

Jodidio, P. (2007). *Calatrava: Complete works 1979–2007*. Cologne: Taschen.

Klatzky, R. L. (1998). Allocentric and egocentric spatial representations: definitions, distinctions, and interconnections. In C. Freksa, C. Habel and K. F. Wender (Eds.), *Spatial cognition: an interdisciplinary approach to representation and processing of spatial knowledge* (pp. 1–17). Berlin: Springer-Verlag.

Krechevsky, I. (1932). "Hypotheses" in rats. *Psychological Review*, **39** (6), 516–532.

Kumaran, D. and Maguire, E. A. (2005). The human hippocampus: cognitive maps or relational memory? *Journal of Neuroscience*, **25**, 7254–7259.

Kumaran, D. and Maguire, E. A. (2006). The dynamics of hippocampal activation during encoding of overlapping sequences. *Neuron*, **49**, 617–629.

Kurrent, F. (Ed.). (1996). *Scale models: houses of the 20th century*. Salzburg: Verlag Anton Pustet.

Landau, B., Gleitman, H. and Spelke, E. (1981). Spatial knowledge and geometric representation in a child blind from birth. *Science*, **213** (4513), 1275–1278.

Leavens, D. A., Hopkins, W. D. and Bard, K. A. (1996). Indexical and referential pointing in chimpanzees (*Pan troglodytes*). *Journal of Comparative Psychology*, **110** (4), 346–353.

Leonard, B. and McNaughton, B. L. (1990). In R. P. Kesner and D. S. Olton (Eds.), *Neurobiology of comparative cognition* (pp. 363–421). Hillsdale, NJ: Erlbaum.

Maguire, E. A., Spiers, H. J., Good, C. D., Hartley, T., Frackowiak, R. S. J. and Burgess, N. (2003). Navigation expertise and the human hippocampus: a structural brain imaging analysis. *Hippocampus*, **13**, 208–217.

Maguire, E. A., Gadian, D. G., Johnsrude, I. S., Good, C. D., Ashburner, J., Frackowiak, R. S. J. and Frith, C. D. (2000). Navigation-related structural change in the hippocampi of taxi drivers. *Proceedings of the National Academy of Sciences*, **97**, 4398–4403.

Miranda, R., Blanco, E., Begega, A., Santín, L. J. and Arias, J. L. (2005). Reversible changes in hippocampal CA1 synapses associated with water maze training in rats. *Synapse*, **59** (3), 177–181.

Morris, T., Spittle, M. and Watt, A. P. (2005). *Imagery in sport: the mental approach to sport*. West Yorkshire: Human Kinetics Europe Ltd.

Nadel, L. (1991). The hippocampus and space revisited. *Hippocampus*, **1**, 221–229.

Nicholson, D. J., Judd, S. P., Cartwright, B. A. and Collett, T. S. (1999). Learning walks and landmark guidance in wood ants (*Formica rufa*). *Journal of Experimental Biology*, **202**, 1831–1838.

Noordzij, M. L., Zuidhoek, S. and Postma, A. (2006). The influence of visual experience on the ability to form spatial mental models based on route and survey descriptions. *Cognition*, **100** (2), 321–342.

O'Keefe, J. and Nadel, L. (1978). *The hippocampus as cognitive map*. Oxford: Clarendon Press.

O'Keefe, J. and Nadel, L. (1979). Precis of O'Keefe & Nadel's *The hippocampus as a cognitive map*. *Behavioral and Brain Sciences*, **2**, 487–533.

Paillard, J. (1991). Motor and representational framing of space. In J. Paillard (Ed.), *Brain and space* (pp. 163–182). New York: Oxford University Press.

Pearson, J., Clifford, C. W. G. and Tong, F. (2008). The functional impact of mental imagery on conscious perception. *Current Biology*, **18**, 982–986.

Poucet, B. (1993). Spatial cognitive maps in animals: new hypotheses on their structure and neural mechanisms. *Psychological Review*, **100**, 163–182.

Price, J. J., Johnson, K. P. and Clayton, D. H. (2004). The evolution of echolocation in swiftlets. *Journal of Avian Biology*, **35**, 135–143.

Rozin, P. (1976). The evolution of intelligence and access to the cognitive unconscious. *Progress in Psychobiology and Physiological Psychology*, **6**, 245–280.

Ryland, F. (1909). *Thought and feeling*. London: Hodder and Stoughton.

Schmahmann, J. D. and Caplan, D. (2006). Cognition, emotion and the cerebellum. *Brain*, **129** (2), 290–292.

Spetch, M. L. and Edwards, C. A. (1988). Pigeons' *Columba livia* use of global and local cues for spatial memory. *Animal Behaviour*, **36**, 293–296.

Spetch, M. L., Cheng, K., MacDonald, S. E., Linkenhoker, B. A., Kelly, D. M. and Doerkson, S. R. (1997). Learning the configuration of a landmark array in humans and pigeons II: generality across search tasks. *Journal of Comparative Psychology*, **111**, 14–24.

Spiers, H. J. and Maguire, E. A. (2006). Thoughts, behaviour, and brain dynamics during n avigation in the real world. *NeuroImage*, **31**, 1826–1840.

Squire, L. R. (1987). *Memory and brain*. New York: Oxford University Press.

Stein, G. (1937/1973). *Everybody's autobiography*. New York: Knopf Publishing Group.

Tolman, E. C. (1932). *Purposive behavior in animals and men*. USA: Appleton-Century-Crofts.

Tolman, E. C. (1948). Cognitive maps in rats and men. *Psychological Review*, **55**, 189–208.

Vanlierde, A. and Wanet-Defalque, M. C. (2004). Abilities and strategies of blind and sighted subjects in visuo-spatial imagery. *Acta Psychologica*, **116**, 205–222.

Vargha-Khadem, F., Gadian, D. G., Watkins, K. E., Connelly, A., Van Paesschen, W. and Mishkin, M. (1997). Differential effects of early hippocampal pathology on episodic and semantic memory. *Science*, **277** (5324), 376–380.

Waller, D., Lippa, Y. and Richardson, A. (2008). Isolating observer-based reference directions in human spatial memory: head, body, and the self-to-array axis. *Cognition*, **106**, 157–183.

Watson, J. and Carr, H. (1908). Orientation of the white rat. *Journal of Comparative Neurology & Psychology*, **18**, 27–44.

Wells, A. J. (2002) Gibson's affordances and Turing's theory of computation. *Ecological Psychology*, **14** (3), 140–180.

Wickens, C. D., Vincow, M. and Yeh, M. (2005). Design applications of visual spatial thinking: the importance of frame of reference. In P. Shah and A. Miyaki (Eds.), *The Cambridge handbook of visuospatial thinking* (pp. 383–425). Cambridge: Cambridge University Press.

Wilson, M. (2002). Six views of embodied cognition. *Psychonomic Bulletin & Review*, **9**, 625–636.

Wood E. R., Dubchenko, P. A. and Eichenbaum, H. (1999). The global record of memory in hippocampal neuronal activity. *Nature*, **397**, 613–616.

Wynn, T. G., Tierson, F. D. and Palmer, C. T. (1996). Evolution of sex differences in spatial cognition. *American Journal of Physical Anthropology*, **101** (S23), 11.

Zoladek, L. and Roberts, W. A. (1978). The sensory basis of spatial memory in the rat. *Animal Learning and Behavior*, **6**, 17–81.

Part I What do animals know and how do they represent external space?

ROBERT W. MITCHELL AND FRANCINE L. DOLINS

2

Psychology and the philosophy of spatial perception: a history, or how the idea of spatial cognition in animals developed

PHILOSOPHY AND PHYSICS

Space has been an object of intense scrutiny and interest in human history, not only in the practical sense of conveniently finding our way through space (see Chapter 20), but also in understanding space through the science of physics and the help of philosophy (Hatfield, 1990; Jammer, 1954/1993; Millar, 2008; O'Keefe and Nadel, 1978). Physics and philosophy have focused on "outer" space, developing "two concepts of space [that] may be contrasted as follows: (*a*) space as positional quality of the world of material objects; (*b*) space as container of all material objects" (Einstein, 1954/1993, p. xv). On a more everyday level, objects take up space, and are parts of a larger space. Thinking about space brings to mind a diversity of ideas: distance, depth, size, shape, volume, extent, position, boundary and dimension.

When, in the middle to late 1800s, many scientists tried to understand the psychology of space, they focused on the perception of space. How were organisms able to perceive space? Did the perception of space develop from sensations, was it already present in them, or did it require some "inborn" faculty? What was needed psychologically (sensorially, perceptually, intuitively, conceptually) to perceive space? These scientists viewed the perception of space as deriving from varying combinations of sensory stimulation, cognition-like entities, and association.

Many of their ideas were the legacy of philosophers attempting to understand the nature of the physical world (nicely detailed in Hatfield, 1990; Morgan, 1977). Descartes (1641/1911), for example, in

Spatial Cognition, Spatial Perception: Mapping the Self and Space, ed. Francine L. Dolins and Robert W. Mitchell. Published by Cambridge University Press.

a highly influential move, distinguished physical material things from mental experiences because only the former had extension. In Descartes' view, mental images of space, such as an imagined triangle, were clearly physical (in that they had extension), whereas thoughts of triangles (the idea of triangles, without instantiation) occurred independent of any medium, and were immaterial. (Although Descartes famously distrusted the senses to provide knowledge of the external world, he assumed that God was good enough to provide us with knowledge by which we could come to know the world.) Henry More (writing in mid to late 1600s), contra Descartes, argued that extension was a property of both matter and spirit, and in fact believed that God and space were identical (Jammer, 1954/1993, pp. 43–47). More's ideas concern what we now think of as absolute space, whereas Descartes' concern corporeal extension. Both ideas about space were developed by Gassendi, and used in Newton's (1687/1999) *Principia*. In Gassendi's (and ultimately Newton's) framework, absolute "space is independent of any substance … Whereas space is infinite, corporeal extension is finite. Space can be occupied by bodies, but corporeal space is impenetrable, subjected to all the vicissitudes of matter, whereas space is unchangeable and immovable" (Jammer, 1954/1993, p. 93). Newton (1687/1999) offered evidence for the reality of absolute space, which was disbelieved by many philosophers. Leibniz posited that absolute space is a fiction and that the concept of space derives from the motion of objects in relation to each other (Jammer, 1954/1993, p. 117). Berkeley (1710/1963, pp. 172–173) believed that the conception of space derived from the contrast between one's experience with one's own body being impeded by another object – in which case one assumed the existence of an object – or not being impeded – in which case one conceived the existence of space. For Kant, space was an intuition of the mind necessary to allow one to experience space at all. But he came to this realization slowly (Hatfield, 1990, p. 87):

> From the 1740s into the 1760s [Kant] held a modified Leibnizian theory of space, rejecting Newtonian absolute space in favor of a relational theory. As he began the line of work that resulted in the first *Critique*, Kant's teaching on space and spatial perception shifted radically. By 1768 he had abandoned the relational theory of space and adopted the Newtonian theory, according to which space constitutes a framework or container distinct from the matter that occupies it. A year later he altered his Newtonian view: although he continued to treat space as something independent from the matter that occupies it, he rejected the doctrine that space has mind-independent existence. He ascribed to

> space the status of a set of a priori laws governing the representations constructed by the human faculty of sensibility.

(Kantian and Newtonian views on space reappear for psychology in O'Keefe and Nadel's [1978] exposition of how cognitive maps are created neuropsychologically.) According to Hatfield's (1990) exposition of philosophical ideas on psychological theory about space, most psychologists misunderstood Kant's ideas, such that German psychology about spatial perception was a muddle. Apparently British psychology, taking Berkeley's ideas as its foundation, was a muddle as well.

PSYCHOLOGY AND PHILOSOPHY

By the late eighteenth century, theorists acknowledged six visual signs of distance (Hatfield, 1990, pp. 37–38):

- accommodation of the eye;
- convergence of the eye;
- apparent size (visual angle);
- "intermediate objects or imagined intervals along the continuous ground space between the observer and a more or less distant object;"
- atmospheric perspective, due to the effect of the intervening atmosphere; and
- clarity of small parts of objects.

These are still used as signs for distance, as are additional sources:

- motion perception;
- binocular disparity; and
- relative size and relative density

(Cutting and Vishton, 1995; other sources of information, such as linear perspective, texture gradients, gravity, etc. are subsumed by these sources [see pp. 94–97; also Kellman, 1995].)

The developing recognition of these signs of distance was stimulated, in part, by the idea that the retinal image was the main source of visual information about space. Using the retinal image to discern space was problematic because, although one can predict what will appear on the retina from knowledge of objects in a field of vision, one cannot predict the physical dimensions of the objects or the field of vision from the retinal image: "while a specific distal event or arrangement is compatible with only a single retinal state, a given retinal state is compatible with countless distal states" (Epstein, 1995, p. 2). Given this difficulty, Berkeley (1709/1963, pp. 32–33) argued that it must be touch that provides us with the idea of space. He believed that the ideas available to touch (which included kinesthesis) were completely different from those available to sight ("we never see and feel one and the same object. That which is seen is one thing, and that which is felt is another" – p. 34), and that it is only through multiple exposures to the diverse relations between the two that one comes to experience sight as providing the experience of objects in space. Berkeley (p. 59) argued that one can, from tactile experiences (which included kinesthesis), develop associations (suggestions) of how to use signs from vision to connect to tactile experiences to know how to predict space from vision (see Hatfield, 1990; Lindsay, 1963; cf. Morgan, 1977, p. 177). In Berkeley's view, the experience of visual space is a cognitive (associational) effect. This view is problematic for several reasons (see, for example, Lindsay, 1963, pp. xix, xvii), but Berkeley's analysis also led him to recognize and elaborate on several of the visual signs of distance detailed above. But these visual signs were, in his view, arbitrarily related to tactile signs of what we, perhaps erroneously, take to be the same.

By the mid-eighteenth century, some philosophers had come to believe that associations between sensations could not all be arbitrary. Particularly disturbed by the view that "ideas" in the mind are unconnected to the real world was Thomas Reid (1785/1969). For Reid, our experience of the world (including space) was part of our nature. He posited that "the mind is so constituted that, from the 'natural signs' delivered by the senses it is able to move directly to 'the thing signified,' which is the object itself" (Robinson, 1982, p. 47). Vision and touch provide us with information about the same things: objects in the world (Reid, 1785/1969, p. 286). The theory that the mind is concerned only with ideas, and not their relation to the world, ignored

> the very principles by which the animal kingdom would have to be regulated if survival were to be possible. The theory was quite literally unnatural. Reid often referred in his essays to creatures such as the caterpillar who would roam over hundreds of different leaves until it settled on the one that was compatible with its needs. The creatures of nature – and man as well – have been fitted by "the mint of Nature" to transact their business with the world they live in, the world as it is. They are guided in their transactions by *the principles of common sense* which they "are under a necessity to take for granted" in all the affairs of life. (Robinson, 1982, pp. 47–48; see Reid, 1785/1969, p. 287)

Important to Reid's view was his distinction between sensations and perceptions.

> A sensation [for example, pain] is a feeling, utterly peculiar to one sense; we can neither define it nor say why it should be one feeling rather than another ... A perception [by contrast] is an act in which we become aware of some object outside ourselves. It is not a sensation *because it is not tied to one particular sense*. The crucial defining feature of a perception is that it can be accomplished by several senses, acting either individually or in concert ... It follows that our ideas of shape, extension, etc. are not sensational, but are in us through nature's stamp. (Morgan, 1977, p. 110)

In addition to providing knowledge about their spatial extent, "objects of sight and touch carry the notion of space along with them; and not the notion only, but the belief for it: for a body could not exist if there were not space to contain it: it could not move if there was no space. Its situation, its distance, and every relation it has to other bodies, suppose space" (Reid, 1785/1969, p. 281).

With a result similar to Reid's assessment, Kant argued that our experiences depend on our natural organization. Kant developed his transcendental methodology (an elaborate edifice we will not describe - see Hatfield, 1990, pp. 91–92) to provide a grounding for the possibility for experience. For example, Kant believed that we must have an intuitive knowledge of space in order to experience it. As Hatfield (p. 89) describes a small part of Kant's argument,

> The perception of space includes a system of (actual or possible) relations among sensations. For these sensations to be represented as outside one another, the ability for spatial representation must already be in place. Thus our fundamental representation of space could not be acquired, for any experience from which it could be acquired (any experience of the "outside and alongside" relation) would already be an experience of space.

From this argument (and others), Kant posited that "our fundamental representation of space is intuitive, not conceptual" (Hatfield, 1990, p. 89), where an intuition concerns only one object, whereas "Concepts ... apply to a variety of individual instances" (p. 90). Space, as an infinite and singular expanse, is thus an intuition. For Kant, then,

> the empirical matter of intuition is constituted by nonspatial sensations varying only in quality and intensity. They are brought to empirical synthesis through the reproductive imagination in accordance with the

> laws of association … Kant believed that the perceptual images of objects are not given from or produced by the senses alone but are constructed by the imagination … [Thus] The imagination constructs perceptual images from the nonspatial sensations that serve as the manifold for the synthesis that underlies empirical intuition. (Hatfield, 1990, pp. 102–103)

The similarities between the conclusions of Kant's and Reid's ideas are evident, though their methods of arriving at them are quite different.

THE INFLUENCE OF BERKELEY

Berkeley's idea that our experience of space derives from tactile experiences had a profound influence on the science of psychology. Consider associationist Bain's (1855/1868, pp. 366–367) elaboration of Berkeleian ideas.

> With regard to these two qualities – namely, (1) the distance of a thing from the seeing eye, and (2) the real dimensions of a body in space – I affirm that they cannot be perceived or known through the medium of sight alone.
>
> Take first the case of Distance, or remoteness. It appears to me that the very *meaning* of this quality – the full import of the fact implied in it – is such as cannot be taken in by mere sight. For what is meant by an object being four yards distant from where we stand? I imagine that, among other things, we understand this – namely, that it would take a certain number of paces to come up to it, or to reduce the distance from four yards, say to one yard. The possibility of a certain amount of locomotion is implied in the very idea of distance … In the case of objects within reach of the hand, the movements of the arm give the measure of distance; they supply the accompanying fact that makes distance something more than a mere visible impression … The actual distance means so many inches, feet, or yards, and of these we have no measure by the eye; they concern the locomotive and other mechanical movements, but not the movements of sight.

Perceived distance certainly depends in part on whether or not we intend to walk it or not; we perceive distance not only in terms of our visual experiences, but also in terms of our goals in relation to the distance and the extent of effort needed to achieve those goals (Witt *et al.*, 2004). Bain, however, wants to state further (seemingly following Berkeley, 1709/1963, pp. 36–37) that the experience of magnitude is dependent on tactile (kinesthetic) experiences: "magnitude is … the extent of movement of the arms or limbs that would be needed to compass the object; and this can be gained in no way but through actual trial by these very organs" (Bain, 1855/1868, p. 371). Magnitude cannot be present with "mere" vision alone (pp. 371–372):

> A certain movement of the eye, as the sweep over a table, gives us the sense of that table's magnitude, when it recalls or revives the extent and direction of arm movement necessary to compass the length, breadth, and height of the table. Previous to this experience, the sight of the table would be a mere visible effect.

Perceptually, an object's form can be discerned by the shape of its resistance to active touch (Weber, 1846/1996, p. 194). But for Bain and other associationists (e.g., Robertson, 1896, pp. 112–113 on "resistance"), the idea of space per se seemed to derive from touch – an idea taken directly from Berkeley. (As Brentano [1874/1973, p. 13] summarized, "the idea of extension and three dimensional space develops from kinesthetic sensations.") For Robertson (pp. 113–114), the first resisting "obstacle" is one's own body. But note how he ignores visual experience and the kinesthetic sense of one's body:

> I have not the slightest doubt that the first object that we become aware of as resisting, and at the same time spread out, is our own body. Of course the child from the very beginning sees as well as touches, but I am putting aside vision for the present, and suppose that we have a child, at first unable to discern a difference between subject and object, beginning to acquire objective experience by touch. And I say ... that the first object it would come to apprehend vaguely is not any other body, but its own. That one object it has always with it; other objects come and go, but it has always the power of touching its own body and thus finding the activity of its own hand impeded.

Even when we passively touch (are touched by) objects, "it comes that we know bodies as extended, *because we know our own body as extended*" (p. 116). So we can (imprecisely – James, 1890, p. 141) match the extent of an object to its touch on our own (extended) body.

It was left initially to Irish philosophers to argue that vision provides the bulk of spatial experience. Abbott (1864, pp. 1–2) begins with satire:

> At the present moment every one who has tasted philosophy ... is firmly convinced that he sees, not persons and things of various bulk, and at divers (*sic*) distances, but merely a variety of colour, or at the best, a flat picture of no perceptible magnitude, and at no perceptible distance. Yet the profoundest metaphysician, when he opens his bodily

> eyes, is mastered by the same belief as the unlearned: he cannot see what he knows he does see, and he cannot help seeing what he knows it is impossible to see. Seeing is, for the time, believing; but on deliberate reflection, sense is fairly overpowered by reason. This is a truly brilliant victory of science. A universal persuasion that nothing really exists, would be scarcely more surprising. It is an ungrateful task to attempt to deprive psychology of its only acknowledged triumph, and to wrest from philosophers the cherished analogy which never failed to support them in bidding defiance to probability and common sense.

Abbott (1864) and Monck (1872) offer diverse support for their belief that vision, and not touch, provides us with the sense of distance and extension, and thus of space. For example, Monck (1872, p. 65) notes that simply seeing color requires seeing extension – an expanse must be present (and is for blind people given sight later in life – see von Senden, 1932/1960, pp. 305–306). For both authors, a significant point of evidence concerns animals. Against the associationist account that visual perception of space is determined by its cognitive connection with tactual perception of space is the evidence that (some) animals are seemingly immediately attuned to the existence of space, specifically through visual experience. "If we find that in animals it is congenital and therefore belongs to the organ of sight itself, it becomes in the highest degree improbable that in man it does not belong to the same organ, but is the result of a different faculty, the gradual product of association" (Abbott, 1864, pp. 167–168; also, Monck, 1872, pp. 13, 42–43; Weber, 1846/1996, pp. 145–147). Knowledge of absolute space, by contrast with spatial experience, derives from intellectual elaboration of experience:

> that the field of distinct vision is always surrounded by other spaces, of which we have a kind of obscure semi-consciousness, renders it impossible for us to see or imagine a limit to space; and when to this we add our positive knowledge of its vastness as something surrounding and containing all that we see, (which experience soon teaches us,) these two elements seem to me to afford everything that is necessary to account for our belief in space infinite. (Monck, 1872, p. 60)

A continuing problem for the Berkeleian idea of the separateness of each source of sensory information was how touch immediately led one to think of visual concomitants, and vice versa (Reid, 1785/1969, p. 284). Bounded spaces, as in objects, allowed for an easy translation between perceptual modalities: "when we are capable of moving our limbs intentionally and consciously, we can, by the movements we must make with our hands about the resistant object, build up a picture of the shape and size of the object" (Weber, 1846/1996, p. 194); as "a man handling an object in the dark ... explores the outlines of the object, he at the same time constructs a visual image of it" (at least "the vast majority of mankind does so") because of the "intimate correlation" between tactile and visual extension (Stout, 1898, p. 382). Similarly, Adam Smith (1759/2004) presented sympathy as deriving from imaginal spatial relocation of one's own body into that of another, a form of kinesthetic–visual matching. (In the early twentieth century, Ryland [1909, pp. 168–169] might have explained such occurrences as resulting from the visual experience of the other activating a motor [or tactile] image of oneself: "attention to the motor image is itself a step towards bringing the movement about [showing a] tendency to turn ideal into real movements ... So when I imagine a movement, the same nerves and muscles are affected as when the actual movement takes place.") Such easy translations from one modality to another argue against Berkeley's view of merely associational correlation: "For inasmuch as association cannot for [Berkeley] be due to any thing in the ideas, for that would imply a common quality in different simple ideas, it must be, as he himself calls it, arbitrary" (Lindsay, 1963, p. xvii).

THE INFLUENCE OF KANT AND REID

Contrary to Berkeley's ideas that the perception of space is derived from one sensory experience and only attached by association to another, the Kantian and Reidian traditions posit that sensory stimulation is not enough to become perceptual experience, but requires something in human nature. Perception, therefore, is not tied to particular sensations, but is a coordination of these accomplished through imagination or intuition. Weber (1846/1996, pp. 145–147) injects Kant's ideas into his scientific investigations:

> *sensations are fundamentally conditions exciting our consciousness ... These conditions do not reveal spatial relationships directly to consciousness but only indirectly, insofar as our soul is capable of representing and interpreting those sensations, and is induced to do so by an innate predisposition or power* ... we feel obliged, for some unknown reason, to order our sensations with respect to the categories of space, time, and number. Were our free will not bolstered by this need when interpreting sensations, we should doubtless never succeed in achieving sensory representations ... experience itself is made possible by our ability to assign our sensations to the appropriate categories of space, time, and number.

Kant's ideas created a problem for empirically driven scientists, in part because our Kantian intuition of space was scientifically unanalyzable. Hatfield (1990) details German scientists' diverse interpretations of Kant's ideas and their various solutions for reconciling Kant's ideas with empirical research, culminating in Helmholtz's attempt to combine empiricism with Kantian transcendental ideas.

As Helmholtz (1878/2003, p. 167) described it, "I believe that one must view the modern era's most essential progress as the resolution of the concept of intuition into the elementary processes of thought … It was the physiological investigations on sense perception, which led us to the final elementary processes in knowledge." Whereas Kant viewed our intuition of space as unanalyzable, Helmholtz believed it is an "unconscious inference" from our experiences of law-like regularities (p. 162; the idea apparently derives from Wundt – see Hatfield, 1990, pp. 198–199).

> if we ask whether there is a marker which is common and perceptible by direct sensation, through which every perception relating to others in space is characterized, then we find, in fact, one such marker in the circumstances that the movement of our body places us in other spatial relationships to perceived objects and thereby also changes the impression which they make on us … If we now make such types of impulses – take a look, move the hands, go back and forth – then we find that the sensations belonging to certain quality circles – namely, those with respect to spatial objects – can be changed; while other mental states of which we are conscious – memories, intentions, wishes, moods – cannot at all … space will also seem physical to us, laden with the qualities of our perceptions of movement, as that through which we move ourselves, through which we are able to look … space would be an innate form of intuition prior to all experience insofar as its perception would be tied to the possibility of the will's motoric impulses, and for which the mental and corporeal ability must be given us through our organization *before* we can have spatial intuition. (Helmholtz, 1878/2003, pp. 158–159, emphasis added)

Helmholtz appears to have his cake and eat it too – Kantian spatial intuitions exist prior to action and experience, but only come to exist through action and experience.

PSYCHOLOGY BIDS ADIEU TO PHILOSOPHY

For William James (1890), neither Berkeleians nor Kantians held the keys to understanding space. In James' view, "*voluminousness* … *discernible in each and every sensation, though more developed in some than in others, is the original sensation of space*, out of which all the exact knowledge about space that we afterwards come to have is woven by processes of discrimination, association, and selection" (pp. 134–135). Summing up late nineteenth-century views on spatial perception, James (pp. 271–272) wrote:

> Really there are but three possible kinds of theory concerning space. Either (1) there is no spatial *quality* of sensation at all, and space is a mere symbol of succession [Berkeley's view]; or (2) there is an *extensive quality given* immediately in certain particular sensations [James' view]; or, finally, (3) there is a *quality produced* out of the inward resources of the mind, to envelop sensations which, as given originally, are not spatial, but which, on being cast in to the spatial form, become united and orderly. This last is the Kantian view. Stumpf admirably designates it the "psychic stimulus" theory, the crude sensations being considered as goads to the mind to put forth its slumbering power.

Because Berkeley's idea that distance is a tactual and not visual experience "has been adopted and enthusiastically hugged in all its vagueness by nearly the whole line of British psychologists who have succeeded him, it will be well for us to begin our study of vision by refuting his notion" (James, 1890, p. 212). He appeals to experience: "It is impossible to lie on one's back on a hill, to let the empty abyss of blue fill one's whole visual field, and to sink deeper and deeper into the merely sensational mode of consciousness regarding it, without feeling that an indeterminate, palpitating, circling depth is as indefeasibly one of its attributes as its breadth" (pp. 212–213). He similarly appeals to experience to refute Kantian ideas about space:

> The essence of the Kantian contention is that there are not *spaces*, but *Space* – one infinite continuous *Unit* – and that our knowledge of *this* cannot be a piecemeal sensational affair, produced by summation and abstraction. To which the obvious reply is that, if any known thing bears on its front the *appearance* of piecemeal construction and abstraction, it is this very notion of the infinite unitary space of the world. It is a *notion*, if ever there was one; and no intuition. Most of us apprehend it in the barest symbolic abridgment: and if perchance we ever do try to make it more adequate, we just add one image of sensible extension to another until we are tired. (p. 275)

> [An] element of constructiveness is present … when we deal with objective spaces too great to be grasped by a single look. The relative positions of the shops in a town, separated by many tortuous streets,

> have to be thus constructed from data apprehended in succession, and the result is a greater or lesser degree of vagueness. (p. 147; cp. quote above from Monck, 1872, p. 60, on infinite space)

Interestingly, constructiveness is present in adaptation to glasses that invert the image normally seen, suggesting that normal seeing develops piecemeal. Morgan (1977, p. 173) describes one subject's experience:

> He found that adaptation proceeded in a haphazard and piecemeal fashion: some objects in the visual field might appear the right way up and others still look wrong. An object could suddenly change its orientation, as when an upside-down candle snapped into the upright when it was lit ... Vehicles were seen correctly as driving on the right-hand side of the road but with their licence (*sic*) numbers reversed, and so on.

This description suggests that various reference frames are present and need to be attuned in any perceptual field (and such seems to be the case for blind person given sight later in life – see von Senden, 1932/1960).

For James, "The theory, stated baldly, is that space is given in the original sensation and that it need not be 'constructed,' 'synthesized,' or 'inferred' by some higher mental process ... *space* was in the stimulus configuration all along!" (Robinson, 1982, pp. 201–202).

> [James] did not "solve" the problem of space perception either in its metaphysical or its purely psychological installments ... Historically, however, James' analysis of perception was a breath of fresh air. His empiricistic arguments outstripped those of Helmholtz and joined with those of Mach in preparing the way for a completely antiseptic experimental psychology of perception. He inspired the scientific psychologists with the reassurance that the facts of experience were the only facts; that all seemingly transcendent [i.e., Kantian] ones *could* be interpreted experientially; that the facts of a *first* sensation need not be interpreted at all. (p. 203)

Undercutting the Kantian idea that Euclidian geometry provides us with knowledge of "space" (see Hatfield, 1990), James shows that our experience of space is variable depending upon the perceptual modality employed. "The interior of one's mouth-cavity feels larger when explored by the tongue than when looked at" (James, 1890, p. 139). "A point [on the body] can only be cognized in its relation to the entire body at once by awakening a *visual* image of the whole body. Such awakening is even more obviously than the previously considered cases a matter of pure association" (p. 161). James explains the existence of geometry itself as resulting from human activity: "The whole science of geometry may be said to owe its being to the exorbitant interest which the human mind takes in *lines*. We cut space up in every direction in order to manufacture them" (p. 150; see Davis, 1986, for a discussion of line-making as the origin of art).

COMPARATIVE PSYCHOLOGY BIDS ADIEU TO CONSCIOUSNESS, AND HELLO TO COGNITION

Knowledge of evolutionary theory caused scientists to revise their thinking about spatial perception to focus on behavioral responses to space, rather than consciousness of it: "the ability to perceive space relations gives animals possessing it an advantage over those without it in the struggle for existence, for it is evident that any organism that is to survive must have a tendency to so move in response to stimuli as to secure those that are advantageous and avoid those that are destructive" (Kirkpatrick, 1901, p. 565). In Kirkpatrick's view, conscious experience is irrelevant:

> The result of a reaction is the important thing biologically and the interesting thing psychologically rather than the sensations experienced while making the movement. (p. 574)

> Experiments in moving the hand as directed by the eye show a remarkable degree of accuracy, but, on the other hand, where the attempt is made to *consciously* compare visual and motor perception of distance the errors are very large. The same is true in comparing visual with tactual in one region, as on the tongue, with visual or tactual in another, as on the finger or the back ... It is clear that, whatever the conscious states of an animal may be, he must, from the first, in order to survive, be a spatially reacting organism. (pp. 567–568)

Spatial perception as experienced is only a small part of a complex coordination between the organism itself and the objects it perceives (p. 566):

> Every spatial reaction that is biologically useful therefore involves the correct relation of (1) a stimulus, (2) the position of the body or one or more of its parts, (3) a movement of the body or a part of it. These are of no value separately but only as related in the reactions ... space perception depends upon the relation of at least three groups of sensations or their images.

Oddly, this seemingly behaviorist analysis suggests that spatial perception is a cognitive phenomenon, rather than a sensory-perceptual one (p. 577):

> There is no such thing as a space sense or a sense of space, for space perception is the cognition of the *relations* of sensation groups to each other as reactions are made in the attainment of practical ends … Con[s]ciousness of space relations is the result of space reactions, not the cause … Ideas of direction, size, distance and shape are gradually formed as the result of the manipulation of objects and the comparison of one with another, while the concept of space as usually discussed is an abstraction from numerous experiences and more or less conscious analysis and synthesis.

The perception of (attention to) space per se seems, thus, strictly human.

When Thorndike (1899) studied the spatial perception of animals, he made clear that he was studying spatial *reactions*: a chick "reacts appropriately in the presence of space-facts, reacts in a fashion which would in the case of a man go with genuine perception of space" (p. 284) and shows "practical appreciation of space facts" (p. 285). "I do not wish to imply that there is in the young chick such consciousness of space-facts as there is in human beings" (p. 284). In his experiments on chicks' "Instinctive reactions to distance, direction, size, etc.," Thorndike provides numbers that support the earlier comments of Abbott (1864) and Monck (1872) on animal space perception:

> If one puts a [four-day-old] chick on top of a box in sight of his fellows below, the chick will regulate his conduct by the height of the box … If the height is less than 10 inches he will jump down as soon as you put him up … At 22 inches he will still jump down, but after more hesitation … At 39 inches the chick *will NOT jump down*. The numerical values given here would, of course, vary with the health, development, hunger and degree of lonesomeness of the chick. [Thus,] at any given age the chick without experience of heights regulates his conduct rather accurately in accord with the space-fact of distance which confronts him. The chick does not peck at objects remote from him, does not, for instance, confuse a bird a score of feet away with a fly near by, or try to get the moon inside his bill. Moreover, he reacts in pecking with considerable accuracy at the very start. (Thorndike, 1899, p. 284)

Playing a Kantian or Reidian note, Thorndike (p. 287) wrote, "one cannot help wondering whether some of the space-perception we trace to experience, some of the coordinations which we attribute to a gradual development from random, accidentally caused movements may not be more or less definitely provided for by the child's inherited brain structure."

Animals' reactions to heights showed an intriguing cognitive element: "hesitation" (Thorndike, 1899, p. 284), "uneasiness when on the edge of a void" (Small, 1899, p. 93), "acting as though he feared to attempt such a dangerous feat" (Watson, 1903, p. 40). In his study of diverse animals' responses to the "sense of support" provided by a surface, Wesley Mills (1898, p. 150) observed that, for many species, "even on the first day of birth they will not creep off a surface on which they rest, if elevated some little distance above the ground. When they approach the edge, they manifest hesitation, grasp with their claws or otherwise attempt to prevent themselves falling, and, it may be, cry out, giving evidence of some profound disturbance in their nervous system." The only animal that seemed to have no hesitation at any height was a water tortoise, likely because it was almost always above water. Such ecological factors clearly influence animals' responses to heights: tortoise species that live in water tend to push off heights regardless of visual experience, those that live on land show great hesitation or resistance to pushing off heights and less likelihood of doing so when blindfolded than when not, and those that live on land and water exhibit behavior either somewhat in between the two other types or like tortoises that live on land (Yerkes, 1904). "And so, strange as it may seem, the 'spatial worth' of sense data, as James would call it, is no more a matter of accurate knowledge than is the development of the sense of space, or [than are] the modes of behavior in different spatial conditions exhibited by the animal" (p. 17). As would be predicted by evolutionary theory (as well as Aristotelian theory – see Wheelwright, 1935/1951), the experience of space, indeed all experiences, must be acted upon (and therefore interpreted) by animals in ways relevant to their needs and expectations. Curiously, many spatial tasks that we would today view as requiring cognitive skill in animals, such as maze learning and detour problems, were not thought of as evincing spatial cognition, but rather associative learning (Small, 1899) or class inferences (Hobhouse, 1901).

In keeping with Kirkpatrick's ideas, Washburn (1908) distinguished five classes of "spatially determined reactions," i.e., "Modification[s] of the behavior of animals with reference to the spatial characteristics of the forces acting upon them" (p. 148). The first class is moving toward or away from a single tactile stimulus. The second is orienting to a stimulus, such as light or gravity. The third is reacting to a moving stimulus (determined by changes in light intensity or contact with different areas of the body). None of these classes requires spatial perception. The fourth class is reacting to an image, where an image is "the perception of simultaneously occurring but differently located stimuli as having certain spatial relations to each other" (p. 193). The fifth class is reactions requiring perception of distance or depth. These two final classes seem dependent on a sense organ that can be moved toward things. "While the sense organ is being moved, it is probable that other reactions of the animal will be suspended" (p. 203).

Washburn (1904) views the final class of spatially determined reactions, requiring perception of distance (or depth), as essential for cognition, in that "discrimination of present experiences and clearly conscious recall of past experiences, have been dependent in part at least upon one factor: the organism's growing power to react to stimuli not in immediate contact with the body" (p. 622). Her argument provides both physiological and psychological reasons. She discerns two "laws" concerning animals' skills in perceptual discrimination (reminiscent of Reid's distinction between perception and sensation):

> First, qualitative discrimination has been developed with reference to stimuli [such as light and sound] that do not immediately hurt or help the organism.
>
> Second, stimuli that are or may be harmful or helpful at the moment of their application have given rise to local discrimination at the expense of qualitative distinctions. (p. 623)

The first law has intriguing consequences for mental representation, as the greater exposure to stimuli that can neither help nor harm the organism (compared to stimuli that can help or harm) results in a more elaborate trace (such as mental imagery) in the nervous system:

> Now what characteristic of a stimulus would determine how thoroughly and deeply it would affect the nervous substance through which it passed? Its intensity, the quantity of energy in it, of course; but still more emphatically the length of time that energy remained in the centers in question, without being drained off into motor paths and transformed into bodily movement ... Delayed reaction, made gradually possible by increasing sensitiveness of the organism to stimuli only indirectly affecting its welfare, is then the source of the image-forming power. (pp. 624–626)

Washburn summarizes: "In a word, upon the possibility of reacting to stimulation that neither harms nor helps the organism at the moment of its operation [i.e., stimuli at a distance], may rest the basis of all higher mental development" (p. 626).

Representations depend on (in fact, are) re-presentations of perceptual images:

> Every image has, in fact, for its physiological basis a very faint form of the same kind of excitation as the corresponding percept has. If I

> imagine a touch on the tip of my nose, the same kind of nervous change takes place there as if I had actually been touched there, only in a much weaker degree. If I imagine a bright light, the retina is affected in the same manner, though in a less degree, as if I actually saw the light. (Ryland, 1909, p. 169)

Delayed reactions, which Washburn viewed as (in part) the cause of mental representation, came to be used in tests as evidence of mental representation (Cole, 1907; Révész, 1924; Tinklepaugh, 1928). In his comparison of monkeys and fowl, Révész (pp. 389–390) notes that the fowl require a visual stimulus to act, and then act immediately upon it, whereas monkeys continue to act when the visual stimulus is hidden:

> The behaviour of the fowl toward objects hidden in its presence shows that it either does not possess any visual representations at all, or if it has such images they are so confused and indefinite that they cannot be fixed and are not to be considered as *motives to action.* [By contrast,] the monkey ... must undoubtedly be governed somewhat or other by the representation of the object. Without this supposition it would be impossible to explain the many and varied methods and ways the monkey tries.

Thus, in Révész's account, at least some animals have both spatial perception and spatial representation. We are, at this point in our history, ready for scientists to recognize that animals might have more complicated representations than where and what a hidden object is. Tolman's (1932) "cognitive map" model isn't far away. We are at the start of the modern era.

REFERENCES

Abbott, T. K. (1864). *Sight and touch: an attempt to disprove the received (or Berkeleian) theory of vision.* London: Longman, Green, Longman, Roberts, & Green.

Bain, A. (1855/1868). *The senses and the intellect.* London: Longman, Green, & Co.

Berkeley, G. (1709/1963). A new theory of vision. In *A new theory of vision and other writings* (pp. 1–86). London: Dent.

Berkeley, G. (1710/1963). A treatise concerning the principles of human knowledge. In *A new theory of vision and other writings* (pp. 87–195). London: Dent.

Brentano, F. (1874/1973). *Psychology from an empirical standpoint* (trans. A. C. Rancurello, D. B. Terrell and L. L. McAlister). London: Routledge & Kegan Paul.

Cole, R. W. (1907). Concerning the intelligence of raccoons. *Journal of Comparative Neurology and Psychology*, **17**, 211–261.

Cutting, J. E. and Vishton, P. M. (1995). Perceiving layout and knowing distances: the integration, relative potency, and contextual use of different information about depth. In W. Epstein and S. Rogers (Eds.), *Perception of space and motion* (pp. 71–118). San Diego: Academic Press.

Davis, W. (1986). The origins of image making. *Current Anthropology*, **27**, 193–215.
Descartes, R. (1641/1911). Meditations on first philosophy [trans. E. S. Haldane and G. R. T. Ross]. *The philosophical works of Descartes* (pp. 185–199). Cambridge: Cambridge University Press.
Einstein, A. (1954/1993). Foreword. In M. Jammer, *Concepts of space: the history of theories of space in physics* (3rd edn, pp. xii–xvii). New York: Dover.
Epstein, W. (1995). The metatheoretical context. In W. Epstein and S. Rogers (Eds.), *Perception of space and motion* (pp. 1–22). San Diego: Academic Press.
Hatfield, G. (1990). *The natural and the normative: theories of spatial perception from Kant to Helmholtz*. Cambridge, MA: MIT Press.
Helmholtz, H. von (1878/2003). The facts of perception. In M. P. Munger (Ed.), *The history of psychology: fundamental questions* (pp. 155–268). Oxford: Oxford University Press.
Hobhouse, L. T. (1901). *Mind in evolution*. London: Macmillan.
James, W. (1890). *The principles of psychology* (vol. 2). New York: Henry Holt & Co.
Jammer, M. (1954/1993). *Concepts of space: the history of theories of space in physics* (3rd edn). New York: Dover.
Kellman, P. J. (1995). Ontogenesis of space and motion perception. In W. Epstein and S. Rogers (Eds.), *Perception of space and motion* (pp. 327–364). San Diego: Academic Press.
Kirkpatrick, E. A. (1901). A genetic view of space perception. *Psychological Review*, **8**, 565–577.
Lindsay, A. D. (1910/1963). Introduction. In G. Berkeley, *A new theory of vision and other writings* (pp. vii–xxiv). London: Dent.
Millar, S. (2008). *Space and sense*. New York: Psychology Press.
Mills, W. (1898). *The nature and development of animal intelligence*. New York: Macmillan.
Monck, W. H. S. (1872). *Space and vision: an attempt to reduce all our knowledge of space from the sense of sight, with a note on the association psychology*. Dublin: William McGee.
Morgan, M. J. (1977). *Molyneux's question: vision, touch and the philosophy of perception*. Cambridge: Cambridge University Press.
Newton, I. (1687/1999). *The principia: mathematical principles of natural philosophy* (trans. I. B. Cohen and A. Whitman). Berkeley: University of California Press.
O'Keefe, J. and Nadel, L. (1978). *The hippocampus as cognitive map*. Oxford: Oxford University Press.
Reid, T. (1785/1969). *Essays on the intellectual powers of man*. Cambridge, MA: MIT Press.
Révész, G. (1924). Experiments on animal space perception. *British Journal of Psychology*, **14**, 388–414.
Robertson, G. C. (1896). *Elements of psychology*. New York: Charles Scribner's Sons.
Robinson, D. N. (1982). *Toward a science of human nature: essays on the psychologies of Mill, Hegel, Wundt, and James*. New York: Columbia University Press.
Ryland, F. (1909). *Thought and feeling*. London: Hodder and Stoughton.
Small, W. S. (1899). Notes on the psychic development of the young white rat. *American Journal of Psychology*, **11**, 80–100.
Smith, A. (1759/2004). *The theory of moral sentiments*. New York: Barnes and Noble.
Stout, G. F. (1898). *A manual of psychology*. London: University Tutorial Press.
Thorndike, E. (1899). The instinctive reaction of young chicks. *Psychological Review*, **6**, 282–291.
Tinklepaugh, O. L. (1928). An experimental study of representative factors in monkeys. *Journal of Comparative Psychology*, **8**, 197–236.

Tolman, E. C. (1932). *Purposive behavior in animals and men*. New York: Appleton Century Crofts.

von Senden, M. (1932/1960). *Space and sight*. London: Methuen & Co.

Washburn, M. F. (1904). A factor in mental development. *Philosophical Review*, **13**, 622–626.

Washburn, M. F. (1908). *The animal mind*. New York: Macmillan.

Watson, J. B. (1903). *Animal education*. Chicago: University of Chicago Press.

Weber, E. H. (1846/1996). *The sense of touch and common sensibility* (trans. of *Tastsinn und Gemeingefühl*, by D. J. Murray). In E. H. Weber, *On the tactile senses* (2nd edn). Hove: Erlbaum (UK) Taylor & Francis.

Wheelwright, P. (1935/1951). *Aristotle: containing selections from seven of the most important books of Aristotle*. New York: Odyssey Press.

Witt, J. K., Proffitt, D. R. and Epstein, W. (2004). Perceiving distance: a role of effort and intent. *Perception*, **33**, 577–590.

Yerkes, R. M. (1904). Space perception of tortoises. *Journal of Comparative Neurology and Psychology*, **14**, 17–26.

KEN CHENG

3

Common principles shared by spatial and other kinds of cognition

INTRODUCTION

Four themes common to spatial cognition and other domains of cognition are reviewed selectively. In some systems of spatial cognition, such as path integration, quantitative values are computed and encoded. This process is probably common to many domains. Generalization and discrimination of spatial locations have been studied recently. Shepard's (1987) law of generalization has been verified for spatial generalization in humans, pigeons and honeybees, and spatial peak shift has been found in pigeons and humans. In spatial cognition, multiple sources of information are often averaged or integrated. Multiple landmarks may dictate landmark-based search. Metric and categorical information may be averaged. And even nonspatial information may be integrated: humans and pigeons can average temporal and positional dictates. Cue competition is found in using landmarks. Both overshadowing and blocking have been demonstrated in the spatial domain, in humans, rats, pigeons, and honeybees. Landmarks and beacons, however, generally fail to block or overshadow the learning of "geometry:" where a target is located in the overall shape of an arena.

COMMON PRINCIPLES SHARED BY SPATIAL AND OTHER KINDS OF COGNITION

Spatial cognition contains elements that are distinctly spatial. The computations necessary for path integration (for example, see Chapter 6) require calculations that do the equivalent of continuous trigonometry. Using geometric properties also has a distinctly spatial flavor (see Cheng and Newcombe, 2005; Chapter 6, this volume). Not all

Spatial Cognition, Spatial Perception: Mapping the Self and Space, ed. Francine L. Dolins and Robert W. Mitchell. Published by Cambridge University Press.

aspects of spatial cognition, however, are special. This chapter describes four themes that spatial cognition shares with many other domains of cognition.

The first theme is that of the encoding of quantities. Much of cognition requires the representation of quantities, and I will argue that this is one of the key jobs for the brain. The second theme is generalization and discrimination. The few studies done on spatial generalization and discrimination concord with general principles found in nonspatial domains. I will examine a universal law of generalization (Shepard, 1987, 1994) and the phenomenon of peak shift in discrimination learning (Hanson, 1959). The third theme is the weighted averaging of multiple sources of information. Dictates of different landmarks may be averaged, metric and categorical information may be averaged, and even dictates of different dimensions of experience, space and time, may be averaged. The fourth theme is cue competition. Here, one predictive cue interferes or competes with the learning or use of another predictive cue. This has been a strong theme in the modern (since the 1960s) analysis of classical conditioning.

REPRESENTATION OF QUANTITIES

Quantities are measures that are typically on a ratio scale. Angles and distances are the prime examples of spatial quantities represented and computed by animals. Angles and directions are circular measures. An angle formed by two walls cannot exceed 360 degrees. And a 360-degree turn in any direction brings one back to the starting point. But both angles and directions are numerical values on which computations can be made. Path integration provides the best example that spatial quantities are represented, and that computations are made on these quantities.

In path integration, an animal keeps track of the straight-line distance and direction from its starting point as it travels over a flat, or sometimes three-dimensional (Wohlgemuth *et al.*, 2001, 2002) surface (Chapter 6, this volume; Wehner and Srinivasan, 2003). At the end of the outbound journey, the animal runs home according to its computed vector. To do this, the animal, from desert ants to rats to humans, must have done the equivalent of adding all the pieces of its outbound journey using vector addition. Models of path integration differ (Maurer and Séguinot, 1995; Müller and Wehner, 1988), but in most cases, computations of quantities representing distances and directions are required. In one connectionist model (Maurer, 1998), representation of distances can be done away with. But the model still possesses many quantities in the form of connection weights.

Representation and computations of quantities are common to other domains of cognition, such as interval timing and classical conditioning. Diverse vertebrate animals can time a short duration of the order of seconds to minutes, an ability called interval timing (review: Meck, 2003). Various models of interval timing, most often tested on rats, pigeons and adult humans, have been proposed over the years. Although the models differ in many ways, quantitative encoding and computations are often found. According to some researchers, an animal encodes a quantity representing the duration of time for an event, such as the duration from a signal onset to the availability of reward (Church, 2003; Church and Broadbent, 1990; Gallistel, 1990; Gibbon and Church, 1984). An internal clock to do the actual timing appears in most models (Church and Broadbent, 1990; Gallistel, 1990; Gibbon and Church, 1984; Killeen and Fetterman, 1988), but not all (see Hopson, 2003). In the most influential theory to date, the Scalar Expectancy Theory (Church, 2003; Gibbon, 1977; Gibbon and Church, 1984), the animal is posited to compare ratios of elapsed time to remembered durations. In a simple connectionist model (Hopson, 2003), weights of connections are quantities that need to be computed.

In classical conditioning, how much quantitative representation takes place depends on the theory. In associative theories (see Pearce and Bouton, 2000), the only quantities represented are associative strengths between stimuli, such as conditioned and unconditioned stimuli. These quantities are updated trial by trial according to parametric rules, and they are the only quantities that govern behavior. Quantitative representation is thus minimal but nevertheless required. A representational theory, on the other hand, is rich with quantities and computations (Gallistel, 2003; Gallistel and Gibbon, 2000, 2002). In this theory, the animal is dealing with rates of occurrences of significant events (the unconditioned stimuli) under different conditions, as differentiated by the conditioned stimuli (e.g., with the tone on vs. without the tone). Numbers of events and temporal durations must be represented and must enter into numerous computations.

It is clear from this brief review that the brains of animals, from ants to humans, must represent and store quantitative values that need to be retrieved at appropriate times to enter into computations for decision-making. As Gallistel (2002) points out, this in turn has profound implications for the neurophysiological mechanisms of memory. Apparently, a thermodynamically stable coding scheme for quantitative information is needed (see Dukas, 1999 and Gallistel, 2002, for thoughtful considerations).

GENERALIZATION AND DISCRIMINATION

If an animal has had success at one place, for example in foraging, it often needs to decide if nearby places are also likely to prove successful. Generalization and discrimination, much studied in traditional learning theory (reviews: Cheng, 2002; Ghirlanda and Enquist, 2003), help with such decisions. In studying generalization, an animal is usually first trained to respond in some way to one stimulus, the S+. It exhibits generalization by responding similarly to stimuli different from but similar to S+. I discuss two principles here: Shepard's (1987) law of generalization, and peak shift.

SHEPARD'S UNIVERSAL LAW OF GENERALIZATION

Shepard's (1987) analysis provides a functional explanation of generalization. He aimed to find universalities by examining the nature of the world. All animals must accommodate to properties of the world. If the properties are universal, then we might expect similar evolved accommodations in all animals. At a suitable level of abstraction, these become universal principles. Shepard's universal law of generalization is an analysis along these lines.

To start with, the problem needs to be cast in the right way. Figure 3.1 illustrates this schematically, using spatial generalization as an example and apt model. Suppose that an animal has found its favorite food in a container at location S+ a number of times; the experimentally minded can think of these as a number of training trials at S+. On the next visit, it finds that the container is at another location, X. The key question is: does the animal "bet" that the container still contains the favorite food?

In this analysis, a distinction between discrimination and generalization is important. In the example, the animal can perfectly well discriminate X from S+. But this is not what is important to the animal. The important question is whether X has the same consequence of interest as S+. The consequence of interest in this case is having the favorite food in the container.

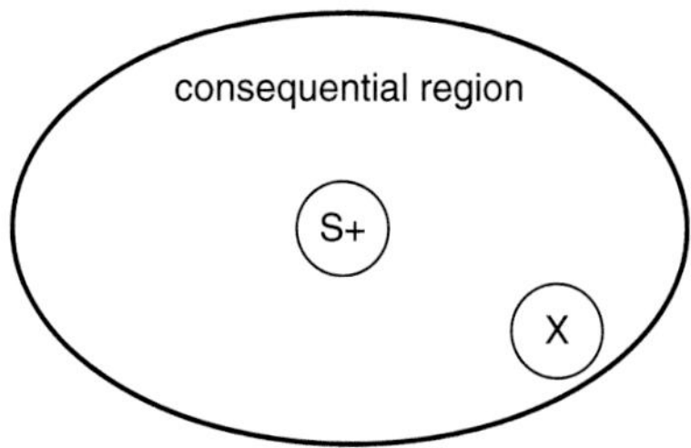

Figure 3.1. Characterization of the problem of spatial generalization. An animal finds favored food in a container located at S+ a number of times. Upon returning this time, it finds the container at X, a discriminably different location. What it has learned from its previous experience is that S+ is in a consequential region, a region in which a consequence of interest (having food in the container) holds. The problem of generalization is now: given that S+ is in the consequential region, what is the probability that X is also in the consequential region? The characterization clearly separates discrimination (telling two locations apart) from generalization (concluding that the two locations have the same consequence of interest). Shepard's (1987) law of generalization is based on such a characterization of the problem.

To continue the analysis in Figure 3.1, the next move is to cast the learning experience (trials with S+) as indicating that some region around S+ is the consequential region, in our example, a region in which the container contains the favorite food. The animal has learned that S+ is in the consequential region. The problem of generalization now becomes: given that S+ is in the consequential region, what is the probability that X is in the consequential region? What Shepard (1987) showed is that under a wide range of assumptions about the nature of the consequential region, the form of the probability distribution is similar. The shape of the distribution, with probability of X being in the consequential region on the y-axis, and psychological distance of X from S+ on the x-axis, comes out concave upward, close to an exponential equation in form. Psychological distance is a dissimilarity metric; it measures how far apart the animal perceives two things to be (see Shepard, 1987, for an account, or Cheng, 2000, 2002, for brief expositions). So far, the data confirm the functional predictions.

Shepard (1987) presented twelve cases of generalization in humans and pigeons that conform to the law. This law applies to spatial generalization as well. The exponential function for response rates (relative to response rate at S+) was confirmed in humans (Cheng and Spetch, 2002), pigeons (Cheng *et al.*, 1997), and in the only inverterbate animal tested to date, the honeybee (Cheng, 1999, 2000). In these studies, a target location was marked in training (the S+). After sufficient training, a number of locations, including the target location, were marked occasionally in testing (but only one location was marked on any one trial). Response rates were taken as a measure of the subject's rating of the probability of reward.

PEAK SHIFT

In the phenomenon of peak shift, two or sometimes more stimuli are used in training. In the typical case, one stimulus, S+, predicts reward, whereas another similar stimulus, S–, predicts no reward. After suitable training, the subject is tested, as in generalization, with a range of stimuli, including S+ and S–. The phenomenon derives its name from the fact that the stimulus that elicits the most responding is often not S+, but a stimulus further from S– than S+ is. Thus, the peak is shifted from where one might expect to find it, that is, shifted from S+. The classic demonstration is Hanson's (1959) work on color wavelength discrimination in pigeons. One wavelength of keylight (S+) signaled that keypecks would be rewarded, while a similar wavelength (S–) signaled that keypecks would go unrewarded. After training with this regime, different wavelengths were presented in unrewarded tests. Pigeons pecked most not to S+, but to wavelengths on the side of S+ away from S–. The training had thus shifted the peak of the generalization function. Since then, it has been found in many species and with various stimulus dimensions (reviews: Cheng and Spetch, 2002; Ghirlanda and Enquist, 2003; Purtle, 1973; Rilling, 1977). Peak shift occurs in the spatial domain in pigeons (Cheng *et al.*, 1997) and humans (Cheng and Spetch, 2002). Spatial locations consisted of marks on a computer monitor. One marked location (S+) signaled the availability of reward, while a nearby location (S–) meant that responses would go unrewarded. Subjects were then presented occasionally with various locations on the monitor, and their responses were measured. Peak responding was sometimes not at S+, but at a location on the side of S+ away from S–.

Do common principles underlie peak shift and generalization in all domains? Many models have been proposed (review: Cheng, 2002), but we have no way at present of deciding among them. It is also unclear whether a single model applies to all domains. Functionally, generalization, in Shepard's (1987) sense, has a different explanation from peak shift. Shepard's law assumes the ability to discriminate stimuli, and the generalization serves to match behavior with probabilistic properties of the world. Peak shift, on the other hand, is very much an attempt to deal with uncertainties in discrimination. The phenomenon is found when S+ and S– are similar and confusable. With imperfect discrimination, a signal detection analysis shows that the animal may judge a shifted stimulus as more likely to be the reward-giving signal than S+ (Lynn *et al.*, 2005).

AVERAGING INFORMATION FROM MULTIPLE SOURCES

Often, when multiple sources of information point to a target location, an animal combines information, in the form of a weighted average of all sources. Different landmarks may be combined. Categorical and metric information may be combined. And even spatial and nonspatial information may be combined. We might ask what purpose is served by using multiple cues. If memory is costly (Dukas, 1999), why not just use one predictive cue? One answer is that spatial memory, like all memory, is not perfectly accurate, and using multiple cues minimizes unsystematic errors. To provide support for this idea, Kamil and Cheng (2001) ran computer simulations of a landmark-based search task. Landmarks were at four corners of a square and the target was somewhere in the middle. Directions to landmarks were coded with an error of ± 1 percent or ± 2 percent. In each case, the more the number of landmarks used, the better the goal can be pinpointed. For accurate localization, using multiple cues may be a necessity, and many cases of use of multiple cues in spatial search exist. This principle for using multiple cues for greater accuracy is likely to be exhibited in many cognitive domains.

AVERAGING LANDMARKS

In the case of landmark-based navigation, an influential model for the averaging of multiple sources is the vector sum model. This model arose in work on gerbils in tasks of locating a hidden goal with respect to cylinders serving as landmarks (Collett *et al.*, 1986). It was followed by my work on landmark use in pigeons (Cheng, 1988, 1989, 1990, 1995). Even though aspects of the model needed to be revised (Cheng, 1994; Cheng and Sherry, 1992), the basic idea of averaging multiple sources of landmark-based information remains (Cheng and Spetch, 1998, 2001a; Cheng *et al.*, 2006).

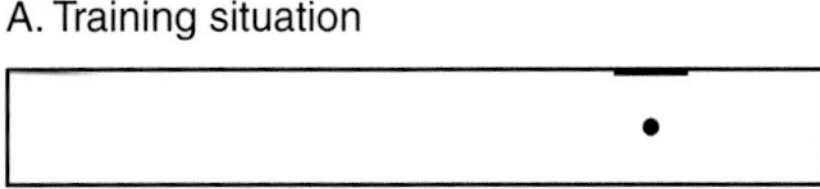

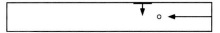

Figure 3.2. An illustration of vector averaging in landmark-based search. A. Hypothetical training situation in a long narrow tunnel. A landmark in the form of a strip on the "top" wall is indicated by the thick bar. The filled circle indicates the location of a hidden target. The problem is to find this location. B. A test situation in which the landmark has been displaced (to the left on the figure). The arrows indicate vector-based information encoded by the animal, the direction and distance to the target from the strip and from the end wall. The form of the argument does not change if more landmarks are included. If the animal averages the dictates of the two vectors indicated, it would search most somewhere between the ends of the two arrows, such as at the open circle in the figure. This model is based on Cheng (1988, 1989).

A sketch of the original model appears in Figure 3.2. The subject's task is to remember where along the tunnel food is found. Suppose that the target distance from the end of the tunnel is x cm. Two landmark sources are available, a distinctive strip on a side wall, and the end of the tunnel. It is not clear whether these constitute two or more landmarks, but to keep things simple, let us say that these are just two different landmarks, as the logic of the argument does not depend on landmark number. To tell whether both landmarks are used in representing the target, the animal is tested in a transformed space, with the strip shifted some distance along the wall, say x1 cm to the left (Figure 3.2b). Various plausible outcomes might arise on such a test. An animal might rely completely on the strip. In that case, its place of peak searching will also be displaced by ~x1 cm in the same direction. An animal might rely completely on the end of the tunnel. In that case, its place of peak searching would be the same place as before, ~x cm from the end. It might rely sometimes completely on the strip, and sometimes completely on the tunnel end. In that case, two search peaks would arise, one at ~x cm from the end, one at ~x + x1 cm from the end. According to the vector sum model, however, the animal averages the two sources of landmark information, and searches most at an intermediate distance between x cm and x + x1 cm from the end (Figure 3.2b). The distance of the peak of search from the tunnel end is determined by the relative weighting of the two sources of landmarks. Indeed, in this model, searching at x cm or x + x1 cm corresponds to weights of 1 or 0 assigned to the landmarks. Such intermediate peaks have been found in a number of experiments on pigeons (Cheng, 1988, 1989; Cheng *et al.*, 2006), in support of the vector sum model.

AVERAGING METRIC AND CATEGORICAL INFORMATION

In humans categorical information might also be averaged with metric information presumably based on landmarks (Huttenlocher and Vasilyeva, 2003; Huttenlocher *et al.*, 1991). An example is the task of remembering the location of a dot inside a circle. Huttenlocher *et al.* found small systematic errors on this task: errors were biased toward the centroid of each quadrant of the circle. Their explanation is that the people tested spontaneously divided the circle into four quadrants, then averaged the categorical code (quadrant to which the dot belongs) with precise metric information. With the further assumption that the category is represented by the centroid (or prototype) of the region, such averaging would produce the observed systematic errors. Similar errors have been found in other tasks with humans of various ages (Huttenlocher *et al.*, 2000; Chapter 23, this volume; Newcombe and Huttenlocher, 2000).

Again, one might ask why humans would average categorical and metric information and produce systematic errors. Huttenlocher *et al.* (1991) showed that if metric encoding is not perfectly precise, it is more accurate to use categorical information in conjunction with metric information, despite the fact that systematic errors are produced as a result. Hence, both pigeons and humans use multiple sources of spatial information to gain accuracy.

AVERAGING SPACE AND TIME

Humans and pigeons can average the dictates of a spatial position and a temporal duration, two different basic dimensions of experience, in guiding behavior (Cheng *et al.*, 1996). The task was to predict when and where a reward would become available. A square on the computer monitor appeared at the left side at the start of each trial. It moved to the right at a steady speed. After a fixed duration of time and, hence, at a fixed spatial position on the monitor, a reward was available to be collected (by keypecking for pigeons, and pressing the keyboard for humans). Responses before the appointed time and place had no effect. The task was thus a modified fixed interval schedule, with position of the square and the duration since its appearance both being perfectly reliable predictors of reward availability.

On occasional unrewarded tests, the speed of the square varied. This put positional and temporal cues in conflict. Suppose that the speed doubled. This would mean that at the correct position, only half the usual time to reward had elapsed, while at the correct temporal duration since trial onset, the square had moved twice as far as usual. Many humans reported using a spatial strategy, attempting to make responses according to the spatial position of the square. Nevertheless, the behavior of both species showed a compromise between the two types of cues. The position of the moving square at peak responding was biased by the temporal dictates: too far at faster-than-normal speeds, too short at less-than-normal speeds. Likewise, the time of peak responding was biased by the positional dictates: too long at slower-than-normal speeds, too short at faster-than-normal speeds. Importantly, the distribution of responding (in time or according to the position of the square) showed single peaks.

The pattern of data shows that a criterion for responding was based on a combination of positional and durational information, an averaging of space and time. The single peak in the response distributions indicates a single criterion for responding. The compromise in the positional and temporal peaks of responding indicates that both durational and positional cues went into calculating this single criterion. Mechanistically, the brain may not only encode quantities, but may also average magnitudes on different fundamental dimensions of experience using a common code for magnitudes across dimensions. The brain might have just one neurophysiological scheme for encoding quantities (Walsh, 2003). Functionally, averaging different dimensions of experience can again increase the accuracy of behavior in typical real-life situations in which perceptual sources of information are not systematically decoupled by experimental manipulation. In human perception and perceptual-motor behavior are many recent examples in which multiple sources of information, within and across modalities, are combined in near optimal fashion to produce greater accuracy (Alais and Burr, 2004; Gepshtein and Banks, 2003; Kersten and Yuille, 2003; Knill and Saunders, 2003; Körding and Wolpert, 2004; review: Cheng *et al.*, 2007; Körding 2007).

CUE COMPETITION

Cue competition has been much investigated in classical conditioning. It refers to situations in which more than one predictive cue (conditioned stimuli) may be used to predict an outcome (the unconditioned stimulus). With more than one conditioned stimulus, the predictors sometimes compete, to the detriment of one or more of the competitors. It is debated whether cue competition is a matter of learning decrement (e.g., Williams, 1996) or retrieval or performance failure (e.g., Denniston *et al.*, 2003). I will use the term "less relied on" to indicate theoretical neutrality on this issue. Overshadowing and blocking are the two most commonly studied forms of cue competition. In overshadowing, two (or more) predictive stimuli are present on learning trials. One of the predictive stimuli (or sometimes both) may be less relied on, compared to the situation in which it is the sole predictor. In blocking, one of the predictive stimuli (the blocking stimulus) is first presented alone as a predictor. In a second phase, the blocking stimulus is presented in conjunction with another predictor, the blocked stimulus. Both stimuli are thus equally good predictors. But of course, the blocking stimulus has had a previous history of being the sole predictor. In the phenomenon of interest, the blocked stimulus is less relied on (compared with appropriate control conditions).

Cases of overshadowing (e.g., Mackintosh, 1976) and blocking (e.g., Kamin, 1969) in classical conditioning are well known. Cases of overshadowing and blocking in the spatial domain are now also well known (reviews: Chamizo, 2003; Cheng and Spetch, 1998, 2001a, 2001b). A good amount of this work was stimulated by the influential theory of O'Keefe and Nadel (1978), which distinguished between a *taxon* and a *locale* system in spatial navigation. Briefly, the taxon system is route-based and S-R in nature, while the locale system is map-based and cognitive in nature. Landmark-based navigation is taken to be part of the locale system. The taxon system shares much with conditioning, while the locale system is taken to be something different. Much of the work on cue competition in spatial learning is an attempt to show that landmark learning obeys general principles found in conditioning. Some examples illustrate the point.

In using landmarks, those nearer to the target should be preferred. The functional reason is that they specify the target location more precisely. In a laboratory study with free-flying honeybees, Cheng *et al.* (1987) showed such a preference. The bees were trained to find a reward in the middle of an array of four colored cylinders (Figure 3.3a).

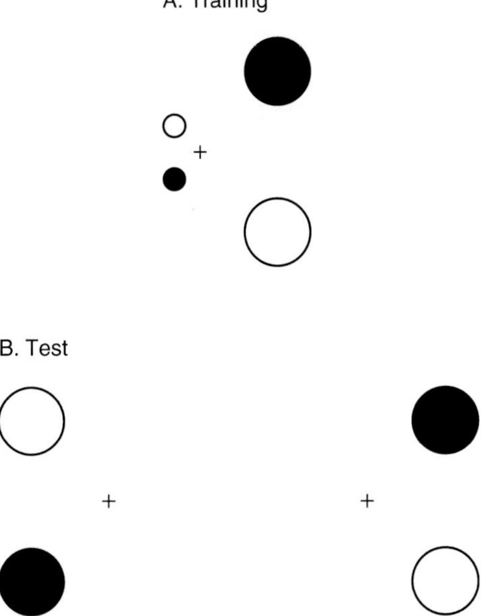

Figure 3.3. Bees prefer to rely on closer landmarks. A. Training situation. The cross indicates target location. Filled and unfilled circles indicate landmarks of different colors. The larger landmarks on the right were three times taller, three times wider, and three times farther away from the target. B. One kind of unrewarded test, with all landmarks of the same (large) size. The crosses indicate theoretical target locations, according to the left and right sets of landmarks. The bees prefer to rely on the landmarks they remember as being closer, the left landmarks in the figure. From Cheng *et al.* (1987).

The landmarks were arranged so that two were much nearer to the target than the other two. To control for the factor of the size that a landmark projected on the eyes of the bees, the farther landmarks were proportionally larger, both in cross-section and in height. At the target, the only perceptible difference between the two sets of stimuli was their distance from the target; bees have the ability to perceive depth or distance based on motion parallax (Lehrer *et al.*, 1988; Srinivasan *et al.*, 1989). On crucial tests after training, the two sets of landmarks (those nearer and farther from the target) were separated from one another (Figure 3.3b). Importantly, the landmark sizes were identical, either all large or all small. The bees searched preferentially at the location defined by the landmarks they *remembered* as closer to the goal (the left side in Figure 3.3b). It must have been memory that was used to determine which landmarks were closer, because the actual distances on the test were identical. This suggests that closer landmarks are weighted more, and might overshadow the use of landmarks farther from the goal.

This point was demonstrated convincingly in a study on humans and pigeons (Spetch, 1995). Subjects searched for an unmarked target on a computer monitor. The target was always at the same place as defined by an array of graphic landmarks on the screen. The target-landmarks array, however, was shifted from trial to trial. In different arrays, the target was the same distance from a crucial landmark. Spetch compared arrays in which the crucial landmark was closest to the target (non-overshadowed landmark) with those in which another landmark was closer to the target than the crucial landmark was (overshadowed landmark). On occasional tests, the crucial landmark was presented alone. On these tests, both humans and pigeons performed worse with overshadowed landmarks than with non-overshadowed landmarks. This demonstrates that a landmark nearer to the target can overshadow a landmark farther from the target.

With respect to discrete landmarks that specify target positions, spatial blocking has also been amply demonstrated in rats (e.g., Biegler and Morris, 1999; Roberts and Pearce, 1999; Rodrigo *et al.*, 1997). In Biegler and Morris' (1999) study, the rats had the task of digging for buried food. When a new landmark (the blocked stimulus) was added in the second (blocking) phase, the animals noticed the landmark and explored it. But their search behavior was not controlled by the blocked landmark. In Roberts and Pearce's (1999) study, the rats had to swim to a submerged platform to escape from water. The platform was indicated by a beacon (a landmark right at the target location) in phase 1. In phase 2, both the beacon and the surrounding landmarks in the room could be used to localize the goal. They found that the beacon blocked the landmark cues in specifying the target location. Honeybees also show blocking of one landmark by another (Cheng and Spetch, 2001b).

The evidence on cue competition in the spatial domain shows that landmark learning obeys some of the principles commonly found in the domain of classical conditioning. Searching in time and searching in space thus has some principles in common (for other principles, see Cheng, 1992). Whether some common underlying mechanism(s) can explain cue competition in both domains remains to be determined.

THE GEOMETRY OF SPACE WITHSTANDS CUE COMPETITION

The cue competition in the spatial domain that I have been reviewing concerns discrete landmarks. In light of all this positive evidence on blocking and overshadowing of landmarks, it is interesting to note when these phenomena fail. The geometric configuration carved out by surrounding surfaces, geometry in short, plays a large role in re-orientation in rats (Cheng, 1986, 2005). In the tasks, rats had to dig for hidden food in an enclosed rectangular arena. The arena, and sometimes the rat as well (Margules and Gallistel, 1988), was rotated so that only cues within the arena could specify target location. In particular, the rats could not inertially keep track of which direction was correct. Two kinds of cues specified the target location. One was the overall geometric shape of the arena, the geometric cues. In a rectangular space, geometric cues leave a two-fold ambiguity about target location. Both the target location, and a rotational error located at 180 degrees rotation through the center from the target, stand in the same geometric relation to the rectangular shape. If, however, other, nongeometric cues are also used to specify the target, then the target location can be unambiguously specified. In my (Cheng, 1986) study, such cues included visual patterns in the corners and on the walls, and olfactory and tactile cues in the corners. The rats made many rotational errors, indicating a primary role for the geometry of the environment in re-orientation. Since then, the role of geometry in re-orientation has been investigated in other species (review: Cheng and Newcombe, 2005).

Geometric information does not suffer from cue competition as landmarks do. The blocking or overshadowing competitor in studies on this topic have mostly been beacons, landmarks right at the target. Beacons are presumably powerful competitors since they specify a target perfectly. One just has to head to the beacon to reach the target; everything else can be ignored. Yet, beacons have failed to overshadow or block the learning of the geometry of environments in pigeons searching for a corner target in a rectangular arena (Kelly *et al.*, 1998), in chicks searching for the center of a square arena (Tommasi and Vallortigara, 2000), and in rats swimming for a hidden platform in a rectangular (Hayward *et al.*, 2003) or triangular pool (Hayward *et al.*, 2003; Pearce *et al.*, 2001). Cues from the room in which a pool was located also failed to disrupt the learning of the geometry of the pool (Hayward *et al.*, 2003). And a salient beacon also failed to block the learning of geometry in a food-searching task in an arena like that used in Cheng's (1986) experiments (Wall *et al.*, 2004). All these species learn readily to use a beacon. The best conclusion is that they learn both beacons and geometry obligatorily in parallel.

In one exception in wild-caught mountain chickadees, a large feature (an entire wall of a different color from the other walls) overshadowed the learning of geometry (Gray *et al.*, 2005). Overshadowing and blocking of geometric cues have also been found with rats (Pearce *et al.*, 2006).

Whether some common principles across domains separate cases of cue competition from failures of cue competition remains to be determined. One possibility is the salience or biological significance of the information. Salient or biologically significant conditioned stimuli typically do not suffer from cue competition (Denniston *et al.*, 1996; Miller and Matute, 1996). Perhaps geometry and beacons are "salient" or "biologically significant" spatial stimuli. At the moment, however, this is merely a circular redescription of the data. Making this point principled requires more theoretical and empirical work.

GRAND THEMES?

Can any grand themes be found in this selective review of common principles underlying spatial cognition and other types of cognition? What I have reviewed are mostly principles, similarities in patterns of data. It is unclear whether there are any deeper similarities or common themes in the mechanisms underlying the principles. On the whole, we are too uncertain about mechanisms at present, even on well studied topics such as generalization and discrimination. And in some cases, for example cue competition, it is not yet clear what are similar and what are different between spatial cognition and other relevant domains such as classical conditioning. Perhaps a small hint of a grand theme comes from the first theme reviewed: quantitative encoding and computations.

The brain computes and encodes information, and some of the information is quantitative. Presumably, some of the quantitative information corresponds to or represents aspects of experience (Gallistel, 1990). The distance to a location and the time elapsed since a significant event are prime examples. What is common is probably the central fact that quantitative dimensions of all kinds are actually encoded. Much else about what is encoded and the computations required will differ. In sensory systems, this point should be obvious.

The same arguments, however, apply when we are further along the stream in the brain. We have to expect many computational differences in learning, memory and cognition because different dimensions of experience have different properties (Gallistel, 2000, 2002). Computations that do not accord with the way of the world, that do not deliver mostly right answers in that domain are not useful (or downright harmful), and will not be selected and maintained in the course of evolution. To take an example, consider path integration again. The mechanism that computes the vector home as the animal undertakes the outbound journey must conform approximately with trigonometric laws; otherwise the job cannot be done and the output is not useful. Other domains of spatial navigation, such as landmark-based navigation, and other nonspatial domains, such as figuring out the return rate at a foraging patch, have different computational demands. Thus, landmark use might depend on computing the distance and direction to a landmark, and foraging optimally might depend on estimating durations between successful searches. Computing quantitative values to represent aspects of the world may be a common theme in cognition. But beyond this fact, we are not likely to find common mechanistic principles across domains.

There is, however, some point in storing computed quantitative information concerning diverse dimensions in a common code, thereby providing a common neurophysiological coding scheme or at least codes that could be readily converted into some common system. Information from different domains often needs to be integrated in making a decision. Quantitative values with different units might need to enter into a computation together. Cheng *et al.*'s (1996) study on the averaging of positional (spatial) and durational (temporal) dictates by pigeons and humans is significant in showing the integration of quantitative information from disparate dimensions of experience. The work suggests that some common code for quantitative values exists (Walsh, 2003), but the evidence is imperfect. Far more mechanistic analyses along these lines, at the behavioral and neurophysiological levels, are required.

SUMMARY

I have reviewed four themes in which principles common between spatial cognition and other domains of cognition are found. The first theme concerns quantitative values, representing dimensions of the world. I argued that it is highly likely that quantitative values are computed and used in some domains of spatial cognition. Path integration is a prime example. I have also suggested that the computational principles used in extracting such information are likely to vary from one domain to another. But the coding scheme for storing such information might be common, a view shared by Gallistel (2002) and Walsh (2003).

The second theme concerns generalization and discrimination. Both are found in the spatial domain as in other domains. A law of generalization (Shepard, 1987) based on a functional analysis has so far been confirmed in spatial generalization. The phenomenon of peak shift is also found in discriminating locations as in discriminating other stimulus dimensions, such as wavelengths of light. Whether there are any common mechanistic principles of generalization and peak shift remains to be seen.

The third theme is that multiple sources of information are sometimes used in navigation, as in other domains. Multiple numbers of landmarks may be averaged (e.g., Cheng, 1988, 1989). Categorical and metric information may be averaged (e.g., Huttenlocher *et al.*, 1991). Even spatial and temporal information may be averaged (Cheng *et al.*, 1996).

Finally, evidence for cue competition in the spatial domain is now plentiful. Both overshadowing and blocking have been found in landmark-based learning, in rats (e.g., Biegler and Morris, 1999; Roberts and Pearce, 1999), pigeons and humans (Spetch, 1995), and honeybees (Cheng and Spetch 2001b). Interestingly, the learning of the geometric shape of an arena (Cheng, 1986) usually does not get blocked or overshadowed.

Overall, while some aspects of spatial cognition may be unique, it appears that there are many aspects common to other forms of cognition across species. Spatial cognition is a blend of specialized principles and general principles of learning. Such a blend should not be surprising. The spatial world that animals live in has properties that are specific to the spatial domain. But the learning animal also encounters some generalities across all domains.

REFERENCES

Alais, D. and Burr, D. (2004). The ventriloquist effect results from near-optimal bimodal integration. *Current Biology*, **14**, 257–262.

Biegler, R. and Morris, R. G. M. (1999). Blocking in the spatial domain with arrays of discrete landmarks. *Journal of Experimental Psychology: Animal Behavior Processes*, **25**, 334–351.

Chamizo, V. D. (2003). Acquisition of knowledge about spatial location: assessing the generality of the mechanism of learning. *Quarterly Journal of Experimental Psychology*, **56B**, 102–113.

Cheng, K. (1986). A purely geometric module in the rat's spatial representation. *Cognition*, **23**, 149–178.

Cheng, K. (1988). Some psychophysics of the pigeon's use of landmarks. *Journal of Comparative Physiology A*, **162**, 815–826.

Cheng, K. (1989). The vector sum model of pigeon landmark use. *Journal of Experimental Psychology: Animal Behavior Processes*, **15**, 366–375.

Cheng, K. (1990). More psychophysics of the pigeon's use of landmarks. *Journal of Comparative Physiology A*, **166**, 857–863.

Cheng, K. (1992). Three psychophysical principles in the processing of spatial and temporal information. In W. K. Honig and J. G. Fetterman (Eds.), *Cognitive aspects of stimulus control* (pp. 69–88). Hillsdale: Erlbaum.

Cheng, K. (1994). The determination of direction in landmark-based spatial search in pigeons: a further test of the vector sum model. *Animal Learning & Behavior*, **22**, 291–301.

Cheng, K. (1995). Landmark-based spatial memory in the pigeon. In D. L. Medin (Ed.), *The psychology of learning and motivation*, vol. 33 (pp. 1–21). New York: Academic Press.

Cheng, K. (1999). Spatial generalization in honeybees confirms Shepard's law. *Behavioural Processes*, **44**, 309–316.

Cheng, K. (2000). Shepard's universal law supported by honeybees in spatial generalization. *Psychological Science*, **11**, 403–408.

Cheng, K. (2002). Generalisation: mechanistic and functional explanations. *Animal Cognition*, **5**, 33–40.

Cheng, K. (2005). Reflections on geometry and navigation. *Connection Science*, **17**, 5–21.

Cheng, K., Collett, T. S., Pickhard, A. and Wehner, R. (1987). The use of visual landmarks by honeybees: bees weight landmarks according to their distance from the goal. *Journal of Comparative Physiology A*, **161**, 469–475.

Cheng, K. and Newcombe, N. S. (2005). Is there a geometric module for spatial orientation? Squaring theory and evidence. *Psychonomic Bulletin & Review*, **12**, 1–23.

Cheng, K. and Sherry, D. F. (1992). Landmark-based spatial memory in birds: the use of edges and distances to represent spatial positions. *Journal of Comparative Psychology*, **106**, 331–341.

Cheng, K. and Spetch, M. L. (1998). Mechanisms of landmark use in mammals and birds. In S. Healy (Ed.), *Spatial representation in animals* (pp. 1–17). Oxford: Oxford University Press.

Cheng, K. and Spetch, M. L. (2001a). Landmark-based spatial memory in pigeons. Cyberchapter in R. G. Cook (Ed.), *Avian visual cognition*. Online, available at: www.pigeon.psy.tufts.edu/avc/cheng/default.htm.

Cheng, K. and Spetch, M. L. (2001b). Blocking in the spatial domain in honeybees. *Animal Learning & Behavior*, **29**, 1–9.

Cheng, K. and Spetch, M. L. (2002). Spatial generalization and peak shift in humans. *Learning and Motivation*, **33**, 358–389.

Cheng, K., Spetch. M. L. and Johnston, M. (1997). Spatial peak shift and generalization in pigeons. *Journal of Experimental Psychology: Animal Behavior Processes*, **23**, 469–481.

Cheng, K., Spetch, M. L. and Miceli, P. (1996). Averaging temporal duration and spatial position. *Journal of Experimental Psychology: Animal Behavior Processes*, **22**, 175–182.

Cheng, K., Spetch. M. L., Kelly D. M. and Bingman, V. P. (2006). Small-scale spatial cognition in pigeons. *Behavioural Processes*, **72**, 115–127.

Cheng, K., Shettleworth. S. J., Huttenlocher, J. and Rieser, J. J. (2007). Bayesian integration of spatial information. *Psychological Bulletin*, **133**, 625–637.

Church, R. M. (2003). A concise introduction to scalar timing theory. In W. H. Meck (Ed.), *Functional and neural mechanisms of interval timing* (pp. 3–22). Boca Raton: CRC Press.

Church, R. M. and Broadbent, H. (1990). Alternative representations of time, number, and rate. *Cognition*, **37**, 55–81.

Collett, T. S., Cartwright, B. A. and Smith, B. A. (1986). Landmark learning and visuo-spatial memories in gerbils. *Journal of Comparative Physiology A*, **158**, 835–851.

Denniston, J. C., Miller, R. R. and Matute, H. (1996). Biological significance as determinant of cue competition. *Psychological Science*, **7**, 325–331.

Denniston, J. C., Savastano, H. I., Blaisdell, A. P. and Miller, R. R. (2003). Cue competition as a retrieval deficit. *Learning and Motivation*, **34**, 1–31.

Dukas, R. (1999). Costs of memory: ideas and predictions. *Journal of Theoretical Biology*, **197**, 41–50.

Gallistel, C. R. (1990). *The organization of learning*. Cambridge, MA: MIT Press.

Gallistel, C. R. (2000). The replacement of general-purpose learning models with adaptively specialized learning modules. In M. S. Gazzaniga (Ed.), *The new cognitive neurosciences* (pp. 1179–1191). Cambridge, MA: MIT Press.

Gallistel, C. R. (2002). The principle of adaptive specialization as it applies to learning and memory. In R. H. Kluwe, G. Lüer and F. Rösler (Eds.), *Principles of human learning and memory* (pp. 250–280). Basel: Birkenhauser Verlag.

Gallistel, C. R. (2003). Conditioning from an information processing perspective. *Behavioural Processes*, **62**, 89–101.

Gallistel, C. R. and Gibbon, J. (2000). Time, rate and conditioning. *Psychological Review*, **107**, 289–344.

Gallistel, C. R. and Gibbon, J. (2002). *The symbolic foundations of conditioned behavior*. Mahwah: Lawrence Erlbaum Associates.

Gepshtein, S. and Banks, M. S. (2003). Viewing geometry determines how vision and haptics combine in size perception. *Current Biology*, **13**, 483–488.

Ghirlanda, S. and Enquist, M. (2003). One hundred years of generalisation. *Animal Behaviour*, **66**, 15–36.

Gibbon, J. (1977). Scalar expectancy theory and Weber's law in animal timing. *Psychological Review*, **84**, 279–335.

Gibbon, J. and Church, R. M. (1984). Sources of variance in an information processing theory of timing. In H. L. Roitblat, T. G. Bever and H. S. Terrace (Eds.), *Animal cognition* (pp. 465–488). Hillsdale: Lawrence Erlbaum Associates.

Gray, E. R., Bloomfield, L. L., Ferrey, A., Spetch, M. L. and Sturdy, C. B. (2005). Spatial encoding in mountain chickadees: features overshadow geometry. *Biology Letters*, **1**, 314–317.

Hanson, H. M. (1959). Effects of discrimination training on stimulus generalization. *Journal of Experimental Psychology*, **58**, 321–334.

Hayward, A., McGregor, A., Good, M. A. and Pearce, J. M. (2003). Absence of overshadowing and blocking between landmarks and geometric cues provided by the shape of a test arena. *Quarterly Journal of Experimental Psychology*, **56B**, 114–126.

Hopson, J. W. (2003). General learning models: timing without a clock. In W. H. Meck (Ed.), *Functional and neural mechanisms of interval timing* (pp. 23–60). Boca Raton: CRC Press.

Huttenlocher, J. and Vasilyeva, M. (2003). How toddlers represent enclosed spaces. *Cognitive Science: A Multidisciplinary Journal*, **27** (5), 749–766.

Huttenlocher, J., Hedges, L. V. and Duncan, S. (1991). Categories and particulars: prototype effects in estimating spatial location. *Psychological Review*, **98**, 352–376.

Huttenlocher, J., Hedges, L. V. and Vevea, J. L. (2000). Why do categories affect stimulus judgement? *Journal of Experimental Psychology: General*, **129**, 220–241.

Huttenlocher, J., Hedges, L. V., Corrigan, B. and Crawford, L. E. (2004). Spatial categories and the estimation of the location. *Cognition*, **93**, 75–97.

Kamil, A. C. and Cheng, K. (2001). Way-finding and landmarks: the multiple-bearings hypothesis. *Journal of Experimental Biology*, **204**, 103–113.

Kamin, L. J. (1969). Selective association and conditioning. In N. J. Mackintosh and W. K. Honig (Eds.), *Fundamental issues in associative learning* (pp. 42–64). Halifax: Dalhousie University Press.

Kelly, D., Spetch, M. L. and Heth, C. D. (1998). Pigeon's encoding of geometric and featural properties of a spatial environment. *Journal of Comparative Psychology*, **112**, 259–269.

Kersten, D. and Yuille, A. (2003). Bayesian models of object perception. *Current Opinion in Neurobiology*, **13**, 150–158.

Killeen, P. R. and Fetterman, J. G. (1988). A behavioral theory of timing. *Psychological Review*, **95**, 274–295.

Knill, D. C. and Saunders, J. A. (2003). Do humans optimally integrate stereo and texture information for judgments of surface slants? *Vision Research*, **43**, 2539–2558.

Körding, K. P. and Wolpert, D. M. (2004). Bayesian integration in sensorimotor learning. *Nature*, **427**, 244–247.

Körding, K. (2007). Decision theory: What "should" the nervous system do? *Science*, **318**, 606–610.

Lehrer, M., Srinivasan, M. V., Zhang, S. W. and Horridge, G. A. (1988). Motion cues provide the bee's visual world with a third dimension. *Nature*, **332**, 356–357.

Lynn, S. K., Cnaani, J. and Papaj, D. R. (2005). Peak shift discrimination learning as a mechanism of signal evolution. *Evolution*, **59**, 1300–1305.

Mackintosh, N. J. (1976). Overshadowing and stimulus intensity. *Animal Learning & Behavior*, **4**, 186–192.

Margules, J. and Gallistel, C. R. (1988). Heading in the rat: determination by environmental shape. *Animal Learning & Behavior*, **16**, 404–410.

Maurer, R. (1998). A connectionist model of path integration with and without a representation of distance to the starting point. *Psychobiology*, **26**, 21–35.

Maurer, R. and Séguinot, V. (1995). What is modelling for? A critical review of the models of path integration. *Journal of Theoretical Biology*, **175**, 457–475.

Meck, W. H. (Ed.) (2003). *Functional and neural mechanisms of interval timing*. Boca Raton: CRC Press.

Miller, R. R. and Matute, H. (1996). Biological significance in forward and backward blocking: resolution of a discrepancy between animal conditioning and human causal judgment. *Journal of Experimental Psychology: General*, **125**, 370–386.

Müller, M. and Wehner, R. (1988). Path integration in desert ants. *Cataglyphis fortis*. *Proceedings of the National Academy of Sciences*, **85**, 5287–5290.

Newcombe, N. S. and Huttenlocher, J. (2000). *Making space: the development of spatial representation and reasoning*. Cambridge, MA: MIT Press.

O'Keefe, J. and Nadel, L. (1978). *The hippocampus as a cognitive map*. Oxford: Oxford University Press.

Pearce, J. M. and Bouton, M. E. (2000). Theories of associative learning in animals. *Annual Review of Psychology*, **52**, 111–39.

Pearce, J. M., Graham, M., Good, M. A., Jones, P. M. and McGregor, A. (2006). Potentiation, overshadowing, and blocking of spatial learning based on the shape of the environment. *Journal of Experimental Psychology: Animal Behavior Processes*, **32**, 201–214.

Pearce, J. M., Ward-Robinson, J., Good, M., Fussell, C. and Aydin, A. (2001). Influence of a beacon on spatial learning based on the shape of the test environment. *Journal of Experimental Psychology: Animal Behavior Processes*, **27**, 329–344.

Purtle, R. B. (1973). Peak shift: a review. *Psychological Bulletin*, **80**, 408–421.

Rilling, M. (1977). Stimulus control and inhibitory processes. In W. K. Honig and J. E. R. Staddon (Eds.), *Handbook of operant behavior* (pp. 432–480). Englewood Cliffs: Prentice-Hall.

Roberts, A. D. L. and Pearce, J. M. (1999). Blocking in the Morris swimming pool. *Journal of Experimental Psychology: Animal Behavior Processes*, **25**, 225–235.

Rodrigo, T., Chamizo, V. D., McLaren, I. P. L. and Mackintosh, N. J. (1997). Blocking in the spatial domain. *Journal of Experimental Psychology: Animal Behavior Processes*, **23**, 110–118.

Shepard, R. N. (1987). Toward a universal law of generalization for psychological science. *Science*, **237**, 1317–1323.

Shepard, R. N. (1994). Perceptual-cognitive universals as reflections of the world. *Psychonomic Bulletin & Review*, **1**, 2–28.

Spetch, M. L. (1995). Overshadowing in landmark learning: touch-screen studies with pigeons and humans. *Journal of Experimental Psychology: Animal Behavior Processes*, **21**, 166–181.

Srinivasan, M. V., Lehrer, M., Zhang, S. W. and Horridge, G. A. (1989). How honeybees measure their distance from objects of unknown size. *Journal of Comparative Physiology A*, **165**, 605–613.

Tommasi, L. and Vallortigara, G. (2000). Searching for the center: spatial cognition in the domestic chick (*Gallus gallus*). *Journal of Experimental Psychology: Animal Behavior Processes*, **26**, 477–486.

Wall, P. L., Botly, L. C. P., Black, C. K. and Shettleworth, S. J. (2004). The geometric module in the rat: independence of shape and feature learning in a food finding task. *Learning & Behavior*, **32**, 289–298.

Walsh, V. (2003). A theory of magnitude: common cortical metrics of time, space and quantity. *Trends in Cognitive Sciences*, **11**, 483–488.

Wehner R. and Srinivasan, M. V. (2003). Path integration in insects. In K. J. Jeffery (Ed.), *The neurobiology of spatial behaviour* (pp. 9–30). Oxford: Oxford University Press.

Williams, B. A. (1996). Evidence that blocking is due to an associative deficit: blocking history affects the degree of subsequent associative competition. *Psychonomic Bulletin & Review*, **3**, 71–74.

Wohlgemuth, S., Ronacher, B. and Wehner, R. (2001). Ant odometry in the third dimension. *Nature*, **411**, 795–798.

Wohlgemuth, S., Ronacher, B. and Wehner, R. (2002). Distance estimation in the third dimension in desert ants. *Journal of Comparative Physiology A*, **188**, 273–281.

EMIL W. MENZEL, JR.

4

To be buried in thought, lost in space, or lost in action: is that the question?

WHAT IS SPACE?

> Time, space, place and motion, being words well known to everybody, I do not define. Yet it is to be remarked, that the vulgar conceive these quantities only in their relation to sensible objects. And hence certain prejudices with respect to them have arisen, to remove which it will be convenient to distinguish them into absolute and relative, true and apparent, mathematical and common, respectively. (Sir Isaac Newton, 1687, as quoted by Mach, 1907, p. 222)

I side with the vulgar commoners. I have no interest in pure, absolute, true, mathematical space. At best, I use the concepts of Euclid, Descartes and Newton as mere tools for examining what animals and people do, so that I might deduce or infer how they perceive and conceive the above quantities in relation to sensible objects. In this chapter I shall review some of my by-now-ancient studies, focusing on my views on spatial perception, spatial cognition, and relations between the two.

A GROUP OF YOUNG WILD-BORN CHIMPANZEES IN A ONE-ACRE FIELD

During my years at the Delta Regional Primate Research Center (now the Tulane Primate Center), between 1965 and 1971, I spent most of my time studying nine young wild-born chimpanzees. I followed them from infancy to puberty. At first they lived in pairs or small groups; later, they were brought together in larger groupings. Some of my work was of a baseline sort – in other words, I did not intentionally introduce extra variables into the situation and tried to see what the animals did when presumably undisturbed. The rest consisted of more than fifty small experiments in which specific variables were introduced into the situation to probe what I thought might be going on under everyday

Spatial Cognition, Spatial Perception: Mapping the Self and Space, ed. Francine L. Dolins and Robert W. Mitchell. Published by Cambridge University Press.

circumstances. Although my studies were both experimental and observational, their approach was similar to that taken by any naturalist who follows in the footsteps of Henry Nissen (1931) and Clarence Ray Carpenter (1964) and sets out to inductively examine a particular group of primates in a particular forest. In almost all observations and experiments the same system of recording was used, and all of my questions were variations on the same themes: *where are the chimpanzees, really? Where will they go next? Why do they go there rather than elsewhere?*

The basic experimental procedure consisted of herding the animals into their indoor sleeping quarters, on the edge of the enclosure, if they were not there already; performing some simple manipulation in the field, or with one or more individuals; returning the pre-selected individuals to the indoors; removing ourselves from the enclosure and climbing onto an observation tower; and then turning the animals loose, usually all of them simultaneously. The prime raw data were ordinary maps of the enclosure, on which my highly skilled, dedicated and invaluable assistant, Palmer Midgett, recorded the simultaneous locations of each animal every time a timer sounded a beep, which in most experiments was every 30 seconds. The enclosure was divided into 351 sectors, most of which were about 3 by 3 m. I myself made additional notes both at the time and afterwards, and numerous movies and photos. The most useful of my notes in the long run were those that I had labeled pure diary, speculation, qualitative summary, or anecdote.

With sufficient practice, most people could flip through my old loose-leaf volumes of maps (or computer analogs thereof) and easily, literally and directly see what the group as a whole, and each individual animal as well, was doing, not only from one minute to the next but also across much longer stretches of time and probably – and most importantly – across the various different conditions to which we exposed the animals. For many purposes the data would be crude if not worthless compared to a continuous sound motion picture. But for my purposes I would not have traded them for any amount of raw film, particularly if someone else had collected that film and had had no clear idea in advance of how it was to be reduced, analyzed and organized.

Everything I have said or hinted at thus far might have been taken right out of an undergraduate textbook on physics (see, in particular, Reif, 1967 and Lindsay and Margenau, 1936, Chapter 2) – except, of course, for my concern for the natural history, prior experiences, and individual identity, psychology and fate of the "physical particles" that I studied. Translating the maps into numbers, the basic data become the Cartesian X, Y, Z spatial coordinates of each of the *N* animals on each of the *T* time intervals, and the basic analytical problem becomes to account for as much as possible of the total variance of these data, using the smallest feasible number of concepts and parameters. To be sure, the *T* time intervals are divided into smaller blocks for some purposes, just as the *N* animals might be divided into subgroups, and the various test objects or test conditions might be divided into classes of objects or events, and the one-acre field might also be divided into blocks and plots, *à la* Sir Ronald A. Fisher's analysis of variation (1930, 1951). Translating these raw numerical data back into visual terms, one of the easiest ways to visualize the data as a whole is to plot a graph on which the ordinate represents the animals' location on one of the three spatial axes of their environment – ignoring the other two axes, at least for the time being – and the abscissa of the graph represents time. But at this point I sense increasing danger of my falling off the deep end into the various statistical and computer ruminations on which I have squandered more years than I care to count; so let me stop here. To clear the air of statistical fire and brimstone, please peruse and ponder the works of Edward Tufte (1982, 1990, 1997). To fog up again, see my other papers, especially Menzel and Menzel (2007). I did, coincidentally, initially have a long, detailed statistical discourse as the introduction to my 1974 paper but I am glad that I dropped it before the paper went to press, thanks to the critical comments of editor Allan Schrier.

Over the course of this chimp research, I made several short trips to Puerto Rico to study the behavior of free-ranging rhesus monkeys in relation to objects in space (to borrow a phrase from Clark Hull, 1952). One result was my zoom lens analogy of perception/cognition (Menzel 1969, 1979; Menzel and Wyers, 1981) and a growing conviction that the term "stimulus as such" is science fiction rather than science. Perceptual objects come in no fixed or invariable sizes or durations. Of course there are some rough minima and maxima, both spatial and temporal, but they are to a considerable extent specific to particular animals in particular locations, with particular perspectives and particular sensory and cognitive capacities and past experiences, etc. It is no wonder that some perceptual problems that seem elementary to us have proved harder for computers to solve than supposedly difficult problems such as (say) chess. The zoom lens analogy has, coincidentally, been reinvented numerous times in a variety of disciplines; it is, I believe, more generic and more powerful than mapping analogies. Its essence is, perhaps, that one does not have to choose, once and for all, whether to be macroscopic, microscopic or telescopic in outlook; one can pick any point within such continua, and also zoom in either direction.

Is a 30.5 × 122 m (slightly less than one acre) enclosure big or small for nine chimpanzees, and this sort of study? As I see it, no area could be too big, and what would be too small is a thorny issue. Given my budget and recording equipment, I felt that the space was adequate, at least while the animals were infants. They outgrew the area very rapidly. On the other hand, they voluntarily spent many hours indoors or even inside a burlap sack or a cardboard box barely big enough to squeeze into, sometimes closing the flaps of the box as well. It is being unable to go anywhere any time, on demand, that makes a space too small, behaviorally speaking, and spaciousness cannot be measured with nothing more than a yardstick or a map. Some people feel trapped and claustrophobic on any island, unless they own a boat or airplane. It would not be difficult to find analogies to this in the migrations and dispersals of nonhumans. The studies on rhesus that I mentioned above led me to wonder also whether any nonhumans perceive and know, as gestalts, such large-scale things as islands, mountains, fields, roads, rivers and entire small forests. Surely some do. (But don't ask me to prove it.) High-flying birds have an obvious and enormous advantage here over any terrestrial animal, for their eyesight is in many ways as good as ours and they are in a position to see some of the objects in question directly, almost at a glance. No infant chimpanzee could possibly see our one-acre enclosure as such, when quadrupedal on ground level, even if the ground had been perfectly flat and without vegetation, which was not the case. Even from the tops of the trees, which some of the chimps did not reach for several months, any chimpanzee would have to scan broadly to view the entire area. I would estimate that it took the animals at least twenty half-hour sessions of initial exploration to learn the area even to a rough first approximation. Do all animals come to recognize a given area as being "enclosed," and do all use the same strategy to discover that? My guess on each question is, no.

WHERE WILL THEY GO NEXT?

The basic locationistic questions, as I have just posed them, first, last, and almost always, almost immediately, gave rise to more specific problems of group behavior because, generally speaking, chimpanzees go where others go. Particularly when the animals were young, more than 90 percent of the variance of any individual's positions in the field enclosure could be predicted if the successive locations of the group as a whole (i.e., the spatial centroid or center of gravity of all *N* animals) were known.

The same questions gave rise to ecological problems, for the group as a whole generally went wherever there were certain classes of objects and environmental features. However, each animal was typically within arm's reach of *some* object or another – if not a companion, then a tree, bush or the fence wire, most often. The amount of time spent in sectors of the enclosure that were relatively barren of objects or shade was usually minimal.

The locationistic questions gave rise to problems of exploration, habituation, learning and memory because, regardless of the class of object or environmental feature, its effect upon spatial adjustments changed with repeated presentations and with the consequences of previous encounters. I do not believe that I have ever tired of watching how animals – especially young ones – respond to novel objects and environments. For humans most of all but also for other animals there seems to be literally no limit as to the amount of information, not to mention the amount of enjoyment or utility, that can be extracted from any given object or environment. Obviously, motivational factors as well as learning are important; without them, why would an animal do anything or go anywhere at all?

The locationistic questions gave rise to problems of perception for many reasons, but one was that it was difficult for us to introduce any object or change, anywhere in our own eye-shot or ear-shot of the enclosure, that was not very quickly detected, and investigated (e.g., Menzel, 1971). The chimps knew their little acre, in their own way, better than we did, and it would not be hard to expand upon that sentiment (C. Menzel *et al.*, 2002). As to how many objects and locations a chimpanzee can remember simultaneously, that obviously depends on what you mean by simultaneously, an object, and a location; but I'd guess it is more than anybody will ever know.

They gave rise to the problem of perceptual organization, as opposed to approach toward or withdrawal from single pre-selected objects, because the animals' movements relative to one another and to inanimate objects appeared to take into account many objects simultaneously, and furthermore to take into account the spatial relationships between the locations of many objects. In the 1960s, some scientists still dismissed the problem of organization as quasi-mystical. I in turn suspected that they were either dodging the problem or trying to translate it into other words, especially if, almost in the next breath, they invoked the concepts of context, spatial gradients or spatial generalization around a given focal object. How can you tell which or what stimuli are at the core or focus of an animal's (as opposed to the experimenter's) attention, and which or what stimuli are context? (Think zoom lens analogy.) That problem has baffled students of perception since the nineteenth century. It brought down the doctrines of empiricism and associationism as they were held by structuralistic psychologists prior to the days of gestalt psychology and of the Gibsons (review: Allport, 1955 – see especially pp. 106–110). For me, circa 1966, core versus context badly shook the doctrines of most learning theories I had been exposed to thus far, as I understood them; and it seemed to me that S-R (stimulus-response) theory in particular was stated, in effect, in terms of two major unknowns. The gestaltists (especially Koffka, 1935 and Köhler, 1929), the Gibsons (1958, 1966, 1969), and Miller *et al.* (1960) made more sense to me. So too did Hediger (1964, 1968); his fascinating descriptions of many different species of wild animals in captivity pay much more than lip service to Psychology's third, and most salient, unknown: organisms.

They gave rise to problems of group structure or social organization because no chimpanzee simply followed the herd; instead, each attended more to certain individuals than to others; and, as a result, there were always stable clusterings and subgroupings of associated individuals within any larger aggregate. I conjecture that in my small groups and small environment every chimpanzee that was awake, healthy and normal was almost constantly cognizant of its location relative to every other chimpanzee, not to mention its own location in the enclosure, and that the first thing it did after any appreciable lapse in attention or consciousness was, in effect, to update itself on this information.

The same locationistic questions gave rise to problems of early experience because the most stable associations between individuals depended on who had been raised with whom: a sort of laboratory counterpart to family relations except that here relations were based on familiarity or long-term spatial propinquity, without genetic relationship, and each chimpanzee was both protector and comforter of, and seeker of protection and comfort from, its peers. [Also, not surprisingly, these wild-born animals did many things that I would not expect from nursery-raised chimpanzees. Köhler might well have complained that critics of his research on chimpanzee intelligence completely ignored the environments and social conditions in which their own animals were raised and housed; I'd say that his complaint was completely justified. Specifically, the chimpanzees he studied were wild-born and presumably mother-reared; those studied by Birch (1945) and Schiller (1957) had been taken from their mothers almost at birth and raised by humans in a nursery; and such maternal deprivation alone is likely to produce chimpanzees that are atypical in many ways, even as adults, and including on Köhler-type learning tasks involving object-manipulation, tool-using, and utilization of space. By the same token I consider Povinelli's recent book (2000) about "the chimpanzee's theory of how the world works" to be naïve, and mistaken in its conclusions. My reasons are more fully spelled out in, for example, Davenport *et al.* (1969, 1973) and Menzel *et al.* (1970).]

They gave rise to problems of social development because as the animals grew older and more skilled at climbing, running, and most activities, they spread out farther and were much less apt to follow each others' every move. Is this increasing social independence or is it more accurately described as the simple ability to perceive or know where your companions are, relative to you, and in what way, and how quickly, the gap could be closed, if necessary? I say it's the latter, if only for the sake of argument. Köhler's most famous claim was that a chimpanzee alone is not a chimpanzee (see Köhler, 1927, p. 282 for the exact quote in Ella Winter's translation; it is somewhat different and less pontifical). I claim, paraphrasing Daniel Defoe's *Robinson Crusoe*, that no living being is ever truly alone.

They gave rise to problems of group coordination and control and individual status, because single individuals were at times capable of leading others to some object or event that only they had seen, and not all individuals were equally capable in this respect. Of particular interest here, more animals followed leaders who were going to a pile of several pieces of hidden food than went to a lesser quantity of food that was directly visible to anyone. Analogously, if two leaders were each shown a pile of hidden food, whichever one had been shown a larger quantity attracted a larger number of followers. So which is more important – perception or memory? And what about the old truisms of a bird-in-hand being better than two in the bush, or out of sight being out of mind? For an informed animal, memory of food in a given place can easily overrule direct perception, if needed. Judgments of the relative quantities of visible versus invisible food are very accurate in individual animals tested in the lab (Menzel and Draper, 1965; Beran, 2004) and this finding did hold up well under quasi-field conditions. So too did the finding that a wide variety of direct and indirect cues to food quantity and location could be used interchangeably, for here the group of chimpanzees also used each other as cues. Menzel and Draper remarked, semi-facetiously, that a bird in hand is worth perhaps 1.02 in the bush. But sometimes it can be worth even less than 1.00 and sometimes it can be worth far more than 2.00, depending on the precise conditions. It is, I suspect, well-nigh impossible to completely separate cognition from perception, or either of these from motivation and action. That seems to me more a linguistic and logical exercise than a biological one.

The problem of extrapolation arose because not only could the animals perceive and remember the distances and directions and locations of objects, but also a sufficient cue for them was for someone else to merely walk or point in a given direction. The shorter the walk we ourselves took toward the location of the hidden food or toy (before putting the leader back with its companions for a delay period and getting ourselves out of the enclosure), the greater the probability that followers could beat the leader to the goal. Our walking and pointing cues (manual or other) were almost equally effective cues of direction, even though the chimpanzees themselves almost never pointed manually toward objects. But why in the world should they point manually? For one thing, chimpanzees are much less inclined to share their goodies than are (some) humans. For another thing, the approximate direction and distance of an animal's goals are by no means difficult to surmise from its behavior even if it does not utter a sound or stand bipedal and raise its arm. My dog, Jip, probably taught me that before I could talk, in which case all I had to do was relearn what human education in my era had induced me to forget.

And, to be sure, the locational questions gave rise to problems of communication, for how else but by some inter-individual interchange can chimpanzees coordinate their movements with respect to each other or achieve group unity? Every organized biological or sociological system presupposes some sort of communication – in the broadest, information-theoretical, definition of that term – and one approach to communication is to start with a case in which a high degree of organization is known to exist, and then work backward to pin down whatever variables happen to account for that organization. Most students of non-human communication follow, instead, what I call a signal-oriented approach; that is, they look for highly specific, molecular classes of presumably innate movement patterns or vocal calls, etc. that are presumed to have evolved for the function of communication, and then they expand out from there. I used to preach on how my way was better, but that sermon would be a digression here.

The chimpanzees' behaviors cast strong doubts upon the empiricist (or perhaps mostly Thorndikean) dogma that slow, gradual learning of new objects and locations is the norm for most animals. I later posed the question more sharply and rhetorically: are learning sets *à la* Harry Harlow learned? (Alternatively, is learning itself learned?) For some marmosets, as for chimpanzees, one-trial learning of new objects and small-scale locations seemed in our test situations to be much closer to the rule than to the exception, unless we posed the problem to the animals poorly or otherwise deliberately set to produce slow, gradual learning. Such a statement would have been flaming heresy in most laboratories fifty or even fewer years ago (see the exchange between Menzel and Juno, 1982, 1984, 1985 and Schrier and Thompson, 1984), but it has by now been endorsed if not confirmed in many field studies and is not often contested any more by laboratory researchers. Of course, one-trial learning never did seem controversial or unexpected to all theorists of the 1960s; Guthrie, Skinner and Lorenz were cases in point. Guthrie and Skinner were fairly staunch empiricists; Lorenz was a fairly staunch nativist. (Nobody is 100 percent of either.) Nature–nurture, for me, has always been a chicken-and-egg problem.

The problem of deception arose because sometimes, instead of following Euclid's rule that the shortest distance between two points is a straight line, an informed leader's path was as devious as that of any animal in a maze or detour problem, and the major "barrier" was not hard to detect if you knew precisely who was on the leader's heels, metaphorically if not physically speaking, and how much food, if any, he would leave for her if he got to it first. My observations on this score went somewhat beyond those of van Lawick-Goodall (1971) but neither of us aimed for, or got, the high publicity achieved later by others, who characterized such behaviors as positively Machiavellian (Byrne and Whiten, 1988). Probably the first to get their priorities straight from the perspective of natural selection were Dawkins and Krebs (1978): why, they asked, would one ever expect deception to be rare or to be found only in humans? Any animal can be devious, and that is true literally – spatially – before it is true metaphorically or introspectively. Ladies and gentlemen of the jury, in the court of Charles Darwin defendants are not assumed to be innocent of deviousness or even intelligent deception, until proved guilty; it is the other way around. The ontogenetic development of deviousness in chimpanzees is slow, gradual and painstaking principally (so I conjecture) when one's test situation is arbitrary and artificial as well as novel. I doubt that one could entirely separate spatial learning from social learning in normally-reared infant chimpanzees, but in either case, "the pursuance of future ends and the choice of means for their attainment are ... the mark and criterion of the presence of mentality in a phenomenon" (James, 1890, Chapter 1; see also Menzel and Menzel, 2007).

The possibility of what might, fancifully speaking, be called rumor-bearing arose because in some tests in which no animal was given precise information as to the location of a distant object, and where (because the object was not entirely hidden) general searching behavior was rewarded, the whole pack sometimes unintentionally led one another in a wrong direction. If one animal for any reason started to move in a different direction and more rapidly than the others, someone would spot this and try to head him off; the instigator of the move would in turn seem to take this as a sign that the other animal knew where to go and try to keep ahead of him; still others would see these two animals and start after them; and so on in a vicious cycle, until the whole pack was racing down the field in a very different direction from the informed leader, checking every likely looking hiding place along the way. If a lead animal stopped, others would stop, and so on, until everyone just stood eyeing each other. The informed leader (to whom we had given a cue of direction) seldom joined in these activities unless it had searched the field in a given direction without finding the object. An alternative description of such events is that the animals not only displayed expectancies of their own, as Otto Tinklepaugh (1928) taught us, but also could acquire expectancies from others.

To be even more colorful, the animals' behavior suggested that they were susceptible to propaganda. Snakes were a common occurrence in the enclosure, and in formal tests as well as everyday observations it seemed clear that piloerection, staring, decelerating locomotion, hooting, picking up and throwing a stick, and so forth, did not generate free-floating emotion on the part of others, but rather they induced staring toward the place at which the informed animal or animals oriented, and all due caution if and when one came close to that place oneself. It was easy for us to induce much the same reactions toward almost any given spot merely by imitating (poorly, in my case) the chimpanzees' cautious "hoo" vocalization or by staring, then heaving a rock, or by pointing a rifle and acting as if we were going to shoot. But of course both the place and the acting had to be natural and plausible for the chimps. Pointing a rifle at (for example) the empty sky or a large, bare patch of ground, or the side of the observation house, produced no reaction; with an old rat-hole in the ground, or tall, thick grass, we were really in business. We could also, just as easily, induce food-searches by displaying interest rather than caution toward a plausible-looking hiding place, and perhaps adding a chimp "food grunt."

One of my most egregious lies was as follows. Whenever the chimps got hold of a water hose they not only played with it but quite often chewed holes in it. On one such occasion, I noticed that the hose was a mottled green and black in color – just like some snake, assuming one were a mediocre zoologist. The next day I accordingly cut off a short piece, all but completely concealed it in tall grass, and then turned the chimps loose while I was still in the enclosure. Just a bit of simulated interest on my part in the general direction of the hose-in-the-grass had every member of the group curious, but also wary. One chimp shortly gave a loud alarm call, at which the action really mounted. (Did it actually see a bit of the "snake"? I certainly thought so, and thought that the other chimps thought so too, but of course that's all just anecdote, don't you think?) They learned better in a matter of minutes, by which time the entire piece of hose was exposed to view; but then too the hose just lay there, did nothing, and got stoned, clubbed and slapped around in the interim. Once more, it is not what "stimuli" fall on the retina but what one thinks or assumes is out there that counts, and it is seldom easy to separate all of the psychological processes in question. To really produce fear in a group of wild chimpanzees, using Kortlandt and Kooij's (1963) stuffed leopard, I would take my clues from Hollywood horror films rather than from ornithology and farmers' scarecrows: in other words, I would leave as much as possible to the animals' imagination, but by all means try to be plausible. Kortlandt and Kooij's fake beast lay indefinitely in the same place, out on open ground in the midday sun, doing nothing more than roll its head in a circle and emit the sounds of the electrical motor that performed that job. Had it, or better yet, just a piece or two of it, been visible for no more than one second, as it disappeared in visually convincing and realistic fashion into a cave or the cover of darkness, it would, I'd bet, have commanded far more respect. One good, realistic growl – in a thoroughly plausible context – would be just as good as the visual display, and possibly even better. I should not, however, have presumed here to judge precisely what constitutes "plausibility" for any given animal, or precisely how non-human primates differ from crows or people; these are, rather, open empirical questions and highly interesting questions, especially if one is concerned with the relations between perception and cognition or (much more specifically) ability to reason. To wax poetical, once the metaphorical leopard disappears into its metaphorical cave, one may chant:

> As from the direct light of the sun to the borrowed reflected light of the moon, so do we pass from the immediate representation of perception, which stands by itself and is its own warrant, to reflection, to the abstract, discursive concepts of reason (Vernunft), which have their whole content only from that knowledge of perception, and in relation to it. (Schopenhauer, 1966, Vol. 1, p. 35).

Precisely where – or, I should say, approximately where, and within what spatial limits – would a chimpanzee assume that a live animal, a banana, a rock and an ice-cube are located, *T* minutes after these items have last been spotted? Are chimps too dumb to realize that live animals may move or hide in the interim; that apples might well disappear if not hidden from other banana-eaters in the vicinity, but do not ordinarily travel far on their own unless they fall from a tree or roll down a hill; that rocks are even less apt to wander unless they are of the perfect size for someone to use for nut-cracking or for breaking a window; and that ice-cubes melt in warm weather? I can attest that my chimps searched over a much broader area for a (totally removed) snake that they had seen in a given locus than for food or toys; hopefully other chapters in this volume will tell us more.

Assuredly too the chimpanzees' behavior toward one another and toward people and other animals suggested that they might have what some psychologists now call a TOM (theory of mind). At this point, however, I prefer to use plain English. I have never really worried over whether nonhumans have theories of mind or even theories of gravity, except in jest or to provoke further observation or reflection. As I see it, events, not Platonic metaphors, are the measure of things real in science. If I wished to push the foregoing metaphors, here's how I'd do it (cf. Menzel, 1964; Menzel and Johnson, 1976). Mother Nature had a theory of space, time, gravity, causality, behavior, mind, soul, etc. before any of her creatures did, even before they were hatched. Says she, the bodies of all my creatures, especially their limbs (if they have any limbs), are designed above all to exploit and cope with the forces of gravity in their chosen niches and to get to them wherever they must go; and eyes and ears are designed above all to scan ahead for whatever might be good to eat or have fun with, or to do one in. Beware, she warns us, of other things that have limbs, eyes, ears, etc. like yours, and can move, especially if their movements seem self-propelled and start to head for you, faster than you yourself can move, and take into account from one instant to the next what you yourself are doing. (See, for example, the studies of ethologists of aversive reactions to whatever looks like an eye, as described by Gould and Gould, 1994, and Gibson's 1969 account of perceptual looming, and my studies of primate responsiveness to objects and object-motion. In antiquity, scholars distinguished animals from plants and inanimate objects on the basis of *motus spontaneus in victu sumendo* – spontaneous movement in the taking of food – and probably every animal with eyesight is sensitive to that distinction.) To be sure, psychologists will be and should be concerned with how sophisticated any given animal's reactions really are, perceptually and cognitively speaking, but please don't throw out the baby with the bath water.

The problems of tool using and invention arose because some of the places the chimps loved most to go were out of, through, or over the fence, and each of their major new strategies, using branches as climbing tools, was increasingly dramatic (Menzel, 1972). Fortunately for us, the chimps never tried digging under the fence as well as going over the top. A list of other apparent anisotropies in the use of space could be quite a long one (Hediger, 1964; Menzel, 1969, 1978). For example, in perceptual psychology there is a well-known phenomenon called the horizontal–vertical illusion: the horizontal and vertical spatial dimensions are not exactly equivalent, and lengths in one direction might be overestimated. In everyday life, try this: if a timid young animal won't come near a tall, novel object, try laying down the object flat on the ground. The change in behavior can be striking, and especially so if you yourself are the test object in question. I do not know whether many students of perception or of cognition would consider such things interesting, but I do.

The problem of memory organization or mapping, as opposed to responsiveness to single objects or locations, independently of further context, arose for reasons hinted at by much of the above. Robert Yerkes tested apes outdoors by walking them around on a leash, showing them which one of two objects (of his own manufacture, and laid side by side) contained food, and then leaving but bringing them back some time later, and letting them pick one of the two objects. Whereas chimps tested indoors in the apparatus of that era did well to remember such an event for a minute, Yerkes' animals had no difficulty even a day or two later. In one more small step for mankind and a giant leap for animal psychology, Tinklepaugh (1932) said: let's make that many pairs of objects, and see how many items can be remembered. He put sixteen pairs of objects in a circle in a big indoor room (think radial maze). Keeping the animal on a leash, he assured that it would be in the center of the circle for each of the sixteen choices, one per each pair of objects or general location. Tinklepaugh, not the subject, dictated which pair of objects the subject went to next, which of course makes his task quite different from, and in some ways less informative than, most radial maze tasks (e.g., Olton and Samuelson, 1976). He offered no evidence that the animals took into account the spatial relations between the sixteen locations, but as I recall, Köhler, Tolman and Yerkes, or at least two of the three, not only were Tinklepaugh's advisors, but also actually sat in on the testing. None of the four, to my recollection, ever said a word about mapping when describing the otherwise very impressive data. Nor would I have done so either.

My innovation was to recognize that the neat circle was superfluous and that man-made objects, including leashes and mazes, sometimes obscure more than they clarify. In other words, I simply carried the animals around outdoors (as they had ridden on their mother's back or belly as infants), allowed them to observe a familiar caregiver placing food in a number of locations, in natural cover, and then examined their relative efficiency in finding the food, as compared to uninformed companions, and, most importantly, the relative overall efficiency of their entire travel route, compared to chance. A different set of locations was used on each test trial. Tolman's neologisms and his psychologese (for example, "cognitive map") were of little importance to me, compared to his more general spatial philosophy (for example, Tolman, 1932, especially pp. 424–426) and straightforward geometry. My 1978 paper on cognitive mapping in chimpanzees summarized my work and thoughts on that topic; C. Menzel *et al.* (2002) is also highly relevant. Drop the word "cognitive" if you really dislike it, and please do recognize that the difference between map (noun: object, picture or other representation) and mapping (present participle of map: process, not object) is important, not only according to any dictionary but also according to human explorers and geographers. Unfortunately, students of animal behavior are seldom clear on the difference, and some mathematicians (for example, Bennett, 1996) are no clearer. I would not call the using of landmarks or "cognitive graphs" (Muller *et al.* 1996) evidence against mapping; they are themselves particular, specialized examples of mapping. As for the more primordial question of what is associated with what, and by what mechanism, S-R associationism *à la* Hull or Guthrie is extinct, as far as I can tell; and S-S associationism *à la* Tolman is beginning to sound as quaint as old-time Idea-Idea association. But you still can't go wrong in studying James (1890). He anticipates almost every student of Mind since his time, regardless of their academic discipline, and his chapter on Association is a gem, especially if read in full context.

The possibility of recall memory as well as recognition arose because sometimes, after having lain on its back apparently asleep for half an hour or longer, an animal sometimes suddenly jumped to its feet and ran straight to a hidden piece of food that we had shown it some time earlier, while all of its uninformed companions simply looked on. Where was the stimulus that triggered this response? Visual information was obviously available once the animal's eyes had opened, but the molar action seemed to commence before that time. Compared to current findings (see particularly C. Menzel, 1999, and Chapter 9, this volume) this is tame as well as anecdotal; I am as amazed and excited by the new data and the new and improved methods as my own colleagues were amazed by mine.

Now then, is it ridiculous to ask all of the above rhetorical questions about fellow-creatures that still or perhaps forever will lack a full-blown human language, let alone the other interests and skills that are necessary to pass the first grade of grammar school? Philosophically speaking, I did not think it ridiculous at all. But how can one prove the point definitively one way or another for a nonhuman, to empirically and experimentally minded scientists who are not familiar with monkeys and apes, especially under natural conditions? I did not know, and still don't. I just tried to cultivate my own garden and to welcome visitors who wish to look for themselves.

As I understand the problems of meta-perception and meta-cognition, they had arisen even before the first day of my outdoor studies, because in the 1960s some investigators were, in effect, claiming that terms such as novelty, exploration, and fear of the dark imply meta-cognition. As they put it, dark is the absence of stimuli, and how can that elicit a response; talk of novelty implies that animals know what they have not yet experienced, and how is that possible; do animals go out to explore what lies ahead, or are they merely trying to avoid past or current conditions, such as hunger or boredom? I was of the new generation in that respect, for the plausibility of the even more controversial term "terra incognita" seemed obvious if one watched how animals respond to even just the inside of a cave, tunnel, barrel, box, bucket, bamboo or even sometimes a hollow straw. There's an old children's song, "The bear went over the mountain ... to see what it could see," and a group of young chimps traversing a one-acre field, or cats in a new house and yard, can certainly convey that same impression, especially in the first hours. In the 1950s and 1960s most students would have seen this as a motivational problem rather than a cognitive one; obviously it is both. It did not occur to me to use the term meta-cognition, largely because the "C" word was racy enough.

The importance of one's own visual perspective as well as one's subjects' momentary vantage points and visual perspectives became even more obvious than I thought it to be at the outset of the study. I intended the title, "a group of young chimpanzees in a one-acre field," to strongly suggest a view of captive animals from the vantage point of a tower, whereas a more truly Tolmanian title, such as "a lone chimpanzee at any given choice-point in the forest," would suggest that one is making every effort to put oneself in all the major possible sorts of situations a chimpanzee might place itself, and to try to view the scene, if not to think, like a chimp, insofar as that is possible. As I think of such contrasts in perspectives now, Tolstoy's philosophy of history, as developed in his *War and Peace*, comes to mind. Is primate intelligence in the head of individuals, or is it instead "distributed," in the jargon of recent times (Johnson, 2001)? My answer to that would be: both, in some sense, but most of what I have said thus far favors the latter view. Who is in the best position to understand individual and group movements and behaviors – emperors, generals, privates, scholars, or illiterate peasants? Tolstoy said it was none of the above, but that perhaps the last should be considered first. If you smell a sermon there, you are right. Implicitly, if not explicitly, I, a non-chimp and self-appointed head caregiver–observer–experimenter–reporter, was granting myself a very privileged status. Without a lot of complementary experience on ground level, in amongst the troops, and peering from the door of their sleeping cage or through the tall grass on all fours, my sense of their psychology would be even poorer than it was. To be a good psychologist, you don't just have to have a good feel for your animals, you also have to have a very good feel for their environments, and I do not know which comes first. My pitiful ability at tree climbing was definitely a handicap, and by all means the metrics of behavioral space are quite different in the trees than they are on the ground. Euclid's geometry is for flatlanders.

And so forth and so on, to the limits of one's perception, imagination, time, motivation or pocket book, whichever expires first.

SO WHAT?

There's a lot more that might be said about spatial perception, spatial cognition, and the relation between the two, but I'll limit myself to two more points.

First of all, perception versus cognition would not be a topic on which I feel any need to take sides, and the same is true of behavior versus cognition and mapping versus association learning. For that matter, whenever I read scientific papers that urge every one to purge certain words from their vocabulary (e.g., Mackintosh, 2002) I wonder why the authors did not follow their own advice, and why they use the same words as blatantly as anyone else ever did. (Answer: it pays to advertise.) How then can we minimize rather than proliferate the number of parameters and concepts we invoke to explain our data? As Einstein said, look at what scientists do, not at what they say.

Second, it is hard for me to believe that people are serious when they (actually the number has been small) tell me that they find spatial approaches to behavior a-theoretical and lacking in depth and scientific direction. Any approach in the hands of some investigators can no doubt be described in this way. But speaking more generally, I beg to differ. Here, almost verbatim, is the last paragraph (p. 149) of my 1974 paper:

> The major advantage of a spatial, geometrical, or mechanical approach to nature is that its questions, methods, and concepts fully utilize the complexity of the real world in their construction, but are themselves extremely simple and general. In principle, the same basic approach is applicable to any overt actions of any species in any situation. Space-like concepts are the common operational ground on which more specialized theories of ecology, sociology, and individual behavior rest, and are the most effective way I know for dealing with several levels of organization simultaneously; i.e., with complex events as a whole. If behavioral science ever develops the tools for studying ecological events as a whole in a really adequate fashion, the old controversies about lab versus field, experiment versus observation, various levels of analysis (ecological, group, individual, intra-individual) and various schools of psychology will all fade into ancient history.

That I call locationism pure and simple (Menzel, 1987). I still sometimes feel nostalgic about it, philosophically speaking. For that matter, I have always had a soft spot for even grander versions of the hope for unity over all domains of science and human knowledge. Compared to Wilson (1998), for example, the dominant theories of behavior in my graduate school days (Hull, Tolman, Guthrie; the dark horses then were Skinner, Lorenz and Hebb) were very timid, not bold or grand. Where psychologists have hesitated to tread since the 1950s, scientists and scholars of other academic disciplines have moved in to do the job for all of us. Physicists are even talking about TOE (theories of everything). Since their TOEs do not yet include anything significant about organic nature, let alone animal or human suffering or psyche, I'd say that we are just getting started.

Whatever unity there is in all the various topics that I have mentioned above is probably easier to convey pictorially or numerically than in words. Ever since Plato (if not ever since cavemen and women gathered around a fire, quite possibly creating and watching shadow-plays as well as drawing on the wall) natural philosophers have known that spatio-temporal, and especially visual, representations can sometimes furnish very powerful and compelling models, explanations, syntheses and generators of new predictions. Maps are an outstanding case in point, especially if one does not limit oneself to animal research or the past century. Microscopes, telescopes, zoom lenses, fisheye lenses, and of course cameras, sound motion pictures and robots came much later in human history than did pictures and maps, which might be why they might seem even more provocative in some ways. In any event, consider now the words of Gustav Fechner, the polymath scientist-humanist who is best known today for what he, and even more so William James (1890), called mere outer, as opposed to true inner, psychophysics:

> The solar system offers quite different aspects as seen from the sun and as seen from the earth. One is the world of Copernicus, the other the world of Ptolemy. It will always be impossible for the same observer to perceive both world systems simultaneously, in spite of the fact that both belong quite indivisibly together and, just like the concave and convex sides of the circle, are only two different modes of appearance of the same matter from different standpoints. Here again one needs but to change the point of view to make evident one world rather than another. (1860/1966, p. 3)

So is the world round or flat? If you say it is round then obviously you are not a psychologist.

Fechner's primordial problems were quite likely these: where am I now, really? Where did I come from? And where am I headed? But I might be projecting, for I myself have pondered them since I was a child, and find them not only deep but bottomless.

And, of course, if you change the pronoun from "they" and "I" to "we" and have any concern for the future of primates in general, not to mention the future of the planet earth, you will have even more to chew on. All in all, spatial problems and metaphors are deeper than they look, and probably deeper than we shall ever know.

ACKNOWLEDGMENTS

This chapter is dedicated to Harriet Anne Menzel in the year of our 56th wedding anniversary, and to all creatures great and small, especially Ian, Rebecca, Emilie, Julian and all their pets. I thank the editors for helping me to reduce my word-count to one-fourth of the first draft, and the publishers for granting me their senior citizen's discount – i.e., an increase in the allowed number of words. Manuscript preparation supported in part by MH-58855 and HD-38051.

REFERENCES

The website http://library.primate.wisc.edu has a fairly complete list of primate publications since 1940.

Allport, F. (1955). *Theories of perception and the concept of structure*. New York: Wiley.

Bennett, A. T. D. (1996). Do animals have cognitive maps? *Journal of Experimental Biology*, **199**, 219–224.

Beran, M. J. (2004). Chimpanzees (*Pan troglodytes*) respond to nonvisible sets after one-by-one addition and removal of items. *Journal of Comparative Psychology*, **118**, 25–36.

Birch, H. G. (1945). The relation of previous experience to insightful problem solving. *Journal of Comparative Psychology*, **38**, 367–383.

Byrne, R. W. and Whiten, A. (1988). *Machiavellian intelligence*. Oxford: Clarendon Press.

Carpenter, C. R. (1964). *Naturalistic behavior of nonhuman primates*. Pennsylvania State University Press, University Park.

Davenport, R. K., Rogers, C. M. and Menzel, E. W. (1969). Intellectual performance of differentially reared chimpanzees: II. Discrimination-learning set. *American Journal of Mental Deficiency*, **73**, 963–969.

Davenport, R. K., Rogers, C. M. and Rumbaugh, D. M. (1973). Long-term cognitive deficits in chimpanzees associate with early impoverished rearing. *Developmental Psychology*, **9**, 343–347.

Dawkins, R. and Krebs, J. R. (1978). Animal signals: information or manipulation? In J. R. Krebs and N. B. Davies (Eds.), *Behavioural ecology: an evolutionary approach* (pp. 282–315). Oxford: Blackwell Scientific Publishers.

Fechner, G. T. (1966). *Elements of psychophysics*. New York: Holt, Rinehart and Winston. (Originally published 1860.)

Fisher, R. A. (1930). *The genetical theory of natural selection*. Oxford: Oxford University Press.

Fisher, R. A. (1951). *The design of experiments* (6th edn). New York: Hafner.

Gibson, E. J. (1969). *Principles of perceptual learning and development*. New York: Appleton-Century-Crofts.

Gibson, J. J. (1958). Visually controlled locomotion and visual orientation in animals. *British Journal of Psychology*, **49**, 182–194.

Gibson, J. J. (1966). *The senses considered as perceptual systems*. New York: Houghton-Mifflin.

Gould, J. L. and Gould, C. G. (1994). *The animal mind*. New York: Scientific American Library.

Hediger, H. (1964). *Wild animals in captivity*. New York: Dover Publications.

Hediger, H. (1968). *Studies of the psychology and behavior of captive animals in zoos and circuses*. New York: Dover Publications.

Hull, C. L. (1952). *A behavior system*. New Haven: Yale University Press.

James, W. (1890). *The principles of psychology* (1st edn, vol. 1). New York: Henry Holt.

Johnson, C. M. (2001). Distributed primate cognition: a review. *Animal Cognition*, **4**, 167–183.

Koffka, K. (1935). *Principles of gestalt psychology*. New York: Harcourt Brace.

Köhler, W. (1927). *The mentality of apes*. New York: Harcourt Brace.

Köhler, W. (1929). *Gestalt psychology*. New York: Liveright.

Kortlandt, A. and Kooij, M. (1963). Protohominid behaviour in primates (preliminary communication). *Symposia of the Zoological Society of London*, **10**, 61–88.

Lindsay, R. B. and Margenau, H. (1936). *Foundations of physics*. New York: John Wiley and Sons.

Mach, E. (1907). *The science of mechanics* (4th edn). Chicago: Open Court.

Mackintosh, N. J. (2002). Do not ask whether they have a cognitive map, but how they find their way about. *Psicológica*, **23**, 165–185.

Menzel, C. R. (1999). Unprompted recall and reporting of hidden objects by a chimpanzee (*Pan troglodytes*) after extended delays. *Journal of Comparative Psychology*, **113**, 426–434.

Menzel, C. R., Savage-Rumbaugh, E. S. and Menzel, E. W. (2002). Bonobo (*Pan paniscus*) spatial memory and communication in a 20-hectare forest. *International Journal of Primatology*, **23**, 601–619.

Menzel, E. W. (1964). Responsiveness to object-movement in young chimpanzees. *Behaviour*, **24**, 147–160.

Menzel, E. W. (1969). Naturalistic and experimental research on primate behavior. In E. Willems and H. Raush (Eds.), *Naturalistic viewpoints in psychological research* (pp. 28–121). New York: Holt, Rinehart & Winston.

Menzel, E. W. (1971). Group behavior in young chimpanzees: responsiveness to cumulative novel changes in a large outdoor enclosure. *Journal of Comparative and Physiological Psychology*, **74**, 46–51.

Menzel, E. W. (1972). Spontaneous invention of ladders in a group of young chimpanzees. *Folia Primatologica*, **15**, 220–232.

Menzel, E. W. (1974). A group of young chimpanzees in a one-acre field. In A. M. Schrier and F. Stollnitz (Eds.), *Behavior of nonhuman primates* (vol. 5, pp. 83–153). New York: Academic Press.

Menzel, E. W. (1978). Cognitive mapping in chimpanzees. In S. H. Hulse, H. Fowler and W. K. Honig (Eds.), *Cognitive processes in animal behavior* (pp. 375–422). Hillsdale: Lawrence Erlbaum.

Menzel, E. W. (1979). General discussion of the methodological problems involved in the study of social interactions. In M. E. Lamb, S. J. Suomi and G. R. Stephenson (Eds.), *Social interaction analysis: methodological issues* (pp. 291–310). University of Wisconsin Press.

Menzel, E. W. (1987). Behavior as a locationist views it. In P. Ellen and C. Thinus-Blanc (Eds.), *Cognitive processes and spatial orientation in animal and man* (pp. 55–72). NATO Conference volume. Dordrecht: Martinus Nijhoff.

Menzel, E. W., Davenport, R. K. and Rogers, C. M. (1970). The development of tool using in wild-born and restriction-reared chimpanzees. *Folia Primatologica*, **12**, 273–283.

Menzel, E. W. and Draper, W. A. (1965). Primate selection of food by size: visible versus invisible rewards. *Journal of Comparative and Physiological Psychology*, **59**, 231–239.

Menzel, E. W. and Johnson, M. K. (1976). Communication and cognitive organization in humans and other animals. *Annals of the New York Academy of Science*, **280**, 131–142.

Menzel, E. W. and Juno, C. (1982). Marmosets *(Saguinus fuscicollis)*: are learning sets learned? *Science*, **217** (4561), 750–752.

Menzel, E. W. and Juno, C. (1984). Are learning sets learned? Or: perhaps no nature-nurture issue has any simple answer. *Animal Learning and Behavior*, **12**, 113–115.

Menzel, E. W. and Juno, C. (1985). Social foraging in marmoset monkeys and the question of intelligence. *Philosophical transactions of the Royal Society of London*, **B308** (1135), 145–158.

Menzel, E. W. and Menzel, C. R. (2007). Do primates plan routes? Simple detour problems reconsidered. In D. A. Washburn (Ed.), *Primate perspectives on behavior and cognition* (pp. 175–206). Washington, DC: American Psychological Association.

Menzel, E. W. and Wyers, E. J. (1981). Cognitive aspects of foraging. In A. C. Kamil and T. Sargent (Eds.), *Foraging behavior: ecological, ethological and psychological approaches* (pp. 355–371). New York: Garland SPTM Press.

Miller, G. A., Galanter, E. and Pribram, K. H. (1960). *Plans and the structure of behavior*. New York: Holt, Rinehart and Winston.

Muller, R. U., Stead, M. and Patch, J. (1996). The hippocampus as a cognitive graph. *Journal of General Physiology*, **107**, 663–694.

Nissen, H. W. (1931). A field study of the chimpanzee: observations of chimpanzee behavior and environment in Western French Guinea. *Comparative Psychology Monographs*, **8** (1), 1–122.

Olton, D. S. and Samuelson, R. J. (1976). Remembrances of places past: spatial memory in rats. *Journal of Experimental Psychology: Animal Behavior Processes*, **2**, 97–116.

Povinelli, D. J. (2000). *Folk physics for apes: the chimpanzee's theory of how the world works*. Oxford: Oxford University Press.

Reif, F. (1967). *Statistical physics*. New York: McGraw-Hill.

Schiller, P. (1957). Innate motor action as a basis for learning. In C. Schiller (Ed.), *Instinctive behavior* (pp. 264–287). New York: International Universities.

Schopenhauer, A. (1966). *The world as will and representation* (3rd edn). New York: Dover Publications.

Schrier, A. M. and Thompson, C. R. (1984). Are learning sets learned? A reply. *Animal Learning and Behavior*, **12**, 109–112.

Tinklepaugh, O. L. (1928). An experimental study of representative factors in monkeys. *Journal of Comparative Psychology*, **8**, 197–236.

Tinklepaugh, O. L. (1932). Multiple delayed reaction with chimpanzee and monkeys. *Journal of Comparative Psychology*, **13**, 207–243.

Tolman. E. C. (1932). *Purposive behavior in animals and men*. New York: Appleton-Century-Crofts.

Tufte, E.R. (1982). *The visual display of quantitative information.* Cheshire, CT: Graphics Press.

Tufte, E. R. (1990). *Envisioning information.* Cheshire, CT: Graphics Press.

Tufte, E. R. (1997). *Visual explanations.* Cheshire, CT: Graphics Press.

Van Lawick-Goodall, J. (1971). *In the shadow of man.* Boston, MA: Houghton.

Wilson, E. O. (1998). *Consilience: the unity of knowledge.* New York: Alfred A. Knopf.

Part II Perception and memory of landmarks: implications for spatial behavior and cognition

CATHERINE THINUS-BLANC, VANESSA CHABANNE, LUCA TOMMASI, PATRICK PÉRUCH AND JACQUES VAUCLAIR

5

The encoding of geometry in various vertebrate species

A wide range of environmental features are likely to be selected and processed to become constitutive elements of spatial representations. Among these features, the geometry of surfaces defined by various types of discrete elements has recently been the target of an increasing number of studies in several species. In the present context, geometry is defined as the relative position (in terms of directions and distances) of several "objects" that define an arrangement of any type (regular or not). These objects can be segments of a perimeter (continuous walls) or non-continuous elements that can also define a surface if one considers the virtual lines joining them. Thus, encoding geometry implies a global processing of the situation, regardless of the intrinsic properties (color, shape, size, texture, etc.) of the discrete elements that define it. In contrast, discrete elements or local cues are individually processed through the perception and encoding of their nonspatial attributes and can be used as isolated (or local) landmarks, even if they are located within an array of landmarks. Geometric encoding corresponds to taking into account the whole situation regardless of local details, whereas non-geometric processing corresponds to the analysis of the specific features of discrete local elements independently of their spatial relationships with others. Thus, the same "object" can be processed as an individual local cue or as a part of an array of objects related to each other by geometric relationships. It is the nature of the task to be performed, the subject's abilities, etc., that determine the predominating type of processing that is implemented, though both of them can be done jointly.

The first experiments designed to specifically address the encoding of geometry were conducted in rats by Cheng and Gallistel (1984) and then by Cheng (1986). In a working memory task, disoriented rats

Spatial Cognition, Spatial Perception: Mapping the Self and Space, ed. Francine L. Dolins and Robert W. Mitchell. Published by Cambridge University Press.

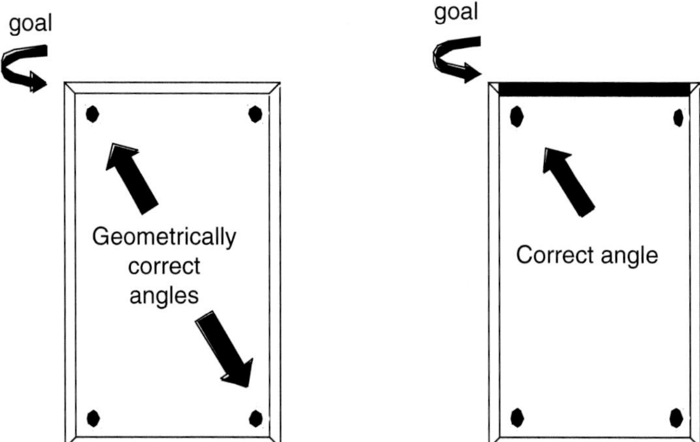

Figure 5.1. Schematic representation of the situation classically used in geometric encoding experiments. Left side: there is no distinctive cue within the apparatus that would allow the subject to differentiate the goal from the empty geometrically equivalent corner. Right side: one of the walls is different (here in black) from the other ones and can be used to find the reward. Other types of local information can be used such as panels of various sizes in the corners, on the lengths or on the widths of the apparatus.

searching for food in one corner of a rectangular enclosure (geometric information) with distinctive local cues in the four corners (non-geometric information), systematically confused equivalent corners (i. e., the reinforced corner and its rotational equivalent, see Figure 5.1). Rats seemed to rely mainly on geometric information even when the local cues differed from each other for a number of features, such as texture, brightness and even smell despite the fact that these would have allowed them to make the distinction between two geometrically similar places, with only one being baited. To account for these results, Cheng proposed that re-orientation in rats depends on a "geometric module," a task-specific, encapsulated system (see Fodor, 1983).

A few years later, Spelke and her collaborators (Hermer and Spelke, 1994, 1996; Wang *et al.*, 1999) examined the same question in a series of studies with human children and adults. They found a similar pattern of data in children (mean age: twenty-two months) whereas adults were able to use both geometric and local cues to guide their search. According to the authors, the joint use of both geometric and local landmarks, whose individual characteristics would have allowed to remove the geometric ambiguity related to the symmetry of the situation, would be tightly related to the mastery of language (spatial language in particular) that would be necessary to "penetrate" the geometric module and to allow for re-orientation by integrating discrete cues.

The notion of modularity and the idea that spatial language is necessary for the joint use of two categories of environmental information have not escaped the interest of researchers in the field of spatial cognition. Children and several animal species have been tested in more or less similar conditions. Though consistent results emerge, some discrepant data lead one to question, within an evolutionary standpoint, the interpretations drawn from the initial studies. It should be underlined that the above-mentioned paradigm tackles the problem of encoding the geometry by presenting the core of the problem as the neglect of local cues versus global geometric features. In addition, this paradigm implies disorientation of the subject before being tested. Other studies have indirectly addressed this issue in a more traditional perspective related to the nature and use of spatial representations for orientation. In the case of piloting, i.e., position-based navigation relying on external landmarks, for instance, the organism must make use of the spatial relations (directions and distances) that exist between landmarks, in order to extrapolate the position of the goal (Thinus-Blanc, 1996). A landmark can be used as a discrete element for guidance (i.e., as a beacon), so that a specific landmark is always located relative to the goal site. In such a case, distinctive properties of landmarks are of major importance. They can also be processed in a configurational way by combining spatial relationships among them and their intrinsic properties. Another level of more abstract configurational processing is to extract the relative direction and distance relationships between objects, regardless of the distinctive properties of the latter.

Such an abstract processing encompasses high-level cognitive computation related to the notions of proportionality of distances, center of an environment, etc. The concept of center is straightforward in human communication because it is one of those linguistic labels that are used to describe unambiguously the spatial structure of our environment. Moreover, the center has a special status in our perceptual world, and much has been written about its uniqueness in human visual perception (Arnheim, 1988). However, we will present studies showing that "the center" is intuitively recognized by rats and chickens searching for a hidden goal in a number of enclosed environments differing both in geometric shape and in size.

It is the aim of the present chapter to briefly review the main data issued from the approach of geometric and local cues processing, and to discuss the possible reasons of divergences, within a broader vision of spatial cognition from a comparative standpoint. Though several studies have been conducted in teleost fish (e.g., Sovrano *et al.*, 2002, 2003; Vargas *et al.*, 2004), we will focus in the present chapter on birds, rodents, monkeys and children.

THE ENCODING OF GEOMETRY IN BIRDS

Birds' spatial abilities offer some of the most impressive examples when talking about comparative spatial cognition. Decades of research on migrating birds and homing pigeons, and on the outstanding memory abilities of food-storing birds have provided us with exemplary instances of adaptations in the domain of spatial cognition (Berthold, 1991; Shettleworth, 1998). The time-honored tradition of field studies on the orientation abilities of avian species has been recently supplemented by research carried out in laboratory-controlled conditions, unveiling many details of spatial representation in birds.

If empirical research using multiple landmarks has elucidated important aspects of the encoding of distance and direction in birds, there has been as much work on birds' spatial memory in enclosed environments. In these types of experiments, the geometric shape of the environment (defined by the solid surfaces delimiting it) is the crucial variable, and the interest of the experimenter is focused on measuring how much it actually controls spatial learning. It can be analysed per se or in association with one or more landmarks present in the enclosure, thus giving rise to interesting dissociations between the role of continuous geometric information (enclosure shape) and discrete information (landmarks).

The two avian species tested so far using this paradigm have been domestic chickens and pigeons. Young chickens trained in a reference memory version of the task rely on geometric information when non-geometric information (flat panels attached to the corners of the enclosure) is absent or removed; they undoubtedly also encode non-geometric information, because when the two types of information are contradictory, the chickens' spatial responses rely on the latter (Vallortigara *et al.*, 1990). Quite similar to young chickens, pigeons trained in this task show learning based on geometric information (Kelly *et al.*, 1998) whether or not they are trained in the presence of non-geometric information (distinctive three-dimensional objects located at the corners). When the two types of information conflict, pigeons rely primarily on non-geometric information. A difference between young chickens and pigeons was found when only the non-geometric information associated with the reinforced corner and its rotational equivalent was removed during tests: in this case chickens made rotational errors, relying exclusively on geometric information (see also Vallortigara *et al.*, 2004), whereas pigeons searched at the correct corner, showing that they had also encoded non-geometric information distant from the goal. When pigeons were transferred to a small-sized replica of the training enclosure, their choice was controlled by non-geometric information (in the geometrically incorrect corner): pigeons, in fact, searched very little in the corners preserving the correct relative geometry (which were associated with incorrect non-geometric information) (Kelly *et al.*, 1998). Because in this last test the two types of information were contradictory, it could not be excluded that relative geometric information had been encoded by pigeons. This latter possibility was investigated by training pigeons in the rectangular enclosure in the absence of non-geometric information, so that learning could be based only on the relative geometry of the enclosed space (Kelly and Spetch, 2001). When they were transferred in two small-sized replicas of the training enclosure (two-thirds of the linear size, half of the linear size), pigeons searched in the geometrically correct corners, showing that they had correctly encoded the relative geometry of the training enclosure, despite the fact that the testing enclosure differed from the training enclosure in terms of absolute geometry. When tested in a square-shaped enclosure, in which both absolute and relative geometry were altered, pigeons searched randomly in the four corners. This result suggests that pigeons encode the geometry of an enclosed environment at least in terms of the relative length of its composing surfaces. But geometry concerns angles, not only distances. So, the question arises, do birds encode angles as well as distances?

Training domestic chicks to search for a food reward located in one corner of a parallelogram-shaped enclosure (a slightly modified version of the "geometric module" paradigm) answered this question (Tommasi and Polli, 2004). Between trials, chicks were passively disoriented and the enclosure was rotated, making re-orientation possible only on the basis of the internal spatial structure of the enclosure. In order to re-orient, chicks could rely on two sources of information: the relative lengths of the walls of the enclosure (associated to their left–right sense order) and the angles subtended by walls at corners.

Chicks learned the task choosing equally often the reinforced corner and its rotational equivalent. Results of tests carried out in novel enclosures, the shapes of which were chosen ad hoc (1) to induce re-orientation based only on the ratio of walls lengths plus right–left sense order (rectangular enclosure), or (2) to induce re-orientation based only on corner angles (rhombus-shaped enclosure), suggested that chicks encoded both features of the environment. In a third test, in which chicks faced a conflict between these geometric features (mirror parallelogram-shaped enclosure), re-orientation seemed to depend on the salience of corner angles. Corner angles seem to be highly salient in the representation of enclosed environments by chicks, and their perceptual abilities might include aspects of the encoding of both geometric and non-geometric information.

Another type of experiment that has revealed interesting aspects of the encoding of geometry in birds, concerns the capability of young domestic chicks in localizing the center of enclosures, while relying only on their geometric shape. In the first of these studies (Tommasi *et al.*, 1997) it was shown that chicks can be trained to find a hidden goal in the geometric center of an enclosure, with remarkable precision. When chicks were trained to localize the center of a square-shaped enclosure and were then transferred in a circular-shaped enclosure or triangular-shaped enclosures (both equilateral and isosceles) roughly the same size, they continued searching in the center. However, in a rectangular-shaped enclosure obtained by doubling one side of the training square-shaped enclosure, the search pattern concentrated in the geometric center and in the centers of the composing squares. These results demonstrate that chicks had encoded the center as defined by the relationships between distances, and as an absolute distance from some selected reference point along the walls. In agreement with this last result, chicks trained to find the center of a square-shaped, a circular-shaped, or a triangular-shaped enclosure and then transferred in an enlarged replica of the training enclosure, showed searching both in the center of the larger enclosure and at a distance from the walls corresponding to the distance from the center to the walls of the training enclosure. When chicks were trained in a large square-shaped enclosure and then transferred in a smaller-sized replica of it (Tommasi and Vallortigara, 2000), however, searching concentrated only in the geometric center of the smaller enclosure, suggesting that encoding of relative rather than absolute distances was favored. Interestingly, similar results have been obtained recently in pigeons using the same task (Gray *et al.*, 2004).

In another series of experiments, chicks were trained to localize the center of a square-shaped enclosure in the presence of a central landmark. Chicks continued searching in the geometric center when the landmark was removed (Tommasi and Vallortigara, 2000, 2001) and when the landmark was displaced to a position farther from the center (Tommasi and Vallortigara, 2001; Tommasi *et al.*, 2003), showing that chicks relied primarily on the global geometric information conveyed by the shape of the enclosure even when the local information provided by the landmark was available. However, recent data (Della Chiesa *et al.*, unpublished results) suggest that using a concentric array of four landmarks which is then shifted during test might take control over chicks' searching response.

In summary, birds such as chickens and pigeons are capable of encoding and using geometric features of the environment. When additional discrete local cues are present and provide information coherent with geometry, they are conjointly used. In general, when there is conflicting information, non-geometric information prevails over that of geometric information.

THE ENCODING OF GEOMETRY IN RODENTS

In a series of reaction-to-change tests, hamsters explored four different objects in a circular open field. After habituation (quantified by the number and duration of contacts with the objects), the shape of the initial object was modified. Such changes induced a renewal of exploratory activity directed either selectively to the displaced objects or to all of them, even if the objects were identical (Poucet *et al.*, 1986; Thinus-Blanc *et al.*, 1987). In contrast, modifying the size of the configuration had no effect though this change induced important changes in the motor patterns of activity. Such data demonstrate that hamsters had spontaneously encoded the geometry of the object arrangement, given that a modification of this feature induced strong reactions, most likely the result of a comparison between a stored representation of the initial arrangement and the perception of the new one. The lack of re-exploration of the same configuration of a different size may correspond to the formation of a geometric category. Similar data have been found in gerbils (Thinus-Blanc and Ingle, 1985), rats (e.g., Save *et al.*, 1992), and mice (Ammassari-Teule *et al.*, 1996; Roullet *et al.*, 1998). Other authors have reached the same conclusion by using different kinds of experimental situations (Greene and Cook, 1997; see also Thinus-Blanc, 1996, for a review).

Rats can localize the central position of an enclosure with a regular geometric shape, relying only on the spatial information provided by the geometry of the environment (Tommasi and Thinus-Blanc, 2004). Rats trained to search for a food reward hidden under sawdust in the center of a square-shaped enclosure designed to force orientation based on the overall geometry of the environment were then tested in a number of enclosures differing in geometric shape and in size (rectangular, double side, square-shaped, equilateral triangular). The rats transferred their acquired knowledge to the novel spatial situation: once the rats learned to localize the center of the environment, they could also transfer such capability to environments differing in shape and/or in size, thus generalizing their central searching behavior across different geometric shapes.

These data suggest that rats, whose spatial abilities were already known to depend strongly on geometrical aspects of the environment (Cheng, 1986; Gallistel, 1990), can learn to localize the central position of an environment and transfer this learned ability to environments differing in shape and in size. In the initial situation (Cheng, 1986; Cheng and Gallistel, 1984), rats searched for food previously hidden in one of the four corners of a rectangular apparatus. After the animals were familiarized with the experimental environment, they were removed from the apparatus, disoriented within a closed box, and returned to the apparatus to search for food. In this reorientation task, rats had to re-establish their position and heading before they engaged in goal-directed behavior. To be oriented again, rats could rely on the shape of the experimental apparatus, on the patterns of panels in the corners, on the odors (if any) or on the brightness of the walls. Rats showed a high rate of search both at the correct corner (with the reward) and at the rotationally equivalent opposite corner, and these two corners on the same diagonal are defined by the same geometric relation within the experimental apparatus (width and length). This search pattern was constant during all experiments conducted by Cheng (1986), in spite of the availability of many cues such as strong distinctive odors and large differences in contrast and luminosity of the walls that could easily differentiate the two symmetrical locations. The replication of this experiment (Margules and Gallistel, 1988) by using two similar apparatuses, one for the exposure, the other for tests, the latter being differently located and oriented compared to the first one, led the authors to a similar conclusion. Rats failed to use the non-geometric information to correctly locate the baited corner. These findings demonstrate that rats use the shape of the environment but seem to neglect its non-geometric properties.

Electrophysiology research on "place cells" in the hippocampus of the rat (neurons whose activity depends on the spatial position of the rat in a given environment) brings a convincing support to behavioral data. In a study by O'Keefe and Burgess (1996), the same place cells were recorded while the rat explored environments of different shapes and sizes. By modeling the observed patterns of activity, the authors called these neurons "boundary vector cells" (Burgess, 2002; Burgess *et al.*, 2000; Hartley *et al.*, 2000) because they responded as a Gaussian function of the distance to the nearest boundary or barrier (along a given allocentric direction, as in these experiments, extra-maze orientation cues were present and unchanged).

The model evoked above does not require any learning. However, the place fields of a single cell recorded while the rat explores two geometrically different enclosures, progressively differentiate over time: after an initial stage of homotopy (during which the neuron fires when the rat views spatially corresponding regions of the two environments) the neuron starts to fire in unrelatable spatial regions of the two enclosures (Lever *et al.*, 2002). Nevertheless, neurons in the subiculum and entorhinal cortex (two subregions of the hippocampal formation) have homotopic place fields even in different environments (Sharp, 1999), suggesting a likely candidate for the neural processes responsible for spatial generalization as it appears to be the case for the center.

In summary, the various approaches that have been conducted so far have unambiguously evidenced rodents' capabilities in encoding geometric environmental features. However, when they could be used conjointly with discrete (non-geometric) cues that would optimize the performance level, the latter are usually neglected when the disorientation procedure is used. This is not always the case in more traditional conditions without disorientation. For instance, in an experiment by Suzuki *et al.* (1980), a different conspicuous landmark was placed at the end of each of the eight arms of a radial maze. After the rat had visited three of the baited arms, the relative positions of the landmarks were changed. This modification greatly affected performance, demonstrating that rats relied both on the whole landmark configuration (i.e., on the geometric properties linking the various landmarks) and on their specific properties. Many other examples of the joint use of geometric and local information are available in the literature (see Thinus-Blanc, 1996, for review), and the peculiarity of the data about geometric processing evoked in the present chapter may be due to the use of the disorientation procedure. However, the reason why this procedure evidences in most of the cases a predominating processing of geometric information is still an unsolved issue.

THE ENCODING OF GEOMETRY IN PRIMATES

Nonhuman primates

The reaction-to-change paradigm has also been used in primates by Gouteux *et al.* (1999). During a phase of habituation, young baboons were individually familiarized with the initial spatial configuration made by four objects affixed to the wall of an outdoor enclosure. A decrease in the duration of the contacts was observed, indicating that the baboons got progressively familiarized with the initial situation. After habituation, animals were tested for their exploratory reactions (contact durations and order of spontaneous visits) to spatial changes made to the initial object configuration. Two kinds of spatial changes were made: a modification of the shape of the configuration (by displacing one of the four objects), and of the spatial arrangement without changing the initial shape (by exchanging the location of two objects). In a second experiment, the four objects were identical, and a modification of the spatial arrangement was performed. Finally, in a third experiment, a substitution of a familiar object with a novel one was performed, without changing the object configuration. Baboons strongly reacted to the spatial modifications of the configuration in that they massively and selectively re-explored only the displaced objects. In contrast, the baboons were less sensitive to changes of the local features (i.e., object replacements) that did not affect the initial spatial configuration, in that they showed no specific spatial re-exploration of the displaced objects. However, the results also suggest that the geometric encoding requires that the various elements that define the geometry of the explored space were not identical and that each of the various elements specifies a location to induce a selective re-exploration directed toward the displaced objects.

Use of both geometric and local information is a common feature of human spatial cognition so much that Hermer and Spelke (1996) propose that language is necessary to bind the geometric module and other, non-geometric modules. Initial work with nonhuman primates argues against this proposal. Tamarins were able to integrate information about the relative position of food above or below a landmark. The monkeys were successful in using the geometric information (i.e., the spatial relationships among various informative elements) and in combining it with non-geometric (i.e., color/shape) features (Deipolyi *et al.*, 2001).

A more elaborate examination tested three young rhesus monkeys under several conditions (Gouteux *et al.*, 2001a). The experimental setting was a rectangular room and an adapted procedure of disorientation was used. When only the geometric features of the experimental rectangular room were available (Experiment 1), monkeys used them and made as many correct responses as rotational errors. In Experiments 2 and 3, a large non-geometric feature covered one of the walls of the apparatus, either associated or dissociated from the reward box. In both cases, monkeys confined their search to the correct corner box with high consistency. In subsequent experiments, the monkeys' ability to use small, distal and proximal cues either indirectly (Experiment 4) or directly (Experiment 5) associated to the corners of the rectangular apparatus were investigated. In both conditions, the monkeys used the geometry only. They were unable to correctly locate the target and made many rotational errors. Finally, the effect of the landmark size on monkeys' orientation abilities was investigated. Whereas a small central cue was not salient enough to allow for a non-ambiguous orientation (Experiment 6), a larger central cue (Experiment 7) or four large different corner cues (Experiment 8) provided the relevant information for the monkeys to correctly orient themselves to find the rewarded box.

Thus, the rhesus monkey, a mammal species phylogenetically close to humans, is able to combine both geometric and non-geometric information to re-orient. This result casts some doubt on the role of language, at least in the particular experimental set-up that was used.

Children

Hermer and Spelke (1994, 1996) renewed the study of the coding of geometric information by human toddlers and adults. In their experiments, toddlers (aged from eighteen to twenty-four months) saw a desired toy that was being hidden in one of the corners of a rectangular homogeneous experimental chamber (four by six feet). After disorientation, the participants were asked to retrieve the toy. In one of the experiments, the chamber contained no distinctive landmark. In another one, a non-geometric feature (blue wall) that broke the symmetry of the experimental apparatus was added. Hermer and Spelke (1994, 1996) found that, when no information other than the shape of the environment was available, children searched equally often in the correct and in the rotationally equivalent corner, and more frequently in these two corners than in the other two remaining corners. However, when non-geometric information (a blue wall or a pair of toys placed in the room) was added, children still divided their searches between the two rotationally equivalent corners and seemed to ignore the added salient cues.

Hermer and Spelke concluded that young children do not use information other than the shape of the experimental environment to re-orient, even when more salient non-geometric information was available and could help them to locate the correct corner. Hermer and Spelke (1994, 1996) also tested human adults within the same experimental set-up. Unlike children, adults were able to use both geometric and non-geometric information to optimize their search. Similar results have been found by Wang *et al.* (1999) in a square room: children (between eighteen and twenty-four months old) used a distinctive geometric cue, but not a colored wall, to locate the hidden object, even though they had been familiarized with the colored wall over multiple training sessions. Further studies (Hermer-Vazquez *et al.*, 2001) revealed that by five to seven years of age, developmental changes for re-orientation occurred and that these changes correlated with the ability to produce phrases involving the words “left” and “right.”

Both the results and the interpretations put forward by the work of Spelke and colleagues have now been challenged. Thus, Learmonth *et al.* (2001) showed that toddlers could correctly locate, after disorientation, the correct corner when (a) the size of the room is big (four times the size of Hermer and Spelke apparatus) and (b) when the landmarks look permanent, e.g., a bookcase or a door, whereas Hermer and Spelke (1994, 1996) used movable landmarks (e.g., a toy) (see also Chapter 23). The role of size was further emphasized in a subsequent study using a within-subjects design: the same children (three years of age) who did not use the landmark in the smaller space did use it in the larger one (Learmonth *et al.*, 2002). Finally, Huttenlocher and Vasilyeva (2003) examined search behavior after disorientation in the rectangular apparatus and also in an isosceles triangular-shaped room providing an additional cue in that one of the corners is unique in angular size. The authors demonstrated that the toddlers could localize the correct corner without re-establishing their original headings and that their use of geometric cues was not limited to surrounding spaces. These latter results suggest that children are able to represent their position relative to the entire space (i.e., outside versus inside).

The experiments reported so far have been conducted in the locomotor space involving a navigational task. Gouteux *et al.* (2001b) examined whether similar search patterns are found in a manipulatory task, using a tabletop model of a rectangular room. Three groups of children (three-, four- and five-year-olds) and one group of adults were tested. The task involved movement of the spatial layout relative to a stationary child. Results show that geometric encoding appears only at four years of age, that is, later than in the locomotor space. The joint use of geometry and local cues emerges at five years of age. These data show that similar types of processing are implemented in both manipulatory and locomotor spaces, but they do not occur at the same time. Being immersed in the environment seems to play a crucial role here (see also Chapters 21 and 22).

DISCUSSION

On the whole, the experiments described in this chapter span a wide range of methods yet demonstrate that the encoding of geometric features of space is a process common to various species. It must be an evolutionary ancient mechanism prone to contribute to the survival of a species. According to Hermer and Spelke (1996) reliance on geometric information for orientation is adaptive because the macroscopic shape of the environment seldom changes. In contrast, the visual appearance may be modified through the course of seasons (snow, absence of leaves, for instance), and new scent markings may interfere with familiar scent trails. With regards to the traditional notion of cognitive maps as developed by O'Keefe and Nadel (1978), for instance, geometric features, i.e., relative distances and angular relationships among environmental elements, are constituent elements of such maps. Together with other types of information, they are helpful for determining the location of non-conspicuous goals, by reference to a configuration of landmarks, for instance (Suzuki *et al.*, 1980; cf. also Thinus-Blanc, 1996, for a review). Other types of geometric processing such as that of proportionality or center location refer to an abstract encoding, generalizable to other isomorphic situations, regardless of the specific characteristics of the objects that define geometric patterns. If this were confirmed, it would mean that spatial representations do have abstract and generalizable properties.

However, a puzzling set of data concerns the predominance of geometric information over local cues (when both are relevant and are not conflicting) and the quasi neglect of the latter observed in many experiments conducted in rectangular enclosures, and that only after a disorientation procedure (e.g., Hermer and Spelke, 1996, in young children). In such cases, an optimal performance level would have been reached by taking jointly into account geometric and local features. Thus, the question at issue is related to some kind of incompatibility between two types of converging information that can be efficiently used separately. Instead of having an additive effect, geometric information seems to induce in some cases the neglect of local cues. To account for this phenomenon, Cheng (1986) and Hermer and Spelke (1994, 1996) have proposed the existence of an encapsulated geometric module, not penetrable to other kind of information until, in humans, mastery of language. Along the adaptive hypothesis by Hermer and Spelke (1996), modularity would preserve the validity and reliability of geometric information from other types of changes such as those affecting the appearance of discrete cues.

But why do some data not fit into this interpretative framework? By definition, modular processing is generalizable to a wide range of situations. Indeed, the literature demonstrates that this is not the case, especially in children. For Newcombe (2002), the available evidence shows that data supporting modularity are found only in very small rooms as used by Hermer and Spelke. More recent work (Learmonth *et al.*, 2001, 2002; see also Chapter 23, this volume) brings no evidence that geometric information is encapsulated in human toddlers as it is indeed integrated with other relevant information about the spatial world, provided that landmarks are permanent. In rodents, the spatial stability of environmental objects has been pointed out as an important requirement for cues to be considered as reliable sources of spatial information (Biegler and Morris, 1993; 1996).

In addition, the size of the discrete cues has been shown in monkeys to have a strong influence on whether they use them or not (e.g., Gouteux *et al.*, 2001a). Perhaps, large landmarks may appear as more stable constituent elements of the situation than small items (unlike Hermer and Spelke's (1994) initial experiment in toddlers, the animals could not see the experimenter manipulating the cues). Data of the same type have been found in chickens and pigeons. Both bird species can jointly use geometric and non-geometric information. In addition, in case of conflict in spatial information, the latter predominates over the former (Kelly *et al.*, 1998; Tommasi and Polli, 2004; Vallortigara *et al.*, 1990). Finally, it is in rodents only that the modularity hypothesis has never been questioned. In Cheng's experiments, landmarks of large size (one entire wall of the apparatus could be different) did not help rats to differentiate the correct corner from the geometrically identical corner diagonally opposite from it. However, as far as we know, experiments of this type have never been replicated with larger enclosures.

In summary, the modular hypothesis cannot account for all of the data obtained so far with the disorientation procedure that requires directional information in symmetrical situations such as rectangular enclosures. When present, local cues provide this information, but it is very close to the subject, particularly in very small experimental rooms. The neglect of proximal cues might be due to the fact that they are less reliable than distal information, the latter of which does not change visually as much when the organism is moving.

Another less speculative explanation has been recently provided by Vargas *et al.* (2004). These authors propose that the phenomena of blocking and overshadowing in the domain of the Pavlovian and the instrumental conditioning would interfere with the learning of some of the cues over the others in a situation that provides redundant spatial information (Chamizo, 2003; Mackintosh, 1974; Rodrigo *et al.*, 1997). Consequently, according to Vargas *et al.* (2004, p. 214):

> it could be expected that the different types of spatial cues would induce in the animals a competitive-like encoding process, such that the control by one group or set of cues could interfere with the potential power of other cues to control behaviour, or that a source of information would be overshadowed by another, more salient cue, but both having the same predictive potential.

In the rectangular enclosure with additional non-geometric information, it may be easier and more reliable for children and rodents, for instance, to predominantly encode geometric information for adaptive reasons (Hermer and Spelke, 1996). This processing would overshadow the other types of cues. The advantage of this interpretative framework is that it emphasizes the effects of differentiated eco-ethological constraints on behavior. It is reasonable to assess that diversified types of environment, of potential predators, etc., together with the evolution of perceptual and cognitive abilities has induced a correlative diversity in the salience of important information to be predominantly encoded. The hypothesis of an effect of blocking and overshadowing is not incompatible with that of the existence of a geometric module. Indeed, they may correspond to two very different ways to envision spatial cognition, both having in common that they meet the criteria of "cognitive economy" when redundant information is available, and hence of adaptive value.

REFERENCES

Ammassari-Teule, M., Tozzi, A., Rossi-Arnaud, C., Save, E. and Thinus-Blanc, C. (1996). Reactions to spatial and nonspatial change in two inbred strains of mice: further evidence supporting the hippocampal-dysfunction hypothesis in the DBA/2 strain. *Psychobiology*, **23**, 284–289.

Arnheim, R. (1988). *The power of the center*. Berkeley: University of California Press.

Berthold, P. (1991). *Orientation in birds*. Basel: Birkhäuser Verlag.

Biegler, R. and Morris, R. G. M. (1993). Landmark stability is a prerequisite for spatial but not discrimination learning. *Nature*, **361**, 631–633.

Biegler, R. and Morris, R. G. M. (1996). Landmark stability – further-studies pointing to a role in spatial-learning. *Quarterly Journal of Experimental Psychology*, **49B**, 307–345.

Burgess, N. (2002). The hippocampus, space, and viewpoints in episodic memory. *The Quarterly Journal of Experimental Psychology, 55A*, **4**, 1057–1080.

Burgess, N., Jackson, A., Hartley, T. and O'Keefe, J. (2000). Predictions derived from modeling the hippocampal role in navigation. *Biological Cybernetics*, **83**, 301–312.

Chamizo, V. D. (2003). Acquisition of knowledge about spatial location: assessing the generality of the mechanism of learning. *Quarterly Journal of Experimental Psychology*, **56B**, 102–113.

Cheng, K. (1986). A purely geometric module in the rat's spatial representation. *Cognition*, **23**, 149–178.

Cheng, K. and Gallistel, C. R. (1984). Testing the geometric power of an animal's spatial representation. In H. L. Roitblat, T. C. Bever and H. S. Terrace (Eds.), *Animal Cognition* (pp. 409–423). Hillsdale: Erlbaum.

Deipolyi, A., Santos, L. and Hauser, M. D. (2001). The role of landmarks in cotton-top tamarin spatial foraging: evidence for geometric and non-geometric features. *Animal Cognition*, **4**, 99–108.

Fodor, J. A. (1983). *The modularity of mind*. Cambridge, MA: MIT Press.

Gallistel, C. R. (1990). *The organization of learning*. Cambridge, MA: MIT Press.

Gouteux, S., Vauclair, J. and Thinus-Blanc, C. (1999). Reaction to spatial novelty and exploratory strategies in baboons. *Animal Learning and Behavior*, **27**, 323–332.

Gouteux, S., Vauclair, J. and Thinus-Blanc, C. (2001a). Rhesus monkeys use geometric and nongeometric information during a reorientation task. *Journal of Experimental Psychology: General*, **130**, 505–519.

Gouteux, S., Vauclair, J. and Thinus-Blanc, C. (2001b). Reorientation in a small scale environment by 3–4 and 5 year-old children. *Cognitive Development*, **16**, 853–869.

Gray, E. R., Spetch, M. L., Kelly, D. M. and Nguyen, A. (2004). Searching in the center: pigeons (*Columba livia*) encode relative distance from walls of an enclosure. *Journal of Comparative Psychology*, **118**, 113–117.

Greene, C. M. and Cook, R. G. (1997). Landmark geometry and identity controls spatial navigation in rats. *Animal Learning & Behavior*, **25**, 312–323.

Hartley, T., Burgess, N., Lever, C., Cacucci, E. and O'Keefe, J. (2000). Modeling place fields in terms of the cortical inputs to the hippocampus. *Hippocampus*, **10**, 369–379.

Hermer, L. and Spelke, E. (1994). A geometric process for spatial reorientation in young children. *Nature*, **370**, 57–59.

Hermer, L. and Spelke, E. (1996). Modularity and development: the case of spatial reorientation. *Cognition*, **61**, 195–232.

Hermer-Vazquez, L., Moffet, A. and Munkholm, P. (2001). Language, space, and the development of cognitive flexibility in humans: the case of two spatial memory tasks. *Cognition*, **79**, 263–299.

Huttenlocher, J. and Vasilyeva, M. (2003). How toddlers represent enclosed spaces. *Cognitive Science*, **27**, 749–766.

Kelly, D. M. and Spetch, M. L. (2001). Pigeons encode relative geometry. *Journal of Experimental Psychology: Animal Behavior Processes*, **27**, 417–422.

Kelly, D. M., Spetch, M. L. and Heth, C. D. (1998). Pigeons' (*Columba livia*) encoding of geometric and featural properties of a spatial environment. *Journal of Comparative Psychology*, **112**, 259–269.

Learmonth, A. E., Nadel, L. and Newcombe, N. S. (2002). Children's use of landmarks: implications for modularity theory. *Psychological Science*, **13**, 3337–3341.

Learmonth, A. E., Newcombe, N. S. and Huttenlocher, J. (2001). Toddlers' use of metric information and landmarks to reorient. *Journal of Experimental Child Psychology*, **80**, 225–244.

Lever, C., Wills, T., Cacucci, F., Burgess, N. and O'Keefe, J. (2002). Long-term plasticity in hippocampal place-cell representation of environmental geometry. *Nature*, **416**, 90–94.

Mackintosh, N. J. (1974). *The psychology of animal learning*. San Diego: Academic Press.

Margules, J. and Gallistel, C. R. (1988). Heading in the rat: determination by environmental shape. *Animal Learning and Behavior*, **16**, 404–410.

Newcombe, N. S. (2002). The nativist-empiricist controversy in the context of recent research on spatial and quantitative development. *Psychological Science*, **13**, 395–401.

O'Keefe, J. and Burgess, N. (1996). Geometric determinants of the place fields of hippocampal neurons. *Nature*, **381**, 368–369.

O'Keefe, J. and Nadel, L. (1978). *The hippocampus as a cognitive map*. Oxford: Clarendon Press.

Poucet, B., Chapuis, N., Durup, M. and Thinus-Blanc, C. (1986). A study of exploratory behavior as an index of spatial knowledge in hamsters. *Animal Learning and Behavior*, **14**, 93–100.

Rodrigo, T., Chamizo, V. D., McLaren, I. P. L. and Mackintosh, N. J. (1997). Blocking in the spatial domain. *Journal of Experimental Psychology: Animal Behavior Processes*, **23**, 110–118.

Roullet, P., Bozec, G. and Carton, N. (1998). Detection of object orientation and spatial changes in mice: importance of local views. *Physiology and Behavior*, **64**, 2, 203–207.

Save, E., Buhot, M. C., Foreman, N. and Thinus-Blanc, C. (1992). Exploratory activity and response to a spatial change in rats with hippocampal or posterior parietal cortical lesions. *Behavioural Brain Research*, **47**, 113–127.

Sharp, P. E. (1999). Complementary roles for hippocampal versus subicular/entorhinal place cells in coding place, context, and events. *Hippocampus*, **9**, 432–443.

Shettleworth, S. J. (1998). *Cognition, evolution and behavior*. Oxford: Oxford University Press.

Sovrano, V. A., Bisazza, A. and Vallortigara, G. (2002). Modularity and spatial reorientation in a simple mind: encoding of geometric and nongeometric properties of a spatial environment by fish. *Cognition*, **85**, B51–B59.

Sovrano, V. A., Bisazza, A. and Vallortigara, G. (2003). Modularity as a fish views it: conjoining geometric and nongeometric information for spatial reorientation. *Journal of Experimental Psychology: Animal Behavioral Processes*, **29**, 199–210.

Suzuki, S., Augerinos, G. and Black, A.H. (1980). Stimulus control of spatial behavior on the eight arm maze in rats. *Learning and Motivation*, **11**, 1–18.

Thinus-Blanc, C. (1996). *Animal spatial cognition*. Singapore: World Scientific Press.

Thinus-Blanc, C. and Ingle, D. (1985). Spatial behavior in gerbils (*Meriones unguiculatus*). *Journal of Comparative Psychology*, **99**, 311–315.

Thinus-Blanc, C., Bouzouba, L., Chaix, K., Chapuis, N., Durup, M. and Poucet, B. (1987). A study of spatial parameters encoded during exploration in hamsters. *Journal of Experimental Psychology: Animal Behavior Processes*, **13**, 418–427.

Tommasi, L. and Polli, C. (2004). Representation of two geometric features of the environment in the domestic chick (*Gallus gallus*). *Animal Cognition*, **7**, 53–59.

Tommasi, L. and Thinus-Blanc, C. (2004). Generalization in place learning and geometry knowledge in rats. *Learning & Memory*, **11**, 153–161.

Tommasi, L. and Vallortigara, G. (2000). Searching for the center: spatial cognition in the domestic chick (*Gallus gallus*). *Journal of Experimental Psychology: Animal Behavior Processes*, **26**, 477–486.

Tommasi, L. and Vallortigara, G. (2001). Encoding of geometric and landmark information in the left and right hemispheres of the avian brain. *Behavioral Neuroscience*, **115**, 602–613.

Tommasi, L., Vallortigara, G. and Zanforlin, M. (1997). Young chickens learn to localize the centre of a spatial environment. *Journal of Comparative Physiology-A: Sensory, Neural and Behavioral Physiology*, **180**, 567–572.

Tommasi, L., Gagliardo, A., Andrew, R.J. and Vallortigara, G. (2003). Separate processing mechanisms for the encoding of geometric and landmark information in the avian hippocampus. *European Journal of Neuroscience*, **17**, 1695–1702.

Vallortigara, G., Pagni, P. and Sovrano, V.A. (2004). Separate geometric and non-geometric modules for spatial reorientation: evidence from a lopsided animal brain. *Journal of Cognitive Neuroscience*, **16**, 390–400.

Vallortigara, G., Zanforlin, M. and Pasti, G. (1990). Geometric modules in animals' spatial representations: a test with chicks (*Gallus gallus domesticus*). *Journal of Comparative Psychology*, **104**, 248–254.

Vargas, J.P., López, J.C., Salas, C. and Thinus-Blanc, C. (2004). Encoding of geometric and featural spatial information by Goldfish. *Journal of Comparative Psychology*, **118**, 206–216.

Wang, R.F., Hermer, L. and Spelke, E.S. (1999). Mechanisms of reorientation and object localization by children: a comparison with rats. *Behavioral Neuroscience*, **113**, 475–485.

TOM COLLETT AND PAUL GRAHAM

6

The visually guided routes of ants

INTRODUCTION

The foraging routes of ants, as displayed by columns of ants following odor trails, have long interested naturalists. Here we focus on the less spectacular visually guided routes of individual ants for what these routes can reveal about the spatial knowledge that ants have acquired of their local environment and the ways in which ants learn and use this information.

Figure 6.1 gives two examples of visually guided routes recorded at the beginning of the last century. The first example comes from Santschi (1913), a Swiss physician who spent most of his life practising medicine in Tunisia, but whose major avocation was the study of ants (see Wehner, 1990 for a biographical sketch). The diagram shows the route followed by a single desert ant (*Cataglyphis bicolor*) between its nest and feeding site, as it weaves through scrub. Although, these ants forage individually and do not lay chemical trails, the details of the multiple nestbound and foodbound paths shown here are intriguingly similar from one trip to the next, suggesting the importance of visuomotor memories in guiding routes.

The second example is taken from a monograph by Cornetz (1910). He was a French civil engineer working in Algeria, who also spent many hours recording the trails of ants in North Africa. He walked behind them, inscribing their paths with a stick in the dust, and afterwards transcribed the tracks onto paper. This particular group of tracks displays the result of a revealing experiment. The first trace is the homeward path of an ant after it had fed. Just before the ant reached the nest, Cornetz caught the ant and carried it in his hat back to the site where it had previously been given sugar. On release, the ant repeated its previous homeward route, presumably guided by the same visual landmark memories as before.

Spatial Cognition, Spatial Perception: Mapping the Self and Space, ed. Francine L. Dolins and Robert W. Mitchell. Published by Cambridge University Press

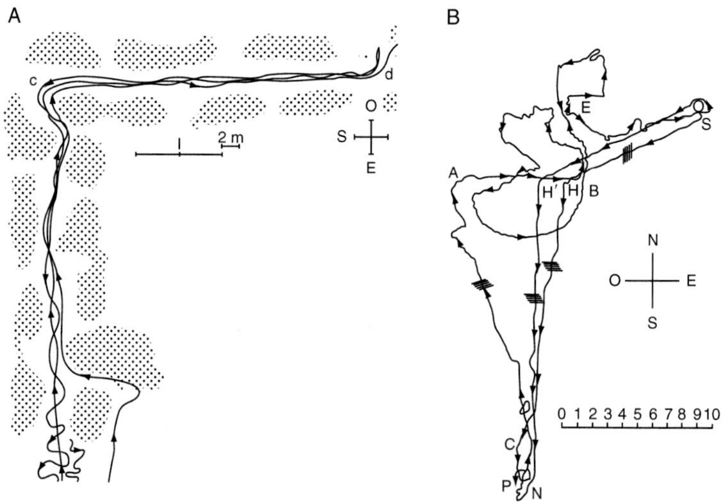

Figure 6.1. Early recordings of ant routes. A, Traces show two outward journeys and one return journey performed by a single desert ant (*Cataglyphis bicolor*). Despite the lack of chemical trails, the routes are similar. Stippling shows vegetation (from Santschi, 1913). B, The path of a desert ant was followed from its nest (N) until the ant reached a food site (S, top right). The ant's subsequent homeward trip is shown as a solid line labelled H. The same ant was then carried back to the food site. The ants second return journey, shown by the dashed line labelled H′, closely resembles its first (from Cornetz, 1910). Routes that are both idiosyncratic and stereotyped are likely to be learnt.

Since then several other species have been shown to follow visually guided routes (e.g., *Formica* sp.: Rosengren, 1971; Fukushi, 2001; *Neoponera apicalis*,: Fresneau, 1985; *Dinoponera gigantean*: Fourcassié et al., 1999; *Melophorus bagoti*: Kohler and Wehner, 2005). Harrison et al. (1989) found that the giant tropical ant, *Paraponera clavata*, follows pheromone trails, when new to a route, but once it is experienced with a particular route, it comes to learn the associated visual cues and to employ them in preference to olfactory ones. Without the need to antennate close to the ground ants following visual cues can travel the same route faster.

We would like to know how such routes are learnt, what navigational information ants acquire, and how that information is used for route guidance. We start by describing the innate mechanisms that make it possible to acquire routes rapidly. We discuss the kinds of processes that underpin an ant's ability to follow established routes, emphasising that navigational knowledge is largely procedural, in the sense that the ant is following a sequence of stored instructions cued by the environment, rather than using an internal map to track its location and plan its route. Thus, in the examples of Santschi and Cornetz, we suppose that the ants recognise landmarks along the route and then recall what to do next.

Route following is a matter of habit. In the simplest case, foraging involves a fixed route to a profitable site, to which an ant may adhere for much of its foraging career (e.g., Fresneau, 1985). In other environments the more scattered distribution of resources may make the ant's task more complex. Although the Mediterranean desert ant, *Cataglyphis fortis*, is primarily a scavenger of dead invertebrates, an individual ant tends to restrict its search to a limited range of directions from the nest (Wehner et al., 1983; Wehner, 1987) and to acquire routes, which gradually increase in length as an ant explores outwards to find new food (Schmidt-Hempel, 1984; Wehner et al., 2004). Insects can also show flexibility in their choice of route. Honeybees will take different routes at different times of day to exploit several temporally limited nectar sources (Wahl, 1932; Ribbands, 1964), and ants in the laboratory learn similar tasks (Harrison and Breed, 1987; Schatz et al., 1999). Changing one's habits can be particularly challenging, we end with a discussion of problems switching from one long-held route to another when a favored food source dries up and a new one is discovered.

SPATIAL KNOWLEDGE FROM PATH INTEGRATION

An established route to a feeding area is the consummation of a learning process that is possible because insects possess innate navigational strategies, which can operate largely without acquired knowledge of their environment and, which generate paths that the insects learn to follow. Ants either find a new site through information supplied by others; for instance, by following a chemical trail. Or, like the desert ant *Cataglyphis*, they explore individually and must then keep track of their outward route so that they can find their way back to the nest, a job which they accomplish using path integration (PI) (for reviews see Collett and Collett, 2000; Wehner and Srinivasan, 2003). This remarkable navigational strategy enables an animal to leave its nest, explore unfamiliar terrain, memorise the location of a newly discovered food site (Wehner et al., 1983; Collett et al., 1999; Wehner et al., 2004) and then return by a direct and often

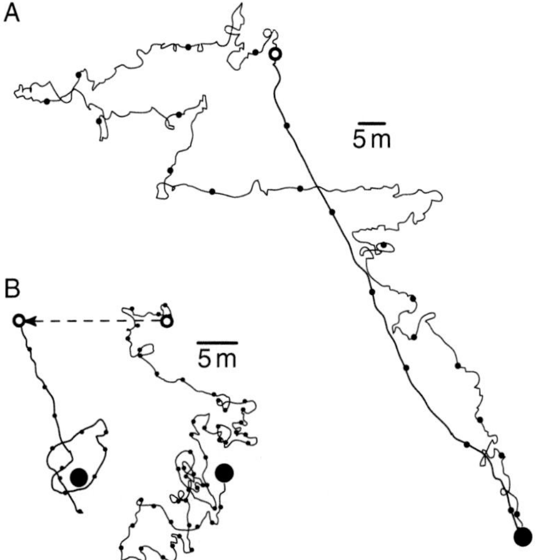

Figure 6.2. Path integration in desert ants. A, The outward journey of a foraging ant (*Cataglyphis* sp.) is shown with marker points every 60 s. After the ant has found a food item, its homeward path is direct (after Wehner, 1990). B, Before the ant can make a return trip, it is displaced some metres to the west. The ant's subsequent homeward path is parallel to and of the same length as the vector from food to nest. This path cannot be the result of landmark guidance, but is instead a vector derived from a linear integration of the outward route. Filled circle denotes nest, open circle food (after Wehner, 1982).

novel path to its nest. Figure 6.2 presents a well-known example from desert ants of a homeward path guided by PI occurring after a roundabout trip to locate food.

PI in ants operates roughly as follows: as an ant follows its path it monitors the direction and distance that it travels using a celestial compass and some kind of pedometer to update an accumulator that holds an estimate of its current direction and distance from the nest. When the ant encounters food, it remembers the PI coordinates of the site relative to the nest and can thus use PI to go back to the food site as well as to return home. The nest is thus the origin of this global PI system and ants seem to set their accumulator to zero when they are inside the nest (Knaden and Wehner, 2006).

The performance of a route using PI involves a process that is equivalent to subtracting the insect's current PI coordinates from those of its goal. The difference between these coordinates specifies the trajectory that the insect takes. This process can be seen at work

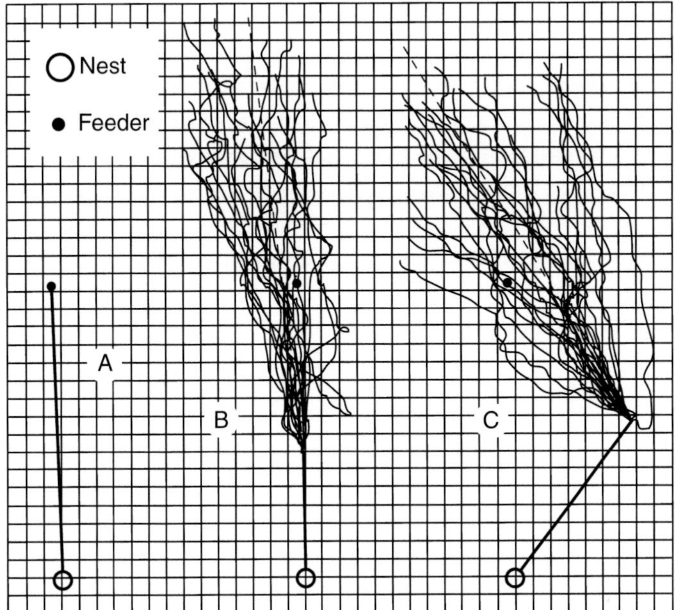

Figure 6.3. Trajectories to a food site guided by path integration. Ants (*Cataglyphis fortis*) were trained to find food at the end of a 16 m channel from which the natural landscape was invisible (A). In tests, the channel was shortened and the trajectories of ants were recorded after they left the channel. B, On exiting a channel of the same orientation as in training, ants head towards the feeder location. When the channel is rotated by 45 ° (C), ants turn through 90 ° at the end of the channel and again head towards the feeder location. Since the landscape on exiting the channel was unfamiliar, the ants' trajectories must have been guided by PI. Grid squares are 1 m across (after Collett et al., 1999).

when ants are forced to make a detour on their outward or homeward trip (Schmidt et al., 1992; Collett et al., 1999). At the end of the detour, they turn directly towards their goal, setting the direction of their new path over unfamiliar ground from their knowledge of their current PI coordinates and those of their destination (Figure 6.3).

We have described PI in terms of coordinates to stress that PI involves the use of positional and metric information. The ant knows its own position and the location of significant sites that it has previously visited and remembered in terms of their PI coordinates relative to its nest. And we have seen that this positional information is employed flexibly to allow novel routes to be performed over unfamiliar terrain. However, at least in ants, this PI information seems to be isolated from other navigational strategies. Familiar visual landmarks along a route do not acquire PI coordinates and the performance of novel paths using PI is only possible if the ant has itself travelled from its nest to the place from where it plans a new route to a known destination. If an insect is passively displaced, its path computed by PI will miss the goal by the magnitude of the displacement. To date, there is no evidence that ants use visual landmarks to reset their PI coordinates (Sassi and Wehner, 1997; Collett et al., 1998, 2003; Andel and Wehner, 2003, Knaden and Wehner, 2005; Collett and Collett, 2009b). When displaced to a familiar location, ants can find their way home either by following a familiar route from that location (Kohler and Wehner, 2005), or by random search until a familiar route is encountered (Wehner et al., 2006). As detailed below, PI helps with the acquisition of routes guided by landmarks, but it does not seem to provide the basis for a metric, 'cognitive' map on which landmarks can be placed.

ROUTE ACQUISITION

The seeds of route learning can be seen in the paths of a naïve desert ant at the start of its foraging career (Wehner et al., 2004). On the first few short trips from the nest, the ant is unconcerned about gathering food. Instead, it probably learns about visual features close to the nest that will later help relocate the nest-hole. Foraging only starts after these initial excursions. In the example shown in Figure 6.4, the ant has several abortive trips in different directions before it discovers an item of food (Figure 6.4, run 9). It then returns straight home and on its next outward trip makes directly for this new site, guided by PI. Although its second visit is fruitless, it makes a third trip to the same site before giving up and looking elsewhere. This persistence after a successful trip both allows the ant to take advantage of a patchy distribution of food and to learn visual landmarks along the repeated path. It means that if food is reasonably abundant, the ant will tend to stick to a single familiar sector around its nest, exploring outwards from its previous find (Schmid-Hempel, 1984). This strategy, when adopted by the whole foraging force, will lead to the efficient exploitation of resources with a variety of spatial distributions.

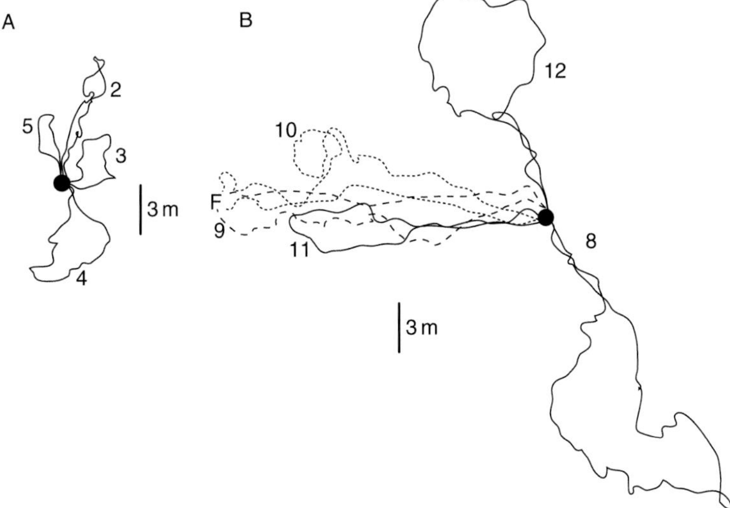

Figure 6.4. A sequence of trips at the start of the foraging career of a desert ant (*Cataglyphis bicolor*). A, Trips 2 to 5. These early runs are short, their possible function is to learn the surroundings of the nest. B, Trips 8 to 12. On run 9, the ant discovers a food item and its subsequent two trips, both unsuccessful, are in the same direction. The disappointed ant then changes direction on trip 12 and after (after Wehner et al., 2004).

While the initial target of an emergent route is set by the target's PI coordinates, the examples in the next section illustrate that the details of the path to the goal are determined by a mix of the ant's inbuilt responses to visual obstacles and landmarks, as well as the state of its PI system at different stages of the route.

Routes guided by extended landmarks

Ants will use extended objects, like walls, for visual guidance. An example from desert ants navigating by means of a vertical barrier is shown in Figure 6.5. The metre high barrier was placed perpendicular to the direct route between nest and feeder. Naïve ants seeing a barrier will often detour around it, frequently aiming directly at the end of the barrier. From there they head straight at the feeder on the way out or towards the nest on the way home, with this latter segment of the path set by PI (Figure 6.5A and C). Through visual learning, this route

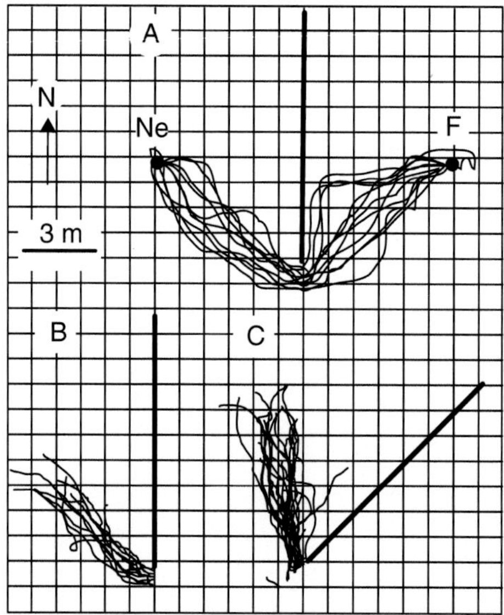

Figure 6.5. Visually guided routes around a barrier. A, Homeward routes taken by desert ants (*Cataglyphis fortis*) around a 10 m long barrier after visiting a feeder (F). B and C. Returning ants about to enter their nest (Ne) after a foraging trip were taken to the end of an identical barrier on a test ground and released with zero home-vector close to the end of the barrier. B, Barrier in training orientation: ants recreate the path that in training took them to the nest. C, Barrier rotated through 45 °: the ants' path also rotates by about 45 °. Thus, the ants' paths are set by visual properties of the barrier and not by the sky compass (after Collett et al., 2001).

becomes tied to the visual properties of the barrier and independent of PI (Collett et al., 2001).

To demonstrate this independence, ants, which are caught when they are close to the nest on their return trip, are placed at the end of a similar barrier on a test ground. Since the ants' homeward trip is complete, the ants have no useful PI information to guide a second trip home, nonetheless the direction and distance of their path is normal (Figure 6.5B). In this test, the barrier could just have triggered the performance of a remembered compass direction. But this explanation cannot account for the ants' behavior when they are placed at the end of a barrier after it has been rotated through 45 ° (Figure 6.5C). The ants' paths also rotate through 45 °, implying that their direction is set principally by the visual information that the barrier supplies, rather than by the sky compass. While navigating by PI, the ant has learnt the appearance of visual landmarks viewed *en route*. The response to the rotated barrier shows that subsequent route performance becomes emancipated from PI, which had initially provided the scaffold for learning.

Further experiments on wood ants have shown how they control their path relative to such a barrier (Graham and Collett, 2002). Ants were trained to go from a start point to a feeding site along a line that was either parallel or oblique to a 20 cm high black wall. To ensure that the ants were guided by the landmark and by nothing else, the wall, the starting point and the feeder were rotated and translated together as a group between each trial. Ants learned to follow the training path relative to the wall irrespective of the wall's position or orientation within the training arena.

How does the wall fix the ants' path? The results of experimental manipulations on ants trained to take a path parallel to the wall imply that they have learned the retinal elevation of the top of the wall (or some other height-related feature) as seen from the trained path, and that they adjust their path so as to keep the top of the wall at the learnt elevation. First, when the wall was made twice as high as in training and an ant was started at its accustomed distance from the wall, its path was directed away from the wall, so bringing the top of the wall towards its usual retinal elevation. But when both the height of the wall and the ants' starting distance were doubled, so that the top of the wall was seen at the expected retinal elevation, the ants' paths were parallel to the wall. When the height of the wall was as in training and ants were started at double their normal distance from the wall, so that the retinal elevation of the top of the wall was unexpectedly low, the ants approached the wall (Figure 6.6).

The point on the barrier closest to the ant has the highest angular elevation. This bump in the elevation profile can be used to control the direction of the route relative to the barrier (Collett, 2009). The bump will be imaged and kept at 90° retinal eccentricity, if the route is parallel to the barrier, and at a greater or smaller angle, if the route is oblique to the barrier. Thus once visual cues are learned, they can then guide the route without support from PI.

Routes guided by isolated landmarks

Isolated landmarks influence routes in a different fashion. Many insects, including ants (Santschi, 1913) and honeybees (von Frisch, 1967; Chittka, 1995), are naturally attracted towards conspicuous upright objects. The insects aim at such 'beacons' (Collett and Rees,

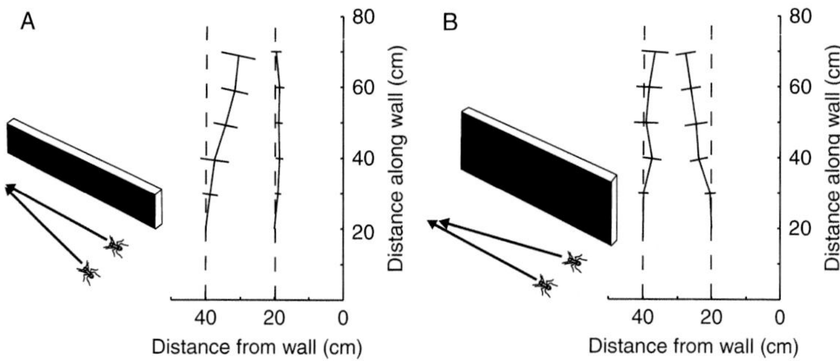

Figure 6.6. Wood ants (*Formica rufa*) guide their path relative to a barrier by its apparent height. A, Ants trained along a path 20 cm from and parallel to a 20 cm high wall are started either 20 cm or 40 cm from the wall. Their path is parallel to the wall when 20 cm from the wall and turns towards the wall at 40 cm from it. B, With a 40 cm high wall, the ants' paths are parallel to the wall if they are started 40 cm away, and they turn away from the wall if they are started 20 cm from the wall (after Graham and Collett, 2002).

1997) provided that the beacons do not divert the insects too far from the direct path to their goal. By exploiting this inbuilt propensity, insects can stick to the same route throughout the learning process, and so speed up their learning of the route.

We show in Figure 6.7 an example from wood ants learning routes indoors (Graham et al., 2003). Ants were trained to go from a starting position to a feeding site in an arena that was empty apart from a single cylinder, which was placed to the side of the direct route; on the left for some ants and on the right for others. Ants were attracted to the cylinder from the beginning of training and their mature routes bowed to the left or to the right according to the position of the cylinder. Tests revealed that the position of the cylinder had become a significant intermediate goal on the ants' route, such that ants learnt additional visual cues from the vantage point of that site. Ants, guided by these extra cues, could then reach the site of the cylinder, when the cylinder was removed and the ants were displaced from their normal start position.

In general, it seems that the shapes of the routes that ants learn are set jointly by PI and by innate responses to visual landmarks. In the two experiments described here, the first part of the route to the end of a barrier, or to a beacon, depends on the position of that visual landmark, while the second part, from the landmark to the goal, is fixed by the difference between the stored PI coordinates of the goal and the state of the PI accumulator when the insect reaches the landmark at the end of its detour.

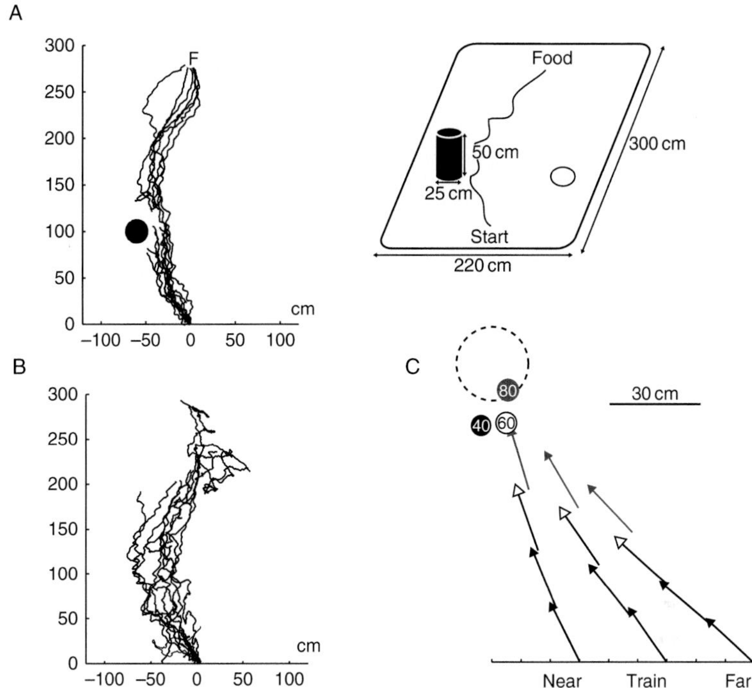

Figure 6.7. The shape of an acquired route is biased by a beacon. Inset: Wood.ants (*Formica rufa*) were trained indoors to a food source with a large black cylinder to the side of the direct route from start to food. A, The routes of trained ants are significantly bowed towards the cylinder and the bowing persists after the cylinder has been removed (B). C, When the cylinder is removed, the paths of ants which start 30 cm either side of their usual position converge on the site of the cylinder. Arrows show mean vectors over successive 20 cm segments. The ringed numbers (40, 60 and 80) show the best estimates of the point of convergence of the three mean vectors after the ants have travelled 40, 60 and 80 cm respectively (after Graham et al., 2003).

Routes between corridors of landmarks

A further example of the structuring of routes by the environment through which an ant passes is seen when it runs along a corridor between objects (Heusser and Wehner, 2002). The ant's innate strategy is to place itself so as to equalise the angular elevations of the two objects (Figure 6.8). A likely benefit of this strategy is to maximize the

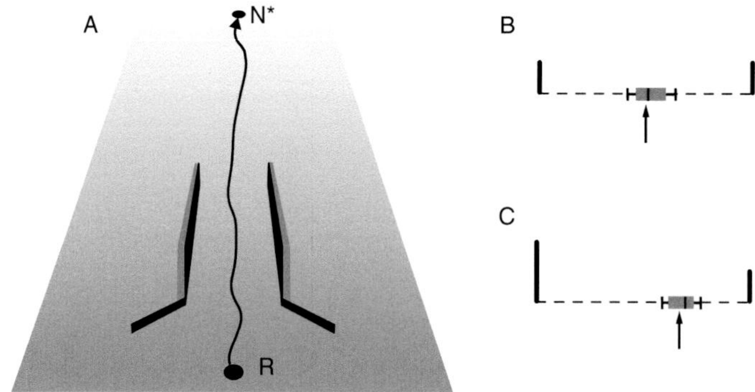

Figure 6.8. Desert ants (*Cataglyphis fortis*) walking between barriers adjust their path to equalize the barriers' apparent height. Ants ran over empty ground between their nest and a food source 20 m away. A, On a test run each ant was taken from the food to a release site close to the end of a pair of barriers 20 or 40 cm high and 150 cm apart with the home-vector pointing along the centre of the corridor. B, Barriers on both sides were 20 cm high. Arrow gives position where barriers have the same angular elevation. Box plot shows median mean position of ant within the corridor and the ranges including 25% and 75% of the data to the left and right of the median. C, Tests where the height of the left barrier is doubled. Paths shift to equalise the angular heights of the barriers (after Heusser and Wehner, 2002).

view of the sky and so improve the information available to the ant's sky compass.

Slow changes in routes

By charting the ontogeny of a route we can see that other less definable processes are also active (Graham and Collett, 2006). The sequence of outbound and homeward routes of several wood ants was recorded in an indoor arena from the ants' first experience of the route to about their 30th foraging trip (Figure 6.9). Outward trips stabilized rapidly (Figures 6.9B and C). The interesting feature here is the slow development of the homeward route, which changed in a consistent manner across all the ants from a strong curve to the right to a path that is only slightly curved (Figure 6.9D). The initial curvature is presumably caused by the ants' innate response to some visual feature of their surroundings. However, unlike a beacon, which makes an insect diverge permanently from a straight route, and then

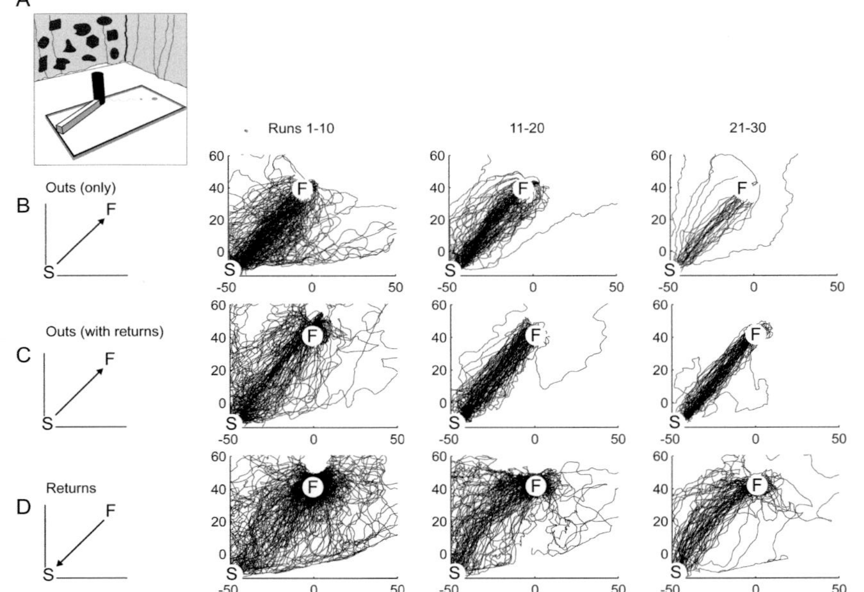

Figure 6.9. The maturation of outward and homeward routes from an ant's first to 30th run. A, Training situation: Wood ants (*Formica rufa*) go from a start point along a narrow corridor, emerging from under a black cylinder into to an area of open arena where there is a feeder. Attempts were made to record all trajectories and they are shown grouped by run number in B to D. The paths of an individual ant are only included if the ant survived to perform at least 10 routes. Low barriers that were used to constrain ants within the half of the arena containing the feeder influenced the paths on initial runs. B, Food-bound trajectories of ants ($n = 14$) that were carried back to the nest after feeding without performing homeward trajectories. C and D, The food-bound and homeward trajectories, respectively, of ants that performed round trips ($n = 12$). After Collett and Graham (2006).

itself becomes an intermediate goal, the curvature of these homeward routes is not sustained. By some unknown process the ants' path gradually straightens.

SPEED OF ROUTE ACQUISITION

Long ago, Tinbergen (1932) found that wasps will learn the location of their nest hole relative to a circle of pine cones during a single survey flight. Since then it has become increasingly clear that insects acquire navigational and sensory memories related to foraging tasks very rapidly. For instance, Menzel (1968) showed that colors signalling food sites are learned in very few trials, and Bisch-Knaden and Wehner (2003) demonstrated similarly rapid learning of an association between a route segment of a given direction and distance and the visual cues triggering that path segment.

The method of route acquisition that we have described encourages rapid learning. The use of a framework provided by a PI controlled trajectory to a goal modulated by reactions to visual features means that the route is specified correctly along all its length, so that the entire route can be revised on every trip. The ant need not discover what to learn as the route is mapped out automatically while the ant follows the PI based path.

In principle, ants could further speed up route learning by acquiring information for guiding a homeward route when traveling to the feeder, and vice-versa. Wood ants do learn something about the route home when running their outward route (Graham and Collett, 2006). Such bidirectional learning can be seen when ants are allowed to perform a sequence of outward trips on a two-leg route (Figure 6.9A), but are always carried home. To test the ants' knowledge of the way home after this procedure, ants are taken singly from the nest to the feeding site (to prevent any complication from PI) and then after feeding each ant returns on its own for the very first time. The ants' paths were very roughly directed along the first leg of the homeward route. It seems likely that they acquire views of the homeward direction while on their food-ward route during periods in which they turn around and briefly retrace their steps towards home. This retracing behaviour is found primarily when wood ants are inexperienced with a route (Judd and Collett, 1998; Nicholson et al., 1999; Graham and Collett, 2006).

Bidirectional learning seems to be a weak process and routes can be learned without it. Ants performing normal homeward routes,

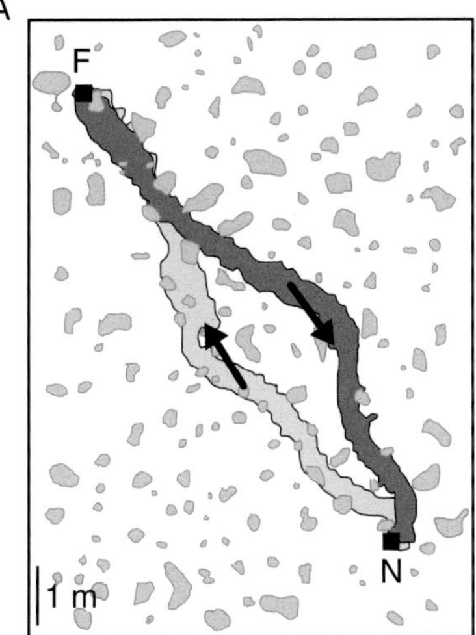

Figure 6.10. Separate food-ward and homeward routes of an Australian desert ant (*Melophorus bagoti*). Shading represents the area covered by five successive outbound and five corresponding inbound runs, light grey blobs indicate tussocks of grass (after Wehner, 2003).

and ants, which are carried home, acquire their outward routes at about the same rate (Figures 6.9A and B). In addition, some species acquire markedly different routes on their outward and homeward journeys (Kohler and Wehner, 2005), and therefore have no opportunity for bidirectional learning (Figure 6.10). One benefit of bidirectional learning may be to aid species, with routes that are guided by a mix of chemical trails and visual cues, develop similar paths in both directions. Such species tend to forage as a group and identical food-ward and homeward paths would allow communication between ants travelling in opposite directions. Another possible benefit of bidirectional learning (see section on Changing routes) is to help ants who have changed feeding sites learn the outward route to a new feeding site while they travel home from it guided by PI.

FAMILIAR VISUAL LANDMARKS AND LEARNT ROUTES

Familiar visual landmarks can specify the path that an ant takes in two very different ways. First, as we saw with extended landmarks, the ant can move to attain particular visual goals. It learns what a landmark or an array of landmarks looks like from the vantage point of its goal or from sites along its route, and subsequently moves to recapture or maintain the learnt scene on its retina. Secondly, landmarks can prime an action that the ant performs. When an ant or bee recognizes a view along a route, it may perform an action that it has associated with that view, such as making a turn or moving in a particular compass direction.

Attraction by learnt views

Evidence from a variety of insects, including flies (Collett and Land, 1975) and water striders (Junger, 1991), in addition to ants and bees (Wehner and Räber, 1979; Cartwright and Collett, 1983), suggests that these insects learn scenes as two dimensional views of their surroundings from particular vantage points and then use the stored view to navigate back to where the view was acquired. By changing the appearance of an insect's surroundings after its acquisition of a view, one can work out what aspects of the surroundings are captured in its memory. Such experiments show that well-trained insects are more concerned with two dimensional than with three dimensional features of the scene.

Figure 6.11A shows the search distribution of wood ants trained to feed at a point midway between two cylinders of the same size (Graham et al., 2004). If one cylinder is enlarged and the other made smaller, the ants' search shifts to where the two cylinders subtend the size normally seen from the goal (Figure 6.11B). If both cylinders are made smaller, there are two possible search positions. Ants then search at both positions (Figure 6.11C), with search driven by the cylinder that the ant is facing (Figures 6.11E and F), suggesting that the ant has a retinotopic memory of the cylinder within its frontal visual field. Ants care more about recapturing the correct appearance of each cylinder as seen from the goal, than about keeping the cylinder at the correct distance.

Identifying landmarks

The example of Figure 6.11 is unusual in that the two landmarks had the same retinal size from the goal, so that ants have no need

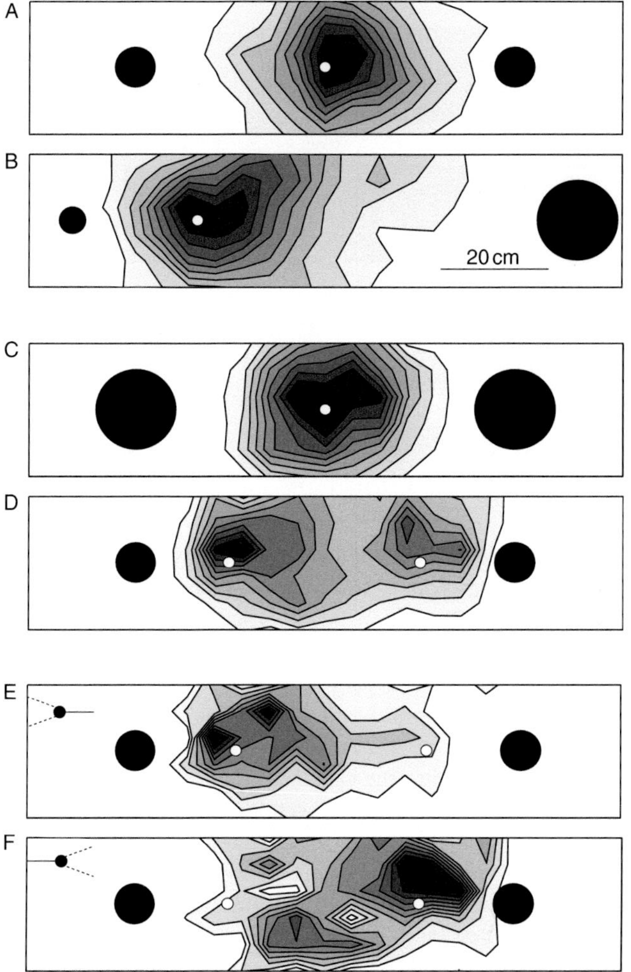

Figure 6.11. Wood ants (*Formica rufa*) search for an absent feeder using a stored two dimensional view or snapshot. A, Ants are trained to find food midway between two cylinders and their search distribution is concentrated on that spot when the feeder is missing. B, Training cylinders are replaced by one cylinder that is smaller (in height and width) and one that is larger. Ants search where the cylinders look the same as they did in training. C, As A for a new group of ants trained with two large cylinders. D, Tested with two small cylinders replacing the training cylinders. Ants search in two locations, one appropriate for each cylinder. E, Subsets of distribution D when ants fixate within ± 20 ° of the centre of one or other cylinder, as shown by the schematic ant to the left of each distribution (from Graham et al., 2004).

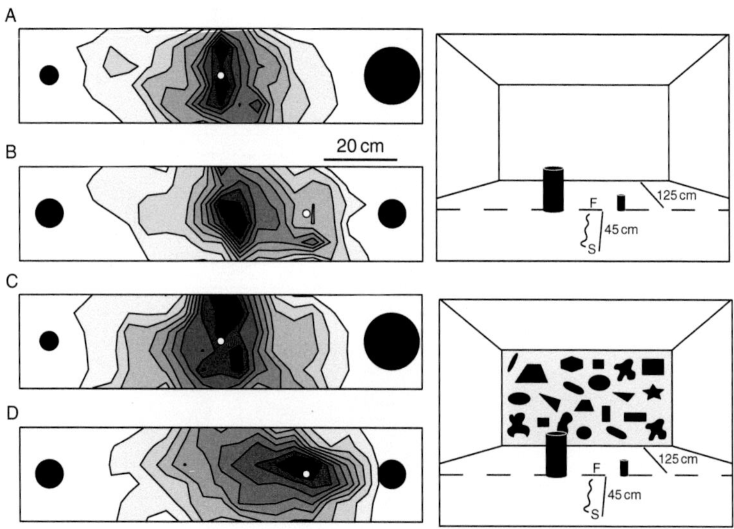

Figure 6.12. Recognizing cylinders by their surroundings. Wood ants (*Formica rufa*) were trained to a feeder midway between a small and a large cylinder (filled circles). In one experiment (A, B) the arena was surrounded entirely by white curtains. In a second experiment (C, D) one wall parallel to the array was patterned. A, C, Search distributions with training array. B, D, Search distributions with two intermediate sized cylinders. Open circles represent position of feeder as predicted by a snapshot in which the identity of each cylinder is known. Ants search correctly in both training conditions. They search correctly in tests if they have been trained and tested with the patterned curtain (D). They fail to search correctly in tests when they were trained and tested with a white curtain (B). The retinal position of the patterned curtain helps ants know which intermediate cylinder represents the small cylinder and which the large, but it cannot itself give locational information, since after each trial cylinders and food are shifted relative to the curtain (after Graham et al., 2004).

to distinguish between the cylinders. More typically, the view from the goal will contain landmarks with a range of apparent sizes. Accurate guidance then requires the memory of each landmark to be matched correctly to its counterpart in the ant's current image. An experimental example of some of the problems of identifying landmarks is shown in Figure 6.12. A food site is placed midway between cylinders of different sizes (Graham et al., 2004), such that the retinal image of the right hand cylinder at the food site is wider and taller than the left. A problem arises, if these training cylinders are replaced by two similar cylinders of intermediate size. Where do the ants search? If ants can identify the right-hand cylinder as the larger one, they would search closer to that cylinder. Ants seemed not to know which cylinder was which and they searched midway between the two cylinders, when they were trained in a uniformly white curtained arena (Figure 6.12B). But they searched near the right-hand cylinder, when one of the curtains was patterned and could provide an additional cue for identifying the cylinders (Figure 6.12D). At the food site, the ants see the curtain with its left eye when it faces the small cylinder and with its right eye when it faces the large cylinder. The cylinders can thus be distinguished by linking a frontal view of each cylinder at the feeding site to the left or right lateral view of the patterned curtain.

Compass direction is another way to disambiguate similar landmarks. Åkesson and Wehner (2002) placed a square array of four cylinders asymmetrically about a desert ant's nest so that the nest was in the southeast corner of the square (Figure 6.13). Food was provided 15 m southeast of the nest. Once ants were accustomed to this arrangement, they were caught at the entrance to the nest on their homeward trip from the feeder and taken to a distant test area with a different panorama. They were then released 2 m to the northwest or southeast of a similar array of four cylinders. After release at a point southeast of the array, ants, as expected, searched for the nest in the southeast corner. But, when released from the northeast, ants searched first in the corner that was closest to the release site and which they reached first. Next they looked in the southeast corner, which they could only pick out from the others through the use of compass cues.

Finally, ants can discriminate between landmarks of different shapes. Desert ants on their way home were accustomed to passing a cylindrical landmark on the right and a triangular landmark on the left (Figure 6.14). The ants demonstrated that they distinguished between these landmarks when they were taken from the feeder and released on a test ground with either a cylindrical or a triangular landmark in the direct line of their home vector. Ants always detoured respectively to the left or right of the test landmark to place the cylinder or the triangular landmark on the expected side.

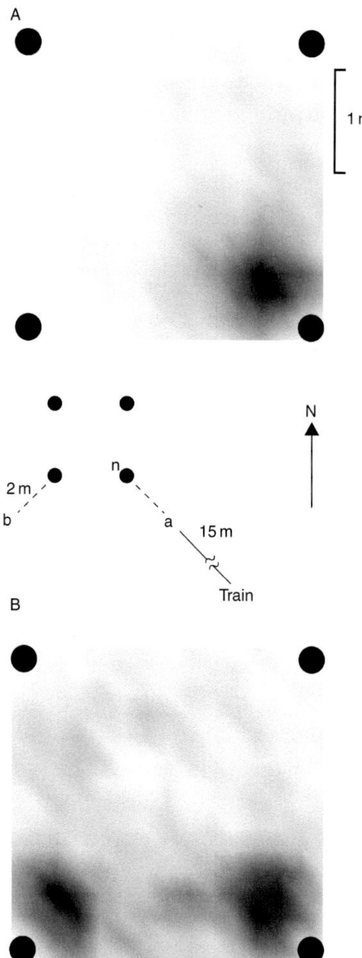

Fig. 6.13. Ants (*Cataglyphis fortis*) use compass cues to aid snapshot guidance. Ants learn to find their nest in the south east (SE) corner of an array of four cylinders and are accustomed to approaching it from the SE. If ants ignore compass information, a visual snapshot recorded at the nest would be matched in all four corners. To see whether ants disambiguate the four possible matching sites, a similar array of cylinders is placed on unfamiliar terrain and ants are released from points around the array (centre panel). A, Ants taken at the nest with zero home vector and released at "a", 2 m SE of the array, searched only at the fictive position of the nest (dark cloud). B, Zero vector ants released at "b", 2 m SW of the array, generated two search peaks. The first peak was close to the point where the ants entered the array. The second, larger peak was at the fictive position of the nest. Both route and compass cues aid snapshot matching (after Åkesson and Wehner, 2002).

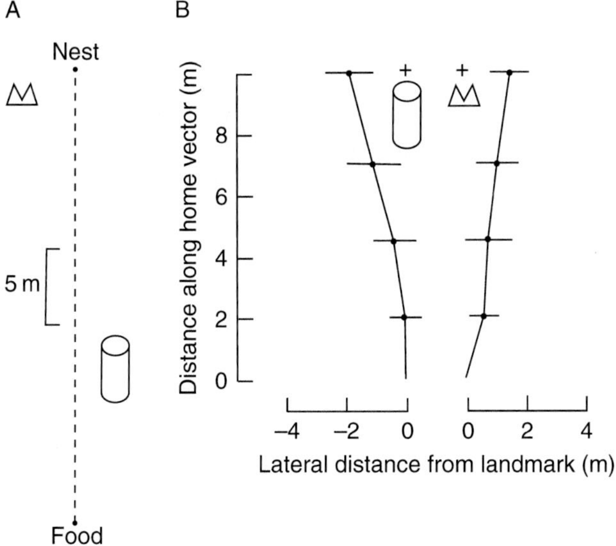

Figure 6.14. Ants recognize the appearance of landmarks along a homeward route. A, Two differently shaped landmarks are placed on either side of the route that is taken by desert ants (*Cataglyphis bicolor*) between a feeder and their nest. On its homeward trip an ant sees the cylinder to its right and the triangles to its left. B, Tests in which single ants have been carried from the feeder to a test area. Either the cylindrical or the triangular landmark is placed on the direct route to the nest, as defined by PI. Ants detour around the landmark so as to place the cylinder in its right visual field and the triangle in its left visual field so matching what they see in training. Data are shown as mean paths with standard deviations (after Collett et al., 1992).

Actions linked to familiar landmarks

Ants, which are initially guided to a goal by a mix of landmarks and path segments directed by global PI, soon learn the directions and distances of the various path segments comprising their route. They then follow these path segments, and can do so with their global PI system in any of a wide range of states. Such local vectors can be linked to familiar landmarks (for bees see Collett et al., 1993, 1996; Srinivasan et al., 1997; Collett et al., 2002; for ants see Collett et al., 1998; Bisch Knaden and Wehner, 2003), to familiar contexts (Knaden et al., 2006, Collett and Collett, 2009a), and to each other (for bees see Collett et al., 1993; Zhang et al., 1996), such that a local vector is triggered on viewing the appropriate landmark or encountering the familiar context.

To be sure that a path segment can be guided by a local vector, the possibility must be eliminated that the direction of the segment is controlled by visual features or by global PI. These conditions were met in tests on desert ants that had been trained on an L-shaped route between nest and feeder. The ant travelled 10 m north from its nest over open ground and then turned through a right angle into an open topped channel with a view of the sky to reach a feeder at the channel's end. It followed that path in reverse to return to the nest. Because the channel was placed in a trench dug into the sand, ants did not see it once they had moved away. In tests the homeward route was recorded on a test ground with another hidden channel (Figure 6.15A). If an ant was allowed to return almost to its nest, so that its home-vector was nulled, and then caught and released at the end of the test channel, the ant travelled down the channel and on leaving it headed roughly south (Figure 6.15B). This directional response is guided neither by global PI, nor by a landmark, because, unlike the barrier in Figure 6.5, the channel cannot be seen. The response seems to be a learnt vector that is triggered by the change in view as the ant goes from the channel to open ground (see Collett and Collett, 2009a and 2009b for a discussion of the distance component of local vectors and how it might be acquired and encoded).

A similar test showed that local vectors take precedence over the global vectors that have their origin at the nest. When the ant was taken from the feeder to a test channel that was half the length of the training channel, its global vector pointed obliquely at the virtual nest and on about a third of trials the ants followed this vector (Figure 6.15C). But more often, the ant, on leaving the shortened channel, performed a short local vector at right angles to the channel before turning and following its global vector (Figure 6.15D). Thus, initially the memorized local vector is favored over the global PI trajectory.

THE INDEPENDENCE OF LANDMARK GUIDANCE AND PI

In the previous section we saw that one particular learned component of a habitual route – a local vector – can operate independently of

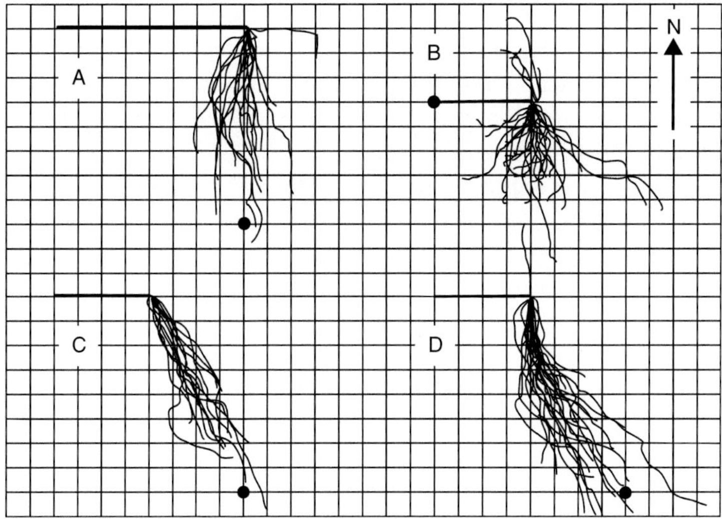

Figure 6.15. Local vectors often take precedence over global PI vectors. Desert ants (*Cataglyphis fortis*) were trained on an L-shaped foraging route. The homeward path consisted of 8 m in an open topped channel (thick line) that was hidden in a trench in the sand followed by 8 m over open ground. A, The 'homing' trajectories of ants taken from the feeder to a test ground and placed at the end of an 8 m channel similar to the training channel. The trajectories are oriented southwards, as they are in training. B, The trajectories of ants allowed to return to the nest before being placed in a 4 m channel on the test ground are also oriented southwards, even though the global PI vector is zero. These trajectories are local vectors triggered by leaving the channel. C, D, Trajectories from ants taken from the feeder to the end of a 4 m channel can be separated into two groups. C, Some ants follow their home-vector from the end of the channel to the fictive nest site (marked with a dot). D, Most ants head due south following a local vector for about 2 m before taking a global vector to the fictive nest. Performance of the global vector is suppressed while the local vector runs (after Collett et al., 1998).

global PI. An indication that the performance of a complete memorized route does not rely on global PI was given by Wehner et al. (1996). They found that the homeward route taken by a foraging desert ant was the same, whether it was performed when the ant left the feeder as normal, or whether the ant was collected at the nest entrance on its way home and replaced on the feeding site with zero home vector. Andel and Wehner (2003) pushed this method further by repeatedly catching desert ants close to the nest after they had performed a homeward route and each time taking the ants back to the feeder from where they returned home. The ants' landmark driven homeward route remained the same on successive trips, unaffected by the state of the global PI vector that changed with each repetition. At the beginning of the ant's first return, the global PI vector pointed as normal at the nest. At the beginning of its second return, the global PI vector started at zero and pointed back to the feeder as the ant travelled home. On the third and later returns, the PI system indicated that the ant had walked beyond the nest. The predicted changes in the global PI vector could be seen in changes in the direction of the home-vector when the ant was released on unfamiliar terrain. Before the first homeward trip, the home-vector was, as it usually is, in the direction of the nest, but on the third trip the home-vector pointed in the opposite direction. Thus, the ant's global PI state is not reset when the ant performs its familiar route and responds to familiar views (see also Collett et al. 2003, Knaden and Wehner, 2005, Collett and Collett, 2009b).

The decoupling of route memories from the global PI system is nicely illustrated in a study on an Australian desert ant, *Melophorus* (Kohler and Wehner, 2005). Homeward bound ants caught either just as they left the feeder, or after they had almost reached the nest, and were displaced to a position in the middle of their habitual route. Despite the differing PI states at the release site, the path taken by the displaced ants was indistinguishable from the last half of their habitual route to the nest (Figure 6.16).

The separation between global PI and visual guidance can also be seen in the learning of places. When ants turn, they tend to overestimate the magnitude of their turn and this error is reflected in systematic errors of path integration (Müller and Wehner, 1988). After following an L-shaped route in a channel all the way to a feeder, the direction of the ants' home vector indicates that the ant 'thinks' it has turned through more than the right angle it was forced to make (see for example Figure 6.17A left). Similar errors are made on outward routes. When an ant turns freely towards the feeder, it overestimates its turn and consequently under-turns towards the feeder. These errors of measurement on the ant's outward trip are not only found when an ant is unexpectedly forced to detour, as in Figure 6.3, but also continue without improvement

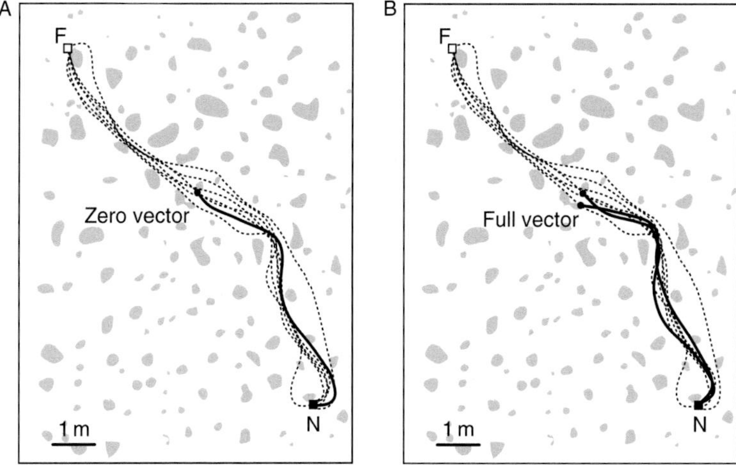

Figure. 6.16. Australian desert ants (*Melophorus bagoti*) rejoin routes after displacement. Dotted lines show normal homeward trajectories of an ant. Black lines show that the ant follows similar paths when (A) it is taken from the feeder (F) to a position midway between F and nest (N), or (B) it runs almost to the nest and is returned with zero home vector to the same midway position. Grey areas depict grass tussocks. Visual route following is not influenced by the state of the PI system (from Kohler and Wehner, 2005).

when ants follow repeatedly an L-shaped route to a feeder over several days (Collett et al., 1999). They continue to search for the feeder in the wrong place indicating that their encoding of the feeder site in global PI coordinates is erroneous (Figure 6.17B left). If the feeder location is habitually marked by a cylinder, ants still under-turn at the L, but when the cylinder comes within view the ants' path bends towards it (Figure 6.17 right). The curved shape of the path to the food site is stable and shows that PI tells the ant that the food is in one location while landmarks indicate another (Collett and Collett, 2009a). Because it seems fundamental to insect navigation, it is worth reiterating a general conclusion that follows from this separation between PI and landmark based navigational strategies. Ants do not possess one often suggested substrate for a cognitive map. Since familiar views do not have global PI coordinates associated with them, ants displaced to a familiar site cannot use PI to plan a novel route from that site to another destination.

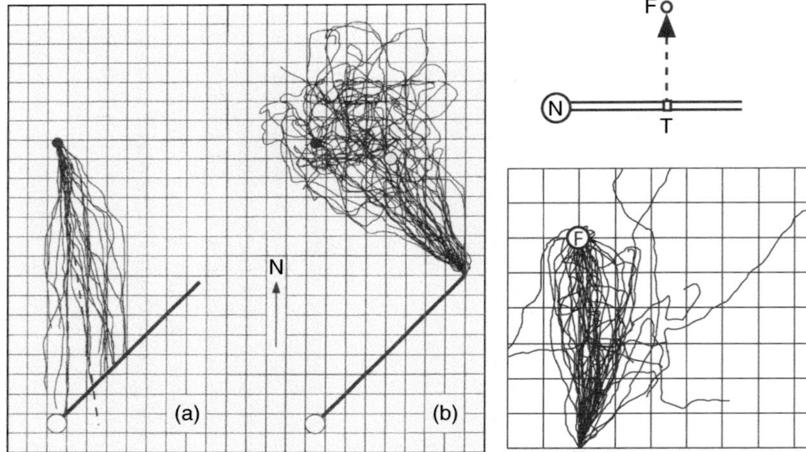

Figure 6.17. Stable errors in the PI encoding of places. The first segment of an L-shaped path to the feeder is within an open topped channel hidden in a trench. The second leg is over open ground. Ants return home directly over open ground all the way to the nest. Left, trajectories in an area with no landmarks in the vicinity. A, Homeward trajectories. B, the segment of food-ward trajectories that is over open ground. The directions of both sets of trajectories miss the nest and the feeding site respectively. Right, Food-ward trajectories in a similar set up but aided by visual features. The feeder is marked with a small cylinder and there are two bushes 20 to 30 m away. The initial path on leaving the channel is again misdirected, probably because of PI errors. But later in the trajectory, the cylinder guides a curving path to the food site. (left, after Collett et al. 1999; right, after Collett and Collett, 2009).

CHANGING ROUTES

When a food site fails and ants explore to find a new site, by what means does the established route adapt? As already described, *Cataglyphis bicolor* explores for food in the immediate neighbourhood of a previously visited site and in the same direction beyond it (Schmid Hempel, 1984). When this strategy is successful, the old route can be kept and just lengthened. But, if the ant's familiar sector is barren, so that the ant explores in a new direction (as in Figure 6.4) how does the ant develop a route to a replacement site?

One potential problem is interference from the old route. PI can lessen the problem on the homeward route. An ant reaches the new feeding site circuitously *via* its original route followed by search. At this site the ant has no route information to interfere with PI instructions for the direct path home. The ant can then build up a new visually guided route along the path of this new home-vector. If an ant has abandoned a foraging direction after a series of unrewarded trips, the prescription for learning the route to a new site in a new direction is equally simple: use PI instructions to find the feeding site and learn visual cues along the way.

The old outward route may be less easy to abandon if the new site is discovered by gradual exploration in a new direction from the old site. With PI navigation suppressed by route memories (Figures 6.15 and 6.16), the old route will persist, newly reinforced, every time that the ant follows it and continues on to the new site. In this case, an additional mechanism seems to be needed. Learning the outward route when traveling home would be one solution. Another possibility is that the discrepancy between the old initial direction and the new PI computed direction to the food induces the ant to abandon its old route and switch to the PI direction of the new site and thus acquire a visually guided route along the new food vector.

Signs of interactions between landmark memories that do slow down a change between two food sites come from a laboratory study of wood ants (Graham et al., 2007). Wood ants were first trained along an 80 cm route from a fixed start position to a feeding site in an arena. Once ants were well trained, food was shifted to a new location so that the vector between the start point and the food site rotated by 45 °, and the ants were encouraged to feed there. The heading of the initial segment of the route began by pointing directly at the old feeding site and gradually rotated over 20–30 trials through intermediate directions until it pointed at the new feeding site (Figures 6.18A and C).

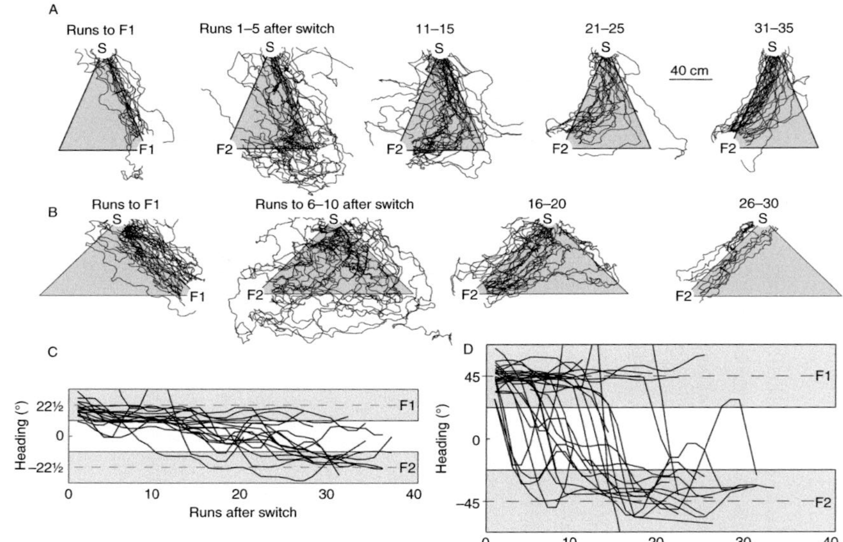

Figure 6.18. The dynamics of changing routes and its dependence on the separation between feeder sites. Wood ants are trained over about 30 trials to a feeder at F1. The feeder is then shifted to F2 and the subsequent 30 to 35 trajectories are recorded. Data are shown for two separations. A, B, Trajectories at the end of training to F1 and during the changeover to F2. Trajectories are grouped by trial number. Grey area shows triangle enclosed by start (S) and F1 and F2. C, D, Time course of changes in the initial direction of the trajectories with each line showing the smoothed data for a single ant. The upper and lower shaded areas represent the range of directions that are categorised as towards F1 or towards F2 respectively. Changes are more gradual and slower for the smaller separation than they are for the larger separation between F1 and F2 (from Graham et al., 2007).

This rotation is not only seen when ants are free to take any route they want to the new feeder site. It is also found when ants are carried from the nest close to the new site and corralled within a small circle centred on the feeder. After feeding the ants make their own way home with a low barrier preventing them from exploring the area between the two feeding sites during their return. Occasional probe trials - in which ants perform a normal outward path from the start position - show that they still acquire a gradually rotating outward route, probably guided by views captured close to the second feeder and on the way home. Over time their memory of this second location slowly comes to dominate their memory of the first. The intermediate routes indicate that for a while the ant's memories of both feeder sites are active and that ants are guided by a combination of the stored views of both locations.

When the direction between two training sites was increased to 90 °, the ants' views from the two sites are more distinct and memories of them do not seem to interact. The initial segment does not rotate but instead switches abruptly between the two sites (Figures 6.18B and D). Perhaps because the views from the two food sites differ more with the larger separation and their memories can be reinforced independently, ants switch routes in about half as many trials (Figure 6.18D). This difference between the two separations indicates that interference between coactivated memories may indeed delay a change of route, particularly if the routes and destinations have visual features in common.

Given the right circumstances, ants can switch readily between several familiar routes. Colonies of the harvester ant *Pogonomyrmex barbatus* may have up to eight distinct trails leading to different destinations (Greene and Gordon, 2007). Between 30 and 50 patroller ants explore their colony's trail system early each morning, and by laying a streak of Dufours secretions on the nest mound, control which particular trails the colony follows that day. Trail following may well employ both visual and chemical cues and in the absence of directional signals from patrollers, foragers tend to pick the trail that they used the previous day (Greene and Gordon, 2007). There is the interesting possibility that the patrollers' recruitment signal exerts a stronger effect in the morning, when route memories used the previous day have been dormant over night. We have seen in many experiments on wood ants that visually guided routes tend to be fragile at the start of each day.

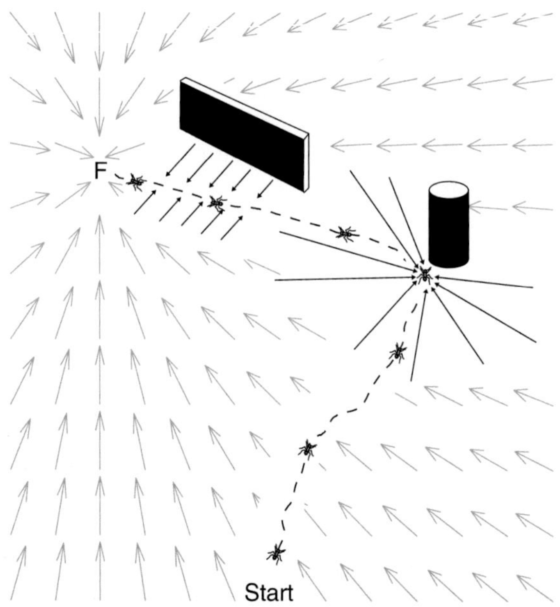

Figure 6.19. Landmark and path integration attractors along a route. Both beacons and extended landmarks draw ants to points or lines on a route, as shown by black arrows. The path integration attractor (grey arrows) can draw ants to the goal from all positions. PI can aid route learning, as shown here by the path relative to the extended landmark.

Routes and attractors

The various learned components of routes can be described through the common language of attractors. Route learning can then be envisaged as the development of visual attractors along a path taken through a PI attractor space (Figure 6.19). Because a visual environment transforms relatively smoothly as the ant passes through it, a stored view can provide an attractor that draws an ant to the site where the view was taken. PI navigation, in which an ant guides its path using the difference between its goal and current coordinates, can also be described as the ant moving towards an attractor state.

Thus, in learning a route, the ant transfers control of its path, which is first set by a PI attractor in conjunction with the ant's intrinsic reaction to visual features of the terrain, to a sequence of visual attractors that with the aid of local vectors take it from one sub goal to the next. These visual attractors make routes robust in that ants are drawn back on to a route if they are displaced or stray from it.

They also make reaching the final destination more certain as any errors in performing a local vector are eliminated each time a visual attractor is reached.

Similarly, the changing routes of wood ants, as they switch from one feeding site to another, can be interpreted as the routes being controlled by two sets of attractors, the strengths of which alter as learning progresses, or with changing contextual signals, like motivational state, time of day and recruitment cues or contextual cues change with time of day, motivational state, or recruitment signals.

CONCLUSIONS

Spatial memories are only of use if they are tied firmly to the physical world. Path integration is grounded by having a fixed origin (the nest), an odometer to measure distances travelled and a sky compass to give the direction of the insect's path. The PI memories that ants and bees store consist of the PI coordinates of significant sites within a coordinate system centred on their nest. Using PI, ants can plan a trajectory to a remembered site by subtracting the PI coordinates of their present position from those of the site. Thus, PI allows an insect to approach a remembered goal from any position it has reached. In its moment-to-moment operation, this global PI system (at least in ants) seems to be insulated from landmark memories, so that errors within the PI system cannot be remedied by consulting local visual cues. There is, thus, good reason for insects to use familiar landmarks and follow routes, when that is possible, while having global PI available as an independent back-up if landmarks fail.

The isolation of landmark memories from global PI means that landmark memories do not have positional coordinates attached to them. Instead they are fixed to the world by virtue of being acquired at a particular location and by being retrieved appropriately, so that they can be used safely for guidance to that location. They can act in two distinct ways, either as attractors bringing an insect to the site where the landmark memory was acquired, or as repellents triggering stored actions that carry the insect in a particular direction away from that spot. These memories are directional and tied to particular routes (Harris et al., 2005; Wehner et al. 2006) so that while they are effective in keeping insects on familiar routes, there is no evidence that they play a more general role in planning new ones.

REFERENCES

Åkesson, S. and Wehner, R. (2002). Visual navigation in desert ants *Cataglyphis fortis*: are snapshots coupled to a celestial system of reference? *Journal of Experimental Biology*, **205**, 1971–1978.

Andel, D. M. and Wehner, R. (2003). Path integration in desert ants, *Cataglyphis*: redirecting global vectors. In N. Elster & H. Zimmerman (Eds.) *Proceedings of the 29th Göttingen Neurobiology Conference* p. 574. Stuttgart: Thieme Verlag.

Bisch-Knaden, S. and Wehner, R. (2003). Local vectors in desert ants: context-dependent landmark learning during outbound and homebound runs. *Journal of Comparative Physiology A: Neuroethology Sensory Neural and Behavioral Physiology*, **189**, 181–187.

Cartwright, B. A. and Collett, T. S. (1983). Landmark learning in bees: experiments and models. *Journal of Comparative Physiology A: Neuroethology Sensory Neural and Behavioral Physiology*, **151**, 521–543.

Chittka, L., Kunze, J., Shipman, C. and Buchman, S. L. (1995). The significance of landmarks for path integration of homing honey bee foragers. *Naturwissenshaften*, **82**, 341–343.

Collett, M. and Collett, T. S. (2000). How do insects use path integration for their navigation? *Biological Cybernetics*, **83**, 245–259.

Collett, M. (2009) Spatial memories in insects. *Current Biology*, 19 R1103-R1108.

Collett, M. and Collett, T. S. (2009a). The learning and maintenance of local vectors in desert ant navigation. *Journal of Experimental Biology*, **212**, 895–900.

Collett, M. and Collett, T. S. (2009b). Local and global navigational coordinate systems in desert ants. *Journal of Experimental Biology*, **212**, 901–905.

Collett, M., Collett, T. S., Bisch, S. and Wehner, R. (1998). Local and global vectors in desert ant navigation. *Nature*, **394**, 269–272.

Collett, M., Collett, T. S., Chameron, S. and Wehner, R. (2003). Do familiar landmarks reset the global path integration system of desert ants? *Journal of Experimental Biology*, **206**, 877–882.

Collett, T. S., Baron, J. and Sellen, K. (1996). On the encoding of movement vectors by honeybees: are distance and direction represented independently? *Journal of Comparative Physiology A: Neuroethology Sensory Neural and Behavioral Physiology*, **179**, 395–406.

Collett, M., Collett, T. S. and Wehner, R. (1999). Calibration of vector navigation in desert ants. *Current Biology*, **9**, 1031–1034.

Collett, M., Harland, D. and Collett, T. S. (2002). The use of landmarks and panoramic context in the performance of local vectors by navigating honeybees. *Journal of Experimental Biology*, **205**, 807–814.

Collett, T. S., Collett, M. and Wehner, R. (2001). The guidance of desert ants by extended landmarks. *Journal of Experimental Biology*, **204**, 1635–1639.

Collett, T. S., Dillmann, E., Giger, A. and Wehner, R. (1992). Visual landmarks and route following in desert ants. *Journal of Comparative Physiology A: Neuroethology Sensory Neural and Behavioral Physiology*, **170**, 435–442.

Collett, T. S., Fry, S. N. and Wehner, R. (1993). Sequence learning by honeybees. *Journal of Comparative Physiology A: Neuroethology Sensory Neural and Behavioral Physiology*, **172**, 145–150.

Collett, T. S. and Land, M. F. (1975). Visual spatial memory in a hoverfly. *Journal of Comparative Physiology A: Neuroethology Sensory Neural and Behavioral Physiology*, **100**, 59–84.

Collett, T. S. and Rees, J. A. (1997). View-based navigation in Hymenoptera: multiple strategies of landmark guidance in the approach to a feeder.

Journal of Comparative Physiology A: Sensory Neural and Behavioral Physiology, **181**, 47–58.

Cornetz, M. V. (1910). Album faisant suite aux trajets de fourmis et retours aux nid. *Mémoires Insitut Général Psychologique Paris*, **2**, 1–67.

Fourcassié, V., Henriques, A. and Fontella, C. (1999). Route fidelity and spatial orientation in the ant *Dinoponera gigantea* (Hymenoptera, Formicidae) in a primary forest: a preliminary study. *Sociobiology*, **34**(3), 505–524.

Fresneau, D. (1985). Individual foraging and path fidelity in a ponerine ant. *Insectes Sociaux*, **32**(2), 109–116.

Fukushi, T. (2001). Homing in wood ants, *Formica japonica*: use of the skyline panorama. *Journal of Experimental Biology*, **204**, 2063–2072.

Graham, P. and Collett, T. S. (2002). View-based navigation in insects: how wood ants (*Formica rufa* L.) look at and are guided by extended landmarks. *Journal of Experimental Biology*, **205**, 2499–2509.

Graham, P. and Collett, T. S. (2006). Bi-directional route learning in wood ants. *Journal of Experimental Biology*, **209**, 3677–3684.

Graham, P., Durier, V. and Collett, T. S. (2004). The binding and recall of snapshot memories in wood ants (*Formica rufa* L.). *Journal of Experimental Biology*, **207**, 393–398.

Graham P., Durier V. and Collett T. S. (2007). The co-activation of snapshot memories in the wood ant. *Journal of Experimental Biology* **210**, 2128–2136.

Graham, P., Fauria, K. and Collett, T. S. (2003). The influence of beacon-aiming on the routes of wood ants. *Journal of Experimental Biology*, **206**, 535–541.

Greene, M. J. and Gordon, D. M. (2007). How patrollers set foraging direction in harvester ants. *American Naturalist*, **170**, 943–948.

Harris, R. A., Hempel-de-Ibarra, N., Graham, P. and Collett, T. S. (2005). Ant navigation-priming of visual route memories. *Nature*, **438**, 302.

Harrison, J. M. and Breed, M. D. (1987). Temporal learning in the giant tropical ant, *Paraponera clavata*. *Physiological Entomology*, **12**, 317–320.

Harrison, J. F., Fewell, J. H., Stiller, T. M. and Breed, M. D. (1989). Effects of experience on use of orientation cues in the giant tropical ant. *Animal Behaviour*, **37**, 869–871.

Heusser, D. and Wehner, R. (2002). The visual centring response in desert ants, *Cataglyphis fortis*. *Journal of Experimental Biology*, **205**, 585–590.

Judd, S. P. D. and Collett, T. S. (1998). Multiple stored views and landmark guidance in ants. *Nature*, **392**, 710–714.

Junger, W. (1991). Waterstriders (*Gerris-paludum* F.) compensate for drift with a discontinuously working visual position servo. *Journal of Comparative Physiology A: Sensory Neural and Behavioral Physiology*, **169**, 633–639.

Knaden, M. and Wehner, R. (2005). Nest mark orientation in desert ants *Cataglyphis*: what does it do to the path integrator? *Animal Behaviour*, **70**, 1349–1354.

Knaden, M. and Wehner, R. (2006). Ant navigation: resetting the path integrator. *Journal of Experimental Biology*, **209**, 26–31.

Knaden, M., Lange C. and Wehner, R. (2006). The importance of procedural knowledge in desert-ant navigation. *Current Biology*, **16**, R916–R917.

Kohler, M. and Wehner, R. (2005). Idiosyncratic route-based memories in desert ants, *Melophorus bagoti*: how do they interact with path-integration vectors? *Neurobiology of Learning and Memory*, **83**, 1–12.

Menzel, R. (1968). Das Gedächtnis der Honigbiene für Spektralfarben. I. Kurzzeitiges und langzeitiges Behalten. *Zeitscrift für vergleichende Physiologie*, **60**, 82–102.

Müller, M. and Wehner, R. (1988). Path integration in desert ants, *Cataglyphis fortis*. *Proceedings of the National Academy of Sciences USA*, **85**, 5287–5290.

Nicholson, D. J., Judd, P. D., Cartwright, B. A. and Collett, T. S. (1999). Learning walks and landmark guidance in wood ants (*Formica rufa*). *Journal of Experimental Biology*, **202**, 1831–1838.

Ribbands, C. R. (1964). *The behaviour and social life of honeybees*. Dover Publications, New York.

Rosengren, R. (1971). Route fidelity, visual memory and recruitment behaviour in foraging wood ants of genus *Formica* (Hymenopterus, Formicidae). *Acta Zoologica Fennica*, **133**, 1–106.

Santschi, F. (1913). Comment s'orientent les fourmis. *Revue Suisse de Zoologie*, **21**, 347–425.

Sassi, S. and Wehner, R. (1997). Dead reckoning in desert ants, *Cataglyphis fortis*: can homeward-bound vectors be reactivated by familiar landmark configuration? In N. Elsner and H. Wässle (Eds.), *Proceedings of the 25th Neurobiology Conference of Göttingen* (p. 484). Stuttgart: Georg Thieme Verlag.

Schatz, B., Lachaud, J.-P. and Beugnon, G. (1999). Spatio-temporal learning by the ant *Ectomma ruidum*. *Journal of Experimental Biology*, **202**, 1897–1907.

Schmid-Hempel, P. (1984). Individually different foraging methods in the desert ant *Cataglyphis bicolor* (Hymenoptera, Formicidae). *Behavioral Ecology and Sociobiology*, **14**, 263–271.

Schmidt, I., Collett, T. S., Dillier, F. X. and Wehner, R. (1992). How desert ants cope with enforced detours on their way home. *Journal of Comparative Physiology A: Sensory Neural and Behavioral Physiology*, **171**, 285–288.

Srinivasan, M. V., Zhang, S. W. and Bidwell, N. J. (1997). Visually mediated odometry in honeybees. *Journal of Experimental Biology*, **200**, 2513–2522.

Tinbergen, N. (1932). Über die orientierung des Bienenwolfes (*Philanthus triangulum* Fabr.). *Zeitscrift für vergleichende Physiologie*, **16**, 305–335.

von Frisch, K. (1967). *The dance language and orientation of bees*. London: Oxford University Press.

Wahl, O. (1932). Neue untersuchungen über das zietgedächtnis der bienen. *Zeitschrift für vergleichende Physiologie*, **16**, 529–589.

Wehner, R. (1987). Spatial organisation of foraging behaviour in individually searching desert ants, *Cataglyphis* (Sahara desert) and *Ocymyrmex* (Namib desert). In J. M. Pasteels and J. L. Deneubourg (Eds.), *From individual to collective behaviour in social insects* (pp. 15–42). Basel: Birkhäuser Verlag.

Wehner, R. (1990). On the brink of introducing sensory ecology: Felix Santschi (1872–1940), Tabib-En-Neml. *Behavioral Ecology and Sociobiology*, **27**, 295–306.

Wehner, R. (2003). Desert ant navigation: how miniature brains solve complex tasks. *Journal of Comparative Physiology A: Neuroethology Sensory Neural and Behavioral Physiology*, **189**, 579–588.

Wehner, R., Boyer, M., Loertscher, F., Sommer, S. and Menzi, U. (2006). Ant navigation: one-way routes rather than maps. *Current Biology*, **16**, 75–79.

Wehner, R., Harkness, R. D. and Schmid-Hempel, P. (1983). *Foraging strategies of individually searching ants* Cataglyohis bicolor (Hymenoptera: Formicidae). Stuttgart and New York: Gustav Fischer Verlag.

Wehner, R., Meier, C. and Zollikofer, C. (2004). The ontogeny of foraging behaviour in desert ants, *Cataglyphis bicolor*. *Ecological Entomology*, **29**, 240–250.

Wehner, R., Michel, B. and Antonsen, P. (1996). Visual navigation in insects: coupling of egocentric and geocentric information. *Journal of Experimental Biology*, **199**, 129–140.

Wehner, R. and Räber, F. (1979). Visual spatial memory in desert ants, *Cataglyphis bicolor*. *Experientia*, **35**, 1569–1571.

Wehner, R. and Srinivasan, M. V. (2003). Path integration in insects. In K. J. Jeffery (Ed.), *The neurobiology of spatial behaviour* (pp. 9–30). Oxford: Oxford University Press.

Zhang, S. W., Bartsch, K. and Srinivasan, M. V. (1996). Maze learning by honeybees. *Neurobiology of Learning and Memory*, **66** (3), 267–282.

SUSAN D. HEALY AND VICTORIA A. BRAITHWAITE

7

The role of landmarks in small- and large-scale navigation

While some animals, like plankton, apparently drift haphazardly through their environment, many animals do not, perhaps even planning to move to a specific destination, such as home or a favored feeding site. There are many possible sources of information to aid their orientation and it is typical for them to use multiple cues. Depending on their current environment and that in which they grew up, animals may access and use magnetic, visual (e.g., stars, sun, landmarks), olfactory, auditory, electrical or tactile cues. The particular cue used will often depend on a preference hierarchy established during development. This flexibility of cue use became strikingly apparent during the peak of the debate over the use by pigeons of olfactory and/or magnetic cues over a decade ago. Although in hindsight it may seem obvious that an animal should use the cue it can access most easily or knows to be the most reliable, both of which will depend on the animal's current circumstance, there continues to be, in the pigeon homing literature at least, an ongoing battle for supremacy of one cue over another (e.g., Holland, 2003; Walker *et al.*, 2002; Wallraff *et al.*, 1999). Without wishing to dwell on this debate, it is of relevance to this chapter in that, in spite of the evidence to the contrary, the two major camps in pigeon homing ignore or deny a role for visual landmarks and continue to focus their dispute on whether olfaction or magnetism is the primary cue. Given the wealth of data on visual landmark use in other animals, it would be interesting if the homing pigeon is one of the few species for whom visual landmarks do not play a major role in navigation.

The aim of this chapter is to examine the use of and memory of landmarks both over relatively short distances, such as flown by honeybees and birds, walked by ants and spiders and swum by fish within in a

Spatial Cognition, Spatial Perception: Mapping the Self and Space, ed. Francine L. Dolins and Robert W. Mitchell. Published by Cambridge University Press.

home range; and over considerably longer distances, such as swum by migratory fish and turtles, flown by migratory birds, or walked by migrating mammals like reindeer and elephants. Landmarks will typically be used in the context of visual landmarks as the vast majority of the work done thus far has concentrated on this modality (although see, for example, Burt de Perera, 2004; Höller and Schmidt, 1996; Tolimieri *et al.*, 2000).

LONG-DISTANCE MIGRATION

The long-distance movements of birds and fish, in particular, can entail journeys of several thousands of kilometers. Both the enormity of these feats and the accuracy with which the return to breeding or feeding grounds is achieved has attracted much attention. However, in spite of this interest and although it is half a century since Kramer (1953, 1961) proposed that the success of these journeys depended on the use of both a map and a compass, it is really only the compass component for which we yet have any significant understanding. This is due, largely, to the relative accessibility of investigating compass use. Whether celestial or magnetic, each of these sources of information can, fairly readily, be manipulated, both in the laboratory and even in the field. "Clock-shifting," for example, by which the animal's internal circadian clock is altered, is generated by changing the onset of light for an animal held in a laboratory. The shift can either advance or put back the animal's circadian clock by several hours and the animal's navigational response to the position of the sun can then be tested. Polarizing filters can be placed on orientation cages or arenas to shift the plane of polarized light (e.g., Hawryshyn *et al.*, 1990). Planetaria are used to manipulate stellar cues (e.g., Emlen, 1970), and fixing magnets to an animal's body allows an assessment of how disruption of magnetic information affects navigational performance (Åkesson *et al.*, 2001).

It is considerably more difficult, however, to determine the role of landmarks in these long-distance journeys. There is a suggestion that both monarch butterflies and birds will follow large landmarks that provide directional information such as long, straight roads or mountain ranges (e.g., Bruderer and Jenni, 1990; see also Wiggett *et al.*, 1989) but it is difficult to manipulate visual landmarks with the ease with which compass cues have been. Manipulating the animal's visual inputs, as Schmidt-Koenig and Walcott (1978) did by putting frosted lenses over the eyes of homing pigeons *Columba livia*, results in determining whether or not the animal does or doesn't use that kind of cue and whether it has some kind of back up navigational system. In the case of pigeon homing, the finding that pigeons flew almost to their home loft diverted pigeon researchers from further investigation of the use of visual landmarks for almost twenty years.

There are hints, however, that landmarks of some description are used for long-distance journeys in addition to a compass. For example, experienced migrating starlings *Sturnus vulgaris* are less likely to lose their way due to either natural or experimental displacements than are starlings making the trip for the first time (Perdeck, 1958). The accuracy of many philopatric birds that return to the same breeding territory year after year (e.g., hooded warblers *Wilsonia citrina* not only return to the same territory, they remember which neighbors shared which boundaries the previous year: Godard, 1991) is also well documented. In the case of the hooded warbler, in particular, it is not obvious that their successful territory relocation can be achieved with a compass alone. Another hint that memory for landmarks may be used to aid return to the correct territory comes from another quarter: the size of the hippocampus, the brain region concerned with processing of spatial information, is larger in experienced migrants than in naïve birds (Healy *et al.*, 1996). Hippocampal volume is not, however, correlated with migration distance (Krebs *et al.*, 1989). These data are consistent with the idea that these migrants may use both a map and a compass, with the map utilized largely for the end of the journey to allow pinpointing of the correct destination.

It is the case, however, that the apparent "needle in a haystack" nature of this feat need not mean that a compass or a feature providing a directional cue alone may not be sufficient for the entire journey. For example, using a directional cue alone may be how migrant fish like the salmonids return home to breed. The fish appears to learn the particular cocktail of waterborne chemicals of its birthplace and then follows up the gradient of these odors to the appropriate destination (Hasler and Scholz, 1983). This was demonstrated experimentally by exposing juvenile salmon to artificial odorants and then decoying these fish to unfamiliar streams scented with the odorants, during their homing migrations (Hasler and Scholz, 1983; Scholz *et al.*, 1976). This is a sufficiently effective mechanism that return to natal streams (or those streams to which very young fish are transplanted) has a success rate of around 98 percent (Hard and Hoard, 1999; Hoard, 1996). It is not clear yet whether or not the fish uses these odors as landmarks or in a map-like way when it nears its destination.

TELEMETRY

The major impediment to investigating the role of landmarks is the difficulty in experimentally manipulating either the visual abilities of the long-distance migrant or the possible landmarks it may pay attention to. Although devices such as radio collars have been used for several decades to keep track of animals, often to investigate ranging behavior and so on, the data have rarely been used to determine in detail how animals may use landmarks to navigate over long distances.

However, there is light at the end of this particular tunnel, in the form of the major technical advances such as the use of Global Positioning Satellite (GPS) telemetry for tracking animals over the length of their journeys. Although the size of the tracking equipment is currently such that only relatively large animals have been feasible subjects thus far, the paths of a number of seabirds, polar migrants and turtles have been recorded (Block *et al.*, 2001; Luschi *et al.*, 1996; Weimerskirch *et al.*, 2002). One relatively common finding to date is the efficiency of the routes taken, sometimes in the form of straight lines (e.g., waved albatrosses flying from breeding grounds in the Galapagos more than 1,000 km to specific upwelling zones off Peru: Mouritsen *et al.*, 2003) or in terms of dealing with geographical barriers (Alerstam, 2001). In the case of the waved albatrosses, the straight-line flights were achieved even when the birds were made to wear strong magnets. The tracking data from the turtles (green and loggerhead), both after capture and release at unfamiliar locations and after being made to carry magnets, has those interested in navigation deep in discussion as to their interpretation. There are simply too few data and the variation among animals is too large to be sure as yet what the turtles are doing. It is, thus, not yet possible to determine what, if any role, landmarks play in the movement patterns of these animals. Nonetheless, this is clearly just the beginning and the turtles do, at least under some circumstances, appear to be able to reach their destination without the aid of magnetic information (Avens and Lohmann, 2003; Luschi *et al.*, 2003).

The detailed examination of tracking data may, eventually, provide us with the kind of insight into landmark use over long distances that we have achieved in understanding substantially shorter journeys. The potential of this was shown by Guilford *et al.* (2004) and Roberts *et al.* (2004) who developed a statistical analysis for examining global positioning satellite data from pigeons homing over a few kilometers. Their analysis of the second-by-second homing tracks provides an assessment of the spatial uncertainty in direction shown by pigeons en route home. When the tracking data were laid over a topographical map of the area, Guilford *et al.* and Roberts *et al.* found that the pigeons' uncertainty was highest at release (even though these were familiar sites). This uncertainty, the authors suggest, may be due to the birds having a relatively poor view of the surrounding landscape at that point. High uncertainty also appeared to occur when the pigeons were flying over complex landscape features such as road junctions or settlements. On the other hand, lowest uncertainty occurred close to the loft (as would be expected), possibly when the birds gained sight of the home loft. Tracking data of this kind also show pigeons following man-made structures such as roads or other visual features such as streams when these are relatively concordant with the pigeon's flight direction (Biro *et al.*, 2002a). As with Monarch butterflies' flight patterns, the pigeons appear to use these landmarks as directional cues.

Currently, GPS use is limited by the size of the "backpack" of equipment that the animal must carry around. If the animal is flying then this issue of weight is even more of a problem. However, tracking of smaller animals is also now possible although, in this case, constrained by the environment rather than the weight of the equipment. Harmonic radar has enabled the tracking of honybees *Apis mellifera* flying from the hive on orientation flights, flights they carry out prior to undertaking foraging flights. Radar tracks show that the structure of the flights change with increasing experience (e.g., Capaldi *et al.*, 2000). For example, although the flights last a similar duration, the bees fly further with each successive flight. Flying faster implies they are flying higher with each flight as bees change their flight speed to maintain a constant speed of image movement across their retina. The increase in height as they move away from the hive would suggest bees see and encode landmarks very close to the hive in some detail but that landmarks further from the hive are perceived and remembered with much less resolution. This gradient of resolution in memory for landmarks might make returning to the hive much easier. One of the downsides to the use of harmonic radar is not only the equipment and manpower currently required, it also only works effectively in open landscapes, such as unwooded fields and deserts.

Irrespective of scale, tracking animals in water brings it own problems but it is now possible to use a system similar to that used on terrestrial animals. Small-scale navigation within a home range has successfully been studied in fish using little transponders inserted into the body cavity (PIT tags). The unique code of each transponder is detected by a series of antennae distributed on the streambed as the fish swims overhead. This system has been used to monitor upstream and downstream, and homing movements of juvenile Atlantic salmon (*Salmo salar*) within a 30 m section of a river (Armstrong *et al.*, 1996, 1997). In order to prevent interference between the antennae they must be separated by at least a meter so this tracking system is currently constrained in the level of accuracy to which the PIT tag will operate. As with the GPS tracking systems used with birds, however, the equipment is becoming smaller, allowing ever greater tracking precision (Armstrong *et al.*, 2001) affording increasingly rich data on local movements by fish, although the downside is that such a system is not readily amenable for long-distance tracking.

NAVIGATION, PILOTAGE AND MAPS

Analysis of the fine detail of homing and migration, as provided by these recent technical developments, seems to offer the possibility that, finally, we may be able to determine the role that landmarks play in journeys, both short and long. The examination of the role of landmarks in migration has been complicated not simply by the difficulty of experimentally manipulating them in the field or testing their use in plausible laboratory settings. There are also several ways in which they may be used. First, an animal may use them in order to determine its current location. This would be done by the animal recognizing some or all of the surrounding environmental features. Such recognition requires the animal to be in a familiar location. A striking example of landmarks being used in this way is the stereotypical flights that bees and wasps fly on departure from and when returning to flowers or the hive. The insect zigzags back and forward in front of the goal before departure and appears to learn both the nature of the landmarks very close to the goal and the distance they are from the goal (retinotopic memories, snapshots: e.g., Cartwright and Collett, 1983; Collett and Baron, 1994; Lehrer, 1993). The precision required for return to the goal is then achieved by apparently matching up the landmark angles and distances (the snapshot). In order that such a system could be useful at considerable distances from home, the suggestion was that the animal would have an album of such snapshots and match each in succession along its journey. Each snapshot of landmarks would be associated with a vector that enabled the animal to move in the appropriate direction to encounter the next snapshot and so on until it reached its goal. Judd and Collett (1998) have shown that the behavior of wood ants to visual landmarks en route from nest to foraging areas is consistent with this model of successive snapshots. Routes to a goal are acquired as the animal moves away from that goal after its first visit (see Chapter 6, this volume). Like bees and wasps, wood ants frequently turn back to look at the goal as well as making short movements toward it.

The possibility that vertebrates may employ something akin to snapshots in making navigation decisions is supported by data from an experiment in which chickens were trained to choose between two objects, behind one of which was a food reward. Dawkins and Woodington (2000) found that the chickens each adopted a particular route around the object and would fixate the object with the same eye from the same position during approach to the object. Data from an experiment on homing pigeons might also be interpreted as support for the use of snapshots by vertebrates. Homing pigeons, given the opportunity to view the surrounding landscape immediately prior to release, home faster if the view is a familiar one (Braithwaite, 1993; Braithwaite and Guilford, 1991; Braithwaite and Newman, 1994; Burt *et al.*, 1997). Biro *et al.* (2002b) dissected this result further by providing pigeons prior to release with a view that was only a 140-degree portion of the surrounding landscape. On subsequent releases pigeons that saw the same 140-degree portion homed faster than birds provided with a new 140-degree view. It seems as if the birds, under these conditions at least, were unable to extrapolate from one portion to another although the effect disappeared with repeated training, perhaps due to the birds building up a sufficiently substantial visual memory of the release site such that there were no longer any novel views.

A second possible use of landmarks is that associated with Kramer's (1953, 1961) concept of "a map and compass." The map component in this case allows for the fact that an animal travelling over familiar terrain may be able to extend this knowledge of the familiar in order to get home from unfamiliar locations, perhaps by recognizing prominent familiar landmarks currently being viewed from a novel angle or distance, or by detecting familiar atmospheric odors. The ability of homing pigeons to return home when displaced by experimenters to unfamiliar release sites as far away as several hundred kilometers (considered to be "true" navigation), has led to an ongoing debate as to how they achieve this feat, in particular, whether or not they detect familiar odors from the unfamiliar site (reviews in Able, 1996; Papi, 1995; Wallraff, 2001).

Once the animal knows where it is, it then needs to determine the correct direction to home. While this could be done using a solar, stellar or magnetic compass, it might also be achieved by third possible use of landmarks, in which compass bearings are taken from the landmarks. This has been referred to as a mosaic map (Wallraff, 1974).

Landmarks themselves might also provide directional information such that an animal moves from one to another in a chain-like fashion. This is referred to as pilotage (Griffin, 1952) or beaconing. Some species of reef fish may use this technique in order to migrate daily from foraging grounds to spawning sites at the tip of their coral reef. If a head of coral from the reef is removed, butterfly fish (family Chaetodontidae) will stop at the site and apparently begin to search for the missing coral head, before swimming on (Reese, 1989). However, if the fish were using just a chain-like method of tracing their route, they would not reach the location of the missing coral head but would, rather, be lost on departure from the previous coral head. Simply moving along a route akin to that of trapliners (hymenoptera and hummingbirds which are thought to repeatedly visit flowers in the same sequence) (see Thomson and Chittka, 2001), is not sufficient evidence for the use pilotage via landmarks. Slightly more compelling is the observation that brown surgeon fish (*Acanthurus nigrofuscus*) are deflected from their swim path when landmarks are displaced (Mazeroll and Montgomery, 1998). Curiously, then, there appear to be few examples that demonstrate the use of pilotage in vertebrates (see Holland, 2003, for a discussion of this issue). On the other hand, there is considerable evidence that flying and walking insects move toward prominent landmarks that lie along their route, even though this may mean that they are not then following the most direct trajectory (see Collett and Zeil, 1998, for review).

There is yet one more landmark-based guidance system: the cognitive map, first proposed by Tolman (1948) to explain the success of rats looking for food in a maze. With no apparent pattern to their movements, the rats rarely revisited sites they had emptied of food. Like the proposal that insects retain retinotopic views of their familiar landscape, the cognitive map hypothesis is a proposal with a neurophysiological substrate underpinning the behavioral mechanism, which is that animals have a neural representation (thought to be in, or involve, the hippocampus) of the spatial relationships among familiar landmarks (O'Keefe and Nadel, 1978). Such a map is thought to bear at least a passing resemblance to a topographical map, in that information on landmark features, distances and angles are all stored, independently of the animal's own location or movement. The apparent benefits of such a neurally expensive investment are that the possessor of such a mapping system would be able to take shortcuts, to execute new routes around obstacles and to plan routes in advance of carrying them out (e.g., Jacobs, 2003; Jacobs and Schenk, 2003).

In the mid-1980s Gould (1986) appeared to have shown that honeybees had such a system when his bees flew to a feeder, to which they had previously only flown from the hive, when released from an unfamiliar release point. But in a subsequent experiment in which Dyer (1991) ruled out the possibility that the bees could detect familiar distant landmarks (the release site was in a quarry), the bees did not fly to the familiar, rewarding feeder but rather either flew home or on a flight path of the appropriate direction for flying to the feeder from the hive (i.e., the trained flight path). Menzel *et al.* (1990) also disputed the use by honeybees of cognitive maps when they showed that the bees preferred to use celestial cues to landmarks.

As with all of the proposed mechanisms of landmark choice, demonstration of the use or not of a cognitive map has proved difficult. Even the honeybee, regardless of the verbal point that the honeybee brain seems to have rather too few neurons to be able to implement a cognitive map, may, in fact be capable of taking shortcuts. Bees displaced from their hive did take novel shortcuts when returning to the hive under some but not other experimental conditions (Menzel *et al.*, 1998). While the authors were cautious with regard to inferences about cognitive mapping in bees, it does seem plausible that the use of such abilities might be sensitive to experimental conditions and therefore it may have been premature to abandon the idea of cognitive mapping in bees.

If taking a novel route is considered a sufficient demonstration of the possession of a cognitive map, then there appear to be several examples of such behavior. Goldfish (*Carassius auratus*) swimming a maze can take a novel route from a previously unvisited start point to reach a goal area (Rodríguez *et al.*, 1994). Likewise, *Portia* spiders make successful detours around obstacles to reach food, and dogs can also make successful shortcuts in the appropriate experimental paradigm (Chapuis and Varlet, 1987; Jackson and Wilcox, 1993). How they are able to do this is still unclear and none of these data are sufficiently compelling to have ended the debate as to the existence of a cognitive map (see, for example, in favour: Pearce *et al.*, 1998; Reid and Staddon, 1998; Yeap and Jeffries, 1999; opposed: Benhamou, 1996; Brown and Drew, 1998).

FEATURELESS LANDSCAPES

Accurate navigation to a goal would seem to require the use of landmarks at some point in the journey, even if only to recognize its successful end. However, such an assumption needs to be considered more carefully in situations when the animals are traveling through apparently featureless landscapes. Path integration, in which the animal keeps continuous track of the distance and direction in which it is traveling on its outward journey allowing it to compute a direct vector for its return journey, has been observed in all animals thus tested (Etienne *et al.*, 1998). In some instances there is very little evidence for the use of landmarks. For example, blind mole rats *Spalax ehrenbergi* assess direction through internal signals and using the earth's magnetic field. This technique is highly accurate and most effectively enables them to avoid unnecessary and expensive digging of tunnels (Kimchi and Terkel, 2003; Kimchi *et al.*, 2004).

Desert ants provide another example of animals living in an apparently featureless landscape. However, although path integration is of major importance to them (Andel and Wehner, 2004), they can and do use landmarks (Collett *et al.*, 1999). In the ants' case, their landscape is only seemingly featureless. It is perhaps worth noting at this point, that the kinds of cues that experimenters find salient or even noticeable may well not be so to the animal (see below for further comments).

Green turtles that make their way across the Atlantic to Ascension Island, loggerheads moving through the Indian Ocean and leatherback turtles moving across the Pacific also move through an environment that may appear to us to be relatively landmark-free. And yet all these species show rather directed movement patterns – either to a particular location or in quite straight lines. Although the directedness of these movements is a remarkable feature of these movements and appears to be based at least to some degree on geomagnetic cues (Lohmann *et al.*, 2008; Luschi *et al.*, 2003) that may be used by the animal in a map-like way, it is much more likely to be the endpoints of the journey that are specified by familiar landmarks, somewhat akin to the use of different kinds of information when we drive along motorways. Our requirement for recognition of particular locations, as opposed to simply following directions, is much greater the closer we are to our origin and destination. How close the animal needs to be remains to be seen although here again tracking of animals is beginning to provide us with some insight. It may also be that one of the more impressive feats, that of the green turtles swimming to Ascension Island, may actually be achieved entirely by directional information such that, once they get sufficiently close, cues from the island itself will guide them in.

We know that animals will locate objects that are significantly smaller even than Ascension Island whilst swimming through the oceans. For example, it has long been known that pelagic fish are attracted to floating objects, both natural and man-made. Fishing industries use this to their advantage in two ways, either by using flotation devices fed out from a boat or by mooring buoys some distance off shore and, indeed, more than half of the catch of the tropical tuna is taken in this way. More recently, Girard *et al.* (2004) tagged yellowfin tuna to determine whether these moored buoys, known as Fish Aggregating Devices (FADs) to fishermen, stimulated fish to aggregate around them or, alternatively, acted as attracting devices, perhaps for navigational use by fish on migration routes. Their tracking data show that fish orient toward a FAD from as far away as 10 km. These fish appear to locate FADs in a way that recalls the ability of those familiar with ocean travel to locate land from distances that to the naïve seem impressive, using cues that are apparently inconspicuous. It is not clear, at this point, what cue(s) the fish use to orient toward a FAD although it is possible that they may use cues emanating from the FADs themselves such as sound or from the other animals that congregate around the FAD. Juvenile fish use sounds emanating from a coral reef, such as the snapping and popping sounds generated by fish and invertebrates, to guide them toward the reef as they settle out from the plankton (Simpson *et al.*, 2004; Tolimieri *et al.*, 2000). The possibility that the fish in Girard *et al.*'s study might use the FADs as a series of sequential landmarks along a route is not supported by the data, as the tuna did not orient in the direction of the next FAD on leaving. Nonetheless, the use of the FADs by these fish does serve as a reminder that animals may pay attention to and use cues when moving around their environment that we do not find salient or conspicuous, over distances that we do not initially appreciate.

CUE PREFERENCE

There has been much effort invested by two, often separate, conglomerations of researchers into the cue(s) by which animals choose to navigate: those interested in how birds migrate and home, specifically in pigeons (see next section), and those addressing questions of goal location in laboratory animals, usually rats and pigeons. In both of these bodies of research the choice of cues has often been thought of as being one of preference rather than of salience. Once preferences have been found, the issue of salience has been raised in rather post hoc fashion. In some cases, the animal's behavior in its natural environment has been used to predict which cues the animal should be particularly attentive to.

Most of the work investigating cues used as landmarks has necessarily, however, been done in the laboratory and by far the largest number of experiments have been carried out on rats running radial mazes or swimming in Morris water mazes. Although most rat strains have rather poor visual capabilities, the conclusion from all of this work is that rats, by and large, pay most attention to distal visual cues, where "distal" usually refers to landmarks (cues) that are provided on the walls of the room in which testing is occurring or on the curtains surrounding the maze (referred to as a "cue-controlled" condition; e.g., O'Keefe and Conway, 1980). They appear to pay rather little attention to cues that are provided close to the goal (although it is also the case that if the submerged platform in a Morris water maze is visible, they swim directly to it on the first trial). Rotating or removing the cues on or around the walls of the room will typically result in a much greater performance deficit than altering the visual features immediately adjacent to the platform's location (e.g., Krimm, 1994; Save *et al.*, 1998).

A similar preference for more distal or global cues has been observed in cue preference tests in food-storing birds. Black-capped chickadees *Parus atricapillus* and dark-eyed juncos *Junco hyemalis* were provided with four feeders, each bearing a different color pattern, one of which contained food. They were allowed to eat part but not all of the food before having to leave the experimental room. Immediately prior to their return, the rewarded feeder was switched with another feeder such that the feeder in the previously rewarded position now bore a different color pattern (see Figure 7.1). When the birds returned to the experimental room, black-capped chickadees typically chose the feeder in the correct location, prior to visiting the feeder with the correct color pattern. Juncos, on the other hand, were just as likely to visit the feeder with the correct color pattern as they were to visit the feeder in the correct location (Brodbeck, 1994). When the array of feeders were moved along the wall such that only the position occupied by the rewarded feeder was shared by the shifted array, the black-capped chickadees preferred to visit this feeder – i.e. the feeder in the correct location relative to the room cues, rather than to the array cues.

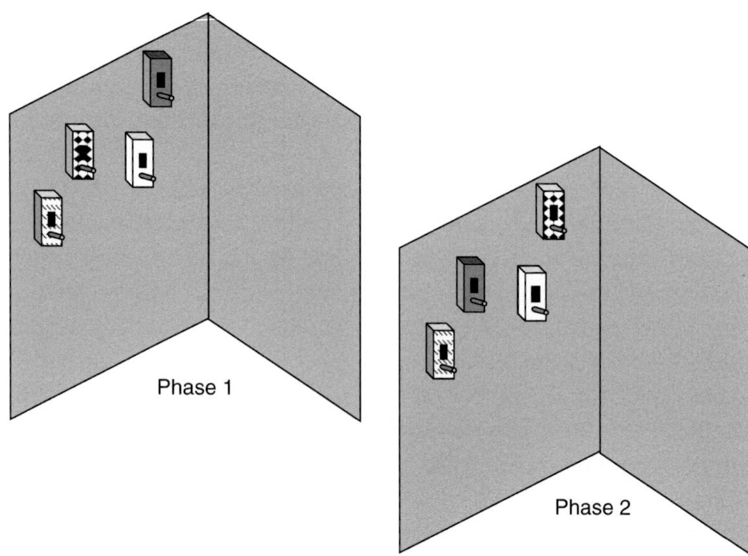

Figure 7.1. Schematic of Brodbeck's (1994) experiment. In phase 1, the black-capped chickadee entered the room and had to search in the feeders for the one that contained a piece of peanut. Each feeder bore a different colour pattern. Before the bird entered in phase 2, the feeder that had contained food was switched with one of the others. Adapted from Brodbeck (1994) with permission from D. R. Brodbeck.

A similar cue preference was shown by black-capped chickadees and juncos when they were tested in an analagous test presented to them on a touch screen (Brodbeck and Shettleworth, 1995). Food-storers apparent preference for spatial cues (in this case, spatial should be taken to mean a configuration of visual cues at some distance from the goal) and a non-storer's indifference between spatial and proximal visual cues was also found in comparisons between marsh tits *Parus palustris* (storers) and blue tits *P. caeruleus* (non-storers) and between European jays *Garrulus glandarius* (storers) and jackdaws *Corvus monedula* (non-storers) (Clayton and Krebs, 1994).

Although this difference in cue use might be due to differences in visual capacities (Macphail and Bolhuis, 2001), an alternative explanation is hinted at by preference for spatial over visual cues by species like rufous hummingbirds and rats. Rufous hummingbirds *Selasphorus rufus*, tested in an analogous experiment to that of Brodbeck's (1994) in the field, strongly preferred to return to the flower in the correct position in the array, rather than to the flower bearing the correct color pattern (Healy and Hurly, 1995; Hurly and Healy, 2002). Cheng

(1986) trained rats to find food in one corner of a rectangular arena, the four corners of which all bore different visual features, and two of the corners had different olfactory cues. Transformations of the arena consistently resulted in the rats choosing the incorrect corner (invariably the one diagonally opposite from the correct corner, for example, top left or bottom right) while ignoring the information provided by the visual and olfactory cues. Neither the hummingbirds nor the rats are food storers (although on some occasions rats may be induced to store) so the preference shown by tits/chickadees and corvids is not one confined to animals that store food. On the other hand, like juncos and jackdaws, both chicks and pigeons either prefer visual cues or have a divided preference between visual and spatial cues (Kelly *et al.*, 1998; Vallortigara *et al.*, 1990), even though it can be shown that both species encode both the visual and the spatial information.

The preference for a particular cue may be dependent on the ecology of the species (e.g., Shettleworth, 2003). Sticklebacks *Gasterosteus aculaetus* from rivers, for example, make choices in mazes based on water flow direction or memory for turns they have made, while sticklebacks from ponds that are stable over the long term, use visual landmarks to guide their decisions (Braithwaite and Girvan, 2003; Girvan and Braithwaite, 1998). However, with the homing pigeon literature as a conspicuous example, it seems likely that there are other influences, such as development, that play a major role in cue preference. The more often an animal visits a stable rewarded location, the more likely it is to use spatial rather than visual (beacon-like) cues to return to it. It may also be that the cue preference a species exhibits is context-dependent for although there is a preference in a number of species, for example, for spatial (geometric) cues over visual cues, like rats in a Morris water maze when the platform is visible, if foodstoring birds and hummingbirds are pointed to the location of food by a conspicuous visual cue, they will preferentially visit that site (Hurly and Healy, 1996; Pravosudov and Clayton, 2001; Sherry and Vaccarino, 1989).

Foodstoring species have proved useful for determining how animals might encode one or more landmarks relative to the goal. In 1989, Cheng proposed that an animal encodes both the distance and direction of the goal (or vector) from each landmark (the vector-sum model). Shifting a local landmark diagonally from the goal, for example, should result in an equal change in distance and direction towards the new landmark position. However, Cheng and Sherry (1992) found that both pigeons and black-capped chickadees moved their search

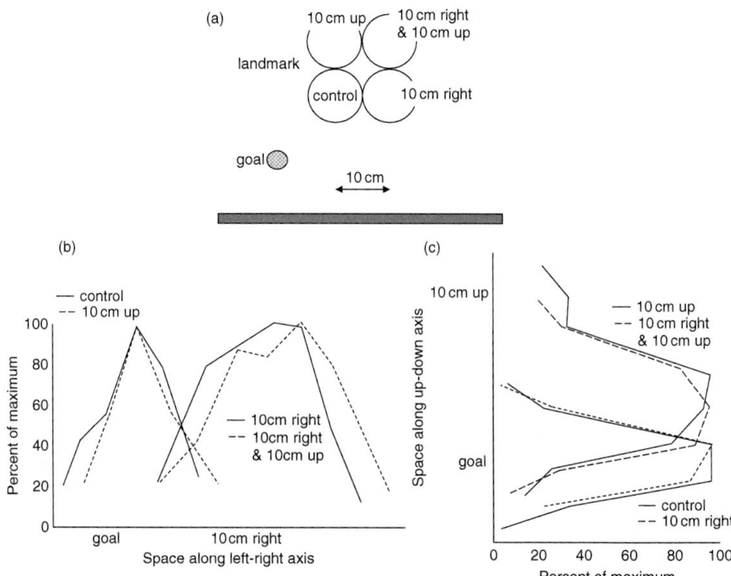

Figure 7.2. (a) Schematic of the relative positions of a large circular landmark to a goal at which birds were to search for reward; control position, landmark moved 10 cm to the right of the control, landmark moved 10 cm up from the control and the landmark moved both 10 cm to the right and 10 cm up. (b) Search distributions of black-capped chickadees along the left–right axis (b) and along the up–down axis (c). The birds search patterns were much more affected by landmark shifts along the left–right axis than by landmark shifts along the perpendicular axis. Adapted from Cheng and Sherry (1992) with permission from K. Cheng.

patterns more in a parallel direction than in the predicted perpendicular direction (see Figure 7.2). Although this result might be interpreted as a violation of the predictions of the vector-sum model, it may also be that the birds used landmarks other than that provided close to the goal. Indeed, the black-capped chickadees were far less affected by the movement of the local landmark than were the pigeons, consistent with the data described above. Gould-Beierle and Kamil (1996), for example, found that although Clark's nutcrackers shifted their search in a parallel direction rather than perpendicularly following a diagonal shift of the landmark they did so to far less a degree than had the black-capped chickadees. A subsequent experiment demonstrated that the location of the birds' searching was much more dependent on global cues (which had changed in the earlier experiment) than on the local cue. Proximity alone is not sufficient to determine which landmark the birds will use: although both Basil (1993) and Bennett (1993) found for Clark's nutcrackers and European jays, respectively, that landmarks that were within 40 cm of the goal were weighted more heavily than those 50–70 cm away, the birds also weighted taller landmarks more heavily than shorter landmarks.

One of the problems with attempting to make sense of all of these results is that the cues used in different experiments are being used by the animals with regard to their salience and not necessarily their proximity to the goal (which may itself, however, be an aspect contributing to a cue's salience). Comparing species within the same experiment is one way to attempt to deal with this although it is only when the same species are compared in more than one experimental design can one begin to claim species differences in preference (e.g., the series of experiments by Brodbeck and Shettleworth).

An alternative is to follow the example of the experiments carried out by Gould-Beierle and Kamil (1999) in which they looked at cue preference in the same species and same experimental set-up but trained different groups of Clark's nutcrackers to search for a hidden reward, using the same landmarks at different distances from the goal. The birds with a cylinder and piece of wood (described as an "edge") set about 10 cm from the goal relied heavily on these cues and moved their search pattern as these two moved within the room while the movement or removal of cues (colored posters) on the walls did not affect their search. Birds trained with the same cylinder and edge set 20–30 cm from the goal moved their search pattern when these cues were moved. When they were removed altogether, the birds moved with the movement of the posters on the walls. Finally, the group with the cylinder and edge set 70–90 cm from the goal were relatively less affected by the movement of these cues and by the movement of the posters, suggesting that they were using the larger room features to localize the goal. This group also took longer to learn the goal location than either of the other groups, which is consistent with a decrease in the accuracy with which the cylinder and edge could be used to specify the goal location. These data are useful in demonstrating that there are differences within the same species in the same experimental setup that depend on the proximity of the objects provided for use as cues to the goal. The experiment also allowed a comparison between the use of two cues close to the goal – the vertical cylinder and the horizontal edge. In all groups the birds paid more attention to the cylinder than to the edge.

It makes intuitive sense that large objects should be more salient with regard to a goal than small objects. However, salience is not always

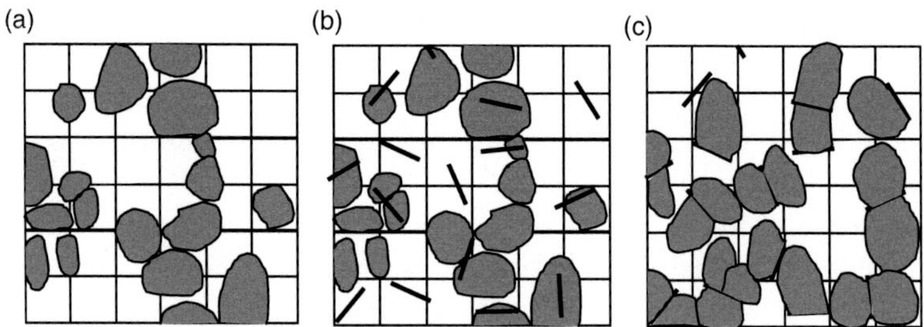

Figure 7.3. (a) Unmanipulated territories (grey patches) on a flat, grassy lawn, of cicada killer wasps *Sphecius speciosus*. (b) Dowels (short dark lines) provided so that they did no coincide with any territory boundaries. (c) The same area the morning after the provision of the dowels. Wasp territories have shifted such that their boundaries coincide with the dowel placements. Reprinted from Eason *et al.* (1999), with permission from Elsevier.

necessarily a matter of object size. The shift in search pattern by the black-capped chickadees in response to a local cue shift appeared to be due to the birds paying attention to the edge of the tray in which the testing was being carried out. Edges seem to play a particularly striking role in the use of landmarks in the development of territorial boundaries. In an experimental investigation of this, Eason *et al.* (1999) put out pieces of dowel (90 cm long, 1 cm diameter) on a lawn on which male cicada killer wasps *Sphecius speciosus*were defending territories. Within twenty-four hours the wasps had moved their territories such that their boundaries coincided with the placement of the dowels (see Figure 7.3). The wasps appear to do this to reduce the time spent disputing a boundary with a neighbor. It appears that blockhead cichlids, *Steatocranus casuarius*, may use landmarks in a similar fashion (Lamanna and Eason, 2003).

SEX DIFFERENCES

There is a well-known source of variation in landmark use within species: a difference between the sexes. In a number of mammals, at least, males seem to use geometric information while females use landmarks (geometric information may be either the use of non-landmark based compasses [e.g., a sun or magnetic compass] or of visual features in such a way as to provide directional information [e.g., the shape of a room]). Typically women when giving directions provide descriptions of the landmarks along the route (e.g., turn left at the corner with the blue house, go along until you reach the church and so on), whereas men describe the route in terms of directions and distances (e.g., go east for 400 feet, then turn north onto High Street for half a mile etc.; Dabbs *et al.*, 1998). A similar difference is found when men and women learn new routes (Galea and Kimura, 1993), i.e., women use the landmarks to learn a route while the men learn and remember the directions and distances (Galea and Kimura, 1993; MacFadden *et al.*, 2003; Montello *et al.*, 1999; Sandstrom *et al.*, 1998; Saucier *et al.*, 2002). Men also learn new routes faster with fewer errors than do women, although it is not clear how the difference in cue use might enable this. Perhaps maps are more readily and accurately constructed if based on directions and distances than with a series of descriptions of landmarks.

The possibility that there may be some more general basis to this sex difference comes from the common observation of male superiority in spatial learning and memory in a number of rodent species (e.g., meadow voles, deer mice, rats). Female rats also appear to rely far more heavily on the landmarks provided in a test arena/room than do males: if the landmarks are removed the number of errors females make increase while males appear unaffected. Changing the shape of the room whilst leaving the landmarks intact, however, has a detrimental effect on male performance but not on females' (e.g., Kanit *et al.*, 1998, 2000; Roof and Stein, 1999; Tropp and Markus, 2001; Williams *et al.*, 1990). This difference in cue use appears to have a hormonal basis: manipulating testosterone levels in males via castration led to those males relying on landmarks while increasing estrogen levels in females resulted in their using geometric cues.

Explanations for the sex difference in performance on spatial tasks, more latterly being correlated with the difference in cue use, have ranged from the evolutionary to the mechanistic: sexual selection on polygynous males ranging over larger areas than their conspecific females and heterospecific monogamous males (e.g., Gaulin and Fitzgerald, 1989), natural selection on males moving over long distances to hunt or to engage in warfare (Geary, 1995; Silverman and Eals, 1992), selection on females to reduce movement during pregnancy and lactation (Sherry and Hampson, 1997), and stress differentially affecting female cognition in novel circumstances (Shors, 1998). The data do not, as yet, allow for definitive exclusion of any of these (see Jones *et al.*, 2003).

HIERARCHICAL USE OF LANDMARKS

Determining which cues animals are paying most attention to can be difficult in both featureless environments and in seemingly crowded ones although for different reasons. In the former, we may have problems detecting the cue(s) at all and, in the latter, we are likely either to assume that animals are using the cues we find most obvious or to provide them with cues most obvious to us. In many cases, animals, as shown by Gould-Beierle and Kamil (1999), can use more than one set of cues. Preference is usually related to spatial proximity of the cues to the goal and perhaps to the accuracy with which the close cues specify the goal's location. A similar effect of distance among landmarks and goal seems to be an explanation for a switch in cue use by rufous hummingbirds foraging on flowers in an array, several of which were rewarded. When the flowers were closer than 40 cm to each other, the hummingbirds visited those in the correct position in the array, when they were further apart, the birds preferentially visited the flowers in the correct global location (Healy and Hurly, 1998).

Hierarchies of cue use have long been part of the homing pigeon literature in spite of the continuing arguments between the leading players as to which cue has the major role. Much of the argument centers around the directional information the birds use and whether this is magnetic or olfactory. The use of visual landmarks or of olfactory landmarks has taken something of a back seat not least because of the difficulty of experimentally manipulating the landmarks a homing pigeon might be paying attention to (hence the value of both the pre-release and tracking data described earlier). A far less contentious example of a cue hierarchy comes from an experimental manipulation by Fauria *et al.* (2004) of landmarks used by solitary bees to relocate their nest. Bees built nests in a set up surrounded by differing landmarks. When the cues immediately surrounding the nest were either reduced in number or removed entirely, the bees took longer to make the correct nest entry decision but none were lost. However, when the more distant cues were manipulated or removed, increasing numbers of bees were lost entirely (see Figure 7.4). These bees had, thus, learned a number of different cues for relocating their nests.

This preference for using cues that do not immediately identify the goal by their proximity has been shown in a number of experiments in which animals have been trained to a particular goal that appears to be marked by very conspicuous landmarks (a subset of which are beacons). The test then is usually to remove the conspicuous landmarks and the somewhat counterintuitive finding has been that the animals reach the goal location nonetheless. Anecdotally, this is seen by pet owners when their pet waits in the correct location for its meal in the absence of its food bowl and North Americans who

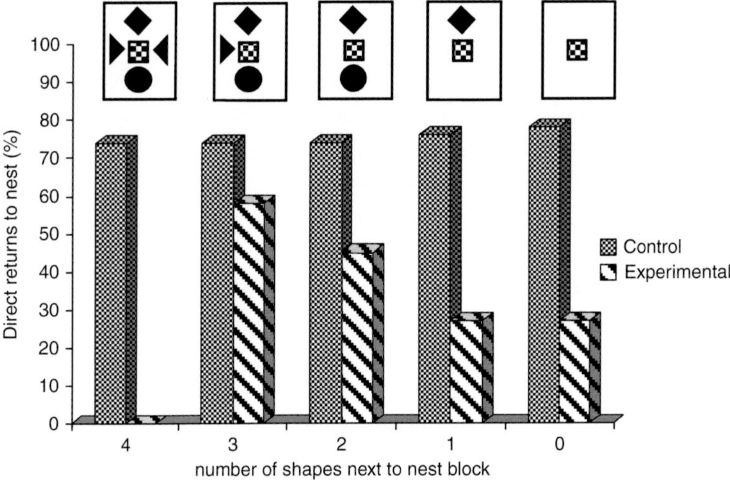

Figure 7.4. The proportion of direct returns to a nest within a nest block (patchwork square within shapes in upper rectangles) by either control bees (all shapes present) or bees with increasing number of shapes surrounding nest block removed. There is a significant difference in the proportion of direct returns between control and experimental when two or more shapes are removed. Reprinted from Fauria *et al.* (2004), with permission from Elsevier.

seasonally put out hummingbird feeders know it is time to do so when a bird appears where the feeder was the previous spring.

It appears as if animals may use the conspicuous, nearby cues when choosing or during initial learning of a reward's location but that with repeated revisiting, it switches to using alternative information. For example, electric fish *Gnatheonemus petersii* trained to swim through an aperture in a divider continued to attempt to swim through the now-closed space when the aperture had been moved (Cain, 1995) and rufous hummingbirds fly to locations previously occupied by rewarding flowers when the array of flowers has been moved some distance away (Healy and Hurly, 1998). Under other circumstances, however, they will return to the correct flowers in an array even when the array has been moved 2 m. Their preference for one landmark or set of landmarks over another can be explained by the phenomenon of overshadowing, i.e., when one landmark takes precedence over others in the control of an animal's behavior. For the hummingbird example, it appears there may be a distance threshold over which nearby landmarks no longer overshadow others. In some experiments, it would appear that distal landmarks may overshadow those nearer to the goal irrespective of the distance (e.g., March *et al.*, 1992; Spetch, 1995). This overshadowing may be due to the reliability of the more distal landmarks. This is particularly likely in instances such as the use of more distal landmarks for food retrieval in foodstoring birds. For some of these animals, such as the black-capped chickadee, it appears that, irrespective of the experimental testing conditions used to date, they will prefer the distal (spatial) landmarks to those immediately at the goal. For others, even for other foodstorers, it appears that they will use whatever landmark more accurately specifies the goal. That this preference is heavily context dependent is shown by the use in the laboratory, by the Clark's nutcracker, of visual landmarks very near to the goal, even though this species in its natural environment stores food for several months, over which time the landmarks at the storage location will change appearance or will disappear altogether.

SUMMARY

There are a number of different ways in which animals may use landmarks to navigate, ranging from the very simple such as the use of a beacon at the goal to head straight to the goal to the representation of a number of landmarks into some kind of a map. The use of landmarks for navigation by animals has been largely explored empirically in a laboratory environment in which it has been possible to investigate the preference for one landmark over another or one cue type over another. It is also in this context that much has been done to investigate variation in landmark use both within and between species, particularly in relation to short-distance goal relocation. The one major exception to this is the work on insect navigation, much of which has been done in the field. For the longer distance navigational problems, such as migration, we have, until recently, had explanations only for the compass directions the animals move in. However, advances in tracking techniques are beginning to provide us with data on the ways in which these animals, too, might use landmarks.

REFERENCES

Able, K. P. (1996). The debate over olfactory navigation by homing pigeons. *Journal of Experimental Biology*, **199**, 121–124.

Åkesson, S., Luschi, P., Papi, F., Broderick, A. C., Glen, F., Godley, B. J. and Hays, G. C. (2001). Oceanic long-distance navigation: do experienced migrants use the Earth's magnetic field? *Journal of Navigation*, **54**, 419–427.

Alerstam, T. (2001). Detours in bird migration. *Journal of Theoretical Biology*, **209**, 319–331.

Andel, D. and Wehner, R. (2004). Path integration in desert ants, Cataglyphis: how to make a homing ant run away from home. *Proceedings of the Royal Society (London) B*, **271**, 1485–1489.

Armstrong, J. D., Braithwaite, V. A. and Huntingford, F. A. (1997). Spatial strategies of wild Atlantic salmon parr: exploration and settlement in unfamiliar areas. *Journal of Animal Ecology*, **66**, 203–211.

Armstrong, J. D., Braithwaite, V. A. and Ryecroft, P. (1996). A flat-bed passive integrated transponder antenna array for monitoring behaviour of Atlantic salmon parr and other fish. *Journal of Fish Biology*, **48**, 539–541.

Armstrong, J. D., Einum, S., Fleming, I. A. and Rycroft, P. (2001). A method for tracking the behaviour of mature and immature salmon parr around nests during spawning. *Journal of Fish Biology*, **59**, 1023–1032.

Avens, L. and Lohmann, K. J. (2003). Use of multiple orientation cues by juvenile loggerhead sea turtles *Caretta caretta*. *Journal of Experimental Biology*, **206**, 4317–4325.

Basil, J. A. (1993). *Neuroanatomical and behavioral correlates of spatial memory in Clark's nutcrackers*. PhD thesis, University of Massachusetts.

Benhamou, S. (1996). No evidence for cognitive mapping in rats. *Animal Behaviour*, **52**, 201–212.

Bennett, A. T. D. (1993). Spatial memory in a food storing corvid. I. Near tall landmarks are primarily used. *Journal of Comparative Physiology A*, **173**, 193–207.

Biro, D., Guilford, T. and Dawkins, M. S. (2002b). Mechanisms of visually mediated site recognition by the homing pigeon. *Animal Behaviour*, **65**, 115–122.

Biro, D., Guilford, T., Dell'Omo, G. and Lipp, H.-P. (2002a). How the viewing of familiar landscapes prior to release allows pigeons to home faster: evidence from GPS tracking. *Journal of Experimental Biology*, **205**, 3833–3844.

Block, B. A., Dewar, H., Blackwell, S. B., Williams, T. D., Prince, E. D., Farwell, C. J., Boustany, A., Teo, S. L. H., Seitz, A., Walli, A. and Fudge, D. (2001). Migratory movements, depth preferences, and thermal biology of Atlantic bluefine tuna. *Science*, **293**, 1310–1314.

Braithwaite, V. A. (1993). When does previewing the landscape affect pigeon homing? *Ethology*, **95**, 141–151.

Braithwaite, V. A. and Girvan, J. R. (2003). Use of waterflow to provide spatial information in a small-scale orientation task. *Journal of Fish Biology*, **63**, 74–83.

Braithwaite, V. A. and Guilford, T. (1991). Viewing familiar landscapes affects pigeon homing. *Proceedings of the Royal Society (London) B*, **245**, 183–186.

Braithwaite, V. A. and Newman, J. A (1994). Visual recognition of familiar release sites allows pigeons to home faster. *Animal Behaviour*, **48**, 1482–1484.

Brodbeck, D. R. (1994). Memory for spatial and local cues: a comparison of a storing and a nonstoring species. *Animal Learning and Behavior*, **22**, 119–133.

Brodbeck, D. R. and Shettleworth, S. J. (1995). Matching location and color of a compound stimulus – comparison of a food-storing and a nonstoring bird species. *Journal of Experimental Psychology: Animal Behavior Processes*, **21**, 64–77.

Brown, M. F. and Drew, M. R. (1998). Exposure to spatial cues facilitates visual discrimination but not spatial guidance. *Learning and Motivation*, **29**, 367–382.

Bruderer, B. and Jenni, L. (1990). Migration across the Alps. In E. Gwinner (Ed.), *Bird migration: the physiology and ecophysiology* (pp. 60–77). Berlin: Springer-Verlag.

Burt, T., Holland, R. and Guilford, T. (1997). Further evidence for visual landmark involvement in the pigeon's familiar area. *Animal Behaviour*, **53**, 1203–1209.

Burt de Perera, T. (2004). A study of spatial parameters encoded in the spatial map of the blind Mexican cave fish (*Astyanax fasciatus*). *Animal Behaviour*, **68**, 291–295.

Cain, P. (1995). Navigation in familiar environments by the weakly electric elephant nose fish, *Gnathonemus petersii* L. (Mormyriformes, Teleostei). *Ethology*, **99**, 332–349.

Capaldi, E. A., Smith, A. D., Osborne, J. L., Fahrbach, S. E., Farris, S. M., Reynolds, D. R., Edwards, A. S., Martin, A., Robinson, G. E., Poppy, G. M. and Riley, J. R. (2000). Ontogeny of orientation flight in the honeybee revealed by harmonic radar. *Nature*, **403**, 537–540.

Cartwright, B. A. and Collett, T. S. (1983). Landmark learning in bees. *Journal of Comparative Physiology A*, **151**, 521–543.

Chapuis, N. and Varlet, C. (1987). Shortcuts by dogs in natural surroundings. *Quarterly Journal of Experimental Psychology*, **39B**, 49–64.

Chapuis, N., Durup, M. and Thinus-Blanc, C. (1987). The role of exploratory experience in a shortcut task by golden hamsters. *Animal Learning and Behavior*, **15**, 174–178.

Cheng, K. (1986). A purely geometric module in the rat's spatial representation. *Cognition*, **23**, 149–178.

Cheng, K. and Sherry, D. F. (1992). Landmark-based spatial memory in birds (*Parus atricapillus* and *Columba livia*): the use of edges and distances to represent spatial positions. *Journal of Comparative Psychology*, **106**, 331–341.

Clayton, N. S. and Krebs, J. R. (1994). One-trial associative memory – comparison of food-storing and nonstoring species of bird. *Animal Learning and Behavior*, **22**, 366–372.

Collett, T. S. and Baron, J. (1994). Biological compasses and the coordinate frame of landmark memories in honeybees. *Nature*, **368**, 137–140.

Collett, T. S. and Zeil, J. (1998). Places and landmarks: an arthropod perspective. In S. D. Healy (Ed.), *Spatial representation in animals* (pp. 18–53). Oxford: Oxford University Press.

Collett, M., Collett, T. S. and Wehner, R. (1999). Calibration of vector navigation in desert ants. *Current Biology*, **9**, 1031–1034.

Dabbs, J. M. J., Chang, E.-L., Strong, R. A. and Milun, R. (1998). Spatial ability, navigation strategy, and geographic knowledge among men and women. *Evolution of Human Behavior*, **19**, 89–98.

Dawkins, M. S. and Woodington, A. (2000). Pattern recognition and active vision in chickens. *Nature*, **403**, 652–655.

Dyer, F. (1991). Bees acquire route-based memories but not cognitive maps in a familiar landscape. *Animal Behaviour*, **41**, 239–246.

Eason, P. K., Cobb, G. A. and Trinca, K. G. (1999). The use of landmarks to define territorial boundaries. *Animal Behaviour*, **58**, 85–91.

Emlen, S. T. (1970). Celestial rotation: its importance in the development of migratory orientation. *Science*, **170**, 1198–1201.

Etienne, A. S., Berlie, J., Georgakopoulos, J. and Maurer, R. (1998). Role of dead reckoning in navigation. In S. D. Healy (Ed.), *Spatial representation in animals* (pp. 54–68). Oxford: Oxford University Press.

Fauria, K., Campan, R. and Grimal, A. (2004). Visual marks learned by the solitary bee *Megachile rotundata* for localizing its nest. *Animal Behaviour*, **67**, 523–530.

Galea, L. A. M. and Kimura, D. (1993). Sex differences in route-learning. *Personality and Individual Differences*, **14** (1), 53–65.

Gaulin, S. J. C. and Fitzgerald, R. W. (1989). Sexual selection for spatial learning ability. *Animal Behaviour*, **37**, 322–331.

Geary, D. C. (1995). Sexual selection and sex differences in spatial cognition. *Learning and Individual Differences*, **7** (4), 289–301.

Girard, C., Benahmou, S. and Dagorn, L. (2004). FAD: fish aggregating device or fish attracting device? A new analysis of yellowfin tuna movements around floating objects. *Animal Behaviour*, **67**, 319–326.

Girvan, J. R. and Braithwaite, V. A. (1998). Population differences in spatial learning in three-spined sticklebacks. *Proceedings of the Royal Society (London) B*, **265**, 913–918.

Godard, R. (1991). Long-term memory of individual neighbours in a migratory songbird. *Nature*, **350**, 228–229.

Gould, J. L. (1986). The locale of honey bees: do insects have cognitive maps? *Science*, **232**, 861–863.

Gould-Beierle, K. L. and Kamil, A. C. (1996). The use of local and global cues by Clark's nutcrackers, *Nucifraga columbiana*. *Animal Behaviour*, **52**, 519–528.

Gould-Beierle, K. L. and Kamil, A. C. (1999). The effect of proximity on landmark use in Clark's nutcrackers. *Animal Behaviour*, **58**, 477–488.

Griffin, D. R. (1952). Bird navigation. *Biological Reviews*, **27**, 359–400.

Guilford, T., Roberts, S., Biro, D. and Rezek, L. (2004). Positional entropy during pigeon homing II: navigational interpretation of Bayesian latent state models. *Journal of Theoretical Biology*, **227**, 25–38.

Hard, J. J. and Hoard, W. R. (1999). Analysis of straying variation in Alaskan hatchery chinook salmon (*Oncorhynchus tshawytscha*) following transplantation. *Canadian Journal of Fisheries and Aquatic Science*, **56**, 578–589.

Hasler, A. D. and Scholz, A. T. (1983). *Olfactory imprinting and homing in salmon*. New York: Springer-Verlag.

Hawryshyn, C. W., Arnold, M. G., Bowering, E. and Cole, R. L. (1990). Spatial orientation of rainbow trout to plane-polarized light: the ontogeny of E-vector discrimination and spectral sensitivity characteristics. *Journal of Comparative Physiology A*, **166**, 565–574.

Healy, S. D. and Hurly, T. A. (1995). Spatial memory in rufous hummingbirds (*Selasphorus rufus*): a field test. *Animal Learning and Behavior*, **23**, 63–68.

Healy, S. D. and Hurly, T. A. (1998). Hummingbirds' memory for flowers: patterns or actual spatial locations? *Journal of Experimental Psychology: Animal Behavior Processes*, **24**, 1–9.

Healy, S. D., Gwinner, E. and Krebs, J. R. (1996). Hippocampal volume in migratory and non-migratory warblers: effects of age and experience. *Behavioural Brain Research*, **81**, 61–68.

Hoard, W. R. (1996). Sequential imprinting in chinook salmon: is it essential for homing fidelity? *Bulletin of the National Research Institute of Aquaculture*, **2** (Suppl.), 59–64.

Holland, R. A. (2003). The role of visual landmarks in the avian familiar area map. *Journal of Experimental Biology*, **206**, 1773–1778.

Höller, P. and Schmidt, U. (1996). The orientation behaviour of the lesser spear-nosed bat, *Phyllostomus discolor* (Chiroptera) in a model roost: concurrence of visual, echoacoustical and endogenous spatial information. *Journal of Comparative Physiology A*, **179**, 245–254.

Hurly, T. A. and Healy, S. D. (1996). Location or local visual cues? Memory for flowers in rufous hummingbirds. *Animal Behaviour*, **51**, 1149–1157.

Hurly, T. A. and Healy, S. D. (2002). Spatial pattern learning by rufous hummingbirds. *Journal of Experimental Psychology: Animal Behavior Processes*, **28**, 209–223.

Jackson, R. R. and Wilcox, R. S. (1993). Observations in nature of detouring behaviour by *Portia fimbriata*, a web-building araneophagic spider (Araneae, Salticidae) from Queensland. *Journal of Zoology, London*, **230**, 135–139.

Jacobs, L. F. (2003). The evolution of the cognitive map. *Brain Behavior and Evolution*, **62**, 128–139.

Jacobs, L. F. and Schenk, F. (2003). Unpacking the cognitive map: the parallel map theory of hippocampal function. *Psychological Reviews*, **110**, 285–315.

Jones, C. M., Braithwaite, V. A. and Healy, S. D. (2003). The evolution of sex differences in spatial ability. *Behavioral Neuroscience*, **117**, 403–411.

Judd, S. P. D. and Collett, T. S. (1998). Multiple stored views and landmark guidance in ants. *Nature*, **392**, 710–714.

Kanit, L., Yilmaz, O., Taskiran, D., Balkan, B., Furedy, J. J. and Pogun, S. (1998). Intersession interval affects performance in the Morris water maze. *International Journal of Neuroscience*, **96**, 197–204.

Kanit, L., Taskiran, D., Yilmaz, O. A., Balkan, B., Demirgoren, S., Furedy, J. J. and Pogun, S. (2000). Sexually dimorphic cognitive style in rats emerges after puberty. *Brain Research Bulletin*, **52** (4), 243–248.

Kelly, D. M., Spetch, M. L. and Heth, C. D. (1998). Pigeons' (*Columba livia*) encoding of geometric and featural properties of a spatial environment. *Journal of Comparative Psychology*, **112**, 259–269.

Kimchi, T. and Terkel, J. (2003). Detours by the blind mole-rat follow assessment of location and physical properties of underground obstacles. *Animal Behaviour*, **66**, 885–891.

Kimchi, T., Etienne, A. S. and Terkel, J. (2004). A subterranean mammal uses the magnetic compass for path integration. *Proceedings of the National Academy of Sciences U.S.A.*, **101**, 1105–1109.

Kramer, G. (1953). Wird die sonnenhohe bei der Heimfindeorientierung verwertet. [Is the elevation of the sun used for home orientation?]. *Journal für Ornithologie*, **94**, 201–219.

Kramer, G. (1961). Long-distance orientation. In A. J. Marshall (Ed.), *Biology and comparative physiology in birds* (pp. 341–371). New York: Academic Press.

Krebs, J. R., Sherry, D. F., Healy, S. D., Perry, V. H. and Vaccarino, A. L. (1989). Hippocampal specialization of food-storing birds. *Proceedings of the National Academy of Sciences (USA)*, **86**, 1388–1392.

Krimm, M. (1994). *Rôle de l'expérience visuelle précoce et des aires visuelles corticales dans le traitement des informations spatiales.* [The role of early visual experience and cortical visual area in the management of spatial information]. Doctoral dissertation. University Aix-Marseille II.

Lamanna, J. R. and Eason, P. K. (2003). Effects of landmarks on territorial establishment. *Animal Behaviour*, **65**, 471–478.

Lehrer, M. (1993). Why do bees turn back and look? *Journal of Comparative Physiology A*, **172**, 544–563.

Lohmann, K. J., Luschi, P. and Hays, G. C. (2008). Goal navigation and island-finding in sea turtles. *Journal of Experimental Marine Biology and Ecology*, **356**, 83–95.

Luschi, P., Hays, G. C. and Papi, F. (2003). A review of long-distance movements by marine turtles, and the possible role of ocean currents. *Oikos*, **103**, 293–302.

Luschi, P., Papi, F. Liew, H. C., Chan, E. H. and Bonadonna, F. (1996). Long-distance migration and homing after displacement in the green turtle (*Chelonia mydas*): a satellite tracking study. *Journal of Comparative Physiology A*, **178**, 447–452.

MacFadden, A., Elias, L. and Saucier, D. (2003). Males and females scan maps similarly, but give directions differently. *Brain and Cognition*, **53**, 297–300.

Macphail, E. M. and Bolhius, J. J. (2001). The evolution of intelligence: adaptive specializations *versus* general process. *Biological Reviews*, **76**, 341–364.

March, J., Chamizo, V. D. and Mackintosh, N. J. (1992). Reciprocal overshadowing between intra-maze and extra-maze cues. *Quarterly Journal of Experimental Psychology*, **45B**, 49–63.

Mazeroll, A. I. and Montgomery, W. L. (1998). Daily migrations of a coral reef fish in the Red Sea (Gulf of Aqaba, Israel): initiation and orientation. *Copeia*, **4**, 893–905.

Menzel, R., Geiger, K., Joerges, J., Muller, U. and Chittka, L. (1998). Bees travel novel homeward routes by integrating separately acquired vector memories. *Animal Behaviour*, **55**, 139–152.

Menzel, R., Chittka, L., Eichmuller, S., Geiger, K., Peitsch, D. and Knoll, P. (1990). Dominance of celestial cues over landmarks disproves map-like orientation in honey-bees. *Zeitschrift für Naturforschaften*, **45C**, 723–726.

Montello, D. R., Lovelace, K. L., Golledge, R. G. and Self, C. M. (1999). Sex-related differences and similarities in geographic and environmental spatial abilities. *Annals of the Association of American Geographers*, **80** (3), 515–534.

Mouritsen, H. (2001). Navigation in birds and other animals. *Image and Visual Computation*, **19**, 713–731.

Mouritsen, H., Huyvaert, K. P., Frost, B. J. and Anderson, D. J. (2003). Waved albatrosses can navigate with strong magnets attached to their head. *Journal of Experimental Biology*, **206**, 4155–4166.

O'Keefe, J. and Conway, D. H. (1980). On the trail of the hippocampal engram. *Physiological Psychology*, **8**, 229–238.

O'Keefe, J. and Nadel, L. (1978). *The hippocampus as a cognitive map*. Oxford: Oxford University Press.

Papi, F. (1995). Recent experiments on pigeon navigation. In E. Alleva, A. Fasolo, H.-P. Lipp, L. Nadel and L. Ricceri (Eds.), *Behavioural brain research in naturalistic and semi-naturalistic settings* (pp. 225–238). Dordrecht: Kluwer.

Pearce, J. M., Roberts, A. D. L. and Good, M. (1998). Hippocampal lesions disrupt navigation based on cognitive maps but not heading vectors. *Nature*, **396**, 75–77.

Perdeck, A. C. (1958). Two types of orientation in migrating *Sturnus vulgaris* and *Fringilla coelebs* as revealed by displacement experiements. *Ardea*, **46**, 1–37.

Pravosudov, V. V. and Clayton, N. S. (2001). Effects of demanding foraging conditions on cache retrieval accuracy in food-caching mountain chickadees (*Poecile cambeli*). *Proceedings of the Royal Society (London) B*, **268**, 363–368.

Reese, E. S. (1989). Orientation behaviour of butterflyfishes (family Chaetodontidae) on coral reefs: spatial learning of route specific landmarks and cognitive maps. *Environmental Biology of Fishes*, **25**, 79–86.

Reid, A. K. and Staddon, J. E. R. (1998). A dynamic route finder for the cognitive map. *Psychological Reviews*, **105**, 585–601.

Roberts, S., Guilford, T., Rezek, I. and Biro, D. (2004). Positional entropy during pigeon homing I: application of Bayesian latent state modelling. *Journal of Theoretical Biology*, **227**, 39–50.

Rodríguez, F., Durán, E., Vargas, J. P., Torres, B. and Salas, C. (1994). Performance of goldfish trained in allocentric and egocentric maze procedures suggest the presence of a cognitive mapping system in fishes. *Animal Learning and Behavior*, **22**, 409–420.

Roof, R. L. and Stein, D. G. (1999). Gender differences in Morris water maze performance depend on task parameters. *Physiological Behavior*, **68**, 81–86.

Sandstrom, N. J., Kaufman, J. and Huettel, S. A. (1998). Males and females use different distal cues in a virtual environment navigation task. *Cognitive Brain Research*, **6**, 351–360.

Saucier, D., MacFadden, A., Elias, L. and Bell, S. (2002). Sex differences in real-world navigation may relate to disorientation and endogenous testosterone levels. *Journal of Cognitive Neuroscience*, **105**.

Save, E., Poucet, B. and Thinus-Blanc, C. (1998). Landmark use and the cognitive map in the rat. In S. D. Healy (Ed.), *Spatial representation in animals* (pp. 119–132). Oxford: Oxford University Press.

Schmidt-Koenig, K. and Walcott, C. (1978). Tracks of pigeons homing with frosted lenses. *Animal Behaviour*, **26**, 480–486.

Scholz, A. T., Horrall, R. M., Cooper, J. C. and Hasler, A. D (1976). Imprinting to chemical cues: the basis for home stream selection in salmon. *Science*, **192**, 1247–1249.

Sherry, D. F. and Hampson, E. (1997). Evolution and the hormonal control of sexually-dimorphic spatial abilities in humans. *Trends in Cognitive Sciences*, **1**, 50–56.

Sherry, D. F. and Vaccarino, A. L. (1989). Hippocampus and memory for food caches in black-capped chickadees. *Behavioral Neuroscience*, **103**, 308–318.

Shettleworth, S. J. (2003). Memory and hippocampal specialization in food-storing birds: challenges for research on comparative cognition. *Brain Behavior and Evolution*, **62**, 108–116.

Shors, T. J. (1998). Stress and sex effects on associative learning: for better or for worse. *The Neuroscientist*, **4**(5), 353–364.

Silverman, I. and Eals, M. (1992). Sex differences in spatial abilities: evolutionary theory and data. In J. Barkow, L. Cosmides and J. Tooby (Eds.), *The adapted mind: evolutionary psychology and the generation of culture* (pp. 533–549). Oxford: Oxford University Press.

Simpson, S. D., Meekan, M. G., McCauley, R. D. and Jeffs, A. (2004). Attraction of settlement-stage coral reef fishes to reef noise. *Marine Ecology Progress Series*, **276**, 263–268.

Spetch, M. L. (1995). Overshadowing in landmark learning – touch-screen studies with pigeons and humans. *Journal of Experimental Psychology: Animal Behavior Processes*, **21**, 166–181.

Thomson, J. D. and Chittka, L. (2001). Pollinator individuality: when does it matter? In L. Chittka and J. D. Thomson (Eds.), *Cognitive ecology of pollination* (pp. 191–213). Cambridge: Cambridge University Press.

Tolimieri, N., Jeffs, A. and Montgomery, J. C. (2000). Ambient sound as a cue for navigation by the pelagic larvae of reef fishes. *Marine Ecology Progress Series*, **207**, 219–224.

Tolman, E. C. (1948). Cognitive maps in rats and men. *Psychological Review*, **55**, 189–208.

Tropp, J. and Markus, E. J. (2001). Effects of mild deprivation on the estrous cycle of rats. *Physiology and Behavior*, **73**, 553–559.

Vallortigara, G., Zanforlin, M. and Pasti, G. (1990). Geometric modules in animals' spatial representations: a test with chicks (*Gallus gallus domesticusi). Journal of Comparative Psychology*, **104**, 248–254.

Walker, M. M., Dennis, T. E. and Kirschvink, J. L. (2002). The magnetic sense and its use in long distance navigation by animals. *Current Opinions of Neurobiology*, **12**, 733–744.

Wallraff, H. G. (1974). *Das Navigation der Vogel*. Munich: Oldenbourg.

Wallraff, H. G. (2001). Navigation by homing pigeons, Part 2. *Ethology, Ecology and Evolution*, 1–48.

Wallraff, H. G., Chappell, J. and Guilford, T. (1999). The roles of the sun and the landscape in pigeon homing. *Journal of Experimental Biology*, **202**, 2121–2126.

Weimerskirch, H., Bonadonna, F., Bailleul, F., Mabile, G., Dell'Omo, G. and Lipp, H.-P. (2002). GPS tracking of foraging albatrosses. *Science*, **295**, 1259–1259.

Wiggett, D. R., Boag, D. A. and Wiggett, A. D. R. (1989). Movements of intercolony natal dispersers in the Columbian ground squirrel. *Canadian Journal of Zoology*, **67**, 1447–1452.

Williams, C. L., Barnett, A. M. and Meck, W. H. (1990). Organizational effects of gonadal secretions on sexual differentiation in spatial memory. *Behavioral Neuroscience*, **104**, 84–97.

Yeap, W. K. and Jefferies, M. E. (1999). Computing a representation of the local environment. *Artificial Intelligence*, **107**, 265–301.

PAUL A. GARBER AND FRANCINE L. DOLINS

8

Examining spatial cognitive strategies in small-scale and large-scale space in tamarin monkeys

INTRODUCTION

In order to effectively track the availability and distribution of distant or widely scattered resources in a tropical rainforest, foragers must monitor their spatial position in relation to salient features in the environment. The question for highly visual animals, such as primates, is how are visuo-spatial features such as river boundaries, forest topography, and other potential landmarks evaluated in spatial decision-making and incorporated into spatial problem-solving strategies?

Controlled laboratory research has shown that in rats, rhesus monkeys and young children, landmarks (distal and local) and the geometry of the space subjects are tested in provide a visual framework for spatial decision-making processes (Benhamou and Poucet, 1998; Cheng, 1986; Gouteux *et al.*, 2001; Hermer and Spelke, 1994). Other studies have found evidence of hierarchies in the saliency of spatial information, quantity information and temporal information used by human and nonhuman primates in selecting feeding sites or travel routes (Byrne and Janson, 2007; Deipolyi *et al.*, 2001; Garber, 1989, 1993a, 2000; Garber and Brown 2006; Garber and Dolins, 1996; Garber and Paciulli, 1997; Hirtle and Jonides, 1985; Menzel, 2005; Menzel and Menzel, 2007; Noser and Byrne, 2007; Poti, 2000; Sheth and Shimojo, 2004). For example, in an experimental field study of wild capuchin monkeys (*Cebus capucinus*) Garber (2000) reports that after a single exposure, individuals using natural spatial cues in their environment returned to the location of five baited platforms and avoided eight platforms that had not previously contained a food reward (all platforms were visually identical, presence or absence of food on a platform was concealed, and olfactory cues were controlled for). However, it required several more trials before these

Spatial Cognition, Spatial Perception: Mapping the Self and Space, ed. Francine L. Dolins and Robert W. Mitchell. Published by Cambridge University Press.

same capuchins preferentially returned to platforms that reliably contained three concealed bananas versus half a concealed banana. Thus, the hierarchy of cues, "place and food" versus "no food" were initially more salient than "more food" versus "less food." In a related experimental field study, Garber and Brown (2006) found that these same capuchins used a combination of two and three landmark arrays (2 m poles) to successfully locate hidden food rewards in small-scale space (local area of the forest of approximately 50 m^2). Recognizing how animals order and re-order information into a hierarchy of sensory cues, social information and landmark arrays contributes to understanding processes of decision-making, rule-based foraging and goal-directed travel in primates (Garber *et al.*, 2009). However, in general little is known concerning how primates internally represent features of the environment and the degree to which visual-spatial information is encoded principally as a topological or metric based map (Garber and Brown, 2006; Menzel *et al.*, 2002, 1999; Normand and Boesch, 2009; Poucet, 1993; Urbani, 2009).

How spatial strategies and internal representations are established, and information encoded and utilized, depends on what spatial information is salient to the traveling animal (discrete landmarks or the shape of the environment) (Cheng, 1986; Gallistel, 1990; Pearce *et al.*, 2001). The primary models used to describe internal spatial representations include strip, topological and metric maps (or "cognitive" maps) (Dyer, 1993; Gallistel, 1990; Muller *et al.*, 1996; O'Keefe and Nadel, 1978; Tolman, 1932). The strip-map representation is, at its simplest, a sequence of interconnecting landmarks that requires an animal to travel solely along the demarcated route between landmarks. Using this type of spatial representation the forager responds selectively to each sequential landmark en route and uses the same sequence of landmarks in reverse order to return. Studies on wood ants (*Formica rufa*) (Collett, 1993; Collett and Collett, 2002; Collett *et al.*, 1992; Nicholson *et al.*, 1999) and desert ants (*Cataglyphis fortis*) (Collett and Collett, 2000; Collett *et al.*, 2001;) are consistent with a strip map. In contrast, a metric spatial representation refers to a system in which landmarks are encoded both veridically and relationally and the animal uses this information to generate flexible spatial behavior and highly accurate, novel travel routes when moving between distant goals (Gallistel, 1990; Muller *et al.*, 1996; Nadel, 1990; O'Keefe and Nadel, 1978, 1979; Tolman, 1932, 1948). Using such a map, the angle and distance between individual points on the landscape are encoded as part of a coordinate-based system.

The topological map, alternatively, can be described as an expanded strip map where nodes (familiar landmarks) connect one set of landmark sequences with another according to a route-based system (Dyer, 1993; Gallistel, 1990). Spatial information encoded topologically is considered not to maintain the true geometric distance, angle and direction between landmarks as would define a "cognitive" (metric) map (Gallistel, 1990; O'Keefe and Nadel, 1978; Tolman, 1932, 1948). Instead, the animals' egocentrically derived information is encoded together with the spatial relationship between landmarks via "distance as effort" (dead reckoning): the resulting internal topological representation maintains somewhat imprecise (exaggerated) distances, angles and directions between known landmarks.

An animal encoding spatial information in a topological map may compute novel travel routes but must use familiar landmarks (nodes) as choice points and beacons (Pearce *et al.*, 2001) to re-adjust or re-compute its course to reach a target destination. One would expect that a free-ranging animal relying on a topological representation would consistently orient to a relatively small set of nodes or switch points in their range and redirect travel toward the next switch point as they move closer to their goal.

In this chapter we examine the ability of two closely related species of New World primates to solve foraging and navigational problems encountered in small-scale and large-scale space. Small-scale space is defined as a local area in which the forager has visual access to all landmarks and spatial information required to directly locate food items. It is analogous to problems primates face when making within-patch foraging decisions. In contrast, large-scale space represents a network of travel routes and feeding sites across a broad landscape. When navigating in large-scale space a forager cannot directly see its target, but rather has to recall and orient to a series of single landmarks or landmark configurations that are associated with the general location of the goal. These landmarks serve to narrow the search area and are analogous to problems primates face in the context of between-patch foraging decisions and navigation to distant resting or refuge sites. Specifically, we compare spatial strategies used by captive cotton-top tamarins (*Saguinus oedipus*) in small-scale space (Dolins, 2009) and free-ranging mustached tamarins (*S. mystax*) and saddleback tamarins (*S. fuscicollis*) in large-scale space. Our goal is to examine whether tamarins systematically apply different types of spatial strategies in these two foraging contexts. Using this theoretical framework to distinguish between spatial strategies and associated behavioral patterns, we address the following questions: (1) what type of spacial strategies do captive tamarins use in small-scale space;

(2) what type of spatial strategies do free-ranging tamarins exhibit when traveling in large-scale space; and 3) in comparing captive and wild tamarins' spatial strategies what does this tell us about the ability of these primates to encode and recall single landmarks, multiple landmarks and landmark arrays in locating resources?

A study of spatial cognition in small-scale space

In two separate experiments, captive cotton-top tamarins (*Saguinus oedipus*) were tested on their ability to localize hidden food items based on the spatial relationships between provided landmarks, when no other cues (e.g., room or olfactory cues) were reliable (Dolins, 2009). In the first experiment, three visual landmark cues making up a triangular array were maintained in a constant spatial relationship with eleven hidden food rewards on a foraging board of sixty-four holes. During each experimental trial the configuration of the landmark+hidden food array was rotated and translated with respect to the perimeter of the foraging board. Performance was measured by the monkeys' efficient localization of the food items hidden in baited holes on the foraging board. This task required reliance on the spatial relationship between the experimentally presented landmark cues. The foraging problem presented to the tamarins via the foraging board was analogous to problems that wild tamarins face when exploring tree holes, crevices in bark, leaf curls, and bromeliad whorls for concealed resources (Garber, 1980, 1984, 1993b; Neyman, 1977).

As indicated in Table 8.1, the monkeys performed significantly above chance prior to and immediately after a rotation of the landmark array. These results are consistent with an ability to encode the spatial relationships among three spatial cues to localize hidden food, and transference of spatial relationships from the previously learned configuration to the rotation configuration. The search patterns used by the tamarins to select holes on the foraging board revealed that their sequence of foraging choices was not significantly different from that expected based on a model of optimal patch choice and is consistent with reliance on a relational spatial strategy (Dolins, 2009).

In a second experiment, we investigated whether the monkeys would search effectively using the relationship between the two cues to localize hidden food items. In this experiment, twelve captive cotton-top tamarins were presented with a simpler array of only two landmarks and four hidden food rewards in which the configuration of cues+food was rotated and translated with respect to the perimeter of

Table 8.1. *Experiment 1: percent success post-rotation*

	Experiment 1: percent success post-rotation*				
Session number	Subject 1 percent success	Subject 2 percent success	Subject 3 percent success	Subject 4 percent success	Subject 5 percent success
1	21.5	14.5	16.0	20.5	25.0
2	22.0	30.0	24.0	21.0	50.0
3	22.5	29.0	19.0	21.5	30.5
4	28.0	18.0	24.0	21.0	50.0
5	18.0	23.0	25.0	20.5	24.0
6	23.0	32.0	23.5	26.0	15.5
7	29.0	29.0	18.0	26.0	28.0
8	25.5	35.5	25.5	45.0	17.0
9	27.5	34.0	26.0	25.0	38.5
10	41.0	28.0	20.5	27.0	18.0
11	43.0	36.5	33.0	26.0	50.0
12	29.0	28.0	32.0	33.5	34.0
13	22.0	29.0	14.5	26.0	36.5
14	38.0	29.5	25.5	32.0	29.5
15	31.0	44.5	24.0	32.0	24.0

* chance level performance post-rotation was 13.11 percent.

the foraging board. In order to eliminate bias associated with the possibility of learned positions of landmarks in relation to external cues, we present data from the first test sessions immediately post-rotation/translation of the cues+food configuration. Two possible strategies might emerge. If the monkeys searched close to cues that provided reliable information in the previous session, then we can conclude they were using an associative/orientation strategy. That is, they encoded each landmark as a discrete item. Alternatively, if they searched the four baited holes between the two landmarks and not around the cues or baited holes this is most consistent with reliance on the spatial relationships between the landmark cues.

In Table 8.2 we present data and analyze the monkeys' performance for the first two sessions of each trial (holes entered per subject directly after a rotation/translation). This was done because in the final part of each testing session, the monkeys were observed to preferentially search previously baited holes. This effect is consistent with a primacy-recency pattern (Dolins, 1994). Using data from the first

Table 8.2. *Experiment 2: mean proportion correct baited hole choices between and outside landmarks out of four possible baited hole choices*

Subjects	Mean proportion correct baited hole choices – BETWEEN landmarks	Mean proportion correct baited hole choices – OUTSIDE landmarks
1	1.6	0.1
2	1.6	0.2
3	1.9	0.1
4	1.8	0.1
5	2.3	0.05

two-thirds of each session, we calculated the proportion of holes searched "between" the cues relative to holes searched "outside" and "surrounding" the cues. In addition, the proportion of incorrect searches was computed for all holes surrounding each cue, simulating a "searching nearby the cue" strategy.

The monkeys' performance searching "between" versus "outside" the cues for each first session after a rotation/translation indicates that they preferentially searched holes positioned between the cues, even when presented with novel transformations of the cues+food configuration on the foraging board. The monkeys' performance was not consistent with searching in the vicinity of one or the other visual landmark. Rather, the evidence points to these tamarins' ability to develop an internal spatial representation using a set of landmark cues relationally to develop an efficient search strategy.

Analyzing the kinds of errors exhibited by the tamarins may offer insight into their learning and cognitive abilities. In Experiment 1 in the first post-rotation session, their success in locating hidden rewards, although significantly greater than chance, was not as high as compared to the final session of that condition. The difficulty of solving a rotation task, which involves inhibiting attention to external cues and attending to the spatial relationships of the landmark array solely, may underlie this decrease in performance. Experience also may play a part in the success of accomplishing this task. Alternatively, these errors may suggest that the rotation of a landmark array rather than viewing a landmark array from new perspectives (due to travel or movement around a location) is a more difficult task for these monkeys.

In human and nonhuman primates, there is evidence that the degree to which an array of landmarks is rotated from its original position, is correlated with the time required to generate an effective solution or mental representation (Corballis, 1988; Shepard and Cooper, 1982; Shepard and Metzler, 1971). In one study, for example, baboons and humans were presented with increased rotation of sample block letters. Both baboons and humans exhibited parity in solving these mental rotation tasks and their response latency increased with an increase in rotation angle (Hopkins *et al.*, 1993). This finding implies that the greater the rotation, the more difficult the process of mental computation of object recognition. The ability to rotate mental images has been documented in several animal species (Garber and Brown, 2006; Hollard and Delius, 1982; Hopkins *et al.*, 1993; Shepard and Cooper, 1982) and facilitates the generation of flexible navigation in response to self-mobility within an environment in which certain features of the landscape are stable and unchanging while other features change seasonally or are less predictable over time. Mental object rotation is consistent with both topological and metric spatial representations, as is recognition of objects and locations from different visual perspectives. It is not consistent with encoding spatial information in the form of a strip map.

The behavioral ecology of wild tamarins suggests that individuals commonly encounter ecological problems in navigating to feeding, sleeping and refuge sites for which proficiency and accuracy in rotating mental images and recognizing locations from varied vantage points is likely to be extremely important. Wild mustached and saddleback tamarins inhabit tropical forests characterized by a dense canopy. These primates travel 1,400–2,000 meters per day and exploit home ranges of 30–120 hectares (Garber, 1993b; Peres, 1993). In many cases they exhibit goal-directed and relatively straight-line travel between changing and dispersed feeding/sleeping sites (Garber, 1988, 1989, 1993a). The monkeys travel pattern and their ability to efficiently locate feeding sites that vary in time and space (e.g., seasonally available ripening fruit, nectar, insects and exudate holes) suggest these primates possess a detailed spatial representation of many points and environmental features across their range (Garber, 1988, 1989). Recognizing and orienting to single landmarks or landmark arrays from different or novel vantage points is consistent with the ability to rotate a recognized mental image. Below we detail information collected during a natural field study of wild Peruvian tamarins in order to examine the types of mental representation used in large-scale space.

Tamarins use of landmarks in large-scale space

The study of spatial cognitive strategies in wild primates provides insight into the relationship between a species' ecology, habitat use, patterns of decision-making, and its cognitive abilities (Garber, 1988, 1989; Milton, 1980). Nonhuman primates exploit home ranges that vary in size from less than one hectare in the pygmy marmoset (*Cebuella pygmaea*) to several square kilometers in the African apes. The ability to navigate efficiently in both large-scale (traveling between feeding sites) and small-scale space (locate resources within a food patch) may require that foragers rely on different search images or mental representations of spatial information (Garber, 2000). When traveling in large-scale space particular sets or arrays of associated landmarks may not be visible from all vantages, and therefore a single or small set of fixed points or features of the environment may be used as a beacon for orientation. A forager in "small-scale space," however, may be able to obtain many views of the same sets of landmarks or perceive an entire spatial array within its immediate visual field. In this way, opportunities to use landmark arrays may be more common in small-scale than in large-scale space.

Mustached (*S. mystax*) and saddleback tamarins (*S. fuscicollis*) inhabit tropical rainforest environments throughout the Amazon Basin. In Peru and Brazil, these primates form a single, large, mixed species troop in which one group of each species feeds, forages, rests and travels together throughout the year (Garber, 1988, 1989, 1993a, 1993b). These associations are extremely stable with individuals of each species responding to each other's predator and alarm calls, cooperatively defending a common home range and major feeding sites, and feeding together in the same trees at the same time. The traveling, dietary and foraging behavior of *S. mystax* and *S. fuscicollis* in the wild (Garber, 1993a, 1993b) closely resembles that of the cotton-top tamarins in the wild (Neyman, 1977; Savage *et al.*, 1996, 1997), and thus provides a strong basis for our comparative analyses.

Methods

Subjects were a mixed species troop of mustached (*Saguinus mystax*) and saddleback tamarins (*S. fuscicollis*) that jointly exploited a common 40 ha home range at our study site (Rio Blanco) in northeastern Peru. The data presented in this study represent a subsection (three months) of a larger dataset on foraging decisions, travel and spatial memory in

Table 8.3. S. mystax *and* S. fuscicollis *quadrat classification and use*

Quadrat classification (2500 m^2)	Quantity of quadrat type in home range
TRAVEL	144
Travel: straight-line	126
Travel nodes: turning (90 degrees or greater)	18
FEEDING/SLEEPING (local/core areas)	6
TOTAL in home range	200

tamarins (Garber, 1989, 1993a, 2000). Given that members of our tamarin mixed-species troop traveled together and fed in the same trees, we combined data for both species in our analysis. The position and behavioral data of focal animals within each social group were recorded every two minutes (30 data points per hour) throughout the day from the time the monkeys left their sleeping tree in the morning until the time they returned to a sleeping tree in the late afternoon. Within the troops' home range, markers placed approximately every 20 m throughout an extensive trail system (over 1,300 marked and mapped points) in the forest provided an accurate assessment of the travel routes and arboreal pathways selected by the tamarins. Using these data, we calculated travel distance and travel direction.

Results: field observations

Maps of the monkeys' home ranges were subdivided into 50 m x 50 m (2500 m^2) boxes that included approximately 200 quadrats in total. Quadrats were categorized in terms of their pattern of use (Table 8.3). For example, there were areas of the forest which the tamarins traveled through, but did not sleep, rest or forage in. These were scored as "travel" quadrats. Areas of the forest in which the tamarins commonly were observed to re-direct travel by changing direction or turning >90 degrees were scored as node or landmark quadrats. Finally, areas of the forest that contained frequently used feeding and sleeping sites were designated as core feeding/resting areas.

Over the course of our three-month study period, the tamarins utilized 144 "travel" quadrats (out of a possible 200). In general, the tamarins moved through an area of 8 ha (thirty-six quadrats) per day. Travel appeared to be "goal oriented" with the monkeys navigating relatively straight-line distances. We defined a travel sequence when the monkeys moved a distance of at least 150 meters (through three quadrats). Of sixty-nine travel sequences, approximately 40 percent (N=29) were used on multiple occasions, whereas the remaining 60 percent were used on only one occasion. Thus, the tamarins were found to navigate in large-scale space using a combination of traditional pathways and novel travel routes that included segments of these traditional pathways. In order to determine the degree to which tamarin travel routes were most consistent with a topological-based or a metric-based spatial representation, we analyzed the frequency of "turning" in the 144 "travel" quadrats. Turning represents a behavioral choice or navigation pattern during which the monkeys did not travel in a relatively straight-line through a quadrat, but turned at a >90-degree angle which re-oriented their travel path. These turning quadrats may have contained critical landmark information attended to by the tamarins and represent nodes or switch points to re-orient travel. Eighteen of the 144 "travel" quadrats were identified as nodes or "turning" areas where the monkeys made 90-degree changes in direction.

The tamarins traveled through each of these eighteen "turning" quadrats an average of 7.2 times during the three-month study period. Four out of every seven visits (57 percent) to these "turning" quadrats resulted in a significant 90-degree change of travel direction. In comparison, the monkeys visited each of the 126 "straight-line" travel quadrats on average 4.2 times. These visits resulted in turns of 90 degrees or greater only 21.4 percent of the time.

We also identified six core areas or quadrats within the tamarin's home range that contained several trees that were used repeatedly as resting and feeding sites across days. These quadrats can be thought of as "local areas" or areas of repeated use (see Garber, 2000). The tamarins were observed to visit these local areas from a variety of different directions and travel routes, which afforded individuals different views of these ecologically important parts of the forest. These different views provided the tamarins with opportunities to encounter and encode several fixed landmark cues from different directions and visual perspectives which could potentially be used to generate a set of landmark arrays in small-scale space that were specific to each area.

Based on patterns of travel and home range use, it appears that in large-scale space, mustached and saddleback tamarins retain spatial representations in which certain quadrats contain landmarks or other environmental features that serve as signposts, beacons or switch points to re-orient travel direction. These “turning” quadrats were located approximately 102 m from the nearest major feeding site. From these turning quadrats the tamarins were able to re-orient travel and reach feeding trees using a variety of direct but alternative routes. We argue that once the tamarins arrived in the general vicinity of commonly used quadrats that contained important feeding and resting sites, they relied on local views, possibly incorporating an array of landmarks to target the location of their specific goal (Garber, 2000).

General discussion

The observations of wild mustached and saddleback tamarins (*S. mystax* and *S. fuscicollis*) provide quantitative evidence that travel is goal-directed, and that individuals navigate to distant feeding and resting sites using relatively straight-line movement (travel segments of 100–300 m). However, we also identified particular areas of the forest in which the tamarins consistently adjusted and re-oriented their direction of travel by turning >90 degrees and then once again traveled along a relatively straight-line segment. Although the tamarins do not re-use the identical routes in these straight-line segments, they do orient to specific quadrats in the forest to redirect travel. We hypothesize that these turning quadrats most likely contain distal, yet familiar, and temporally stable landmarks which are re-used for orientation forming a non-veridical internal topological internal representation of the salient points in the home range.

A topological representation of nodes interconnected in a route-based system provides a more parsimonious explanation for the tamarin’s goal-directed and efficient navigational skills in large-scale space than an internal metric map. If tamarins relied on internal metric representations they would travel in straight lines computing the shortest distance between the initial point of departure and the goal location (i.e., triangulation). Instead, our evidence points to reliance on internal topological maps such that the monkeys have imposed a sequential node+route-based organization to navigate within their home range. Using such a representation, space is segmented into a large number of “straight line” quadrats and a small number of “turning” quadrats, which are grouped together to form travel routes. Unfortunately, we were not able to determine which specific landmarks in a given choice point area the monkeys may be attending to. It can only be inferred that they were using some type of identifiable and reliable cues as landmarks in these specific sectors of their home range.

An animal traveling in large-scale space may utilize a number of directional strategies to achieve efficient navigation between distant sites; in real terms, an animal exhibiting straight-line travel between distant sites may be relying on either an internal topological or metric map. However, a route-based system, such as a topological representation, provides a simpler explanation for animals' goal-directed and efficient navigational skills in large-scale space than that explained by a "cognitive map." In contrast, when an animal reaches a well-known feeding site (as in small-scale space) it may shift its general directional strategy to one that incorporates more specific and detailed spatial information about that area, utilizing spatial information encoded geometrically as in metric representations. Shifting to a metrically based strategy affords an animal efficient location and exploitation of resources located within a local dependent area (Poucet, 1993). Such a shift appears to characterize tamarins in our captive study, which we feel is consistent with foraging problems primates face in small-scale space. These problems include not only finding resources but equally as important, maintaining access to resources also sought by other group members (Garber *et al.* 2009; Giraldeau and Caraco, 2000). Individuals who arrive at a feeding patch first may benefit by gaining a "finder's advantage" over others who arrive later. This is analogous to the concept of "scramble" feeding competition and may have played an important role in shaping tamarin cognition and feeding ecology

Small- and large-scale space present different ecological and problem-solving challenges to a foraging animal. These challenges reflect an animal's spatial strategies and the way in which landmarks are perceived and encoded in memory. While some types of cues may become encoded in an internal topological map, others may be stored in memory for concurrent use in coordinating navigation to and recognition of specific locations (as in a metrically based map) (Gallistel, 1990). Thus, while distal cues in an array may act as points within a cognitive framework providing means to achieve broader navigational and behavioral goals, other cues may become more salient for precise direction and localization en route (Byrne and Janson, 2007). In large-scale space, a forager would then represent extensive areas in one kind of array, while simultaneously sectoring other areas into small-scale units using recognizable landmarks. Moreover, it may be simpler or more reliable for a forager to orient to a set of landmarks than a single landmark (Garber and Brown, 2006; Kamil and Cheng, 2001). This relates to the fact that a forager orienting to a single landmark+goal will need to maintain a large number of different mental representations because the relative spatial relationship between the landmark+goal will vary depending on the direction it approaches the goal. In contrast a forager relying on a set of landmarks that bound the goal, regardless of travel direction, only needs to take a path between the landmarks to reach the goal.

De Lillo *et al.* (1997) found that primates "group" visual objects, thereby imposing a representational spatial structure on their environment, that is, sectoring space. In forming search strategies, for instance, captive brown capuchins (*Cebus apella*) organize the search space into clusters or chunks, reducing the demands on recalling discrete items or landmarks used to locate feeding sites by creating smaller spatial sectors from the larger environment. Wild white-faced capuchins also appear to group landmarks in small-scale space into arrays (Garber and Brown, 2006). Chunking spatial information allows animals to organize a search space, large or small, creating efficient foraging and navigational patterns (Menzel *et al.*, 2002). Chunking objects in large-scale space where objects are more distantly located from one another (where landmarks are not able to be perceived in one visual scan) may represent a very different and perhaps more difficult cognitive task in comparison to chunking landmarks in small-scale "localized" space (Poucet, 1993). In this regard, by orienting to a relatively smaller set of predictable nodes or switch points in large-scale space, the forager can more easily locate these "stable and familiar" areas of the range and use landmarks in these areas to re-orient travel to locate more ephemerally used feeding and resting sites.

In summary, we feel that data on wild mustached tamarin and saddleback tamarin travel and range use in large-scale space support a model of navigational strategies that rely on an internal topological representation. We also presented the results of a series of experimental studies of captive cotton-top tamarins in small-scale space that are consistent with an ability to use landmarks relationally to locate hidden food rewards. Taken together these data indicate that tamarins, and presumably other primates, have the ability to employ topological and metric spatial strategies depending on the information available and the nature of the task. Ranging data on two species of free-ranging lemurs (*Lemur catta* and *Propithecus verreauxi verreauxi*) (Dolins, unpublished data) similarly support a model of navigational strategies in large-scale space that rely on topological representations. In humans, topological strategies are most heavily utilized when navigating long distances, whereas in localized space metric strategies predominate in both children and adults (Miller and Baillargeon, 1990; Newcombe, 1988; Piaget and Inhelder, 1967).

Finally, in order to better understand how nonhuman primates and humans internally represent, sector, localize and navigate to goals in space requires further comparative examinations of spatial problem-solving and navigation in both captive and free-ranging individuals. In particular, studies need to focus on the set of conditions in which landmarks are used singly or in an array, and the specific features of landmarks that are predictable and salient for individuals in large- and small-scale space. An ability to experimentally identify and distinguish the components of topological and geometric representations used by primates is a critical first step in defining their spatial strategies.

REFERENCES

Benhamou, S. and Poucet, B. (1998). Landmark use by navigating rats (*Rattus norvegicus*): contrasting geometric and featural information. *Journal of Comparative Psychology*, **112** (3), 317–322.

Byrne, R. W. and Janson, C. H. (2007). What wild primates know about resources: opening up the black box. *Animal Cognition*, **10**, 357–367.

Cheng, K. (1986). A purely geometric module in the rat's spatial representation. *Cognition*, **23**, 149–178.

Collett, M. and Collett, T. S. (2000). How do insects use path integration for their navigation? *Biological Cybernetics*, **83** (3), 245–259.

Collett, T. S. and Collett, M. (2002). Memory use in insect visual navigation. *Nature Reviews Neuroscience*, **3** (7), 542–552.

Collett, T. S., Collett, M. and Wehner, R. (2001). The guidance of desert ants by extended landmarks. *Journal of Experimental Biology*, **240** (9), 1635–1639.

Collett, T. S., Dillmann, E., Giger, A. and Wehner, R. (1992). Visual landmarks and route following in desert ants. *Journal of Comparative Physiology A*, **170**, 435–442.

Corballis, M. C. (1988). Recognition of disoriented shapes. *Psychological Review*, **95** (1), 115–123.

Deipolyi, A., Santos, L. and Hauser, M. D. (2001). The role of landmarks in cotton-top tamarin spatial foraging: evidence for geometric and non-geometric features. *Animal Cognition*, **4** (2), 99–108.

De Lillo, C., Visalberghi, E. and Aversano, M. (1997). The organization of exhaustive searches in a patchy space by capuchin monkeys *(Cebus apella)*. *Journal of Comparative Psychology*, **11** (1), 82–90.

Dolins, F. L. (1994). *Spatial relational learning and foraging in cotton-top tamarins (Saguinus oedipus oedipus)*, University of Stirling, Scotland, unpublished PhD thesis.

Dolins, F. L. (2009). Captive cotton-top tamarins' (*Saguinus oedipus oedipus*) use of landmarks to localize hidden food items. *American Journal of Primatology*, **71**, 316–323.

Dyer, F. C. (1993). Large-scale spatial memory and navigation in honey bees. In: *Proceedings of the Royal Institute of Navigation: Orientation and Navigation In Birds, Humans, and Other Animals*, Oxford.

Dyer, F. C. (1991). Bees acquire route-based memories but not cognitive maps in a familiar landscape. *Journal of Animal Behavior*, **41**, 239.

Gallistel, C. R. (1990). *The organization of learning*. London: The MIT Press.
Garber P. A. (1980). Locomotor behavior and feeding ecology of the Panamanian Tamarin (*Saguinus oedipus geoffroyi*). *Int J Primat* **1**, 185–201.
Garber, P. A. (1984). Use of habitat and positional behavior in a neotropical primate, *Saguinus oedipus*. In P. Rodman and J. Cant (Eds.), *Adaptations for foraging in non-human primates* (pp. 112–133). New York: Columbia University Press.
Garber, P. A. (1988). Foraging decisions during nectar feeding by tamarin monkeys (*Saguinus mystax* and *Saguinus fuscicollis*, Callitrichidae, Primates) in Amazonian Peru. *Biotropica*, **20** (2), 100–106.
Garber, P. A. (1989). Role of spatial memory in primate foraging patterns: *Saguinus mystax* and *Saguinus fuscicollis*. *American Journal of Primatology*, **19** (4), 203–216.
Garber, P. A. (1993a). Seasonal patterns of diet and ranging in two species of tamarin monkeys: stability versus variability. *International Journal of Primatology*, **14**, 145–166.
Garber, P. A. (1993b). Feeding ecology and behaviour of the genus *Saguinus*. In A. B. Rylands (Ed.), *Marmosets and tamarins: systematics, ecology and behaviour* (pp. 273–295). Oxford: Oxford University Press.
Garber, P. A. (2000). The ecology of group movement: evidence for the use of spatial, temporal, and social information in some primate foragers. In S. Boinski and P. A. Garber (Eds.), *On the move: how and why animals travel in groups* (pp. 261–298). Chicago: University of Chicago Press.
Garber, P. A. and Brown, E. (2006). Use of landmark cues to locate feeding sites in wild capuchin monkeys (*Cebus capucninus*): an experimental field study. In A. Estrada, P. A. Garber, M. Pavelka and L. Luecke (Eds.), *New perspectives in the study of mesoamerican primates: distribution, ecology, behavior and conservation* (pp. 311–332). New York: Kluwer.
Garber, P. A. and Dolins, F. L. (1996). Testing learning paradigms in the field: evidence for use of spatial and perceptual information and rule-based foraging in wild moustached tamarins. In M. A. Norconk, A. L. Rosenberger and P. A. Garber (Eds.), *Adaptive radiations of neotropical primates* (pp. 201–216). New York: Plenum Press.
Garber, P. A. and Paciulli, L. (1997). Experimental field study of spatial memory and learning in wild capuchin monkeys (*Cebus capucinus*). *Folia Primatologica*, **68**, 236–253.
Garber, P. A., Bicca-Marques, J.-C. and Azevedo-Lopes, M. A. O. (2009). Primate cognition: intregrating social and ecological information in decision-making. In P. A. Garber, A. Estrada, J. C. Bicca-Marques, E. Heymann and K. B. Strier (Eds.), *South American primates: comparative perspectives in the study of behavior, ecology, and conservation* (pp. 365–385). New York: Springer.
Giraldeau, L. A. and Caraco, T. (2000) *Social foraging theory*. Princeton: Princeton University Press.
Gouteux, S., Thinus-Blanc, C. and Vauclair, J. (2001). Rhesus monkeys use geometric and nongeometric information during a reorientation task. *Journal of Experimental Psychology: General*, **130** (3), 505–519.
Hermer, L. and Spelke, S. S. (1994). A geometric process for spatial reorientation in young children. *Nature*, **370**, 57–59.
Hirtle, S. C. and Jonides, J. (1985). Evidence of hierarchies in cognitive maps. *Memory & Cognition*, **13**(3), 208–217.
Hollard, V. D. and Delius, J. D. (1982). Rotational invariance in visual pattern recognition by pigeons and humans. *Science*, **218**, 804–806.

Hopkins, W. D., Fagot, J. and Vauclair, J. (1993). Mirror-image matching and mental rotation problem solving by baboons (*Papio papio*): unilateral input enhances performance. *Journal of Experimental Psychology; General*, **122**, 61–72.

Kamil, A. C. and Cheng, K. (2001). Way-finding and landmarks: the multiple-bearings hypothesis. *Journal of Experimental Biology*, **204**, 103–113.

Menzel, C. R. (1999). Unprompted recall and reporting of hidden objects by a chimpanzee (*Pan troglodytes*) after extended delays. *Journal of Comparative Psychology*, **113**, 426–434.

Menzel C. R. (2005). Progress in the study of chimpanzee recall and episodic memory. In H. S. Terrace and J. Metcalfe (Eds.), *The missing link in cognition: Origins of self-reflective consciousness*. New York: Oxford University Press. pp. 188–224.

Menzel, C. R., Savage-Rumbaugh, E. S. and Menzel, E. W. (2002). Bonobo (*Pan paniscus*) spatial memory and communication in a 20-hectare forest. *International Journal of Primatology*, **23**, 601–619.

Menzel E. W. and Menzel C. R. (2007). Do primates plan routes? Simple detour problems reconsidered. In D. A. Washburn (Ed.), *Primate perspectives on behavior and cognition* (pp. 175–206). Washington, DC: American Psychological Association.

Menzel, E. W., Juno, C. and Garrud, P. (1985). Social foraging in marmoset monkeys and the question of intelligence. *Philosophical Transactions of the Royal Society of London B; Biological Sciences*, **308**, 145–158.

Miller, K. F. and Baillargeon, R. (1990). Length and distance: Do preschoolers think that occlusion brings things together? *Developmental Psychology*, **26**(1), 103–114.

Milton, K. (1980). *The foraging strategy of howler monkeys: a study in primate economics*. New York: Columbia University Press.

Muller, R. U., Stead, M. and Pach, J. (1996). The hippocampus as a cognitive graph. *Journal of General Physiology*, **107**, 663–694.

Nadel, L. (1990). Varieties of spatial cognition: psychobiological considerations. In A. Diamond (Ed.), *The development and neural bases of higher cognitive functions* (pp. 613–636). New York: Annals of the New York Academy of Sciences, Vol. 608.

Newcombe, N. (1988). The paradox of proximity in early spatial representation. *British Journal of Developmental Psychology*, **6**(4), 376–378.

Neyman, P. (1977). Aspects of the ecology and social organization of free-ranging cotton-top tamarins (*Saguinus oedipus*) and the conservation status of the species. In D. G. Kleiman (Ed.), *The biology and conservation of the Callitrichidae*. Washington, DC: Smithsonian Institution Press.

Nicholson, D. J., Judd, S. P., Cartwright, B. A. and Collett, T. S. (1999). Learning walks and landmark guidance in wood ants (*Formica rufa*). *Journal of Experimental Biology*, **202** (13), 1831–1838.

Normand, E. and Boesch, C. (2009). Sophisticated Euclidean maps in forest chimpanzees. *Animal Behaviour*, 1–7.

Noser, R. and Byrne, R. W. (2007). Travel routes and planning of visits to out-of-sight resources in wild chacma baboons, *Papio ursinus*. *Animal Behaviour*, **73**, 257–266.

O'Keefe, J. and Nadel L. (1979). Precis of O'Keefe and Nadel's "The hippocampus as a cognitive map". *Behavioral and Brain Sciences*, **2**, 487–533.

O'Keefe, J. and Nadel, L. (1978). *The hippocampus as a cognitive map*. Oxford: Clarendon Press.

Pearce, J. M., Ward-Robinson, J., Good, M., Fussell, C. and Aydin, A (2001). Influence of a beacon on spatial learning based on the shape of the test environment. *Journal of Experimental Psychology: Animal Behavior Processes*, **27** (4), 329–344.

Peres, C. A. (1993). Diet and feeding ecology of saddle-back (*Saguinus fuscicollis*) and moustached (*S. mystax*) tamarins in an Amazonian *terra firme* forest. *Journal of Zoology London*, **230**, 567–592.

Piaget, J. and Inhelder, B. (1967). *The child's conception of space*. New York: Norton.

Poti, P. (2000). Aspects of spatial cognition in capuchins (*Cebus apella*): frames of reference and scale of space. *Animal Cognition*, **3**, 69–77.

Poucet, B. (1993). Spatial cognitive maps in animals: new hypotheses on their structure and neural mechanisms. *Psychological Review*, **100**, 163–182.

Savage, A., Giraldo, L. H., Soto, L. H. and Snowdon, C. T. (1996). Demography, group composition, and dispersal in wild cotton-top tamarins (*Saguinus oedipus*) groups. *American Journal of Primatology*, **38**, 85–100.

Savage, A., Shideler, S. E. Soto, L. H., Causado, J., Giraldo, L. H., Lasley, B. L. and Snowdon, C. T. (1997). Reproductive events of wild cotton-top tamarins (*Saguinus oedipus*) in Colombia. *American Journal of Primatology*, **43**, 329–337.

Shepard R. N. and Cooper, L. (1982). *Mental images and their transformations*. Boston: MIT Press.

Shepard, R. N., and Metzler, J. (1971). Mental rotation of three-dimensional objects. *Science*, **171**, 701–703.

Sheth, B. R. and Shimojo, S. (2004). Extrinsic cues suppress the encoding of intrinsic cues. *Journal of Cognitive Neuroscience*, **16** (2), 339–350.

Tolman, E. C. (1932). *Purposive behavior in animals and men*. USA: Appleton-Century-Crofts.

Tolman, E. C. (1948). Cognitive maps in rats and men. *Psychological Review*, **55**, 189–208.

Urbani, B. (2009). Spatial mapping in wild white-faced capuchin monkeys (*Cebus capucinus*). Ph.D. thesis. Urbana, IL: University of Illinois.

CHARLES MENZEL

9

Spatial learning and foraging in macaques

In this chapter, I present some of the questions about spatial learning that arise from a comparative and naturalistic perspective on macaque foraging. I then describe experimental studies of foraging and learning, focusing on experimental studies of macaques in tasks that require locomotion. These studies address two main issues in spatial learning. The first issue concerns the types of ecological associations that macaques learn in real, three-dimensional environments. The second issue concerns the types of experience that macaques require before they are capable of one-trial learning of novel distributions of food. Finally, I will comment on some of the questions that remain unanswered about nonhuman primate spatial cognition and memory.

NATURALISTIC BACKGROUND

Fifty years ago, few reputable students of nonhuman primate learning took field research seriously. Thorndike's (1898) negative opinions regarding studies conducted outside the laboratory were still accepted by some as definitive, and even one of the most prominent students of primate learning went so far as to claim "Our own data show without question that the monkeys we raise in the laboratory are ... brighter ... than any monkeys ever raised in a wild or feral state" (Harlow and Mears, 1979, p. 5) and that the best, not the worst, animal learning occurs in the idealized conditions of the laboratory (Harlow and Mears, 1979, p. 47). Today we are more aware of the relevance of field research to primate learning, and we can document our claims more rigorously and quantitatively than was the custom in the distant past (e.g., Janson, 1998; Janmaat *et al.*, 2006a, 2006b).

In the past decade or two, careful and detailed studies of animals in their natural habitats have indicated that primates can move toward

Spatial Cognition, Spatial Perception: Mapping the Self and Space, ed. Francine L. Dolins and Robert W. Mitchell. Published by Cambridge University Press.

resource locations from great distances, before they can see the locations. Tamarins move along fairly direct routes from one ripe-fruit-bearing tree to another tree of the same species, despite the fact that the tropical forests in which they live are dense and visibility is limited; and a social group may approach a given tree from different angles on different occasions (Garber, 1989). The long-distance movements of baboons also can be oriented toward resource locations that are out of view. Yellow baboons at Ruaha National Park, Tanzania, usually encounter baobab fruit and *Combretum obovatum* after a period of direct and rapid travel over a long distance. These are the main plant foods of the baboons that are spatially predictable and that provide a large number of grams per minute of handling time (Pochron, 2001). The baboons' direct and rapid travel prior to encounters with spatially predictable, economical foods differs from the indirect, slower travel that precedes their encounters with other types of food that are less predictable and less economical (Pochron, 2001). To give two additional examples, in the arid Erer region of Ethiopia, hamadryas baboons walk faster before reaching a feeding place even if it is not yet in sight, where they will stay longer than fifteen minutes (Sigg and Stolba, 1981; Kummer, 1995), and mangabeys in African rainforests approach distant, non-visible trees with fruit more frequently, and faster, than they approach trees without fruits (Janmaat *et al.*, 2006a).

A longstanding aim in behavioral research is to identify the specific environmental cues that animals follow when they search for food. If one can define the start point and goal in advance of the animal's movement, this provides a more powerful sort of test than if the start point and goal are unknown to the investigator. Encounters between macaques and ripe fruits have been arranged experimentally in the field, permitting detailed study of where macaques go next after finding an initial food item. The reaction of Japanese macaques after finding a piece of experimentally introduced, native akebi fruit was different from that after discovering other, equally preferred foods such as chocolate or non-native banana (C. Menzel, 1991). Animals who found ripe akebi fruit stared upwards and entered and inspected distant trees containing akebi vines, even at times of year when the naturally occurring akebi fruits were not ripe. In contrast, animals who found chocolate returned at later times, sometimes on the following day, to the precise location of the find and confined much of their visual and manual searching to a small area of ground. Animals who found nothing, a control condition, did not enter akebi vines or search in the vicinity of their starting locations. In other words, animals who found a piece of artificially introduced akebi fruit seemed to put more weight on reliable, established information about multiple past finds than on the location of one recent find, as if their memory evoked a search "routine" for this recurrent seasonal fruit. Macaques who found chocolate seemed to put more weight on the location of a single recent find than on past finds of native plant foods.

The Japanese macaques' behavior suggested the use of long-term memory in foraging and a degree of flexibility in organizing a search routine (C. Menzel, 1991). Their memory appeared to be of something specific. Analogous performances have been reported in other primate species and situations. For example, captive orangutans that encountered a few pieces of banana in a single location in a 1,900 m^2 outdoor enclosure subsequently moved to and inspected other locations that had contained bananas on previous days, rather than locations that had contained grapes (Scheumann and Call, 2006).

TWO VIEWPOINTS REGARDING SPACE

As the foregoing section suggests, a detailed examination of how individual primates navigate and find food in their particular ecological contexts gives rise to questions about spatial perception and memory. The issue of how "space" is structured from the standpoint of a nonhuman primate arises, because the amount of space used, and the types of environmental features used as physical supports and reference points during travel, feeding and social interactions can vary sharply across species or even across individuals of the same species (Hediger, 1968; E. Menzel, 1969; C. Menzel, 1986). The issue of memory arises, because the environmental features that serve as goals, local signs, choice points and landmarks are not always within range of seeing or hearing at the time of response (Altmann, 1998; Janson, 1998; Dominy *et al.*, 2001). Animals often appear to fill in gaps in the information provided by direct visual perception.

To illustrate the operation of learning and memory capabilities in the context of foraging, consider where a long-tailed macaque goes after it discovers one or a few initial food items. Long-tailed macaques are diverse feeders (Aldrich-Blake, 1980; Wheatley, 1980; Lucas and Corlett, 1991). Many of the types of food they use are rapidly depleted and hard to detect from a distance (Van Schaik *et al.*, 1983). Like many animals in nature, they face constantly changing foraging problems and do not always have a large number of consecutive chances to perfect their responses on a given problem (Gass, 1985). To forage efficiently, macaques should be capable of forming new search strategies immediately for new occurrences of food. That is, they should be innovative. If a macaque finds one or a few initial food items, how does it find more of the same food type? Where does it go and what locations does it inspect?

One hypothesis about how animals use the available space for foraging is the "structure guided" hypothesis (C. Menzel, 1996b). According to this hypothesis, macaques organize their search by the spatial proximity of food items to environmental structures. The type of structure that the animal inspects might vary depending on where the animal discovered an initial item. For example, if a macaque finds a piece of highly preferred food within a streambed, it might restrict its subsequent search to locations along the same streambed. Alternatively, if it finds a new type of highly preferred food at the base of a fern, then it might inspect other ferns, in preference to rocks, trees or logs. In the natural habitat of macaques, foods can be associated spatially with visible borders such as stream edges or forest edges, or can be found in discrete structures, such as trees of a particular species (Wheatley, 1980).

A competing hypothesis is that macaques organize their search by the distance from the food location (C. Menzel, 1996b). According to this "pure spatial gradients" hypothesis, available space is homogeneous and unstructured. It is an undifferentiated area around a reinforcer. The animal allocates its search effort to concentric rings around a food item according to some decreasing function of the Euclidean distance from the item. After finding food, the animal might slow its movement, increase its rate of turning, and inspect any location that lies within a reasonable distance of an initial find, regardless of the position of visible environmental structures. For example, a carrion crow will search on a beach in a more or less random walk, up to about 2 meters from an encountered eggshell (Tinbergen *et al.*, 1967), and other birds, fish and insects show increased turning and area-restricted search after finding a resource (Bell, 1991). Formal analyses of two-dimensional spatial gradients in learning and navigation can be found in Hull's (1952) classic study and in Reid and Staddon (1998).

STRUCTURE-GUIDED FORAGING

To test the structure-guided hypothesis, I studied captive-born long-tailed macaques in a 880 m^2 outdoor enclosure at the Bokengut field

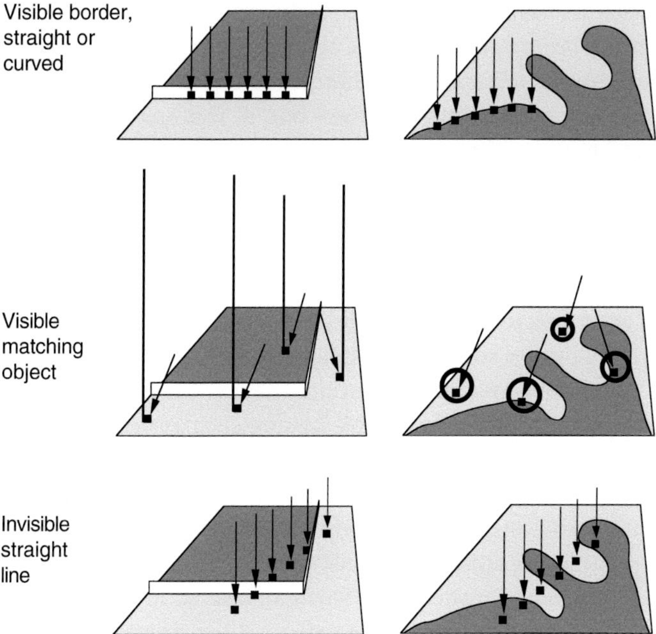

Figure 9.1. Sample layouts of hidden food in three experimental conditions. (a) Visible environmental border: the food (black squares) lies along either a low cement wall between a grass field and a sand field (left); or along the border between a stone field and a grass field (right). (b) Visible matching objects: food next to bases of vertical poles (left); or next to wooden crates (right). (c) Invisible line: the line cuts across grass and sand fields (left); it cuts across stone and grass fields (right). Drawn by Lorenz Gygax, based on photographs in C. Menzel (1996b).

station, University of Zürich (C. Menzel, 1996a, 1996b). The enclosure contained tree trunks, grass fields, rock fields, elevated walkways and visual barriers. The hypothesis tested was that macaques would extrapolate their search within dimensions of environmental structure that, in the macaques' natural habitat, normally would be correlated with the distribution of food, such as along visible borders, within visible patches (surface areas), within particular height levels or within discrete visible objects of the same type. It seemed possible that even captive-born macaques, who had had strictly limited experience searching for food, would follow visible structures.

While the macaques were confined indoors and could not see what I was doing, I hid food in the macaques' enclosure according to one of several different types of rules. The rules are shown in Figure 9.1.

Thus, food was hidden either next to discrete visible objects of the same general type (matching objects), or along a visible borderline, or along an arbitrarily oriented, invisible straight line. When the food was hidden along an invisible straight line, it was spaced at intervals of either 1 m or 3 m. To get the macaques started on each trial, I also left behind one to three piles of visible food that were distributed according to the rule, and of course additional food piles were hidden according to the rule. Each trial used a different set of locations. In some trials, I hid high-preference food according to one rule and low-preference food according to a different rule, to determine whether the macaques would follow the rule for the high-preference food.

Results were that if a macaque detected a single visible pile of food next to a continuous visible border, such as next to a borderline between a grass field and a sand field, or next to a borderline between a sand field and a stone field, the animal restricted much of its subsequent manual inspections to other locations along the same border. Importantly, the animals searched along visible borders from Trial 1 of the experiment. Furthermore, when an animal found food next to a single discrete visible object (e.g., a log, a stone, a vertical pole) it might walk several meters to another object of the same general type and inspect it manually. Extension of search from a single baited object to other, similar-looking objects was observed from Trial 1 of the experiment. Still further, if the macaques found high-preference food, banana, next to a single object of type A (e.g., a yellow stone) and found low-preference food, carrot, next to a single object of type B (e.g., a low post), they inspected other locations of type A more often than they inspected other locations of type B.

Figure 9.2 shows the time course of food acquisition within the trial by the group as a whole, for trials on which the animals were presented with the types of food distributions shown in Figure 9.1. It may be seen from the cumulative curves in Figure 9.2 that the macaques found hidden food items much more quickly when the items were hidden according to a "natural" rule, along a visible border or next to visible matching objects, than when the food was hidden according to a relatively "unnatural" rule, that is, at regular intervals of 1 m or 3 m along an arbitrarily oriented, invisible straight line. These results are in strong agreement with the structure-guided hypothesis.

In the macaques' natural habitat, food items can be associated with directional changes in a visible environmental gradient, such as a directional change in vegetation type or color. The captive long-tailed macaques varied the direction of their search according to some purely

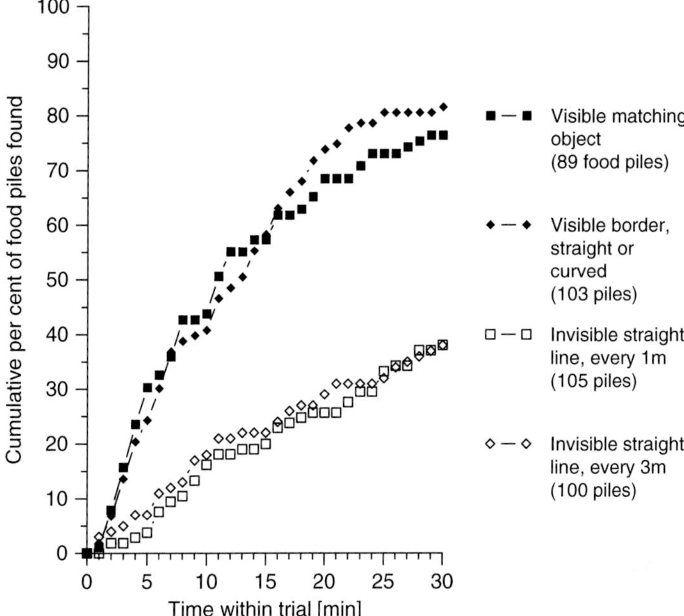

Figure 9.2. Cumulative percentage of food piles found as a function of time within a trial and the various food distributions presented in the experiment. Black symbols are results when food was hidden next to visible environmental structures; open symbols are results when food was hidden along invisible straight lines. The numbers in parentheses are the total numbers of food piles hidden in fifteen trials. Reprinted from C. Menzel (1996b).

spatial characteristics of the food distribution (Hemmi and Menzel, 1995). Five male macaques were tested individually and were presented with three piles of food at 1-meter intervals along an invisible, arbitrarily oriented straight line, within a large sand field. The position and orientation of the line varied across trials. The three food piles differed primarily in visibility, and they decreased in visibility along the line. The first pile was a small but easily visible piece of banana and five visible raisins placed together on the ground. The second pile consisted of eight visible raisins, and the third pile consisted of four visible raisins and four raisins hidden under the sand.

After finding the food, the macaques immediately started searching. They stood bipedally, and they moved aside objects manually while examining the ground visually. From Trial 1, they restricted their searching almost entirely to the sand field, rather than to the surrounding terrain. Out of 216 instances of bipedal standing,

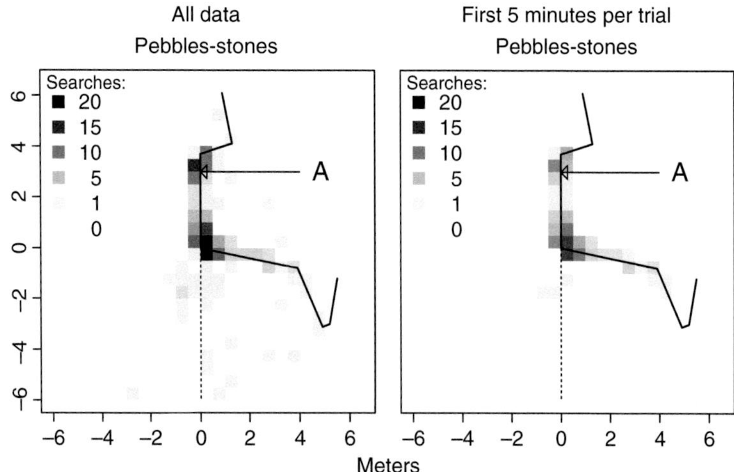

Figure 9.3. The locations that six long-tailed macaques (*Macaca fascicularis*) inspected manually, following their discovery of food along a straight-line segment of a border between a field of stones and a field of pebbles. "A" = location of visible banana. A small amount of visible raisins also was present at 1-m intervals along the border, from "A" (x, y coordinates 0, 3) to the "corner" (x, y coordinates 0, 0). It may be seen in the figure that the animals also searched along this section of the border.

215 occurred within the sand field, as did 531 of 537 instances of manual searching. Restriction of search to the sand field could not be explained by the amount of time spent inside versus outside the sand field. When the macaques left the sand field they typically stopped searching, climbed to an elevated perch in the outdoor enclosure, and remained there until the end of the trial. Furthermore, the macaques searched approximately along the invisible line, in the direction in which food had decreased in visibility. The findings suggested that the macaques did not simply react to each food pile they discovered on its own, but also to some aspect of the spatial relationships of the cues. The animals did not simply follow the gradient of increasing food reinforcement. Instead, they searched in the direction of decreasing cue visibility, i.e., in the direction in which they could not decide at a glance whether there was more food (Hemmi and Menzel, 1995).

In a further experiment, two food distribution rules were pitted against each other to determine which rule the macaque would follow. Six male macaques were tested as dyads and presented with four piles of food that decreased in visibility, at 1-meter intervals along a straight section of a visible environmental border. At the location of the least visible cue, the border curved sharply. Figure 9.3 shows an example of one of the types of borders used, one between a field of stones and a field of pebbles. Most of the stones were 8 cm to 35 cm in diameter. The question was where the macaques would go if they searched beyond the cues: would they make a turn and continue to search along the border, as predicted by the structure-guided hypothesis, or would they go straight and continue to search along the invisible straight line defined by the four initial piles? Figure 9.3 shows that the macaques searched along the full spatial range defined by the four cue piles of food (from point "A" to the corner). More importantly, they extended their search beyond the cues. They made a turn and searched more frequently along the visible border than along the invisible straight line, particularly in the first five minutes of the trial. This outcome was in agreement with the structure-guided hypothesis.

In the macaques' natural habitat, different species of fruit, fungus or insects can be found at characteristic heights ranging from ground level to the upper canopy. A further experiment examined the relative importance of two dimensions of environmental structure, height level and type of discrete object, in guiding foraging. The captive-born macaques were tested individually or in small subgroups in the outdoor enclosure and were presented with a small amount of visible food in a single location. The visible food was located on a vertical pole at one of two height levels (either low or high) and next to one of two types of containers (either a bag or a can). Numerous other exemplars of both types of containers were distributed at both height levels throughout the outdoor enclosure. The question was whether the macaques would restrict their ensuing search by height level, by object type, by both dimensions combined, or by neither dimension.

We found that the animals strongly followed the type of object that had contained the visible food. They were influenced to a lesser but statistically significant degree by the baited height level. If animals found food at the lower level, they tended to restrict their search to other low objects of the same type; if they found food high, they tended to inspect other objects of the same type at both height levels. In sum, both the height level and the type of object in which food was discovered guided foraging, and object type was the more important factor in this experimental situation.

In other tests (C. Menzel, 1996a), the long-tailed macaques made far more visits to locations where they had obtained food on the previous day than to locations they had merely inspected on the previous day without finding food. Social factors also were important. The macaques inspected locations where another animal had searched and obtained food more often than they inspected locations where another animal had merely searched without finding food. They appeared to detect the outcome of another animal's search visually, from a distance of several meters.

The experimental findings on wild Japanese macaques and captive-born long-tailed macaques, taken together, indicate that macaques are sensitive to a large number of different aspects of a food's distribution in space. They show a degree of innovation in their searching, and they show excellent memories for locations that contained food in the past (C. Menzel, 1997). The immediate use of matching visual cues in trial-unique problems by experimentally naïve macaques (C. Menzel, 1996a) seems far more efficient than would be expected from many prior laboratory studies but is consistent with wild macaque foraging behavior (van Schaik *et al.*, 1983). The findings on structure-guided foraging suggest that for macaques, "space" is not perceived or remembered as an empty container. Instead, it is defined largely by the relative positions of objects. Pure spatial gradients are disrupted by the presence of visible objects. Specifically, the close spatial proximity of an initial food item to a visible environmental feature was an important determinant of where the macaques searched. The dimensions of environmental structure that animals followed included discrete visible objects, visible borderlines, surface areas, height levels, and directions associated with a change in food visibility.

I have used the concept of association in a neutral, non-psychological sense to refer to spatial pairings of food and structures in the environment. A reader might ask whether the macaques were therefore using an "associative strategy," or whether their search strategies, as structure-guided, could be characterized as cognitive. The fact that animals need to learn *about* ecological associations does not necessarily mean that they learn *by* association (Gibson, 1979) or that their behavior is non-cognitive. The following two points seem relevant. First, as I have already emphasized, the experimentally naïve macaques showed systematic, reward-producing search patterns on Trial 1 in trial-unique experimental tests. They did not require a large number of consecutive chances on the same problem to show effective foraging. They did not simply repeat the movement patterns that had preceded earlier food finds. Nor did they only re-visit the specific locations or classes of objects that had contained food on recent trials. They extended their search from one or a few initial items to new locations that had never contained food previously. Second, if one were to suggest that an organism's behavior is "cognitive," one might expect the organism to be capable of remembering many different specifics of events and locations, and more details than another individual that had not witnessed the events and locations personally. The possibility remains open that macaques are sensitive to many additional types of past and current cues during their foraging, that they maintain knowledge of the locations and directional bearings of group members that are out of view, and that they rely on a fairly detailed and personalized memory of events and distant places in selecting where to go.

CONCLUSIONS

The naturalistic background and experimental findings reviewed in this chapter raise the question of what more there is that nonhuman primates take into account and anticipate during their foraging. How well can a macaque, tamarin or ape recall events, environmental features and social partners located completely out of sight and hearing of the general navigational or search area? In most animal memory studies, an animal's retention and its retrieval are tested within the same spatial situation in which learning originally occurred. In this respect, most animal studies address recognition memory rather than recall memory. With apes that have been reared with exposure to an artificial language, one can study the types of information that animals can recall and report about areas that are outside the immediate sensory environment (C. Menzel, 1999, 2005). I discuss this topic in a separate chapter (Chapter 24, this volume). The "big question" is, of course, how representative such animals are of their species as a whole. I doubt very much that they are brighter than any primates ever reared in the natural habitat. Instead, I suggest that they serve to show how limited our understanding is of the "upper limits" of primate cognitive capabilities in both captivity and the field.

ACKNOWLEDGMENTS

I thank Lorenz Gygax, Nerida Harley, Jan Hemmi, Ernst Krusi and Marion Maag for assistance with the experiments on long-tailed macaques and Hans Kummer and Emil Menzel for discussions. Research on macaques was supported by NSF grant INT-8603379, a Sigma-Xi Grant-in-Aid, Swiss National Science Foundation Grant 31.27721.89 and

National Research Service Award NS07973 from the US Public Health Service. Manuscript preparation was supported by HD-38051 and HD-056352.

REFERENCES

Aldrich-Blake, F. P. G. (1980). Long-tailed macaques. In D. J. Chivers (Ed.), *Malayan forest primates* (pp. 147–165). New York: Plenum Press.

Altmann, S. A. (1998). *Foraging for survival.* Chicago: University of Chicago Press.

Bell, W. J. (1991). *Searching behaviour.* New York: Chapman and Hall.

Dominy, N. J., Lucas, P. W., Osorio, D. and Yamashita, N. (2001). The sensory ecology of primate food perception. *Evolutionary Anthropology*, **10**, 171–186.

Garber, P. (1989). Role of spatial memory in primate foraging patterns: *Saguinus mystax* and *Saguinus fuscicollis*. *American Journal of Primatology*, **19**, 203–216.

Gass, C. L. (1985). Reaching for an integrated science of behavior. *Behavioral and Brain Sciences*, **8**, 337–338.

Gibson, J. J. (1979). *The ecological approach to visual perception.* Boston: Houghton Mifflin.

Harlow, H. F. and Mears, C. (1979). *The human model: primate perspectives.* Washington, DC: V. H. Winston; New York: Wiley.

Hediger, H. (1968). *The psychology and behaviour of animals in zoos and circuses.* New York: Dover Publications.

Hemmi, J. and Menzel, C. R. (1995). Foraging strategies of long-tailed macaques *Macaca fascicularis*: directional extrapolation. *Animal Behaviour*, **49**, 457–464.

Hull, C. L. (1952). *A behavior system.* New Haven: Yale University Press.

Janmaat, K. R. L., Byrne, R. W. and Zuberbühler, K. (2006a). Evidence for a spatial memory of fruiting states of rainforest trees in wild mangabeys. *Animal Behaviour*, **72**, 797–807.

Janmaat, K. R. L., Byrne, R. W. and Zuberbühler, K. (2006b). Primates take weather into account when searching for fruits. *Current Biology*, **16**, 1232–1237.

Janson, C. H. (1998). Experimental evidence for spatial memory in foraging wild capuchin monkeys, *Cebus apella*. *Animal Behaviour*, **55**, 1229–1243.

Kummer, H. (1995). *In quest of the sacred baboon: a scientist's journey.* Princeton: Princeton University Press.

Lucas, P. W. and Corlett, R. T. (1991). Relationship between the diet of *Macaca fascicularis* and forest phenology. *Folia Primatologica*, **57**, 201–215.

Menzel, C. R. (1986). Structural aspects of arboreality in titi monkeys (*Callicebus moloch*). *American Journal of Physical Anthropology*, **70**, 167–176.

Menzel, C. R. (1991). Cognitive aspects of foraging in Japanese monkeys. *Animal Behaviour*, **41**, 397–402.

Menzel, C. R. (1996a). Spontaneous use of matching visual cues during foraging by long-tailed macaques (*Macaca fascicularis*). *Journal of Comparative Psychology*, **110**, 370–376.

Menzel, C. R. (1996b). Structure-guided foraging in long-tailed macaques. *American Journal of Primatology*, **38**, 117–132.

Menzel, C. R. (1997). Primates' knowledge of their natural habitat: as indicated in foraging. In A. Whiten and R. W. Byrne (Eds.) *Machiavellian intelligence II: extensions and evaluations* (pp. 207–239). Cambridge: Cambridge University Press.

Menzel, C. R. (1999). Unprompted recall and reporting of hidden objects by a chimpanzee (*Pan troglodytes*) after extended delays. *Journal of Comparative Psychology*, **113**, 426–434.

Menzel, C. R. (2005). Progress in the study of chimpanzee recall and episodic memory. In H. Terrace and J. Metcalfe (Eds.), *The missing link in cognition: origins of self-reflective consciousness* (pp. 188–224). New York: Oxford University Press.

Menzel, E. W., jr. (1969). Naturalistic and experimental approaches to primate behavior. In E. Willems and H. Rausch (Eds.), *Naturalistic viewpoints in psychological research* (pp. 78–121). New York: Holt, Rinehart & Winston.

Menzel, E. W., Jr. and Menzel, C. R. (2007). Do primates plan routes? Simple detour problems reconsidered. In D. A. Washburn (Ed.), *Primate perspectives on behavior and cognition* (pp. 175–206). Washington, DC: American Psychological Association.

Pochron, S. T. (2001). Can concurrent speed and directness of travel indicate purposeful encounter in the yellow baboons (*Papio hamadryas cynocephalus*) of Ruaha National Park, Tanzania? *International Journal of Primatology*, **22**, 773–785.

Reid, A. K. and Staddon, J. E. R. (1998). A dynamic route finder for the cognitive map. *Psychological Review*, **105**, 585–601.

Scheumann, M. and Call, J. (2006). Sumatran orangutans and a yellow-cheeked crested gibbon know what is where. *International Journal of Primatology*, **27**, 575–602.

Sigg, H. and Stolba, A. (1981). Home range and daily march in a Hamadryas baboon troop. *Folia Primatologica*, **36**, 40–75.

Thorndike, E. L. (1898). Animal intelligence: an experimental study of the associative processes in animals. *Psychological Monographs*, **2** (8).

Tinbergen, N., Impekoven, M. and Franck, D. (1967). An experiment on spacing-out as a defence against predation. *Behaviour*, **28**, 307–321.

van Schaik, C. P., van Noordwijk, M. A., de Boer, R. J. and den Tonkelaar, I. (1983). The effect of group size on time budgets and social behaviour in wild long-tailed macaques (*Macaca fascicularis*). *Behavioral Ecology and Sociobiology*, **13**, 173–181.

Wheatley, B. P. (1980). Feeding and ranging of East Bornean *Macaca fascicularis*. In D. G. Lindburg (Ed.), *The macaques* (pp. 215–246). New York: Van Nostrand Reinhold.

Part III Evolutionary perspectives on cognitive capacities in spatial perception and object recognition

THOMAS WYNN

10 The evolution of human spatial cognition

INTRODUCTION

Human spatial cognition has evolved. We impose regular geometric shapes on virtually all of our objects and living spaces, for example, and have developed formalized algorithms for determining our precise location in space and for guiding our long-distance movement. No other animal does these things. Some of these abilities are culturally enabled (Global Positioning System is one example), but the underlying understanding of spatial relations is an aspect of human cognition that has evolved over the five-plus million years that separate us from the ancestor we share with chimpanzees. An important piece to any understanding of a complex organic phenomenon is the understanding of how it evolved. We know, for example, that there is a robust and reliable difference in the performances of men and women on certain tests of spatial cognition. Why? Is it the result of natural selection tied to wayfinding (Eals and Silverman, 1994; Silverman *et al.*, 2000; Silverman and Eals, 1992), or a byproduct of some other evolutionary development (Wynn *et al.*, 1996)? Answers to such questions not only satisfy a natural curiosity about where we have been, but also help us understand the nature of the modern mind itself.

Studying the evolution of spatial cognition is more difficult than studying the evolution of bipedalism, but it is not impossible. Fossils, unfortunately, are little help. Fossil crania with good preservation supply information about brain size and even gross brain anatomy from endocasts. These can be useful in documenting general brain evolution, but they do not inform us about specific cognitive abilities, at least not at our present stage of understanding. A second approach, taken by evolutionary psychologists, looks at modern humans and eschews direct evidence of evolution entirely. Evolutionary psychologists argue that evolutionary circumstances are preserved in the cognitive

Spatial Cognition, Spatial Perception: Mapping the Self and Space, ed. Francine L. Dolins and Robert W. Mitchell. Published by Cambridge University Press.

architecture of the modern mind, and that an adequate description of what the mind is designed to do will identify the selective pressures that produced it (Thornhill, 1997). Paleoanthropologists are skeptical of sole reliance on descriptions of the modern mind. We do, in fact, have direct evidence of the products of prehistoric minds in the form of archaeological remains. Cognitive archaeology is a relatively new branch of prehistoric archaeology that employs the methods of archaeology and the theories of cognitive science to reconstruct the evolution of the human mind, including the evolution of spatial cognition (Nowell, 2001).

Archaeology is primarily an observational discipline that studies traces of past action. These traces are occasionally obvious and even spectacular, as in the ancient Mexican city of Teotihuacan, or the cave paintings at Chauvet. But far more often archaeology studies traces of the mundane, everyday activities of preparing food, making and using tools, constructing shelter, and so on. Because these everyday activities were organized by active minds, their traces can be used to reconstruct something of the minds themselves. The difficulties are methodological. Cognitive archaeologists use two basic approaches. The first relies on current cognitive theory to identify patterns in the archaeological evidence that reflect specific cognitive abilities. The persuasive power rests on the explanatory power of the specific theories and the definition of the relevant variables for analysis. The second approach relies on the experimental replication of the prehistoric activities themselves, and then uses the modern participants as surrogates for the prehistoric actors, on the premise that successful reproduction of an action sequence should reproduce something of the cognitive underpinnings. The most successful of these latter have employed brain imagery techniques to study stone tool manufacture (Stout *et al.*, 2000). The two approaches are not mutually exclusive, and indeed cognitive archaeologists often rely on both (Wynn, 2002).

The nature of preservation biases the archaeological record in two important respects. First, not all materials preserve equally well. Stone preserves well, but metal does not. The bones of animals often preserve well in trash middens, but the remains of plant foods usually do not. As a result of such differences in preservation, we do not have equal access to the entire range of past activities. Second, there is a sliding scale of resolution inherent to the archaeological record. The more remote in time, the less we have to study. There are fewer older sites than recent sites because erosion and other natural processes have had more opportunity to work. The state of Colorado contains thousands of archaeological sites dating to the last millennium, for example, but the entire world boasts fewer than ten sites that are older than two million years.

Because many human activities organize action in space it is one domain that is amenable to cognitive analysis. Archaeologists have access to two varieties of spatial patterns. The first includes patterns of sites on the landscape and patterns of activities internal to the sites themselves. The former relies on fairly precise chronological control, and as a result is applicable only to recent time periods. Internal site patterning relies on good context, i.e., it is necessary that the site patterns have been little disturbed by erosion, and this requirement limits its usefulness for the deep past. The second variety of spatial patterning is inherent to individual artifacts. When prehistoric humans made artifacts they organized their actions on the spatial field of the artifact. Even though this is a narrow window, it is one of the most reliable. Archaeologists have excavated and described millions of stone tools dating to virtually the entire range of human evolution. As a result, stone tools are the subject of most analyses of the evolution of spatial cognition, generally considered, including both spatial cognition and shape recognition and imposition.

STONE KNAPPING

Knapping is the process of flaking stone, and has been in the repertoire of hominid behaviors for over 2.5 million years. The basic action of knapping involves striking a mass of rock, termed a core, with another object, termed a hammer. If the knapper strikes the core with enough force, and with an appropriate angle of delivery, the core will break. Given the limits of human strength, it is almost always necessary to strike the core near an edge, which breaks off a small piece termed a flake (Figure 10.1). This flake will have sharp edges that can be used for a variety of tasks; it can also be further modified into a large range of edge shapes for performing more specialized tasks (a chisel-like edge for engraving or a thicker edge for scraping). The core itself may also have sharp edges, making it potentially useful. Not all types of stone are equally useful to the stone knapper, and some raw materials are especially prized (flint, chert, obsidian). The vast majority of technological history (over two million years) consists of developments in stone knapping. Hominids acquired new techniques (e.g., striking the core against an anvil), employed new materials (e.g., use of a soft hammer such as antler), and an increasingly large range of final products.

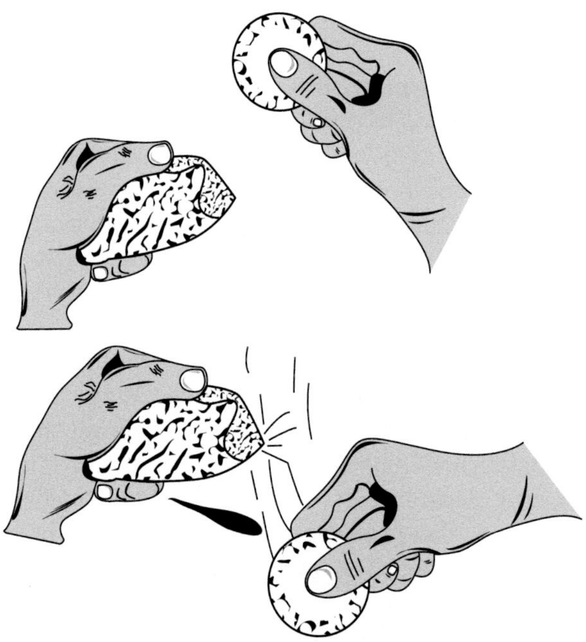

Figure 10.1. The basic action of stone knapping.

Basic stone knapping is primarily a spatial task. Recently Stout *et al.* (2000) have conducted a pilot study of basic stone knapping using positron emission tomography (PET). The subject was a single, experienced knapper, who performed the simple task of knocking single flakes off cores. The result showed highly significant activation in several brain regions. Much of this activation is what one would expect from any skilled, motor task using hand–eye coordination (primary motor and somatosensory cortex), but there was also significant activation of the superior parietal lobes.

> The superior parietal lobe consists of what is referred to as "multi-modal association cortex" and is involved in the internal construction of a cohesive model of external space from diverse visual, tactile, and proprioceptive input. (p. 1220)

In other words, simple stone knapping is a "complex sensorimotor task with a spatial-cognitive component" (p. 1221).

APES

This basic action of stone knapping is within the cognitive and motor abilities of apes. This was initially demonstrated by Wright (1972), who taught an orangutan to flake stone, and use the sharp flake to open the reward box. In the 1990s Toth and Savage-Rumbaugh (Schick *et al.*, 1999; Toth *et al.*, 1993) corroborated and expanded on Wright's results when they taught the bonobo Kanzi to flake stone. Kanzi is able to strike flakes from cores, and use the sharp flakes to cut cords securing a reward box. However, Kanzi is not as adept as human knappers. "[A]s yet he does not seem to have mastered the concept of searching for acute angles on cores from which to detach flakes efficiently, or intentionally using flake acars on one flake of a core as striking platforms for removing flakes from another face" (Toth *et al.* 1993, p. 89). These abilities are basic to modern knapping and, more telling, are evident in the stone tools made two million years ago. Toth *et al.* suggest that this represents a significant cognitive development, though they do not specify just what cognitive ability has evolved (see below, pp. 219–221). Kanzi is also not very accurate in delivering blows, which could simply be a matter of biomechanical constraint (i.e., he does not have the necessary motor control), or it could result from an inability to organize action on the small spatial field of the core. It is the organization of such action, fossilized as patterns of flake scars, that developed significantly during the two million years following the appearance of stone tools.

Even though apes can be taught basic stone knapping, there are no ethological examples of an ape lithic technology. Chimpanzees do use stone hammers to break open nuts (Boesch, 2003; Boesch and Boesch, 1990), and this activity occasionally fractures the hammers, producing flakes (Mercader *et al.*, 2002). It would not require a great evolutionary leap to begin using these pieces. Apes have not done so, probably because have never had the need to (they have very effective canine teeth and incisors). Sometime before 2.5 million years ago a group of hominids, who did not have large incisors or canines, began to produce stone flakes and use them.

THE OLDEST LITHIC TECHNOLOGY

Oldowan is the term archaeologists have given to the earliest stone tools. As the name suggests, these tools have been found at numerous sites in Olduvai Gorge, Tanzania, where they date to about 1.7–1.8 million years ago (Ma). Oldowan tools have been recognized throughout much of East Africa, and there are possible sites in North Africa and the Middle East (Schick and Toth, 2001). The earliest known sites date back to 2.5 Ma (Harris, 1983; Semaw, 2000; Semaw *et al.*, 1997), and late sites date to

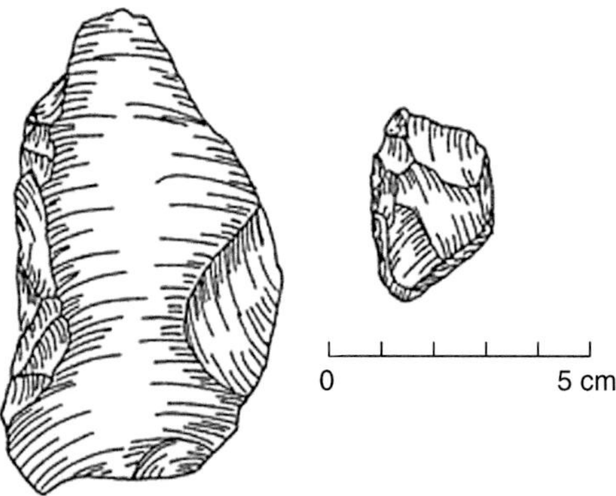

Figure 10.2. Two-million-year-old flake tools.

perhaps 1.5 Ma (de laTorre *et al.*, 2003). In localities where Oldowan sites have been carefully excavated, they are invariably found in association with permanent water, usually next to lakes. The typical habitat was gallery forest woodland. In these sites the tools are usually associated with animal bone, and analysis of cut marks on the bones indicates that the tools were often used for butchery and occasionally for breaking into bones for marrow. The specific body parts represented, and the presence of carnivore gnawing marks, indicate that a form of scavenging was more likely than hunting. The sites have no evidence for structures, or use of fire; indeed, there is no reason to suppose they were anything other than relatively protected spots to which the hominid scavengers carried meat and stone for tools. Paleoanthropologists most often assume the responsible hominid was an early form of *Homo*, though the contemporary form of robust *Australopithecus* might also have been responsible.

The tools themselves are very simple. Most consist of simple flakes removed from pebble cores (Figure 10.2). The hominids used a variety of raw material, with lavas, cherts and quartzites being the most common. These materials have good fracture qualities, a fact clearly appreciated by the hominids. Often, the knappers removed multiple flakes from the core, producing a core tool that was itself useful for breaking into bones for marrow (Figure 10.3). The hominids' primary concern, however, appears to have been the production of flakes with sharp edges (Schick and Toth, 2001; Toth, 1985). The hominids appear not to have been concerned with shape. Neither the core tools nor the flakes have any modification that implies that the knapper had a goal shape in mind. In some late Oldowan assemblages there are some flakes whose edges have been trimmed into a projection

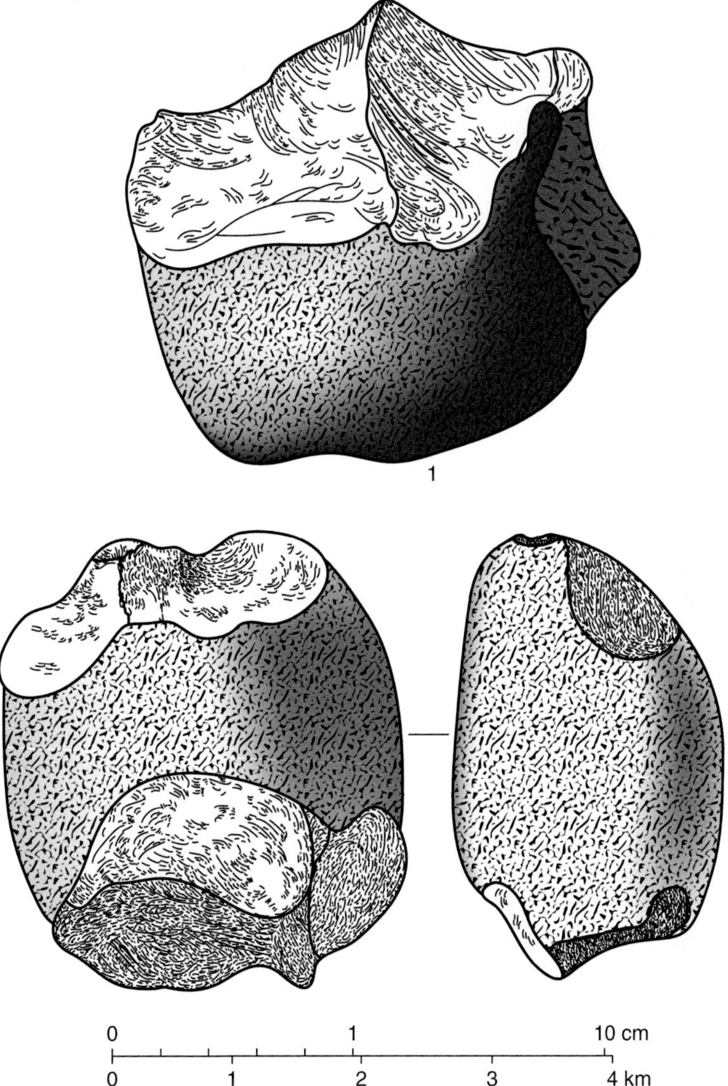

Figure 10.3. Two-million-year-old core tools.

(Figure 10.4) (Leakey, 1971), and at the very late site of Peninj ST one core was knapped in a way that suggests the hominids tried to control the configuration of the flaking surface (de laTorre *et al.*, 2003).

The spatial cognitive requirements for these tools are modest. At a descriptive level several abilities are evident. The knappers were clearly able to direct hammer blows to fairly specific locations on the

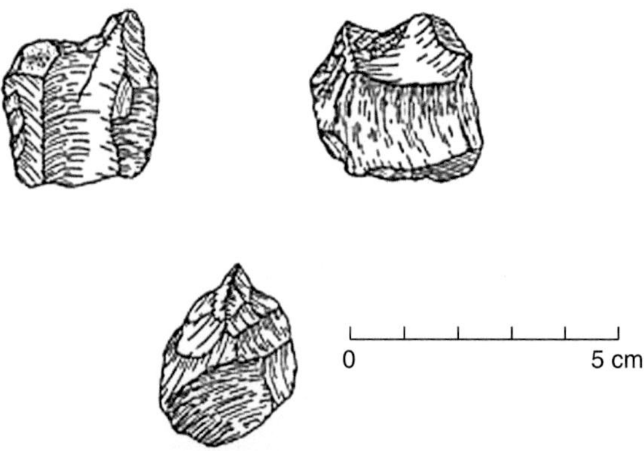

Figure 10.4. Two-million-year-old flake tools with trimmed projections.

cores. They did not simply bash randomly, but understood the nature of knapping, and the optimal places for delivering blows to produce the large, sharp flakes they preferred. They were also able to place blows in relationship to one another, especially when producing core tools with cutting edges (Figure 10.3). Here they placed blows next to previous blows (proximity in a topological sense) and were able to place blows on opposite sides of a continuous edge (boundary in topological terms). On the late examples of modified flakes the knappers delivered trimming blows in a sequence that maintained a direction (boundary + order topologically). We have already seen that apes are capable of knapping stone, though they are not as adept as these early hominids clearly were. Did these early hominids possess spatial abilities that were different from those of apes? If they did, it is not obvious. The cognitive underpinnings of these actions in space still appear very ape-like. Directing action on the spatial field of an object, following a spatial boundary, using proximity, and even ordering action in space are all in the spatial repertoire of apes (Wynn and McGrew, 1989). Toth *et al.* (1993), based on their work with Kanzi, have suggested that simple stone knapping represents a significant cognitive advance over the abilities of apes, but do not stipulate just what it is that changed. In terms of spatial cognition, it is possible that these early hominids were better at detecting "flakeable" edges on the complex background of a core. This is akin to the problems of "spatial visualization" posed by experimental psychologists studying spatial cognition. If true, this would represent a modest development in spatial cognition. However, the issues of skill and anatomical constraint complicate the issue (Stout, 2002). Modern apes are not able to produce directed, powerful blows because of the structure of their arms, hands and shoulders (Toth, personal communication, 1994). Their relative ineptitude may be simply biomechanical, not cognitive (though there is no sharp divide between these two).

The ape-like nature of these early tools has several important implications for the current argument. First, even though hominids had been present on the evolutionary stage for at least two million years prior to the appearance of stone tools, their spatial ability apparently remained at an ape grade. Second, even after the advent of lithic technology, few, if any, evolutionary developments in spatial cognition occurred for the next one million years. Ape spatial thinking was fully capable of directing hominid action for at least three million years. But about 1.5 million years ago the situation changed.

THE FIRST SHAPED TOOLS

Homo erectus (a.k.a. *Homo ergaster*) appeared in East Africa about 1.7–1.8 million years ago. Presumably an evolutionary descendant of earlier *Homo*, it possessed a number of characteristics that distinguish it clearly from its predecessor. Based on the remarkably complete "Turkana boy" from Nariokotome, it is clear that *Homo erectus* was taller than earlier *Homo*, had a larger cranial capacity, and a body well-adapted to strenuous activity in a hot climate (Walker and Leakey, 1993). *Homo erectus* also occupied a different adaptive niche. No longer confined to woodlands near standing water, *Homo erectus* moved out into hot, open savannas, and into higher, cooler elevations (Cachel and Harris, 1995). Indeed, the presence of *Homo erectus* fossils in European Georgia and Southeast Asia by 1.6 Ma suggests that *Homo erectus* expanded rapidly, relying on a very flexible adaptive niche. There is strong evidence for the use, and perhaps control, of fire, and continued reliance on meat, though evidence for hunting is equivocal. Anthropologists generally interpret this evidence in terms of greater reliance on learned, cultural activities. A key component of *Homo erectus* culture was lithic technology.

By 1.4 Ma a new form of lithic technology arose in Africa, for which archaeologists have given the term Early Acheulean. The defining feature of this technology was a large stone tool known generically as a "biface" (Figure 10.5). The first step in making a biface was usually the production of a large flake using a two-handed hammering technique to remove the flakes from a boulder-sized core (large flakes could weigh several pounds, but most were not this large). The knappers then trimmed the edges of the large flake "bifacially," which simply means

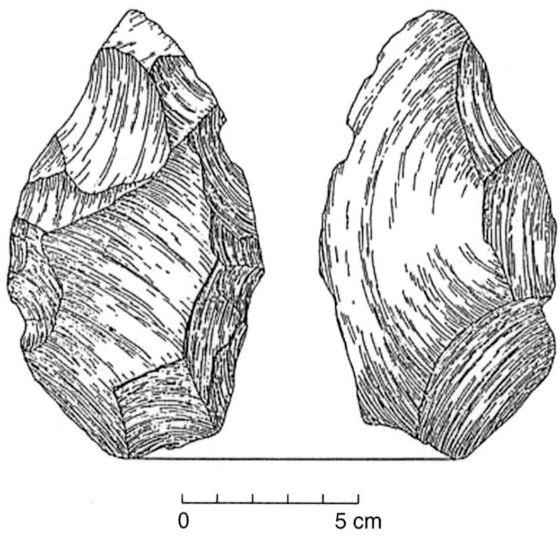

Figure 10.5. 1.4-million-year-old handaxe.

that smaller trimming flakes were struck onto both faces of the flake. Often, the knappers trimmed the entire edge of the flake bifacially, producing a large tool with an extensive edge. Bifaces come in two basic varieties, handaxes with converging sides forming a tip, and cleavers with a blade-like end (Figure 10.6). Bilateral (reflectional) symmetry is often apparent, and the position of trimming indicates that the symmetry was intended by the knappers (Wynn, 1995, 2002). Some early Acheulean assemblages have discoids, which are bifacially trimmed core tools with a round shape. There is some disagreement about the functions of the large bifaces. Based on experimental work using replicas, Nick Toth has argued that they were effective butchery tools, and this is the opinion held by most archaeologists (Schick and Toth, 2001; Toth, 1987). The neuroscientist William Calvin has argued that the handaxes were projectiles, and an important component in the evolution of aimed throwing, which he sees as key to the evolution of the human brain (Calvin, 1993). Beyond the bifaces, the stone tools of the Early Acheulean look very like earlier Oldowan tools – trimmed and untrimmed flakes, cores and amorphous core tools.

Homo erectus organized its action in space differently from earlier hominids, and this represents the first clear indication of evolution away from an ape grade of spatial cognition. At a descriptive level, the clearest development lies in the imposition of shape on objects. While earlier hominids modified the "business" edges of stone tools, they made no effort to impose an overall shape. Early Acheulean knappers

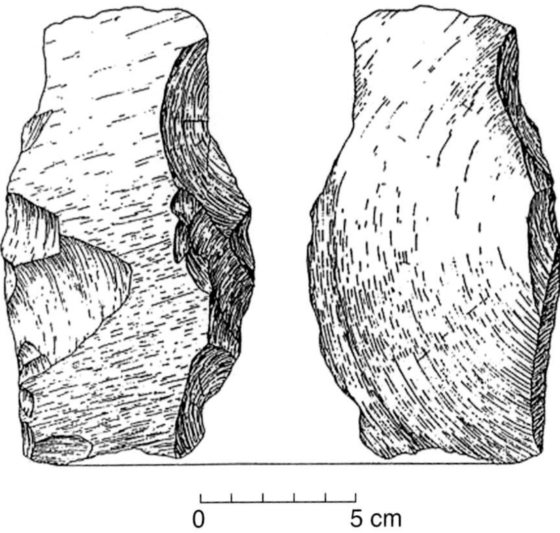

Figure 10.6. 1.4-million-year-old cleaver.

produced bilaterally symmetrical handaxes and cleavers, and round discoids. True, the symmetry is not a precise congruent symmetry, but the knappers clearly tried to mirror the shape of one lateral edge with the other. The shape is not a simple consequence of bifacial trimming. Moreover, there is no functional benefit to the symmetry; any large bifacially trimmed flake, of whatever shape, will do the task of butchery as well.

Specific cognitive interpretations of this development depend largely on one's theoretical framework. From a cognitive neuropsychological perspective the important breakthrough is the coordination of two heretofore separate neural pathways or networks – shape recognition and spatial cognition. Shape recognition is largely controlled by the ventral visual pathway (Albright *et al.*, 2000), and centered primarily in the left temporal region. Recall that a PET study of basic knapping revealed no significant activation of this region, but significant activation of the dorsal pathway, especially the right parietal; basic stone knapping is a spatial activity (Stout *et al.*, 2000). Recognition of faces, which have a symmetrical component, takes place in this left temporal region. Many vertebrates are adept at detecting symmetry, for good evolutionary reasons, but do not use the symmetry recognition network to guide specific motor actions. Early Acheulean knappers appear to have done so. *Homo erectus* introduced shape recognition considerations into the knapping task, which suggests either greater neural interconnectivity between these areas of the brain, or more access via frontal lobe functions. It is not necessary that the Acheulean knappers generated a fully conceived image of the final product before they began a knapping task. In fact the minimum requirement is that they attend to symmetry while knapping, and this could have been accomplished with implicit, even pre-attentive, application of mirroring (Wynn, 2002, with commentaries).

Alternative theoretical orientations might yield alternative explanations (see Wagman (2002) for a Gibsonian perspective, and Wynn (1989) for Piagetian interpretation), but all will need to account for the imposition of shape in an otherwise spatial task.

EUCLIDEAN SPACE

One of the most complex and least understood times in human evolution was the period between 500 and 200 thousand years ago (500–200 Ka). Prior to 500 Ka, *Homo erectus* grade fossils and archaeological remains (and spatial cognition) had characterized the hominid record for over one million years; indeed some have suggested that this was a period of stasis in human evolution (Gould, 2002). Whether this is an accurate description, or an over-interpretation, it does appear to have been a time of little change. Certainly no technological progress is evident. But after 500 Ka this monotonous picture fragmented. Hominid fossils from this period demonstrate a great deal of regional variability, along with the advent of a number of anatomical features of the face and cranium that are sapiens-like (Rightmire, 1998). This confusion of evidence has elicited a handful of names from paleoanthropologists: *Homo antecessor*, *Homo heidlebergensis*, *Homo helmii*, *Homo erectus* (in Asia), and archaic *Homo sapiens*. The behavioral picture from archaeology is little less confusing. A number of "firsts" characterize this period: first occupation of Northern Europe (Roebroeks *et al.*, 1992), first spears (Thieme, 1997), first extensive use of ochre (d'Errico, 2001), first clear evidence of large mammal hunting (Roberts and Parfitt, 1999; Singer *et al.*, 1993; Thieme, 1997), first evidence for possible regional differences in artifact shapes (Wynn and Tierson, 1990), and the earliest known carved object – the enigmatic Berekhat Ram "figurine," which is a modified lump of lava that is not convincingly depictive, but is certainly provocative (d'Errico and Nowell, 2000). The specific thread of human evolution is hard to trace through this knot of evidence; we do not know how many, or which, populations contributed to modern gene pools. However, from the

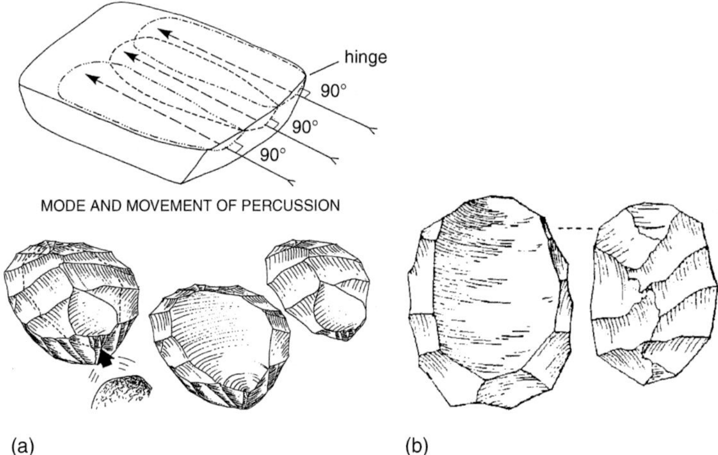

Figure 10.7. Levallois prepared core technique (after Boeda 1994).

softer focus of general evolutionary trends, it is clear that hominids emerged from this period much more modern in terms of anatomy and behavior than they entered it. One of the features that assumed modern status was spatial cognition. Many other aspects of behavior remained archaic: there are no organized campsites and no evidence for structured seasonal movement; hunting and gathering remained opportunistic and, perhaps most telling, there is no convincing evidence for true symbolic culture – no art, no ritual, no personal ornaments (Gamble, 1999).

During this period hominids perfected a number of techniques for flaking stone that are truly impressive, and which are among the most difficult for modern knappers to master. The best known of these is Levallois, which is a multi-step technique for producing large, thin flakes (Figure 10.7). The technique requires preparing a core to a fairly specific shape that facilitates the controlled removal of the final products. While there is no need to visualize a specific final flake shape, the entire procedure appears to have been guided by an asymmetric concept of the core mass that archaeologists have termed a "volumetric concept" (Boeda, 1995). By 200 Ka this "volumetric concept" was a key element in the repertoire of stone knappers. However, despite the complexity of prepared core techniques, the clearest evidence for spatial concepts comes from the bifaces.

Hominids in Africa and Europe continued to make handaxes and cleavers – they were clearly still effective tools – but the resulting artifacts are often much more regular in shape and "finer" in

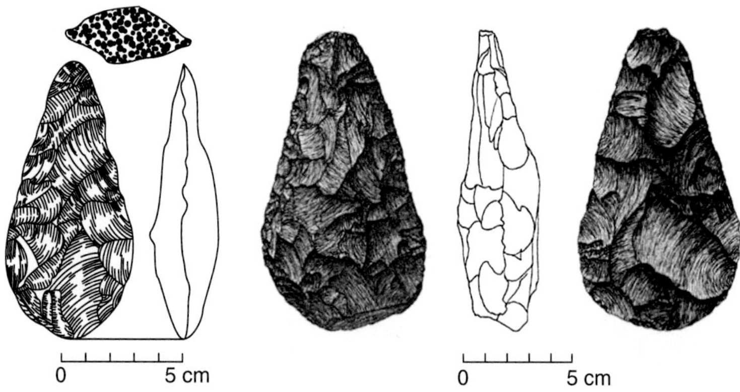

Figure 10.8. 300,000-year-old handaxes.

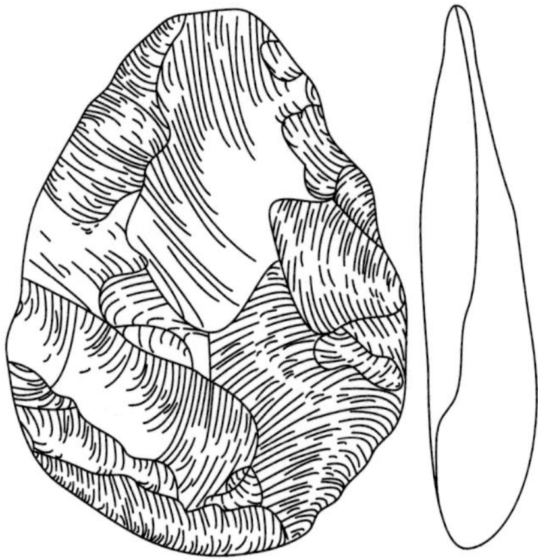

Figure 10.9. Handaxe with "twisted" profile.

appearance (Figure 10.8). Several specific developments are apparent. First, the symmetry was now much more precise, and in fact appears to have been guided by some appreciation of congruency, i.e., mirroring that preserves spatial size as well as overall configuration. Second, the symmetry imposed on bifaces was often three-dimensional – symmetry in plan, profile and cross-section – the result being some truly beautiful artifacts. Third, there appear to have been violations of symmetry, that is, intentional breaking of symmetry for visual effect. "Twisted profile" handaxes are the best examples (Figure 10.9) but there are cleavers whose axis appears "bent" (Figure 10.10). Not all bifaces from this period are beautiful, and some entire assemblages are dominated by unrefined, ugly bifaces. But there are other sites, whose knappers had

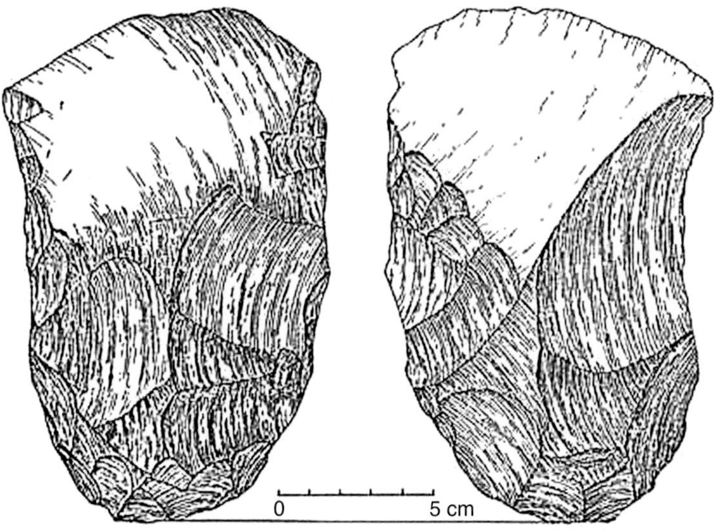

Figure 10.10. Cleaver with "bent" long axis.

access to high quality raw material, where fine examples were the norm. Two good examples are Kalambo Falls in Zambia (Clark, 2001) and Ma'ayan Baruch in Israel (Steklis and Gilead, 1966).

The stone tools suggest that the knappers relied on a number of spatial abilities not used by earlier *Homo erectus*. First, the three-dimensional symmetry required allocentric perception. The knapper needed to control the shape of the pieces from many different angles and points of view, some of which were not physically available (e.g., cross-sections that could not be directly sighted). Second, the congruency of shape suggests a component of spatial quantity in the notion of shape. Here a fairly specific spatial concept has been coordinated in the imposition of shape. From a cognitive neuropsychological perspective this suggests a more sophisticated coordination of the dorsal and ventral visual pathways. Indeed, it appears necessary to invoke visual imagery, in the sense described by Kosslyn (Kosslyn, 1994; Kosslyn *et al.*, 1994, 1997). The knappers of the fine three-dimensionally symmetrical bifaces almost certainly had a visual image of end product that guided their action.

The knappers appear to have guided their action using a Euclidean understanding of spatial relations, one in which spatial quantity is held as a constant and objects and actors move within an invariant spatial frame (Wynn, 1989). Certainly, animals have evolved to perceive such a visual world, but these hominids appear to have understood space in this way, and used this conception to control their action.

DISCUSSION

The most significant conclusion one can draw from this archaeological analysis is that human spatial cognition evolved long ago in conditions very different from those of the modern world, different even from those of modern hunters and gatherers. The first development away from an ape grade of spatial cognition occurred about 1.5 million years ago in the time of early *Homo erectus*. Despite being much more human-like anatomically than earlier hominids, *Homo erectus* had a significantly smaller brain than modern humans, and a concomitantly simple technology. Associated with this technology was the first evidence for coordination of spatial cognition and shape recognition. The second significant development in spatial cognition occurred about one million years later, and encompassed coordination of allocentric perception and size constancy into a Euclidean understanding of spatial relationships. Spatial cognition had attained its modern state. Yet the hominids responsible, likely an early form of *Homo sapiens*, were still very unlike modern humans. In particular, they exhibit no evidence for the rich symbolic culture that characterizes all modern humans. If we want to understand why humans possess their unique set of spatial understandings, we cannot simply study what spatial cognition can do in the modern world. The ability to rotate a three-dimensional solid mentally, and pick a solution from an array of alternatives, is not what spatial cognition is *about*.

In evolutionary analysis there are two different answers to why questions. The first posits that the character in question, e.g., Euclidean thinking, evolved to solve an evolutionary problem. In evolutionary terms, it was an adaptation. The second posits that the character evolved as a by-product of natural selection operating on another character or suite of characters. Given the complex developmental interrelationship of most characteristics, especially neural characteristics, it is difficult for selection to favor narrowly defined characteristics without affecting others (Gould, 2002). In order to conclude that a characteristic is the result of selection, it is necessary that it match closely the adaptive problem it supposedly solves. In this sense, documenting selection is "onerous," to use Williams' (1966) well-known term. The "default setting" is not that a characteristic is the result of selection. Selection must be demonstrated.

SELECTION FOR DOMAIN SPECIFIC ABILITIES

There are several current hypotheses that propose selection for fairly narrow, domain specific, features of human spatial cognition. The best developed is that of Silverman and colleagues (Eals and Silverman, 1994; Silverman *et al.*, 2000; Silverman and Eals, 1992), who have proposed a selective hypothesis for the well-established sex difference in spatial cognition, a hypothesis that also has implications for the evolution of human spatial cognition in general. The hypothesis suggests that the inherent division of labor in hunting and gathering societies has selected for different specific spatial abilities. Like most arguments in evolutionary psychology, it is based on the theoretical stance that the mind is massively modular and that natural selection has designed each module to solve a particular evolutionary problem (Thornhill, 1997). In the most recent version of this argument, Silverman *et al.* (2000) identify the narrow ability of "space constancy" as underpinning the sex differences in spatial cognition, and argue that the male need for effective wayfinding while hunting selected for the difference. They support their hypothesis with experimental evidence for a male advantage in wayfinding, and the statistical correlation between wayfinding and other tests of space constancy, mental rotation in particular. The research makes no reference to the paleoanthropological record, nor to the ethnographic record of wayfinding by modern hunters and gatherers. This apparent lacuna is consonant with a basic tenet of evolutionary psychology – that the evolutionary history of a trait is preserved in its design. If Silverman and colleagues can demonstrate that "space constancy" has been designed to solve the evolutionary problem of wayfinding, then, according to evolutionary psychology, no further support is necessary. Even though this hypothesis focuses on the sex difference in spatial cognition, it is also implicitly a selective hypothesis for a specific spatial ability – i.e., hunting selected for wayfinding ability and space constancy.[1] As such, the hypothesis can be checked against the paleoanthropological record.

The archaeological record lends some support to Silverman's emphasis on space constancy, but challenges features of his general model, especially the nature of the "environment of evolutionary adaptedness" (EEA). Space constancy encompasses "all of the processes involved in maintaining the stability of the surrounding environment while in locomotion" (Silverman *et al.*, 2000, p. 205). It includes maintenance of constant size and shape of objects under transformation, as in the construction of an image of the object from an alternative viewpoint. The archaeological evidence indicates that such space constancy was in the repertoire of the hominids who made the fine later Acheulean handaxes, but was not in the repertoire of early *Homo erectus*. Silverman *et al.* have identified an important element in the evolution of human spatial cognition, one that appears to have been associated with the advent of *Homo sapiens*. The archaeological record even supplies some weak support for their contention that hunting was part of the selective milieu – good evidence for hunting appears at the same time. Other features of Silverman's hypothesis do not fare as well in the presence of the archaeological record. A key feature of most evolutionary psychological arguments is the EEA, the environment of evolutionary adaptedness, which is modeled on presumed features of a hunter-gatherer past. The flaw here lies in the assumptions. Even if the characterization accurately described modern hunter-gatherers (and it probably doesn't), this form of adaptation is a very recent phenomenon in human evolution, perhaps no more that 12,000 years old (Wynn and Coolidge, 2003). The EEA for hominids at the time of the evolution of space constancy was actually very little like that of modern hunter-gatherers. In particular, there is no evidence at all for long-distance hunting trips, managed landscape hunting, or even large-scale cooperative hunts. Yes, the hominids hunted, but very opportunistically. We have no reason to conclude that they used advanced wayfinding. Much later, after 50 Ka or so, it is clear that humans did use such wayfinding, but this is long after the appearance of space constancy in the hominid cognitive repertoire.

SELECTION FOR DOMAIN GENERAL ABILITIES

Evolutionary psychology advocates a massively modular theory of cognition that grants little, if any, role to general abilities, Spearman's "g," or intelligence. There are of course alternative models of cognition that grant varying degrees of importance to general abilities, and these too can be invoked to make sense of the archaeological evidence. Indeed, at first blush the archaeological record makes more sense from a domain general perspective than an extreme modular perspective. After all, the first move away from ape spatial abilities appears to have been the coordination of information from two different neural pathways – shape recognition and spatial cognition – and this would seem to imply some central processing. In fact, Kosslyn (1994) makes just this argument for mental imagery. In defense of the evolutionary psychologists, it must be noted that a massively modular mind does not logically require a brain that is massively localized (Samuels, 2000). In theory at least, a cognitive module could be neurologically diffuse. Nevertheless, the cognitive neuropsychological literature does appear more amenable to modularity in the original Fodorian sense of some encapsulated input/output systems, and some central control (Albright *et al.*, 2000; Fodor, 1983). One of the most influential domain general cognitive theories of the last twenty years has been Baddeley's model of working memory, and it provides a perspective on the archaeological evidence that is rather different from that provided by evolutionary psychology.

Working memory (WM), as described by Baddeley (Baddeley, 1986, 2001; Baddeley and Logie, 1999), is the mechanism by which the mind holds in attention and processes information provided by other neural sources, including perception and long-term memory. In Baddeley's formulation WM is a hierarchical system in which a central executive coordinates inputs from two slave systems, a phonological loop for temporary storage of phonological based information (e.g., words) and a visuospatial sketchpad (also known as visuospatial working memory) for the temporary storage of visuospatial information (Baddeley and Logie, 1999). Features of the basic model remain controversial, especially the status of the central executive, but the distinction between the phonological and visuospatial components has been confirmed many times. Moreover, brain imaging studies indicate that phonological and visuospatial tasks of WM are processed by separate neural networks, both of which include prefrontal lobe activation (Gruber and von Cramon, 2003; Miyake *et al.*, 2001). Working memory in general appears to be the cognitive system underpinning executive functions, a suite of behavioral abilities long associated with frontal lobe function (Barkley, 2001; Kane and Engle, 2002), and even general intelligence (Kyllonen, 1996). It is a prime candidate for evolutionary understanding, and its role in human evolution may have been pivotal (Coolidge and Wynn, 2001; Wynn and Coolidge, 2003).

The nature and role of the visuospatial sketchpad (VSSP) is less well researched and understood than that of the phonological loop, but experimental work by Logie (1995) has identified some basic features. As the term implies, the VSSP functions to hold visual (e.g., images) and spatial information (e.g., location and relative position) in temporary storage and attention where they can be further processed. Experiments using interference tasks (e.g., tracing a path while performing mental rotation) indicates that visual and spatial information are processed separately by the VSSP, though, interestingly, a recent fMRI study indicates a unitary neural substrate for the VSSP (Gruber and von Cramon, 2003). The VSSP also appears more difficult than the phonological store to isolate experimentally from central executive functioning (Miyake *et al.*, 2001). The apparent separation of visual and spatial information has interesting evolutionary implications, as does the close association with central executive functions.

Barkley (2001) has proposed an evolutionary model for executive functions in general. In particular, he has argued that abilities in self-regulation (e.g., placing oneself in future social scenarios) provided a selective advantage in the social context of early hominid behavior. Coolidge and Wynn (2001) have made the more specific proposal that enhanced executive abilities in contingency planning and response inhibition were the key developments in the evolution of modern humans, and have recently recast this argument in terms of enhanced working memory (Wynn and Coolidge, 2003). In both Barkley, and Coolidge and Wynn, the emphasis was on phonological based information, with little discussion of visuospatial abilities, largely because the more recent evolutionary developments, as recognized archaeologically, appear to have been in the domain of language, symbol use, and narrative (Coolidge and Wynn, 2005).

The archaeological evidence suggests that the visuospatial component or working memory may be much older, in an evolutionary sense, than the phonological component. The evidence from early Acheulean bifaces indicates that an ability to coordinate visual (shape recognition) and spatial (stone knapping; flake removal positions) was in the hominid repertoire by 1.5 million years ago. Given that this coordination is currently performed by the VSSP of working memory, it is parsimonious to conclude that this piece of WM evolved long ago with the advent of *Homo erectus*. The emergence of Euclidean spatial ability 400,000 years ago is more difficult to formulate in working memory terms. It could simply have been the result of enhanced capacity of the VSSP (which would fit nicely into Silverman's hypothesis for space constancy), or be the result of the evolution of a nascent phonological loop where shapes and norms could be held and processed as semantic, declarative categories. The variety and sophistication of contemporary stone-knapping procedures, including the Levallois technique in which the careful core preparation determines the final shape of one flake, indicate that these hominids, presumably early *Homo sapiens*, were capable of fairly complex plans of action that were almost certainly the domain of working memory. In other words, the archaeological evidence suggests that some aspects of modern working memory had evolved by 400,000 years ago, a scenario in which spatial abilities per se need not have been specifically selected.

NOTE

1. For purposes of this chapter I have avoided discussion of the sex difference issue. Elsewhere (Wynn *et al.* 1996) I have suggested that the paleoanthropological record supports a scenario in which the sex difference evolved as a by-product of the evolution of the timing of fetal development, fetal androgens in particular. However, the cause or causes behind the sex difference in spatial cognition are not well understood; neurological, cognitive, hormonal and life experience factors have all been implicated. For a recent review see Halpern and Collaer (2005).

REFERENCES

Albright, T. D., Jessell, T. M., Kandel, E. and Posner, M. (2000). Neural science: a century of progress and the mysteries that remain. *Cell*, **100**, S1–S55.

Baddeley, A. (1986). *Working memory*. London: Clarendon.

Baddeley, A. (2001). Is working memory working? *American Psychologist*, November, 851–864.

Baddeley, A. and Logie, R. (1999). Working memory: the multiple-component model. In A. Miyake and P. Shah (Eds.), *Models of working: mechanisms of active maintenance and executive control* (pp. 28–61). Cambridge: Cambridge University Press.

Barkley, R. (2001). The executive functions and self-regulation: an evolutionary neuropsychological perspective. *Neuropsychology Review*, **11** (1), 1–29.

Boeda, E. (1994). *Le concept Levallois: variabilite des methodes*. Paris: CNRS Editions.

Boeda, E. (1995). Levallois: a volumetric construction, methods, a technique. In H. L. Dibble and O. Bar-Yosef (Eds.), *The definition and interpretation of Levallois technology*. Madison: Prehistory Press.

Boesch, C. (2003). Is culture a golden barrier between human and chimpanzee? *Evolutionary Anthropology*, **12** (2), 82–91.

Boesch, C. and Boesch, H. (1990). Tool use and tool making in wild chimpanzees. *Folia Primatologica*, **54**, 86–89.

Cachel, S. and Harris, J. (1995). Ranging patterns, land-use and subsistence in *Homo erectus* from the perspective of evolutionary biology. In J. Bower and S. Sartono (Eds.), *Evolution and ecology of Homo erectus* (pp. 51–66). Leiden: Pithecanthropus Centennial Foundation.

Calvin, W. (1993). The unitary hypothesis: a common neural circuitry for novel manipulations, language, plan-ahead, and throwing? In K. Gibson and T. Ingold (Eds.), *Tools, language, and cognition in human evolution* (pp. 230–250). Cambridge: Cambridge University Press.

Clark, J. D. (Ed.) (2001). *Kalambo Falls prehistoric site, vol. III*. Cambridge: Cambridge University Press.

Coolidge, F. and Wynn, T. (2001). Executive functions of the frontal lobes and the evolutionary ascendancy of *Homo sapiens*. *Cambridge Archaeological Journal*, **11**, 255–260.

Coolidge, F. and Wynn, T. (2005). Working memory, its executive functions, and the emergence of modern thinking. *Cambridge Archaeological Journal*, **15** (1), 5–26.

de laTorre, I., Mora, R., Dominguez-Rodrigo, M., de Luque, L. and Alcala, L. (2003). The Oldowan industry of Peninj and its bearing on the reconstruction of the technological skill of Lower Pleistocene hominids. *Journal of Human Evolution*, **44**, 203–224.

d'Errico, F. (2001). Memories out of mind: the archaeology of the oldest memory systems. In A. Nowell (Ed.), *In the mind's eye: multidisciplinary approaches to the evolution of human cognition* (pp. 33–49). Ann Arbor: International Monographs in Prehistory.

d'Errico, F. and Nowell, A. (2000). A new look at the Berekhat Ram figurine: implications for the origins of symbolism. *Cambridge Archaeological Journal*, **10**, 123–167.

Eals, M. and Silverman, I. (1994). The hunter-gatherer theory of spatial sex differences: proximate factors mediating the female advantage in recall of object arrays. *Ethology and Sociobiology*, **15**, 95–105.

Fodor, J. (1983). *The modularity of mind: an essay on faculty psychology*. Cambridge, MA: MIT Press.

Gamble, C. (1999). *The Palaeolithic societies of Europe*. Cambridge: Cambridge University Press.

Gould, S. (2002). *The structure of evolutionary theory*. Cambridge, MA: Belknap.

Gruber, O. and von Cramon, D. Y. (2003). The functional neuroanatomy of human working memory revisited: evidence from 3-T fMRI studies using classical domain-specific interference tasks. *NeuroImage*, **19** (3), 797–209.

Halpern, D. and Collaer, M. (2005). Sex differences in visuospatial abilities: more than meets the eye. In P. Shah and A. Miyake (Eds.), *The Cambridge handbook of visuospatial thinking* (pp. 170–212). Cambridge: Cambridge University Press.

Harris, J. W. K. (1983). Cultural beginnings: Plio-Pleistocene archaeological occurrences from the Afar, Ethiopia. *African Archaeological Review*, **1**, 3–31.

Kane, M. and Engle, R. (2002). The role of prefrontal cortex in working-memory capacity, executive attention, and general fluid intelligence: an individual-differences perspective. *Psychonomic Bulletin and Review*, **9** (4), 637–671.

Kosslyn, S. (1994). *Image and brain: the resolution of the imagery debate*. Cambridge, MA: MIT Press.

Kosslyn, S. M., Thompson, W. L. and Alpert, N. M. (1997). Neural systems shared by visual imagery and visual perception: a positron emission tomography study. *Neuroimage*, **6**, 320–334.

Kosslyn, S. M., Alpert, N. M., Thompson, W. L., Chabris, C. F., Rauch, S. L. and Anderson, A. K. (1994). Identifying objects seen from different viewpoints: a PET investigation. *Brain*, **117**, 1055–1071.

Kyllonen, P. (1996). Is working memory capacity Spearman's g? In I. Dennis and P. Tapsfield (Eds.), *Human abilities: their nature and measurement* (pp. 49–76). Mahwah, NJ: Lawrence Erlbaum.

Leakey, M. (1971). *Olduvai Gorge Vol. 3*. Cambridge: Cambridge University Press.

Logie, R. (1995). *Visuo-spatial working memory*. Hillsdale: Lawrence Erlbaum.

Mercader, J., Panger, M. and Boesch, C. (2002). Excavation of a chimpanzee stone tool site in the African rainforest. *Science*, **296**, 1452–1455.

Miyake, A., Friedman, N. P., Rettinger, D. A., Shah, P. and Hegarty, M. (2001). How are visuospatial working memory, executive functioning, and spatial abilities related? A latent-variable analysis. *Journal of Experimental Psychology: General*, **130** (4), 621–640.

Nowell, A. (Ed.) (2001). *In the mind's eye: multidisciplinary approaches to the evolution of human cognition*. Ann Arbor: International Monographs in Prehistory.

Rightmire, G. P. (1998). Human Pleistocene in the Middle Pleistocene: the role of *Homo heidelbergensis*. *Evolutionary Anthropology*, **6** (6), 218–227.

Roberts, M. A. and Parfitt, S. A. (1999). *Boxgrove: a Middle Pleistocene hominid site at Eartham Quarry, Boxgrove, West Sussex*. London: English Heritage.

Roebroeks, W., Conard, N. J. and Kolfschoten, T. v. (1992). Dense forests, cold steppes, and the Palaeolithic settlement of northern Europe. *Current Anthropology*, **33** (5), 551–586.

Samuels, R. (2000). Massively modular minds: evolutionary psychology and cognitive architecture. In P. Carruthers and A. Chamberlain (Eds.), *Evolution and the human mind: modularity, language, and meta-cognition* (pp. 13–46). Cambridge: Cambridge University Press.

Schick, K. and Toth, N. (2001). Paleoanthropology at the millennium. In T. D. Price and G. Feinman (Eds.), *Archaeology at the millennium: a sourcebook* (pp. 39–108). New York: Plenum.

Schick, K., Toth, N., Garufi, G., Savage-Rumbaugh, E., Rumbaugh, D. and Sevcik, R. (1999). Continuing investigations into the stone tool-making capabilities of a Bonobo (*Pan paniscus*). *Journal of Archaeological Science*, **26**, 821–832.

Semaw, S. (2000). The world's oldest stone artefacts from Gona, Ethiopia: their implications for understanding stone technology and patterns of human evolution between 2.6–1.5 million years ago. *Journal of Archaeological Science*, **27**, 1197–1214.

Semaw, S., Renne, P., Harris, J., Feibel, C. S., Bernor, R., Fesseha, N., *et al.* (1997). 2.5-million-year-old stone tools from Gona, Ethiopia. *Nature*, **385**, 333–336.

Silverman, I. and Eals, M. (1992). Sex differences in spatial abilities: evolutionary theory and data. In J. Barkow, L. Cosmides and J. Tooby (Eds.), *The adapted mind: evolutionary psychology and the generation of culture* (pp. 487–503). New York: Oxford University Press.

Silverman, I., Choi, J., Mackewn, A., Fisher, M., Moro, J. and Olshansky, E. (2000). Evolved mechanisms underlying wayfinding: further studies on the hunter-gatherer theory of spatial sex differences. *Evolution and Human Behavior*, **21**, 201–213.

Singer, R., Gladfelter, B. G. and Wymer, J. J. (1993). *The Lower Paleolithic Site at Hoxne, England*. Chicago: University of Chicago Press.

Steklis, M. and Gilead, D. (1966). *Ma'ayan Barukh: a lower palaeolithic site in Upper Galilee*. Jerusalem: Jerusalem Center for Prehisoric Research.

Stout, D. (2002). Skill and cognition in stone tool production. *Current Anthropology*, **43** (5), 693–722.

Stout, D., Toth, N., Schick, K., Stout, J. and Hutchins, G. (2000). Stone tool-making and brain activation: positron emission tomography (PET) studies. *Journal of Archaeological Science*, **27**, 1215–1223.

Thieme, H. (1997). Lower Palaeolithic hunting spears from Germany. *Nature*, **385**, 807–810.

Thornhill, R. (1997). The concept of an evolved adaptation. In G. Bock and G. Cardew (Eds.), *Characterizing human psychological Adaptations* (pp. 23–38). New York: John Wiley & Sons.

Toth, N. (1985). Archaeological evidence for preferential right-handedness in the Lower and Middle Pleistocene, and its possible implications. *Journal of Human Evolution*, **14**, 607–614.

Toth, N. (1987). Behavioral inferences from Early Stone Age artifact assemblages: an experimental model. *Journal of Human Evolution*, **16**, 763–787.

Toth, N., Schick, K., Savage-Rumbaugh, S., Sevcik, R. and Savage-Rumbaugh, D. (1993). Pan the tool-maker: investigations into the stone tool-making and tool-using capabilties of a bonobo (*Pan paniscus*). *Journal of Archaeological Science*, **20**, 81–91.

Wagman, J. (2002). Symmetry for the sake of symmetry, or symmetry for the sake of behavior. *Behavioral and Brain Sciences*, **25** (3), 422–423.

Walker, A. and Leakey, R. (Eds.) (1993). *The Nariokotome Homo erectus skeleton.* Cambridge: Harvard University Press.

Williams, G.C. (1966). *Adaptation and natural selection.* Princeton: Princeton University Press.

Wright, R. (1972). Imitative learning of flaked tool technology: the case of an orangutan. *Mankind*, **8**, 296–306.

Wynn, T. (1989). *The evolution of spatial competence.* Urbana: University of Illinois Press.

Wynn, T. (1995). Handaxe enigmas. *World Archaeology*, **27**, 10–24.

Wynn, T. (2002). Archaeology and cognitive evolution. *Behavioral and Brain Sciences*, **25** (3), 389–438.

Wynn, T. and Coolidge, F. (2003). The role of working memory in the evolution of managed foraging. *Before Farming*, **2** (1), 1–16.

Wynn, T. and McGrew, W.C. (1989). An ape's view of the Oldowan. *Man*, **24**, 283–298.

Wynn, T. and Tierson, F. (1990). Regional comparison of the shapes of later Acheulean handaxes. *American Anthropologist*, **92**, 73–84.

Wynn, T., Tierson, F. and Palmer, C. (1996). Evolution of sex differences in spatial cognition. *Yearbook of Physical Anthropology*, **39**, 11–42.

LUCIO REHBEIN, STEVE SCHETTLER, RONALD KILLIANY AND MARK B. MOSS

11

Egocentric and allocentric spatial learning in the nonhuman primate

INTRODUCTION

Visuospatial abilities, including spatial perception and spatial memory, are essential in our performance of activities of daily living. We rely on spatial functions to navigate successfully through our environment. Success in guiding our behavior is dependent on visual landmarks in the environment, i.e., allocentric spatial cues, as well as those that use ourselves as the frame of reference, i.e., egocentric cues. In most instances, it is presumed we rely more on one source than the other, and certainly in some instances, the two may be in conflict. Learning one's way driving in a new environment, as demonstrated in a recent novel study of taxicab drivers (Newman *et al.*, 2007) points out this competitive spatial dynamic.

Unfortunately, the study of the neurobiological basis of spatial behavior in humans has not received the same level of attention in the literature as other domains of higher cortical function. For example, much more is known about the neurobiological bases of memory and executive functions than of spatial functions. Indeed, following brain trauma, insult or disease, spatial cognition is often neglected in the symptomatic work-up.

By contrast, research over many years using animal (often rodent) models in behavioral and single unit studies has implicated several brain regions, including the frontal and parietal association areas, the basal ganglia, and the hippocampal formation, in the neural bases of spatial function. Among these, it is the hippocampal formation that appears to play a pivotal role in the encoding and mapping of one's location in space (O'Keefe and Nadel, 1978). This chapter will focus primarily on the role of the hippocampal formation in spatial cognition in nonhuman primates, most often generalizing from studies on

Spatial Cognition, Spatial Perception: Mapping the Self and Space, ed. Francine L. Dolins and Robert W. Mitchell. Published by Cambridge University Press.

the rhesus monkey (*M. mulatta*) as an animal model of human spatial cognition. Prior to discussing the role of the hippocampal formation in spatial cognition, a brief summary of non-limbic structures that may also play a role is briefly outlined in a separate section below. Following sections on the studies of the hippocampal formation, sections are devoted to the sparse literature on the effects of age and gender differences on spatial function.

NON-LIMBIC REGIONS AND SPATIAL COGNITION

Spatial cognition is subserved by several regions of the brain. Jacobsen (1935) was perhaps the first to demonstrate that selective brain damage in nonhuman primates may produce an impairment in spatial learning. He showed that baboons with damage to the prefrontal cortices failed on delayed response, a task that included a significant spatial component (for further suport, see Goldman-Rakic, 1998; and Owen *et al.*, 1996). Pohl (1973) extended the role of association cortices in spatial learning when he demonstrated an anatomical dissociation between spatial learning in egocentric space (relying on one's own body as the reference to guide behavior) from that of allocentric space (using visual cues in the external world to guide behavior). Dorsolateral prefrontal cortical damage impaired performance on a left–right egocentric discrimination task, but did not affect performance on an allocentric landmark task. Conversely, posterior parietal damage impaired performance on the landmark discrimination task but not on the left–right egocentric discrimination task. Evidence that activity in neurons recorded from the monkey posterior cingulate cortex is more closely related to a target's position in a room than the target's position relative to the monkey (Dean and Platt, 2006) provides further support for the involvement of parietal association areas in mediating spatial learning in allocentric space, as suggested by Pohl's lesion study. Other regions of the brain, including the anterior cingulate cortex, have been shown to play a less direct role in spatial learning (Meunier *et al.*, 1997).

THE HIPPOCAMPUS AND SPATIAL COGNITION IN NONHUMAN PRIMATES

Although the role of the primate hippocampal formation in higher cognitive function is becoming better understood (e.g., Murray *et al.*, 2000; Suzuki and Amaral, 2004), its contribution to the representation and memory of spatial location, an inherent component of several aspects of cognition, has yet to be clearly defined. O'Keefe and Nadel's (1978) seminal model of the hippocampal formation as a key neuroanatomic locus for spatial cognition provides a provocative framework for the discussion studies of spatial cognition in monkeys with damage to the hippocampal formation. According to O'Keefe and Nadel, there are two general forms of space perception: absolute, or non-egocentric, and relative, or egocentric. They defined absolute space as a "unitary, objective space in which the position of the organism [does] not affect the distribution of objects represented with that space" (p. 381). They defined egocentric or relative space as "a variety of spaces defined by the relation between the organism and external objects, or between different parts of the organism" (p. 381). Their idea of relative space includes not only what we term "egocentric" space, but also allocentric space and the space objects occupy. According to their model, the hippocampus is concerned with the non-egocentric, or absolute, space; it receives, but does not store, inputs from systems concerned with relative space. In their terms, the definition of absolute space "embodies the notion of a framework or container within which material objects can be located but which is conceived as existing independently of particular objects or objects in general" (p. 7). The studies of hippocampal function described below are considered in this context accordingly.

Ablation studies of the hippocampal formation

Early studies

Perhaps the first compelling work on spatial learning following damage to the hippocampal formation came from studies by Mahut (1971, 1972). In a series of tasks designed to parallel those used in the clinic (Penfield and Milner, 1958; Scoville, 1954; Scoville and Milner, 1957), Mahut showed that monkeys with medial temporal lobe resections were impaired on the spatial delayed alternation task. In this task, monkeys were required to choose, alternately, either the left or the right of two food-wells, each covered by identical black plaques, to obtain reward. Monkeys with medial temporal lobe lesions were markedly impaired in the performance of this task. But even these findings have to be viewed with some caution regarding a possible brain region that may be pivotal in spatial cognition. Since the delayed alternation task is confounded by a demand on memory, albeit with only a short delay, it is difficult to attribute with confidence the deficit to one of spatial cognition alone.

In order to address the potential confound, Mahut and her students chose a discrimination reversal task that could be administered both in a spatial as well as a nonspatial form. Successful performance on this task requires the monkey to obtain a reward under the left or right of two identical plaques. These investigators were the first to show that adult rhesus monkeys with either hippocampal, fornix or entorhinal damage, were severely impaired on reversal learning using spatial cues as the discriminative stimuli, but were unimpaired when a nonspatial object paradigm was used (Mahut and Zola, 1976; 1977; Moss *et al.*, 1981). Hence this was the first demonstration that the hippocampal formation, and its major afferent and efferent systems, may play a selective role in spatial learning, though the results do not dissociate the attributes of absolute or relative spatial cognition as proposed by O'Keefe and Nadel.

Landmark tasks

The above findings from spatial reversal learning suggested strongly that monkeys with selective damage to the medial temporal lobe, specifically the hippocampal formation and both of its major afferent (entorhinal cortex) and efferent (fornix) systems, plays a role in spatial learning and specifically in successfully using egocentric cues in the environment in guiding learned behavior. In an attempt to expand on this initial finding, a different approach was used to see whether the spatial learning of surgically manipulated monkeys would improve if the location of the correct choice of one of two identical objects could be discerned relative to an allocentric referent, rather than an egocentric one. In a series of experiments conducted by Rehbein (1985a, 1985b; Rehbein and Mahut, 1983), monkeys that received lesions of the hippocampus or transactions of the fornix early in life (approximately at two months of age) were tested as adults on two new spatial tasks that differed from the delayed alternation task in that they could be solved by the use of an alternative spatial strategy. With the delayed alternation and spatial discrimination learning task, monkeys had to rely on their ability to discern the left from the right food well in reference to their own body (egocentric cues). Therefore, in a second experiment, a single distinct object (landmark) was placed off-center on the testing board, and hence provided an external referent (allocentric cue) to help discern left from right. It was presumed that monkeys that otherwise could not guide their responses relative to their own body, could now do so with the presence of an allocentric cue. This task required the monkeys to shift their responses from the left to the right one of two identical plaques, presumably with no additional cues available other than the orientation of their previous response relative to their own body (egocentric cues).

In a second experiment, paralleling that of Pohl (1973), a three-dimensional object (landmark) was added onto the testing tray. The landmark was expected to guide the monkeys' responses without the need for them to depend, exclusively, on an egocentric frame of reference. However, since in Pohl's (1973) study, the landmark was always in physical contact with one of two identical plaques, it is possible that monkeys viewed the display as a complex visual configuration, which could have masked the spatial nature of the task. In order to avoid this possible confound, Rehbein (1985a, 1985b) used a testing tray with eighteen food wells and moved the location of two identical objects and the landmark relative to one another from one trial to the next. The object to the right (or the left) of the landmark was to be the positive one, regardless of the visual configuration of the three elements on any given trial. An illustration of the testing tray is presented in Figure 11.1. If the spatial performance deficit of monkeys with hippocampal or fornix damage was linked with an inability to utilize an egocentric frame of reference, then no impairment was expected on the landmark discrimination task because the egocentric framework would not play a role. Additionally, in order to discover whether the visual landmark was in fact responsible for lack of impaired performance, we decided to assess the effect of removing the landmark during an additional testing session, after learning criterion was reached.

All monkeys learned the landmark task rapidly, within two to thirteen sessions, and no significant group differences in performance were found. However, monkeys in Group H (hippocampal lesions) obtained significantly higher error scores on transfer trials during the additional session without the landmark in which the egocentric framework plays an important role in successful performance. These data are summarized in Table 11.1.

The landmark was expected to help monkeys in the hippocampal group to perform as well as did those in the normal group by, presumably, allowing them to respond under visual guidance in reference to an allocentric cue. However, in hindsight it was apparent that, while it was true that the positive object was always to the right (or the left) of the landmark as intended, it was also, unfortunately, always to the left (or the right) of the monkey. Therefore, the task shared an important

Table 11.1. *Learning scores obtained by normal and operated monkeys on the spatial (left–right) discrimination task with landmark.*

Groups	n	Mean trials	Mean errors	Errors on transfer
Normal control	5	96	26	0.6
Hippocampal	7	158	36	3.2*

*Operated monkeys made significantly more errors in transfer than did monkeys in the normal control group ($p < 0.05$).

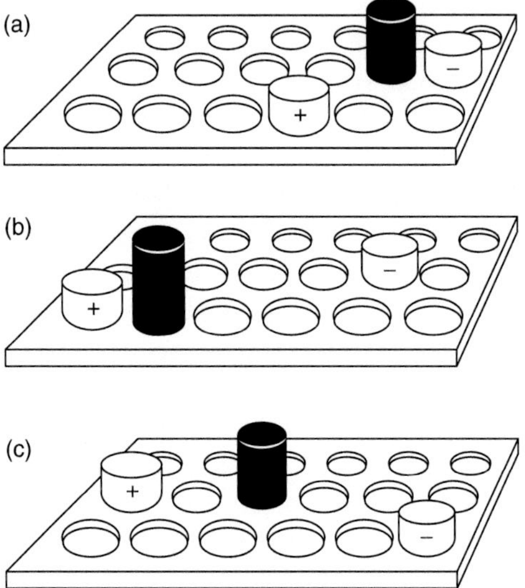

Figure 11.1. Illustration of the procedure used in the landmark discrimination task, with examples of three trials. A: the landmark and the two objects to one side of the center of the tray. B: the landmark, together with one of the objects, on one side of the center of the tray and the other object on the opposite side. C: all three objects dispersed across the tray. "+": denotes presence of food reward. "–": denotes absence of food reward.

egocentric component with the spatial reversal and delayed alternation tasks. However, that the landmark did contribute, at least in part, to more accurate performance was indicated by the deterioration in the accuracy of performance of monkeys in the hippocampal group in its absence. However, this contribution was not strong enough to readily allow monkeys to develop a guidance strategy in allocentric space as evidenced by their inability to perform at the same level of accuracy of normal control animals.

ASYMMETRIC LANDMARK DISCRIMINATION TASK

In the next task in this series of studies, a task was designed to provide a more effective means of guiding the monkey in allocentric spatial learning. Unlike the cylindrical landmark used in the preceding experiment that had no distinct sides but those extrapolated by the observer, it was decided to provide a new landmark that was asymmetric in shape, so that it would have its own orientation in space. This condition was achieved by using a landmark in the shape of an arrow, mounted on a rotating tray, so that no matter how the visual array was rotated, the relationship between the head and body of the arrow remained unchanged. This time the landmark would have its own changing orientation, independent of the observer's point of view, and one of two identical objects would truly be to *its* right (or *its* left), rather than to the right (or left) of the animal's body (see Figure 11.2). Under these modified conditions, it was predicted that monkeys with ablation of the hippocampus would not evidence impaired performance.

In fact, this time monkeys in the hippocampal group (Group H) were not only unimpaired, but also made significantly fewer errors than did those in the normal control group (Group N). These results showed that the two groups of operated monkeys were effectively helped by the presence of an asymmetric visual landmark in allocentric space. A summary of these data is presented in Table 11.2.

An analysis of error patterns showed that the observed facilitation in performance by operated monkeys could be accounted for by at least two factors: (1) normal monkeys spent, on average, four days

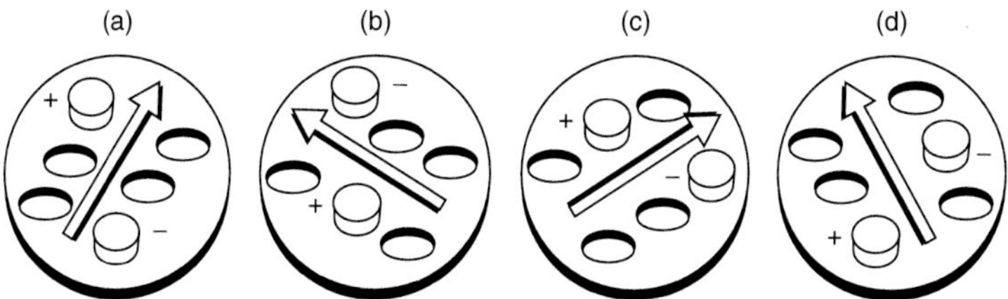

Figure 11.2. Illustration of the procedure used in the spatial discrimination task with asymmetric landmark with examples of four trials. Letters A, B, C and D: denote landmark orientations set at 1, 10, 2 and 11 o'clock, respectively. "+": denotes presence of food reward. "–": denotes absence of food reward.

Table 11.2. *Learning scores obtained by normal and operated monkeys on the spatial discrimination task with* asymmetric landmark.

Groups	n	Mean trials	Mean errors
Normal control	5	288	69.80
Hippocampal	7	185	34.28*

* Operated monkeys made significantly fewer errors to learn the task than did monkeys in the normal control group ($p < 0.05$).

approaching the task with a readily available, but ineffective, egocentric, near-far spatial hypothesis; (2) the great majority of monkeys in the two experimental groups did not adopt a systematic egocentric near–far, or left–right, spatial strategy, but were able to develop a guidance strategy, presumably based on the cues provided by the landmark.

The results obtained with the asymmetric landmark were unequivocal. Monkeys in the hippocampal group, without a single exception, learned the task in fewer trials and errors than did normal monkeys. The critical factor in this task seemed to have been the structural asymmetry of the landmark itself which lent it an orientation in space and which did not depend upon the superimposition or extrapolation of an egocentric frame of reference as seemed to occur with the cylindrical (symmetric) landmark used in the first landmark experiment.

The present study constituted the first attempt to assess the role of visual landmarks in the spatial learning of monkeys with selective ablations of the hippocampus. The ability to perform successfully on the landmark task appears to rely on spatial information that is allocentrically, rather than egocentrically, based in a schema of relative space, and hence, this capacity appears preserved in the absence of an intact hippocampus. A more recent study by Hampton *et al.* (2003) supports the view that monkeys with hippocampal ablations are able to use allocentric cues to find reward in an open field test condition.

Take together, these findings are consistent with one prediction derived from O'Keefe and Nadel's (1978) theory of hippocampal function: namely that animals deprived of the hippocampus can still learn about space by using guidance (or orientation) route learning strategies that fall within a reference system mediated by structures outside of the hippocampal formation. Findings from two earlier studies in which landmarks were used (Pohl, 1973; Ungerleider and Brody, 1977) indicate a posterior parietal cortical locus may represent one of these extrahippocampal areas that mediate landmark discrimination learning. However, in view of the experimental sophistication of monkeys in the present study, and of the differential effects of early and late ablations of the hippocampus on other behavioral functions, the notion that that the mediation of allocentric vs. egocentric space is anatomically dissociable in the adult must await the postoperative administration of the landmark task to older hippocampectomized monkeys.

One additional note from a developmental aspect of spatial learning in the monkey is worthy of mention. Lavenex and Lavenex (2006) described an experiment to assess spatial and nonspatial relational memory in freely moving nine-month-old and adult (eleven to thirteen-year-old) monkeys. Using tasks that employed the presence or absence of local spatial cues, they found that spatial memory processes characterized by an allocentric representation of the environment are present as early as nine months of age in these monkeys. This is of interest given that monkeys with hippocampal damage sustained earlier in life are impaired on a spatial task that requires the use of allocentric cues (Rehbein, 1985a) supporting the view that the hippocampus may play an important role in this circuitry.

Spatial non-matching to sample

As mentioned above, Rehbein (1985b) found that monkeys with ablation of the hippocampal formation were initially impaired, and later recovered, the ability to learn a spatial discrimination on the basis of egocentric cues (spatial reversal, delayed alternation). He also had found no impairment, and even facilitation, in these monkeys' spatial learning when distinctive allocentric cues were provided (landmark discrimination tasks). Taken together, these findings support the view of O'Keefe and Nadel that functions of relative space can take place in the absence of an intact hippocampal formation. It would follow, therefore, that performance on a task by monkeys that required aspects of relative space to be mapped on to the construct of absolute space, should be impaired in monkeys with damage to the hippocampal formation. Hence, a next step was to devise a task, which could not be solved with the use of either egocentric (i.e., left–right, distal–proximal, up–down), or allocentric (i.e., guidance, orientation) spatial strategies. In order to avoid providing monkeys with the benefit of a constant relationship between a given spatial location to be discerned and their body or a salient sensory cue, on or outside the tray, the new task required the use of a trial-unique location in testing space. Similar to the object recognition memory task these monkeys had seen before (Mahut and Moss, 1985), this task would assess the monkey's capacity to distinguish, after a single previous experience, a familiar from a novel position of two identically marked positions on the tray. Thus, performance on the position recognition task depended on the availability of a representation (map) of the testing space, so that responses not given before, nor prescribed by the representation, could be correctly generated. As already mentioned, the position recognition memory task contained important spatial as well as memory component, and therefore, it was particularly appropriate to assess spatial function in monkeys with bilateral section of the fornix which, we knew, were unimpaired on the object recognition memory test (Mahut and Moss; 1985; Rehbein, 1985b; 1991).

As was the case in previous testing, monkeys were tested in a modified Wisconsin General Testing Apparatus. The same tray containing three rows of six wells, used in first landmark task, was used. Two identical aluminum discs, 4 cm in diameter, mounted on brown plastic rims were used as food covers. As illustrated in Figure 11.3, any given trial consisted of covering the bait contained in one of the eighteen wells (sample position) and allowing the monkey to retrieve it by displacing the disc. This was followed, 10 seconds later, by the opportunity to choose between the sample position (now containing no food reward) and a novel position (now containing the food reward). An illustration of the testing situation is presented in Figure 11.3.

Five different sequences of twenty pairs of positions were semi-randomly selected from a listing of all possible combinations of two

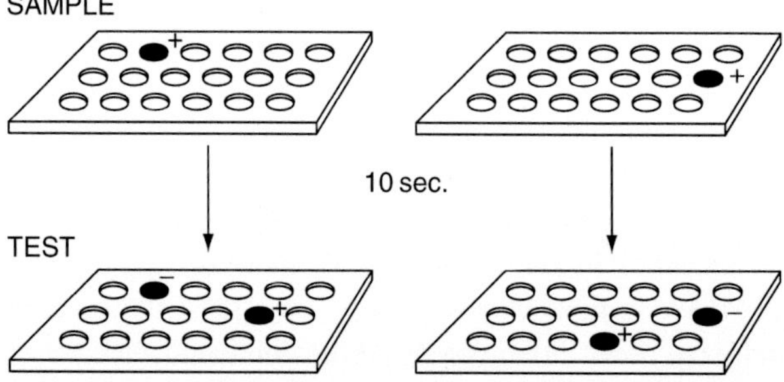

Figure 11.3. Illustration of the procedure used in the position recognition memory task with examples of two trials. "+": denotes presence of food reward. "–": denotes absence of food reward.

positions on the tray. Near–far and left–right sample positions vis-à-vis novel positions were counterbalanced within each session. Twenty trials a day were given and monkeys were trained until a learning criterion of ninety correct responses in 100 consecutive trials was met, or to a maximum of 1,000 trials. Following the basic task with 10-second delays, intervals between sample and recognition test trials were increased in stages to 30, 70 and 130 seconds, for a total of 140 trials with each delay. Control monkeys learned the task in 400 to 720 trials with an average of 148 errors. In contrast, all monkeys in Group H failed to learn the task within the limits of testing, and reached an average of only 72 percent correct response level in the last 100 trials. With delays of 30, 70 and 130 seconds, monkeys in Group H (i.e., those with hippocampal lesions) performed at significantly lower levels of accuracy than did monkeys in Group N (the normal group). These findings would appear to support a key point that an intact hippocampal formation is essential for the performance on a task that makes demands on the ability to retain the capacity for some aspect of absolute space, a principal attribute ascribed to the hippocampus by O'Keefe and Nadel.

Delayed Recognition Span Task (DRST): positions vs. objects

In view of the long-standing impairment shown by monkeys in Group H on either nonspatial, or spatial, trial-unique memory tasks (Mahut and Moss, 1985; Rehbein, 1985b), a different recognition memory test was designed which offered two novel features. First, unlike in most tasks presently in use to assess memory capacity in the monkey, it was not the experimenter who, by increasing the delay or adding items to a list, would set the difficulty of the task. Instead, in the new task, it would be, presumably, the monkey's own immediate memory capacity that would determine the limit of performance. Monkeys would have to identify the appearance of a new stimulus among an array of increasing, but familiar stimuli. According to O'Keefe and Nadel, the ability to detect a new stimulus in otherwise familiar space is subserved by the CA1 subfield of the hippocampus. Second, by using constantly changing sequences of items, the new task would allow us to test memory for the incidental repetition of one and the same sequence of stimuli.

TESTING WITH POSITION SERIES

Testing took place in a modified Wisconsin General Testing Apparatus, and the same eighteen-well tray used in the previous experiment was used. Identical aluminum discs, 4 cm in diameter, mounted on brown plastic rims were prepared as food covers. From the eighteen possible positions on the tray, two complementary sets of nine target positions were selected such that at no axis would they be adjacent to each other, except at the diagonal axis. On the basis of these two sets of positions, twenty-five different sequences of nine positions each were built by using a table of random digits. On any given day monkeys were shown the first position of a series with one of the aluminum discs covering an M&M chocolate candy, or other highly preferred food. Following the retrieval of the bait, the opaque screen was lowered, the same position was re-covered and the second position in the series was baited and covered with another disk. The same procedure was used for the subsequent positions in the series with an interval of approximately 5 seconds between the covering of two successive positions. Presentation of positions within a series was terminated at their first error (i.e., the first time the monkey failed to uncover the most recently covered position). Each daily session consisted of a single presentation of two series with an interval of 15 seconds between series. The second of the two series presented on every other day of testing was always the same as the first of the two (repeated). The initial two series, and series in between the repeated ones, were always different (non-repeated). Thus, three non-repeated series were presented, followed by a fourth series which was the repeat of the third series, and this sequence (three non-repeated series followed by a repetition of the last series, i.e., ABCCDEFF etc.) was continued for a total of thirty-two sessions, or for forty-eight non-repeated and sixteen repeated series.

TESTING WITH OBJECTS SERIES

After thirty-two days of testing with positions, a similar procedure was followed using a set of fifteen small, distinct objects to cover the wells. Twenty-five series of positions, different from the ones used in the spatial form of the task, were prepared. Objects were numbered and pre-assigned to the sequence of positions in each trial by using a random digit table. Thus, each sequence consisted of specific objects in specific positions, and these were changed from one trial to the next, except (as in the testing with positions series above) for every fourth, repeated, series in which order of objects and positions were always the same as the previous series. Monkeys were tested for thirty-two days on this version of the task, or for a total of forty-eight non-repeated and sixteen repeated series.

RE-TEST WITH POSITION SERIES

Immediately following the recognition memory span test with objects, monkeys were re-tested for an additional ten days with non-repeated position series. This was to ensure that the impaired performance on the initial span test for position was not a function of task order, particularly in view of the improved performance on the subsequent span task with objects. When tested with position series, monkeys in Group N reached a mean position recognition span of 3.72 in non-repeated, and 4.61 in repeated series. In contrast, monkeys in Group H reached a mean position recognition span of 2.35 and 2.96 with non-repeated and repeated series, respectively. A posteriori comparisons made between each group's initial and final performance showed the position recognition span of monkeys in Group N increased significantly during testing with non-repeated and repeated series; whereas no significant differences were found between the initial and final position recognition span of monkeys in Group H.

The mean object recognition span of monkeys in Group N was 6.80 with non-repeated, and 6.57 with repeated object series. In contrast, monkeys in Group H attained a mean object recognition span of 3.51 and 3.20 with non-repeated and repeated object series, respectively. Similarly to what was observed with position series, the operated groups of monkeys attained a significantly smaller object recognition span than did the control group of monkeys. Comparisons between each group's performance on the first and the last twelve days of testing showed that monkeys in Group N had significantly increased their object recognition span with training on non-repeated and repeated object series; whereas no significant increase was found in the object recognition span of monkeys in Group H as a result of training both with non-repeated and repeated object series (Rehbein, 1985b, 1991).

The results obtained with the recognition memory span task demonstrated that normal monkeys were able to increase their recognition memory span in all four conditions. Overall, this increase was less pronounced with position than with object series. Also, normal monkeys showed a significantly greater recognition span increase with repeated, than with non-repeated position series. However, no differential increase was observed in their recognition span with repeated and non-repeated object series.

Finally, unlike normal monkeys, those in the hippocampal group did not increase their recognition span either as a function of training, repetition of the same series, or the use of multidimensional objects instead of identically marked locations. Thus, this experiment confirmed and extended the previous findings of a severe deficit in spatial memory following hippocampal damage. These findings provide support for the view by O'Keefe and Nadel that one role of the hippocampal formation (specifically subfield CA1), is to subserve the "misplace system" that signals the presence of new spatial stimuli in an otherwise familiar setting. These data must be viewed with some caution however, given that monkeys with hippocampal damage were also impaired on the object recognition span version of the task that precludes the use of spatial cues. It is plausible that the deficit in spatial recognition may be more general in nature (e.g., short-term "load") rather than one of spatial cognition per se.

These findings were replicated in a study by Beason-Held *et al.* (1999). She found that, relative to monkeys in a normal control group, those with bilateral ibotenic acid lesions of the hippocampal formation were impaired on all conditions of the DRST. Again, as in the studies by Rehbein cited above, unlike normal monkeys, those with damage to the hippocampal formation failed to evidence an improvement in performance on the repeated trial sequences. It is of interest that this finding obtained even though damage to the hippocampus in each case was incomplete. This failure to benefit from repeated trials embedded in a trial unique paradigm raises questions about the role of the hippocampus in the weighting of stimulus-response associations, an issue which warrants further investigation.

Of note, adaptation of the recognition memory span task has been successfully used to assess (1) spatial learning and memory in patients with different degenerating brain diseases (Moss *et al.*, 1986); (2) memory in aging monkeys (Moss *et al.*, 1997); and (3) developmental changes of recognition memory span in children (Rehbein *et al.*, 2009). Accordingly, such translational behavioral tasks may facilitate our ability to assess the role of the hippocampal formation in spatial cognition in human studies.

Electrophysiological studies of the primate hippocampal formation

Recent data from single-cell recording studies in the monkey are also now providing some insight into how space is represented in the primate hippocampus and how this relates to memory and spatial functions. Rolls and his colleagues (Rolls *et al.*, 2005a, 2005b; Rolls and Xiang, 2006) have now identified neurons in the hippocampus that respond to places viewed in an open space ("spatial view" neurons), as well as cells that respond to objects in space ("object cells") and the combination of objects present at particular locations ("object-place" cells) that would be important for object-place memory. In a recent study, Rolls and Xiang (2006) found that about 10 percent of neurons responded differently to different objects independently of location; 13 percent of neurons responded to the spatial view independently of which object was present at the location; and about 12 percent responded to a combination of a particular object and the place where it was shown in the room. Together with the findings from lesion studies, those from Rolls *et al.* would suggest that the hippocampal neurons are concerned with several aspects of spatial mapping that would include allocentric representations. Of course the activity of such units may reflect the convergence of inputs from hippocampal cortical and subcortical afferents as might be posited by O'Keefe and Nadel.

SPATIAL LEARNING IN NORMAL AGING

Whereas the assessment of spatial learning and memory has been considered in the context of selective damage to the limbic system in the adult and infant, recent studies have begun to focus on the effect of aging on spatial learning and memory. As part of a study on executive function and age, Lai *et al.* (1995) assessed the effect of age on both spatial and object reversal learning in a group of aged rhesus monkeys ranging between eighteen and twenty-five years of age. Performance was compared to a group of young adult monkeys between the ages of five and nine years using the experimental paradigm as that described above. The data showed convincingly that as a group, aged monkeys evidenced impairment in spatial reversal learning, but not in object reversal learning. This was perhaps the first study to demonstrate a selective impairment in spatial learning as a consequence of aging in the nonhuman primate, a finding that parallels that found following damage to the hippocampal formation (Mahut, 1971).

In a related study, Moss *et al.* (1988) assessed the performance of a group of twelve early senescent rhesus monkeys (ages twenty to twenty-five years) compared to that of fourteen young adult monkeys (ages five to nine years) on the delayed recognition memory span task (described above). Aged monkeys were impaired on the spatial DRST task relative to young adults with a span of 1.95 as compared to 2.32, respectively. Moreover, a linear relationship was observed between age and degree of impairment. Of interest, the early senescent monkeys, as well as young controls, increased significantly (2.44 and 3.27) for the repeated sequences. The fact that early senescent monkeys achieved a significantly lower mean recognition span than young adult monkeys on the spatial and object conditions, but not the color condition, of the DRST suggest that aged monkeys are impaired on memory load for spatial (including object) information, but not for nonspatial information.

In subsequent studies, we have found that a selective impairment in spatial learning was found in monkeys of advanced age (ages twenty-five to twenty-nine; Moss *et al.*, 1997). Not surprisingly, monkeys who represent the oldest of the old (age twenty-nine and above, equivalent to humans in the ninth and tenth decades of life) were also impaired in spatial recognition memory while, in most instances, evidencing relative sparing of object recognition memory (Herndon *et al.*, 1997).

Finally in an attempt to assess the effect of age on allocentric spatial demands (Killiany *et al.*, 1994), the performance of aged rhesus monkeys (mean age twenty-nine years) was compared to that of young adult monkeys (mean age seven years) on an allocentric task using the apparatus similar to that illustrated above (Figure 11.2). The task required monkeys to choose which of two identical objects was to the left or right of an arrow centered between them. The stimulus array was mounted on a circular tray that could be rotated 360 degrees. In this manner, the correct stimulus changed left–right orientation for egocentric space, but remained constant with regard to allocentric space (e.g., correct object always to the left of the arrow). The task was administered in an acquisition and a test phase. During the acquisition phase, monkeys reached a criterion level of 90 percent correct performance on four arrow orientations. Once criterion was reached, the test phase in which the arrow successively changed orientation from "12:00 to 1:00 to 2:00 etc." was given. Monkeys in the aged group were significantly impaired on both the acquisition and test phases of this task relative to monkeys in the younger group. These findings suggested that aged monkeys not only have difficulty in using egocentric cues in performing spatial tasks as demonstrated by the findings with spatial reversals (Lai *et al.*, 1995), but also evidence difficulty when required to use allocentric spatial information. Whether different neural circuits that may be differentially affected with age subserve these two functions is yet unclear. Regardless, the findings would be consistent with the notion that one of the circuits susceptible to age changes, as evidenced by increased difficulty in navigating egocentric space, is the hippocampal formation.

GENDER DIFFERENCES IN SPATIAL LEARNING

Few studies using nonhuman primates have been conducted to assess the effect of sex in cognition. One recent study assessing the recognition span using the DRST described above, however, found that young adult female monkeys, when compared directly to young adult males, obtained significantly lower spatial spans (Lacreuse *et al.*, 2005). What was surprising in this study was that, although both male and female monkeys evidence impairment in spatial learning with age (Moss *et al.*, 1988), males show a disproportionate decline to the extent that the advantage conferred on male monkeys in spatial learning is lost.

CONCLUSIONS

Taken together, the findings from studies of spatial learning and memory in the monkey suggest that, as in humans, both egocentric and allocentric modes are employed in processing and acting on the spatial environment. Monkeys can perform tasks that rely on learning and remembering stimuli that were defined in space by their relationship to the monkey (delayed alternation and spatial reversal learning). Data from lesion studies suggest that the medial temporal lobe, and specifically the hippocampal formation, may play a pivotal role in mediating spatial learning and memory primarily for egocentric modes. Monkeys with damage to the hippocampal formation are impaired on tasks of delayed alternation, spatial reversal learning, and spatial recognition, all of which rely on the efficient processing of egocentric space. Evidence also suggests that monkeys, in the absence of the hippocampus, can perform at normal, or near normal levels of efficiency, with the availability allocentric cues in the environment. Another important conclusion to be drawn from the initial set of studies discussed in this chapter is that the improved performance on left–right discrimination tasks of monkeys with early hippocampal damage, and their facilitated spatial learning in the presence of allocentric landmarks, did not reflect a recovery of their general spatial capacity. This was demonstrated by their profound impairment on the trial-unique position recognition task and on the position recognition span task. This finding was not surprising in view of the hippocampal monkeys' known impairment on other trial-unique memory tasks (Mahut and Moss, 1985; Rehbein, 1985b). However, the mnemonic deficit reported with the object recognition test in those studies appeared to be greatly enhanced by the spatial nature of the stimuli used.

Unlike in left–right discrimination learning or in landmark-guided behavior, performance on the position recognition task monkeys was not facilitated by either egocentric or allocentric cues. Instead, correct performance in the position recognition and recognition span tasks depended on the availability of a representation (map) of the eighteen positions on the testing tray, so that responses that were not foreseen nor prescribed by the representation (i.e., responses to novel positions within a given trial) could be correctly generated. Thus it would appear that the trial-unique position recognition and the position recognition span studies represent an exploration of a possible third mode of spatial cognition that differs from those involving "egocentric" or "allocentric" spatial strategies. It is this capacity for establishing a representation of the spatial features of the environment which has been proposed as the primary function of the hippocampus by O'Keefe and Nadel (1978). Interestingly, it was also this capacity which appeared most affected in our monkeys with hippocampal damage, in spite of the sophistication arising from six years of nearly uninterrupted training on a wide variety of tasks.

This evidence supports at least two important predictions derived from O'Keefe and Nadel's (1978) theory of hippocampal function (i.e., hippocampal damage left intact the use of guidance hypothesis in the landmark experiments, but impaired spatial mapping in the experiments that required trial-unique position recognition). However, we have reasons to question the exclusively spatial role of the hippocampus advanced by this theory. For example, it has been consistently found that damage to the hippocampus also produces impairment on nonspatial tasks with trial-unique stimuli (Mahut *et al.*, 1982; Rehbein, 1985a). These findings suggest a more encompassing, cognitive-mnemonic involvement of the hippocampus than that proposed by O'Keefe and Nadel (1978). The findings by Rolls and his colleagues identifying the existence of neurons in the hippocampus that respond to objects in space, locations in space, and the relationship of specific objects at specific locations in space would support this possibility. In a slightly different theoretical approach, most of our findings can be best understood in reference to the ideas advanced by Hirsch (1974, 1980). In his view, there are at least two neural systems that can mediate complex learning. One is a cognitive learning system that uses conditional information arising from stimulus configurations, motivational states or items stored in memory, the operation of which depends on the integrity of the hippocampus, a system that fits into the schema of O'Keefe and Nadel. The other is an associative learning system based on the gradual strengthening of S-R bonds as a function of repeated reinforcement for both objects alone as well as for the locations of objects in space, or absolute space alone, the operation of which depends on neural systems other than the hippocampus (see Hirsch and Krajden, 1982, for a characterization of the two systems). While this is not in direct contradiction with the notion posited by O'Keefe and Nadel that spatial learning can occur outside the hippocampus, it extends their view to include a nonspatial role in cognition by the hippocampus.

Another conclusion that can be drawn from the work reviewed in this chapter is that the neural circuitry and processing of spatial cognition still poses a major question to the neurosciences. Despite the explosive growth in this area of research during the past few years, it is still unclear how various brain regions interact to integrate and process information leading to spatial behavior. What is clear however is that when adaptive behavior is performed, several egocentric-allocentric transformations are required and therefore, it is likely that several different brain areas need constant coding and re-coding (updating) in multiple frames of reference. Precisely which brain structures or regions subserve the attention, perception, intention and action in frames of reference that allow an organism to interface its actions with objects in the outside world is yet poorly understood and awaits further investigation.

REFERENCES

Beason-Held, L., Rosene, D. L., Killiany, R. J. and Moss, M. B. (1999). Hippocampal ibotenic acid lesions produce memory deficits in the rhesus monkey. *Hippocampus*, **9**, 562–574.

Dean, H. L. and Platt, M. L. (2006). Allocentric spatial referencing of neuronal activity in macaque posterior cingulate cortex. *Journal of Neuroscience*, **26**, 1117–1127.

Goldman-Rakic, P. S. (1998). The prefrontal landscape: implications of functional architecture for understanding human mentation and the central executive. In A. C. Roberts, T. W. Robbins and L. Weiskrantz (Eds.), *The prefrontal cortex* (pp. 87–102). Oxford: Oxford University Press.

Hampton, R. R., Hampstead, B. M. and Murray, E. A. (2003) Selective hippocampal damage in rhesus monkey impairs spatial memory in an open-field test. *Hippocampus*, **14**, 808–818.

Herndon, J. G., Moss, M. B., Rosene, D. L. and Killiany, R. J. (1997). Patterns of cognitive decline in aged rhesus monkeys. *Behavioral Brain Research*, **87**, 25–34.

Hirsch, R. (1974). The hippocampus and contextual retrieval of information from memory: a theory. *Behavioral Biology*, **12**, 421–444.

Hirsch, R. (1980). The hippocampus, conditional operations, and cognition. *Physiological Psychology*, **8**, 175–182.

Hirsch, R. and Krajden, J. (1982). The hippocampus and the expression of knowledge. In R. L. Isaacson and N. E. Spear (Eds.), *The expression of knowledge* (pp. 213–241). New York: Plenum Press.

Jacobsen, C. F. (1935). An experimental analysis of the frontal association areas in primates. *Archives of Neurology and Psychiatry*, **33**, 558–569.

Killiany, R. J., Lai, Z. C., Lai, J. C., Rosene, D. L. and Moss, M. B. (1994). Impairment in the acquisition and use of allocentric spatial information in aged rhesus monkeys. *Society for Neuroscience Abstracts*, **20**, 388.

Lacreuse, A., Kim, C. B., Rosene, D. L., Killiany, R. J., Moss, M. B., Moore, T. L., Chennareddi, L. and Herndon, J. G. (2005). Sex, age, and training modulate spatial memory in the rhesus monkey (*Macaca mulatta*). *Behavioral Neuroscience*, **119**, 118–126.

Lai, Z. C., Moss, M. B., Killiany, R. J., Rosene, D. L. and Hendon, J. G. (1995). Executive system dysfunction in the aged monkey: spatial and object reversal learning. *Neurobiology of Aging*, **16**, 947–955.

Lavenex, P. and Lavenex, P. B. (2006). Spatial relational memory in 9-month-old macaque monkeys. *Learning and Memory*, **13**, 84–96.

Mahut, H. (1971). Spatial and object reversal learning in monkeys with partial temporal lobe ablations. *Neuropsychologia*, **9**, 409–424.

Mahut, H. (1972). A selective spatial deficit in monkeys after transactions of the fornix. *Neuropsychologia*, **10**, 65–74.

Mahut, H. and Moss, M. B. (1985). The monkey and the sea horse. In R. L. Isaacson and K. H. Pribram (Eds.), *The hippocampus* (vol. 4, pp. 241–279). New York: Plenum Press.

Mahut, H. and Zola, S. (1976). Effects of early hippocampal damage in monkeys. *Society for Neuroscience Abstracts*, **2**, 829.

Mahut, H. and Zola, S. (1977). Ontogenetic time-table for the development of three functions in infant macaques and the effect of early hippocampal damage upon them. *Society for Neuroscience Abstracts*, **3**, 428.

Mahut, H., Zola-Morgan, S. and Moss, M. (1982). Hippocampal resections impair associative learning and recognition memory in the monkey. *Journal of Neuroscience*, **2**, 1214–1229.

Meunier, M., Bachevalier, J. and Mishkin, M. (1997). Effects of orbital frontal and anterior cingulate lesions on object and spatial memory in rhesus monkeys. *Neuropsychologia*, **35**, 999–1015.

Moss, M. B., Mahut, H. and Zola-Morgan, S. (1981). Concurrent discrimination learning of monkeys after hippocampal, entorhinal, or fornix lesions. *Journal of Neuroscience*, **1**, 227–240.

Moss, M. B., Rosene, D. L. and Peters, A. (1988). Effects of aging on visual recognition memory in the rhesus monkey. *Neurobiology of Aging*, **9**, 495–502.

Moss, M. B., Albert, M. S., Butters, N. and Payne, N. (1986). Differential patterns of memory loss among patients with Alzheimer's disease, Huntington's disease, and alcoholic Korsakoff's syndrome. *Archives of Neurology*, **43**, 239–246.

Moss, M. B., Killiany, R. J., Lai, Z. C., Rosene, D. L. and Herndon, J. G. (1997). Recognition span in rhesus monkeys of advanced age. *Neurobiology of Aging*, **18**, 13–19.

Murray, E. A., Bussey, T. J., Hampton, R. R. and Saksida, L. M. (2000). The parahippocampal region and object identification. *Annals of the New York Academy of Science*, **911**, 166–74.

Newman, E. L., Caplan, J. B., Kirschen, M. P., Korolev, I. O., Sekuler, R. and Kahana, M. J. (2007) Learning your way around town: how virtual taxicab drivers learn to use both layout and landmark information. *Cognition*, **104**, 231–253.

O'Keefe, J. and Nadel, L. (1978). *The hippocampus as a cognitive map*. Oxford: Clarendon Press.

Owen, A. M., Evans, A. C. and Petrides, M. (1996). Evidence for a two-stage model of spatial working memory processing within the lateral frontal cortex: a positron emission tomography study. *Cerebral Cortex*, **6**, 31–38.

Penfield, W. and Milner, B. (1958). Memory deficit produced by bilateral lesions in the hippocampal zone. *Archives of Neurology and Psychiatry*, **79**, 475–497.

Pohl, W. (1973). Dissociation of spatial discrimination deficits following frontal and parietal lesions in monkeys. *Journal of Comparative and Physiological Psychology*, **82**, 227–239.

Rehbein, L. (1985a). *Long-term deleterious effects on visual and spatial memory functions after damage to the hippocampus in infancy*. Dissertation presented to the Department of Psychology in partial fulfillment of the requirements for the doctoral degree in the field of Psychology, Northeastern University, Boston, USA.

Rehbein, L. (1985b). Hipocampo y discriminacion espacial en monos. [Hippocampus and spatial discrimination in monkeys.] *Archivos de Biología y Medicina Experimentales*, **18** (2), R-160.

Rehbein, L. (1991). Efectos a largo plazo de lesions tempranas en el hipocampo del mono. [Long-term effects of early lesions in the hippocampus of the monkey.] *Estudios Psicológicos*, **1** (1), 119–149.

Rehbein, L. and Mahut, H. (1983). Long-term deficits in associative and spatial recognition memory after early hippocampal damage in monkeys. *Society for Neuroscience Abstracts*, **9**, 639.

Rehbein, L., Barría, I., Masardo, M., Oyarzún, M. and Schade, N. (2009). Ontogenetic changes in the recognition memory span of children. In preparation.

Rolls, E. T. and Xiang, J. Z. (2006) Spatial view cells in the primate hippocampus and memory recall. *Reviews in the Neurosciences*, **17**, 175–200.

Rolls, E. T., Xiang, J. and Franco, L. (2005a). Spatial view cells and the representation of place in the primate hippocampus. *Journal of Neurophysiology*, **94**, 833–844.

Rolls, E. T., Xiang, J. and Franco, L. (2005b). Object, space and object-space representations in the primate hippocampus. *Hippocampus*, **9**, 467–480.

Scoville, W. (1954). The limbic lobe in man. *Journal of Neurosurgery*, **11**, 64–66.

Scoville, W. and Milner, B. (1957). Loss of recent memory after bilateral hippocampal lesions. *Journal of Neurology, Neurosurgery and Psychiatry*, **20**, 141–146.

Suzuki, W. A. and Amaral, D. G. (2004) Functional neuroanatomy of the medial temporal lobe memory system. *Cortex*, **40**, 220–222.

Ungerleider, L. G. and Brody, B. A. (1977). Extrapersonal orientations: the role of posterior parietal, anterior frontal and inferotemporal cortex. *Experimental Neurology*, **56**, 265–280.

ROBERT W. MITCHELL AND ESMÉ HOBAN

12

Does echolocation make understanding object permanence unnecessary? Failure to find object permanence understanding in dolphins and beluga whales

Objects are spatial. Understanding the nature of objects begins with perception. For most mammals, vision and touch are the primary means of discovering spatial properties of objects (including the body, here aided by kinesthesis). Mental representations of objects, in which their spatial aspects are encoded, are believed to develop from perceptual modalities available to the organism, allowing it to recognize that objects continue to exist even when they disappear from perception (e.g., Mitchell, 1994; Piaget, 1954; Reisberg and Heuer, 2005; Ryland, 1909). Spatial cognition about objects, including imagery and constructive memory, derives from perception of objects (Finke, 1980, 1986). Spatial cognition employs representations of objects, requiring understanding that these representations concern real objects that can be acted upon. Spatial cognition allows an organism to discern where an object is located on the basis of remembered landmarks, remember an object's properties such as size and shape, and know that objects do not cease to exist when not apparent.

Prey are essential objects in a predator's life. If prey can hide or otherwise become imperceptible, predators cannot rely on perception to discern them. Whereas perception informs about real objects in their surrounding space, cognition is required when objects become imperceptible. Unless a predator has some understanding of the continued existence of imperceptible prey, it should cease to act to obtain that prey. Thus, some level of understanding of object permanence seems an essential cognitive skill for predators of hiding prey (and perhaps for foragers of hidden food – "extractive foragers" – as well [Parker and Gibson, 1977]).

Spatial Cognition, Spatial Perception: Mapping the Self and Space, ed. Francine L. Dolins and Robert W. Mitchell. Published by Cambridge University Press.

Some mammals, such as cetaceans, have an additional perceptual modality for informing them about real objects: echolocation (Au, 1993; Thomas *et al.*, 2002). Echolocating animals use auditory signals rebounding from objects to locate these objects and discern their properties. Echolocation is comparable to seeing, in that dolphins discern properties of objects such as size and shape from these acoustic echoes (DeLong, 2004; Harley *et al.*, 2003; Pack and Herman, 1995). Just as we might find a hologram extremely realistic, dolphins classify echoes from acoustically identical sounds of phantom objects in the same way they classify real objects (Aubauer *et al.*, 2000).

Because of echolocation, dolphins and other cetaceans may develop a different understanding of objects than that of other mammals. For example, some authors speculate that "For the dolphin an object is apparently more of an object when it is heard than when it is seen" (Jerison, 1986, p. 158). To examine the nature of object understanding by cetaceans that use sonar, we used object-permanence tasks to test Atlantic bottlenosed dolphins (*Tursiops truncatus*) and beluga whales (*Delphinapterus leucas*) on understanding of objects.

In an object-permanence task, an organism is allowed to watch a desired object move behind one or more obstacles that prevent the organism from perceiving the object, and then the organism is allowed to act. The organism is expected to recognize that an object continues to exist when not perceived, infer the position of the object from its previous direction, and, if the object is desirable, obtain it by circumventing or removing the obstacle. An organism's responses in an object-permanence task provide evidence of its understanding of objects. Piaget's (1954) depiction of the development of understanding of object permanence during the sensorimotor period in human children proceeds through six stages (see also Doré and Dumas, 1987; Flavell, 1963, pp. 129–135). In the first two stages, infants do not recognize that objects continue to exist when they move outside perception. In stage three, infants can predict the future position of a moving object from its trajectory and can remove an obstacle to obtain an object when part of it is visible, but do not actively search for fully hidden objects. In stage four, infants actively search for hidden objects. At the beginning of stage four, the hidden object is obtained only when the child's search movement begins when the object disappears; at the end of stage four, the object can be obtained when it has been hidden repeatedly behind the same obstacle, but not when a new obstacle is used. In stage five, the child can obtain the object when it is hidden behind a new obstacle, and later can also obtain the object, when hidden behind successive obstacles, by circumventing the last obstacle. At stage five, the child recognizes the object upon seeing it again, but does not recall it when out of sight (Doré and Goulet, 1998). By the end of stage six, the child can predict the future position of an object from the trajectory of something containing the object when the container moves behind an obstacle. At stage six, the child has developed an ability for mental representation and symbolic functioning not present at stage five: the child knows that objects continue to exist when outside perception because he or she mentally represents these objects (see Doré and Goulet, 1998; for alternative interpretation, Thomas and Walden, 1985). (Although according to Piaget object permanence understanding develops over the first year of an infant's life, some researchers suggest that this understanding is present soon after birth [Baillargeon *et al.*, 1985]. However, once confounding influences are removed, infants' early understanding is no longer evident [Rivera *et al.*, 1999].)

Understanding of object permanence has been tested in diverse species (for overview, see Doré and Dumas, 1987; Doré and Goulet, 1998; Etienne, 1984; Heishman *et al.*, 1995). At present, the only non-human species scientists agree achieve stage six are great apes – chimpanzees (*Pan troglodytes*), bonobos (*P. paniscus*), gorillas (*Gorilla g. gorilla*), and orangutans (*Pongo pygmaeus*) (Barth and Call, 2006; Beran and Minihan, 2000; Call, 2001; Collier-Baker *et al.*, 2004; Collier-Baker and Suddendorf, 2006; de Blois *et al.*, 1999; Mathiew *et al.*, 1976; Natale *et al.*, 1986; Wood *et al.*, 1980). Initial studies were suggestive of stage six understanding in capuchin monkeys (*Cebus capucinus*), rhesus monkeys (*Macaca mulatta*), squirrel monkeys (*Saimiri sciurea*), common marmosets (*Callithrix jacchus*), cotton top tamarins (*Saguinus oedipus*) and dogs (*Canis familiaris*), and of stage five in Japanese monkeys (*Macaca fuscata*), wooly monkeys (*Lagothrica flavicauda*) and cats (*Felis catus*) (Doré, 1986; Dumas, 2000; Dumas and Doré, 1989; Fiset *et al.*, 2003; Gagnon and Doré, 1992; Goulet *et al.*, 1994; Mendes and Huber, 2004; Natale *et al.*, 1986; Neiworth *et al.*, 2003; Triana and Pasnak, 1981; Vaughter *et al.*, 1972; Wise *et al.*, 1974). However, evidence using more refined testing supports the idea that non-ape mammals are at best capable of stage five (Collier-Baker *et al.*, 2004; Doré and Goulet, 1998). Other mammals, such as golden hamsters (*Mesocricetus auratus*), attain only stage four (Thinus-Blanc and Scardigli, 1981). Some psittacine birds – African grey parrot (*Psittacus erithacus*), macaw (*Ara maracana*), parakeet (*Melopsittacus undulatus* and *Cyanoramphus auriceps*), cockatiel (*Nymphicus hollandicus*) – appear to attain stage six (Funk, 1996; Pepperberg and Funk, 1990), though implementation of more refined tests recently employed with mammals may demote their status (Collier-Baker *et al.*, 2004; Doré and Goulet, 1998). Magpies (*Pica pica*) attain stage five and perhaps stage six (Pollok *et al.*, 2000), whereas ring doves (*Streptopelian risoria*) attain only stage four (Dumas and Wilkie, 1995). Other animals have not been tested on object permanence tasks per se, though evidence suggestive of stage five exists for crows, and of stage four for cocks and rabbits (species unidentified: Krushinskii, 1960).

One might expect bottlenosed dolphins to achieve stage six understanding on object-permanence tasks. They are intelligent mammalian predators (Herman, 1980, 2002), and one expects intelligent animals to understand object permanence. According to Etienne (1984, p. 309), "Object permanence is ... a prerequisite for anticipatory or insightful behavior and thus is attained by more intelligent species." Dolphins understand symbolic reference when either sounds or human hand gestures are used to indicate objects, actions and object-action sequences (Herman, 1980). Dolphins can also remember and report whether or not they have observed an object being placed into their tank, and report (after searching) whether or not an object is in their tank (Herman and Forestell, 1985). This evidence implies to some researchers that dolphins know that objects continue to exist outside their perception (Herman and Forrestell, 1985, pp. 678–679), but this implication is unwarranted: that dolphins respond distinctively to absence of an object is not evidence that they make reference to the object or know that it exists in extraperceptual space (Hoban, 1986, pp. 135–136, 140–141).

Dolphins' skills in imitation (Herman, 2002; Tayler and Saayman, 1973) and self-recognition (Marten and Psarakos, 1994, 1995; Reiss and Marino, 2001) suggest that, at least on Piaget's (1945/1962) imitation scale, they have mental representation (i.e., stage six understanding) (Mitchell, 1987). However, both of these skills employ a representation of a specific type – kinesthetic–visual matching – that is unnecessary for stage six understanding of object permanence, which usually precedes self-recognition (Mitchell, 1993, 2007). In this view, stage six understanding of object permanence might be useful for some species in developing self-recognition, but the two can develop independently. If dolphins' development across domains is like humans', they should attain stage six understanding of object permanence. They don't.

According to Doré and Goulet (1998), noting unpublished research presented by Doré *et al.* (1991), two captive bottlenosed dolphins, one of whom had provided the evidence described above of both symbolic reference and reporting on the presence or absence of objects, failed to exhibit stage six understanding when tested on a standard object-permanence task, showing at best stage five: the dolphins "fail invisible displacement problems but can retrieve an object that was visibly moved in a succession of screens" (Doré and Goulet, 1998). However, unpublished research performed about a decade earlier with one of these dolphins (Phoenix) and her poolmate (Akeakamai) when they were about three years old failed to provide evidence of stage four understanding of object permanence (Mitchell *et al.*, 1982; cited in Mitchell, 1995). In this research, upon observing their usual food (freshly thawed fish) being tossed into floating upturned frisbees (familiar objects in their tank), the dolphins did not attempt to overturn the frisbees to obtain the fish (which were partly visible hanging over the edge of the frisbees), but rather tended to submerge in the direction the fish were thrown and move their heads back and forth, apparently looking or echolocating for a fish beneath the frisbee into which it had been thrown. Such behavior suggests that these dolphins attained only stage three understanding of object permanence. That they (surprisingly) did not attempt to overturn a frisbee to obtain partially visible fish suggests that they had not attained stage four. Once the dolphins observed people overturning the frisbees containing fish, however, they ate the now available fish and proceeded to overturn any frisbees in their tank regardless of whether they contained fish or not.

The dolphins' initial failure to overturn the frisbees is not attributable to visual deficits, as dolphins "can see quite well in air" (Madsen and Herman, 1980, p. 130; Dawson, 1980). The dolphins' failure also cannot be explained by suggesting that fish are not desirable objects. Not only did the dolphins eat fish from the overturned frisbees, but these same dolphins when tested preferred to have more fish rather than less fish or none at all (Mitchell *et al.*, 1985). One dolphin's apparent lack of stage four understanding when first tested, followed by apparent skill at stage five understanding when tested years later, may be attributable to development, learning, or both. Contrariwise, the methods used to test the dolphins may have inflated the stage of their knowledge. Tests of dogs' object permanence understanding performed by Doré and colleagues at about the time they were testing the dolphins showed evidence of stage six understanding (Gagnon and Doré, 1992, 1993, 1994), but later, more refined testing by these authors and others supported a lack of stage six understanding in dogs (Collier-Baker *et al.*, 2004; Doré *et al.*, 1996; Fiset *et al.*, 2003).

Regardless of discrepancies between the two studies of the dolphins, neither study provided evidence of stage six understanding of object permanence that one would expect of dolphins based on their previously reported skill at tasks that apparently require mental representation of objects.

These observations suggest that dolphins do not mentally represent objects, and do not fully understand object permanence. Perhaps they do not need to (Mitchell, 1995). Understanding object permanence is presumed to be an adaptation that allows an organism "to attribute continued existence to disappearing objects and to localize ... objects in space" (Dumas and Doré, 1989, p. 191), and is particularly important when predators "are dealing with the behavior of prey that has disappeared while being pursued" (Dumas, 2000, p. 232). Dolphins may neither have nor require such an adaptation because they have another adaptation that fulfills nearly all the same functions: echolocation. Echolocating animals can keep track of objects (e.g., fish) they cannot see or touch. Thus, if fish hid in vegetation, a dolphin could discover them through echolocation. Echolocation provides a perceptual system that informs the organism about what many mammals need cognition (i.e., understanding of object permanence) to tell them. Most mammalian predators would have to recognize that prey, though hidden from perception (i.e., disappearing), still exist in order to continue trying to obtain the prey. Dolphins and other cetaceans echolocate while hunting fish (Herzing, 2004; Herzing and dos Santos, 2004), so may never experience the perceptual disappearance of prey, and thus have no need to cognize about them. Consequently, understanding object permanence may be irrelevant for dolphins. According to Warshall (1974, p. 134), dolphins are able to "see," via echolocation, inside some objects. Such perceptual skill seems to obviate the need for stage six understanding of object permanence in almost all situations a dolphin would encounter. Echolocation could also assist where an understanding of object permanence could not, such as in following quickly moving fish through murky water. In some cases of low visibility, cetaceans use passive listening ("passive sonar") rather than echolocation ("active sonar") to detect prey movements or calls (dos Santos and Almada, 2004, p. 403). Such audition, like echolocation, provides perceptual information about prey, again obviating the need for an understanding of object permanence.

An alternative hypothesis is that Phoenix and Akeakamai avoided overturning frisbees because their training taught them, when being fed, to touch a frisbee only when requested. By this account, their behavior resulted not from the lack of stage four understanding of object permanence, but from interference from training. This explanation raises a problem for research with all captive cetaceans, because these organisms are trained to be highly responsive to human cues. The explanation also conflicts with the dolphins' behavior of searching for fish under the frisbees into which they had been thrown: if they knew the fish were *in* the frisbees, they need not have searched for them *underneath*. To evaluate whether echolocation precludes fully developed object permanence understanding, we performed object permanence tasks with other bottlenosed dolphins and another echolocating cetacean, the beluga whale.

EXPERIMENT 1

Method

Participants and context. Two female Atlantic bottlenosed dolphins (Mimi, Nina), and two female beluga whales (Kela, Naku), were tested in December of 1987 at Mystic Aquarium in Mystic, Connecticut, for their understanding of object permanence. The dolphins were about ten years old, and the beluga whales were about seven years old. Mimi had lived at Mystic for twenty-five months, Nina for seventeen months, and the beluga whales for twenty-eight months. The dolphins had been trained at other facilities prior to their arrival at Mystic. The beluga whales were brought within a week of their capture to Mystic. All animals had been trained, using fish rewards, to perform publicly, and thus were familiar with some gestural signals (such as pointing, used to request retrieval of an object).

Each animal was tested separately in the large tank in which performances were held. While each was being tested, dolphins were present in a holding tank to the left of the testing area, and beluga whales were present in a holding tank to the right of the area. Both holding tanks were separated from the performance tank by metal gates that allowed the animal being tested to maintain perceptual contact with her cagemates (to avoid distress at being separated).

Apparatus and procedure. Thawed fish (capelin, their usual food) were used as desired objects, and opaque plastic bowls were used as obstacles into which fish were thrown or placed. During the experiment, we fed each dolphin 0.45 kg of fish, and each beluga whale 0.68 kg of fish. Four bowls (all unfamiliar to the animals, according to trainers) were available: two dog-food bowls (diameter = 24.3 cm; depth = 7.7 cm), one light blue and one grey; and two four-quart bowls (diameter = 30.8 cm; depth = 8.3 cm), one yellow and one dark brown. The depth of these bowls, when floating, was such that fish inside would not be visible to the animal.

During each session, three unfamiliar people were visible to the animals. One man retrieved the bowls when they had moved too far away for fish to be easily thrown into them. (A current tended to move floating bowls about.) Another man interacted with the animals as "the experimenter." And a woman recorded the behavior of the animals and the experimenter, specifically noting into which bowl a fish was placed, whether the animal searched for the fish underwater, which bowl(s) she approached, and what she did with the bowl.

Animals were tested in this order: Kela, Naku, Mimi and Nina. First, each animal was separated, and bowls used for testing were placed in the water for the animal to explore. To support exploration, objects were pointed to, moved about and/or pushed underwater by the experimenter until the animal contacted a bowl or seemed to scan it by echolocation. In this way, animals were provided with exposure to unfamiliar objects, after which the study began. In all phases of the experiment, the experimenter made every attempt to act toward the objects only when the animal was attending to him. All phases tested the animal's comprehension of object permanence at stage four. The number of times the experimenter performed the actions in each phase is presented in Table 12.1. Note that water current was strong and quickly moved objects about the tank.

During Phase I trials, the experimenter threw a fish into a nearby floating bowl. During Phase II trials, the experimenter held a fish directly over a bowl and dropped the fish into the bowl. The experimenter's action during Phase II was intended to make the connection between fish and bowl more salient to the animal than did his action during Phase I. During Phase III trials, within a few seconds the experimenter took a bowl out of the water, held it perpendicular to the water surface, moved a fish held in his other hand into the bowl, moved the bowl such that its bottom was parallel to the water surface, and then put bowl into water. Displacement of the bowl following "disappearance" of the fish inside tested the animal's ability to discern the location of the fish from the trajectory of an object containing the fish rather than from the trajectory of the fish itself. During Phase IV trials, which occurred only for the two dolphins, Phase I trials were repeated to see if previous testing influenced responses. (The term "fish-container" below indicates a bowl with a fish inside.) The animal

Table 12.1. *Sequences of experimenter's actions and of animals' responses in Experiment 1*

Human actions	Kela's actions	Mimi's actions	Nina's actions
Phase I	1. I	1. S, F	1. S, F
	2. I	2. F	2. S
	3. O, I		3. S
			4. S
			5. S
Phase II	1. P, C	1. F	1. F
	2. C		
	3. O, P, A		
Overturn			1. A
Phase III	1. O, A	1. F	1. O, A
Phase IV		1. F, O", I	1. O, A, O*, O*, O*
			2. O*
			3. F

Note: Abbreviations indicate responses of the animals on each (numbered) trial: S = searched for fish underwater; A = ate fish from overturned fish-container; C = contacted (or pushed) fish-container; F = retrieved ("fetched") fish-container to or near experimenter; I = ignored containers (or ignored fish if fish-container was overturned); O = overturned fish-container; O* = overturned empty bowl; O" = accidentally overturned fish-container; P = played with a fish by repeatedly spitting it out.

was always allowed at least thirty seconds to react before another trial took place. Throughout each session, fish were occasionally fed to the animal in order to maintain attention.

Two exceptions to the general sequence of Phases occurred in an attempt to make the fish-containers salient (see Table 12.1). After Phase II with the dolphin Nina, the experimenter pushed a fish-container underwater such that the fish floated out of it. During Phase III with the beluga whale Naku, the experimenter twice pointed to a bowl containing fish.

Results

Results for Kela, Mimi and Nina are presented in Table 12.1 and detailed below. (Beluga whale Naku always ignored the fish-containers, and hence her responses are not presented.) The animals' *initial* responses are summarized in Table 12.2, which shows that the

Table 12.2. *First response to objects by animals across trials in Experiment 1*

	Phase I			Phase II			Phase III			Phase IV		
Animal:	Kela	Mimi	Nina	Kela	Mimi	Nina	Kela	Mimi	Nina	Kela	Mimi	Nina
Initial response	Number of trials											
searched underwater	0	1	5	0	0	0	0	0	0	–	0	0
contacted f-c	0	0	0	2	0	0	0	0	0	–	0	0
retrieved f-c	0	1	0	0	1	1	0	1	0	–	1	1
overturned c	0	0	0	0	0	0	0	0	0	–	0	1
overturned f-c	1	0	0	1	0	0	1	0	1	–	0	1
no response	2	0	0	0	0	0	0	0	0	–	0	0
total	3	2	5	3	1	1	1	1	1	–	1	3

Note: The abbreviation "f-c" refers to a fish-container, and the abbreviation "c" alone refers to a container that holds no fish. Also, responses following the initial one are depicted in Table 12.1 and in the text.

animals' first response was rarely to overturn a fish-container. Only the beluga whale Kela overturned a fish-container in each of the first three phases. The dolphin Nina overturned a fish-container in Phases III and IV, but only after the experimenter overturned the fish-container following Phase II to release the fish inside. And dolphin Mimi accidentally overturned a fish-container while retrieving it in Phase IV. Note that the most successful animal, the beluga whale Kela, appeared the least desirous of fish, whereas both dolphins retrieved and ate all fish thrown into the water and during Phase I searched underwater for the fish near the bowls. All animals frequently looked at the human observers while or before responding to a container.

Kela. When introduced to the bowls, Kela ignored them at first, but later she twice grabbed the same bowl in her mouth and overturned it. When fish were thrown into the water, Kela once ignored a fish and once captured another but then dropped it. However, she ate five fish that were individually placed in her mouth. Twice she played with a fish, spitting it out to the experimenter until he placed it headfirst into her mouth, at which time she ate it. During Phase I, Kela overturned one fish-container, but ignored this fish. (The bowl she overturned was the same bowl she had overturned twice during the exploration phase.) During Phase II, Kela hit a fish-container on two trials, and overturned a fish-container on one trial. (She caught the fish, but spit it out repeatedly, eating it only after the experimenter put it headfirst into her mouth.) During Phase III, Kela overturned the fish-container and ate the fish. Of the five trials on which she responded to the containers, she hit the fish-container on two trials and overturned the fish-container on three trials.

Mimi. When introduced to the bowls, Mimi nudged and overturned a bowl. During Phase I, Mimi searched under the fish-container on the first trial, and retrieved this fish-container and, on the next trial, another one. During Phase II, Mimi retrieved the fish-container. During Phase III, Mimi retrieved the fish-container. During Phase IV, Mimi accidentally overturned the fish-container while retrieving it (the front of the bowl overturned as she pushed it from behind), but ignored the fish. In all, she retrieved five fish-containers in five trials, and only once (accidentally) overturned one.

Nina. When introduced to the bowls, Nina ignored them at first, then pushed one bowl slightly, and scanned a bowl when the experimenter pushed it underwater with its inside toward her. During Phase I, Nina searched underwater near the containers, but only once retrieved the fish-container. During Phase II, Nina retrieved the fish-container. At this point, Nina had retrieved two fish-containers in six trials. When the trainer overturned the last fish-container, Nina ate the fish that was released. During Phase III, Nina overturned the fish-container and ate the fish. During Phase IV, Nina overturned the first bowl into which a fish was thrown, then overturned the three empty bowls in the water. On the next trial she overturned the bowl closest to her after the experimenter threw a fish into the bowl farthest from her. Next, she retrieved a fish-container. During the four trials of Phases III and IV, her initial response on two trials was to overturn fish-containers.

Discussion

The beluga whale Kela's repeated obtaining of the fish from fish-containers suggests that she had attained at least stage four understanding of object permanence, in that she apparently discerned the future position of a fish both from its trajectory and from movement of a bowl containing a fish, and circumvented the obstacle to the fish. Kela's overturning fish-containers following three differently orchestrated placements of fish in bowls implies that the manner of placement was not significant for her retrieval of fish. However, the first fish-container she overturned was the same bowl she had overturned previously when it contained no fish. This fortuitous overturning may have influenced her subsequent responses.

The dolphin Mimi's responses suggest that she had attained stage three understanding of object permanence, in that she once scanned underwater for a fish that landed in a bowl. The fact that she consistently retrieved fish-containers might indicate that she knew where the fish were, and believed the task was to retrieve fish-containers. An alternative interpretation is that she responded to a person's acts of throwing a fish toward a bowl (in Phases I and II) and of throwing a bowl containing a fish (in Phase III) as akin to a person's pointing to an object, in that all these actions indicated to her which bowl to retrieve rather than which bowl contained a fish.

The dolphin Nina's responses are the most complex and are very similar to reactions of the dolphins in the initial study with frisbees. Her repeated underwater searching for fish that landed in bowls in Phase I suggest that she had attained at least stage three, and her failure to overturn fish-containers to obtain fish suggests that she did not attain stage four. Though her overturning the fish-container in Phase

III suggests attainment of stage four, it may be that she simply learned, after observing fish being released when the experimenter overturned the container, that overturning containers could result in fish. Support for this interpretation is her overturning empty bowls and her overturning of the bowl closest to her after the experimenter had thrown a fish into the bowl farthest from her, both of which suggest that she had learned to associate overturning any bowl with obtaining a fish. Thus, her actions likely reflect little understanding of continued existence of fish, and more understanding of reward contingencies associated with obtaining a fish.

A further complication in interpretation is that the human participants who were visible to the animals throughout could see where fish were, and may have inadvertently cued the animals. Another experiment (described next) avoided potential cueing, and also tested animals' responses to fish that were visible in a clear container (Hoban and Mitchell, 1990a, 1990b).

EXPERIMENT 2

Method

Participants and context. The study was again performed at Mystic Aquarium, in July of 1989, nineteen months after Experiment 1. Subjects were the beluga whale Kela and the dolphin Nina, as well as a beluga whale named Aurora. At the time of the study, Aurora was about eight years old, and had lived at Mystic for fifty-nine months. She had been brought within a week of her capture to Mystic. Like the other beluga whales, she was trained to perform publicly. Context was the same as in Experiment 1, except that this time the water current did not move objects around in the tank and, thus, they retained their relative position in the water.

Apparatus and procedure. Thawed fish (capelin) were used as desired objects, and plastic containers were used as obstacles. Each animal received 0.91 kg of fish. Three opaque rectangular plastic containers were used: one blue, one rose and one white (length: 25.1 cm; width: 14 cm; depth: 10.5 cm). The depth of all opaque containers was such that fish thrown into them would not be visible to the animal. Two other plastic containers were used: a transparent rectangular container the size of the opaque containers, and an opaque, white, rounded bowl (diameter of top: 15.2 cm; diameter of base: 7.6 cm; depth: 10.2 cm). Prior to testing, the order of presentation and the bowl into which the fish was to be thrown on each trial were semi-randomly generated, and the recorder remained blind to this information. During each session, two people were visible to the animals. To avoid possible cueing, the "experimenter" performed all the actions in a trial, and then turned away from the animal, at which time the "recorder" observed and recorded the animal's responses. Recorder and experimenter alternated roles across animals.

Animals were tested in this order: Kela, Aurora and Nina. Each animal was tested separately. As in Experiment 1, animals were allowed to explore containers, after which the study began. In all phases, the experimenter made every attempt to perform the required actions only when the animal was attending. Phases I and IV tested the animal's stage four understanding of object permanence; Phase II tested the animal's stage six understanding of object permanence; and Phase III tested the animal's ability to obtain a fish visible inside an object, to test that fish were desirable objects and that objects per se were not influencing the animals' behavior (see Table 12.3).

Phases I, II and IV each had four trials. Each trial began with all three opaque rectangular containers being placed in the water in a predetermined but variable order; which container was to be the obstacle to the fish was also predetermined (see Table 12.3). During Phase I, the experimenter threw a fish into a container. During Phase II, the experimenter placed a fish into the round bowl and dropped this fish-container into one of the three containers. In the two trials in Phase III (performed only with Aurora and Nina), the transparent object only was placed in the water, and the experimenter threw a fish into it. During Phase IV, Phase I was repeated to see if previous testing influenced the animal's responses. Animals were always allowed at least thirty seconds to react before another trial took place. Throughout each session, fish were occasionally, and at predetermined times, fed to the animal to maintain her attention.

Results

Results for the two beluga whales Aurora and Kela are described in Table 12.3 and detailed below. (Dolphin Nina did not contact any fish-containers, not even the transparent one, during any phase; her behavior is not presented.) Animals' *initial* responses are summarized in Table 12.4, which again shows that the animals' first response was rarely to overturn a (non-transparent) fish-container. Both beluga whales readily ate all fish thrown into the tank or released from a

Table 12.3. *Order of placement of rectangular blue, rose and white containers, placement of the fish, and animals' responses in Experiment 2*

	Placement of bowls	Aurora's actions	Kela's actions
Phase I	R(f) W B	1. I	1. O(R), A
	W B(f) R	2. O(B), A	2. I
	B R W(f)	3. +(W), >(R), >(B)	3. O*(B)
	R B(f) W	4. I	4. >>(R), C*(R)
Phase II	W(f) R B	1. S, ++(W), C*(R)	1. ++(W), C*(R)
	W B R(f)	2. >>(W)	2. >(W)
	B(f) W R	3. O(B), A	3. +(B), O*(W)
	R B(f) W	4. I	4. >(R), >(W)
Phase III	Transparent container	1. O, A	
	Transparent container	2. O, A	
Phase IV	B(f) R W	1. ++(B), C*(W)	1. +(B), O*(W)
	W(f) B R	2. O(W), A	2. O*(B), C*(B)
	R W B(f)	3. O*(W), C*(W)	3. O*(R)
	B W(f) R	4. O(W), A	4. O*(B)

Note: B, R or W represent the colors of the containers (blue, rose and white, respectively). The sign (f) means that the fish (or, in Phase II, the fish-container) was put into the container whose sign precedes the (f). Capitalized letters indicating actions are the same as in Table 12.1. Additional responses are indicated by the following signs: + = approached fish-container; ++ = tried to overturn fish-container; > = approached empty container; C* = contacted or pushed empty container; >> = tried to overturn empty container. When B, R or W in parentheses follows an action sign, it represents the color of the container toward which the designated action was directed.

fish-container, and both also frequently looked at the recorder while or before responding to a container.

Aurora. When introduced to the containers, Aurora explored them by scanning or looking at them. During Phase I, she overturned one fish-container and approached (but did not overturn) another. During Phase II, she overturned one fish-container, and attempted (but failed) to overturn a fish-container on one trial and an empty container on another. During Phase III, she overturned the transparent container on both trials. During Phase IV, she overturned two fish-containers and one empty container, and attempted (but failed) to overturn another fish-container. Of the twelve trials in which she

Table 12.4. *First response to objects by animals across trials in Experiment 2*

	Phase I		Phase II		Phase III		Phase IV	
Animal:	Aurora	Kela	Aurora	Kela	Aurora	Kela	Aurora	Kela
Initial response	Number of trials							
searched underwater	0	0	1	0	0	–	0	0
approached f-c	1	0	0	1	0	–	0	1
approached c	0	0	0	2	–	–	0	0
tried to overturn f-c	0	0	0	1	0	–	1	0
tried to overturn c	0	1	1	0	–	–	0	0
overturned f-c	1	1	1	0	2	–	2	0
overturned c	0	1	0	0	–	–	1	3
no response	2	1	1	0	0	–	0	0
total	4	4	4	4	2	–	4	4

Note: see note for Table 12.2.

responded to a container during Phases I, II and IV, her initial responses in seven trials were to the fish-container and she overturned this object in four trials. In two instances her action of overturning a fish-container began immediately upon a fish being thrown into a container: she moved toward the fish as it fell into the container, and her overturning seemed a continuation of her action to obtain the visible fish (consistent with stage four understanding). She also initially overturned one empty container and attempted to overturn another. If her four successful and two unsuccessful attempts to overturn a fish-container are accepted as evidence of her knowledge of the location of the fish in the twelve trials of Phases I, II and IV, her "success" rate is six out of twelve trials. Assuming a chance probability of one in three for selecting the correct container, this rate is not statistically significant (Binomial, $N = 12$, $x = 6$, *ns*). Note, however, that if the three trials in which she failed to respond to any container are not included, and her four successful and two unsuccessful attempts to overturn a fish-container are again used as evidence of her knowledge of the location of the fish, her "success" rate is six out of nine trials, which is statistically significant (Binomial, $N = 9$, $x = 6$, $p < 0.05$).

Kela. When introduced to the containers, Kela explored the rectangular rose container and the round container, wearing the rectangular container on her head as she swam around the tank. During Phases I, II and IV, she exhibited a right preference when she selected a container: in ten of the eleven trials in which she responded to containers, she chose the rightmost container. In only one instance – the first trial – did she overturn a fish-container, and in three other trials she approached and/or contacted a fish-container but did not overturn it. In five trials she failed to overturn any container, and in seven trials she initially responded to an empty container.

Discussion

Kela's overall failure in Experiment 2 suggests that her more successful performance in Experiment 1 resulted from cueing from the observers. Aurora's performance is in keeping with an (at least) stage four understanding of object permanence, though there are difficulties with this interpretation. Her "success" rate depends upon conflating her performance on stage four (Phases I and IV) and stage six (Phase II) tasks, as well as including as "successes" her failed attempts to overturn fish-containers. Even if one grants Aurora success, it might be attributable to her recognizing, as such, sounds or vibrations caused by the fish landing in the container, though if such a perceptual signal were available it is unclear why she did not overturn the correct container in all instances. The fact that Aurora overturned the transparent containers and ate the fish inside indicates that fish were desirable objects for her.

Suggestions and conclusions

Clearly, problems arise when testing echolocating cetaceans' understanding of object permanence. Cetaceans are usually unavailable for protracted intrusive experimentation, and the conditions of their captivity may cause them (because of extensive training) to have expectations that their behavior in the presence of "trainers" (humans with fish) should be contingent upon signals from those trainers.

These expectations could be utilized in testing object permanence understanding. One could begin by training the animal to respond to a particular signal from the experimenter by retrieving a specific object (the retrieval-object). The animal could then be taught to overturn containers (apparently easily learned by dolphins and beluga whales). Finally, while the animal watched, the retrieval-object as well as two other same-size unique objects could each be placed simultaneously into one of three opaque, non-transportable but overturnable containers near the water, and the animal could be requested to get the retrieval-object. The experiment could be enacted in two contexts: placing the retrieval-object into an object the animals cannot "see" through (via echolocation) in one, and (as a check) placing it in an object they can "see" through in the other. Retrieval of the retrieval-object in both contexts would indicate that the animal knew that the retrieval-object continued to exist in extraperceptual space. Note, however, that the training involved in this proposed study introduces conditioning into a display of object permanence understanding; if object permanence understanding must be "spontaneously displayed" without conditioning (Dumas and Doré, 1989, p. 191), this experimental design is unacceptable. An alternative might be to perform Experiment 2 again, but place the fish (or in Phase II the fish-container) into one of three *familiar* containers with which the animals have had extended but unconditioned interaction, and *then* place the containers in the water simultaneously. This elaboration would avoid acoustic and vibratory cues from the fish landing in a container, provide the animal with extensive experience to learn that the containers are available for overturning, and permit spontaneous responding.

At present, the available evidence suggests that echolocating animals such as dolphins and beluga whales have stage four and perhaps stage five understanding of object permanence. There is no evidence that they have stage six understanding of object permanence, and hence no evidence that they mentally represent external (non-self) objects. Indeed, the failure of dolphins to exhibit stage six understanding of object permanence argues against the idea (suggested by Suddendorf and Whiten, 2001) that secondary representation is the basis for such understanding. Cetaceans' understanding of objects clearly requires further study, and other echolocating animals such as bats should also be studied to discern whether they too fail to exhibit a fully developed understanding of object permanence. In this way, whether or not echolocation is an adaptation that preempts the cognitive capacity for understanding object permanence can be determined.

ACKNOWLEDGMENTS

Initial exploratory study occurred at Kewalo Basin Marine Mammal Laboratory. Neil Overstrum, David Merrit and Tim Binder coordinated studies at Mystic Aquarium, and John Wood, Doris Luther and Jim Rice assisted. They, along with Pearl Yao, Peter Sherman, Peter Scheifele,

Kurt Horton, Jenny Flaherty, Linda Phillips, Nicholas Thompson, Shyamala Venkataraman, David Miller and Fran Dolins offered intelligent and stimulating ideas, questions and criticisms. We are grateful for the assistance of all.

REFERENCES

Au, W. W. L. (1993). *The sonar of dolphins*. New York: Springer-Verlag.

Aubauer, R., Au, W. W. L, Nachtigall, P. E., Pawloski, D. A. and DeLong, C. M. (2000). Classification of electronically generated phantom targets by an Atlantic bottlenose dolphin (*Tursiops truncatus*). *Journal of the Acoustical Society of America*, **107**, 2750–2754.

Baillargeon, R., Spelke, E. S. and Wasserman, S. (1985). Object permanence in five-month-old infants. *Cognition*, **20**, 191–208.

Barth, J. and Call, J. (2006). Tracking the displacement of objects: a series of tasks with great apes (*Pan troglodytes*, *Pan paniscus*, *Gorilla gorilla*, and *Pongo pygmaeus*) and young children (*Homo sapiens*). *Journal of Experimental Psychology: Animal Behavior Processes*, **32**, 239–252.

Beran, M. J. and Minahan, M. F. (2000). Monitoring spatial transpositions by bonobos (*Pan paniscus*) and chimpanzees (*P. troglodytes*). *International Journal of Comparative Psychology*, **13**, 1–15.

Call, J. (2001). Object permanence in orangutans (*Pongo pygmaeus*), chimpanzees (*Pan troglodytes*), and children (*Homo sapiens*). *Journal of Comparative Psychology*, **115**, 159–171.

Collier-Baker, E., Davis, J. M. and Suddendorf, T. (2004). Do dogs (*Canis familiaris*) understand invisible displacement? *Journal of Comparative Psychology*, **118**, 421–433.

Collier-Baker, E. and Suddendorf, T. (2006). Do chimpanzees (*Pan troglodytes*) and 2-year-old children (*Homo sapiens*) understand double invisible displacement? *Journal of Comparative Psychology*, **120**, 89–97.

Dawson, W. W. (1980). The cetacean eye. In L. M. Herman (Ed.), *Cetacean behavior*, (pp. 53–100). New York: Wiley.

de Blois, S. T., Novak, M. A. and Bond, M. (1999). Can memory requirements account for species' differences in invisible displacement tasks? *Journal of Experimental Psychology: Animal Behavior Processes*, **25**, 168–176.

DeLong, C. M. (2004). Object-centered representations in echolocating dolphins: evidence from acoustic analyses of object echoes and a human listening study. *Dissertation Abstracts International: Section B: The Sciences and Engineering*, **64**, 5245.

Doré, F. Y. (1986). Object permanence in adult cats (*Felis catus*). *Journal of Comparative Psychology*, **100**, 340–347.

Doré, F. Y. and Dumas, C. (1987). Psychology of animal cognition: Piagetian studies. *Psychological Bulletin*, **102**, 219–233.

Doré, F. Y. and Goulet, S. (1998). The comparative analysis of object knowledge. In J. Langer and M. Killen (Eds.), *Piaget, evolution, and development* (pp. 55–72). Mahwah: Lawrence Erlbaum.

Doré, F. Y., Goulet, S. and Herman, L. M. (1991). *Permanence de l'objet chez deux dauphins* (*Tursiops truncatus*) [Object permanence in two dolphins]. Paper presented at the XXIIIème Journées d'étude de l'Association de Psychologie Scientifique de Langue Française, Rome.

Doré, F.Y., Fiset, S., Goulet, S., Dumas, M.C. and Gagnon, S. (1996). Search behavior in cats and dogs: interspecific differences in spatial cognition. *Animal Learning and Behavior*, **24**, 142–149.

Dos Santos, M.E. and Almada, V.C. (2004). A case for passive sonar: analysis of click train production patterns by bottlenose dolphins in a turbid estuary. In J.A. Thomas, C.F. Moss and M. Vater (Eds.), *Echolocation in bats and dolphins* (pp. 400–403). Chicago: University of Chicago Press.

Dumas, C. (2000). Flexible search behavior in domestic cats (*Felis catus*): a case study of predator-prey interaction. *Journal of Comparative Psychology*, **114**, 232–238.

Dumas, C. and Doré, F.Y. (1989). Cognitive development in kittens (*Felis catus*): a cross-sectional study of object permanence. *Journal of Comparative Psychology*, **103**, 191–200.

Dumas, C. and Wilkie, D.M. (1995). Object permanence in ring doves (*Streptopelian risoria*). *Journal of Comparative Psychology*, **109**, 142–150.

Etienne, A.S. (1984). The meaning of object permanence at different zoological levels. *Human Development*, **27**, 309–320.

Finke, R.A. (1980). Levels of equivalence in imagery and perception. *Psychological Review*, **87**, 113–132.

Finke, R.A. (1986). Mental imagery and the visual system. *Scientific American*, **254**, 88–95.

Fiset, S., Beaulieu, C. and Landry, F. (2003). Duration of dogs' (*Canis familiaris*) working memory in search for disappearing objects. *Animal Cognition*, **6**, 1–10.

Flavell, J.H. (1963). *The developmental psychology of Jean Piaget*. New York: Van Nostrand.

Funk, M.S. (1996). Development of object permanence in the New Zealand parakeet (*Cyanoramphus auriceps*). *Animal Learning and Behavior*, **24**, 375–383.

Gagnon, S. and Doré, F.Y. (1992). Search behavior in various breeds of adult dogs (*Canis familiaris*): object permanence and olfactory cues. *Journal of Comparative Psychology*, **106**, 58–68.

Gagnon, S. and Doré, F.Y. (1993). Search behavior of dogs (*Canis familiaris*) in invisible displacement problems. *Animal Learning and Behavior*, **21**, 246–254.

Gagnon, S. and Doré, F.Y. (1994). A cross-sectional analysis of object permanence development in dogs (*Canis familiaris*). *Journal of Comparative Psychology*, **108**, 220–232.

Goulet, S., Doré, F.Y. and Rousseau, R. (1994). Object permanence and working memory in cats (*Felis catus*). *Journal of Experimental Psychology: Animal Behavior Processes*, **20**, 347–365.

Harley, H.E., Putman, E.A. and Roitblat, H.L. (2003). Bottlenose dolphins perceive object features through echolocation. *Nature*, **424**, 667–668.

Heishman, M., Conant, M. and Pasnak, R. (1995). Human analog tests of the sixth stage of object permanence. *Perceptual and Motor Skills*, **80**, 1059–68.

Herman, L.M. (1980). Cognitive characteristics of dolphins. In L.M. Herman (Ed.), *Cetacean behavior* (pp. 363–429). New York: Wiley.

Herman, L.M. (2002). Vocal, social, and self-imitation by bottlenosed dolphins. In K. Dautenhahn and C.L. Nehaniv (Eds.), *Imitation in animals and artifacts* (pp. 63–108). Cambridge, MA: MIT Press.

Herman, L.M., and Forestell, P.H. (1985). Reporting presence or absence of named objects by a language-trained dolphin. *Neuroscience and Biobehavioral Reviews*, **9**, 667–681.

Herzing, D.L. (2004). Social and nonsocial uses of echolocation in free-ranging *Stenella frontalis* and *Tursiops truncatus*. In J.A. Thomas, C.F. Moss and M. Vater

(Eds.), *Echolocation in bats and dolphins* (pp. 404–410). Chicago: University of Chicago Press.

Herzing, D.L. and dos Santos, M.E. (2004). Functional aspects of echolocation in dolphins. In J.A. Thomas, C.F. Moss, and M. Vater (Eds.) *Echolocation in bats and dolphins* (pp. 386–393). Chicago: University of Chicago Press.

Hoban, E. (1986). *The promise of animal language research.* PhD dissertation, University of Hawaii at Manoa. University Microfilms: Ann Arbor, Michigan.

Hoban, E. and Mitchell, R.W. (1990a). *Object permanence and cetaceans.* Paper presented at the 18th Annual Conference of the International Marine Animal Trainers Association, Illinois.

Hoban, E. and Mitchell, R.W. (1990b). *Failure of bottlenosed dolphins and beluga whales to exhibit stage 4 understanding of object permanence?* Paper presented at the National Animal Behavior Society Meeting, New York.

Jerison, H.J. (1986). The perceptual worlds of dolphins. In R.J. Schusterman, J.A. Thomas and F.G. Wood (Eds.), *Dolphin cognition and behavior: a comparative approach* (pp. 141–166). Hillsdale: Lawrence Erlbaum.

Krushinskii, L.V. (1960). *Animal behavior: its normal and abnormal development.* New York: Consultants Bureau.

Madsen, C.J. and Herman, L.M. (1980). Social and ecological correlates of cetacean vision and visual appearance. In L.M. Herman (Ed.), *Cetacean behavior* (pp. 101–147). New York: Wiley.

Mathiew, M., Bouchard, M.-A., Granger, L. and Herscovitch, J. (1976). Piagetian object permanence in *Cebus capucinus, Lagothrica flavicauda, Pan troglodytes. Animal Behavior*, **24**, 585–588.

Mendes, N. and Huber, L. (2004). Object permanence in common marmosets (*Callithrix jacchus*). *Journal of Comparative Psychology*, **118**, 103–112.

Marten, K. and Psarakos, S. (1994). Evidence of self-awareness in the bottlenose dolphin (*Tursiops truncatus*). In S.T. Parker, R.W. Mitchell and M.L. Boccia (Eds.), *Self-awareness in animals and humans* (pp. 361–379). New York: Cambridge University Press.

Marten, K. and Psarakos, S. (1995). Using self-view television to distinguish between self-examination and social behavior in the bottlenose dolphin (*Tursiops truncatus*). *Consciousness and Cognition*, **4**, 205–224.

Mitchell, R.W. (1987). A comparative-developmental approach to understanding imitation. In P.P.G. Bateson and P.H. Klopfer (Eds.), *Perspectives in ethology*, vol. 7: Alternatives (pp. 183–215). New York: Plenum Press.

Mitchell, R.W. (1993). Mental models of mirror-self-recognition: two theories. *New Ideas in Psychology*, **11**, 295–325.

Mitchell, R.W. (1994). The evolution of primate cognition: simulation, self-knowledge, and knowledge of other minds. In D. Quiatt and J. Itani (Eds.), *Hominid culture in primate perspective* (pp. 177–232). Boulder: University Press of Colorado.

Mitchell, R.W. (1995). Evidence of dolphin self-recognition and the difficulties of interpretation. *Consciousness and Cognition*, **4**, 229–234.

Mitchell, R.W. (2007). Mirrors and matchings: imitation from the perspective of mirror-self-recognition, and why the parietal region is involved in both. In K. Dautenhahn and C.L. Nehaniv (Eds.), *Imitation and social learning in robots, humans and animals* (pp. 103–130). Cambridge: Cambridge University Press.

Mitchell, R.W., Yao, P. and Sherman, P. (1982). *Are dolphins stupid?* Paper presented at the Northeast Regional Meeting of the Animal Behavior Society, Massachusetts.

Mitchell, R.W., Yao, P., Sherman, P. and O'Regan, M. (1985). Discriminative responding of a dolphin (*Tursiops truncatus*) to differentially rewarded stimuli. *Journal of Comparative Psychology*, **99**, 218–225.

Natale, F., Antinucci, F., Spinozzi, G. and Potí, P. (1986). Stage 6 object concept in nonhuman primate cognition: a comparison between gorilla (*Gorilla gorilla gorilla*) and Japanese macaque (*Macaca fuscata*). *Journal of Comparative Psychology*, **100**, 335–339.

Neiworth, J.J., Steinmark, E., Basile, B.M., Wonders, R., Steely, F. and DeHart, C. (2003). A test of object permanence in a new world monkey species, cotton top tamarins (*Saguinus oedipus*). *Animal Cognition*, **6**, 27–37.

Pack, A.A., and Herman, L.M. (1995). Sensory integration in the bottlenosed dolphin: immediate recognition of complex shapes across the senses of echolocation and vision. *Journal of the Acoustical Society of America*, **98**, 722–733.

Parker, S.T. and Gibson, K.R. (1977). Object manipulation, tool use and sensorimotor intelligence as feeding adaptations in cebus monkeys and great apes. *Journal of Human Evolution*, **6**, 623–641.

Pepperberg, I.M. and Funk, M.S. (1990). Object permanence in four species of psittacine birds: an African Grey parrot (*Psittacus erithacus*), an illiger minii macaw (*Ara maracana*), a parakeet (*Melopsittacus undulatus*), and a cockatiel (*Nymphicus hollandicus*). *Animal Learning and Behavior*, **18**, 97–108.

Piaget, J. (1945/1962). Play, dreams and imitation in childhood (trans. C. Gattegno and F.M. Hodgson). New York: Norton. (Original work: J. Piaget (1945). *La formation du symbole chez l'enfant*.)

Piaget, J. (1954). *The construction of reality in the child*. New York: Basic Books.

Pollok, B., Prior, H. and Güntürkün, O. (2000). Development of object permanence in foodstoring magpies (*Pica pica*). *Journal of Comparative Psychology*, **114**, 148–157.

Reisberg, D. and Heuer, F. (2005). Visuospatial images. In P. Shah and A. Miyake (Eds.), *The Cambridge handbook of visuospatial thinking* (pp. 35–80). New York: Cambridge University Press.

Reiss, D. and Marino, L. (2001). Mirror self-recognition in the bottlenose dolphin: a case of cognitive convergence. *Proceedings of the National Academy of Sciences*, **98**, 5937–5942.

Rivera, S.M., Wakeley, A. and Langer, J. (1999). The *drawbridge* phenomenon: representational reasoning or perceptual preference? *Developmental Psychology*, **35**, 427–435.

Ryland, F. (1909). *Thought and feeling*. London: Hodder and Stoughton.

Suddendorf, T. and Whiten, A. (2001). Mental evolution and development: evidence for secondary representation in children, great apes, and other animals. *Psychological Bulletin*, **127**, 629–650.

Tayler, C.K. and Saayman, G.S. (1973). Imitative behaviour by Indian Ocean bottlenose dolphins (*Tursiops aduncus*) in captivity. *Behaviour*, **44**, 286–298.

Thinus-Blanc, C. and Scardigli, P. (1981). Object permanence in the golden hamster. *Perceptual and Motor Skills*, **53**, 1010.

Thomas, J.A., Moss, C.F. and Vater, M. (Eds.) (2002). *Echolocation in bats and dolphins*. Chicago: University of Chicago Press.

Thomas, R.K. and Walden, E.L. (1985). The assessment of cognitive development in human and nonhuman primates. In E.S. Watts (Ed.), *Nonhuman primate models for human growth and development* (pp. 187–215). New York: Alan R. Liss, Inc.

Triana, E. and Pasnak, R. (1981). Object permanence in cats and dogs. *Animal Learning and Behavior*, **9**, 135–139.

Vaughter, R. M., Smotherman, W. and Ordy, J. M. (1972). Development of object permanence in the squirrel monkey. *Developmental Psychology*, **7**, 34–38.
Warshall, P. (1974). The ways of whales. In J. McIntyre (Ed.), *Mind in the waters* (pp. 110–118, 120–121, 124–131, 133–140). New York: Charles Scribner's Sons.
Wise, K. L., Wise, L. A. and Zimmerman, R. R. (1974). Piagetian object permanence in the infant rhesus monkey. *Developmental Psychology*, **10**, 429–437.
Wood, S., Moriarity, K., Gardner, B. T. and Gardner, R. A. (1980). Object permanence in child and chimpanzee. *Animal Learning and Behavior*, **8**, 3–9.

ANGELO ARLEO AND LAURE RONDI-REIG

13

Multimodal sensory integration and concurrent navigation strategies for spatial cognition in real and artificial organisms

INTRODUCTION

Spatial cognition involves the ability of a navigating agent (be it an animal or an autonomous artifact) to acquire spatial knowledge (e.g., spatio-temporal relations between environmental cues or events), organize it properly, and employ it to adapt its motor behavior to the specific context (e.g., performing flexible goal-oriented behavior to solve a navigation task). Similar to other high-level brain functions, spatial cognition calls upon parallel information processing mediated by multiple neural substrates that interact, either cooperatively or competitively, to promote appropriate spatial behavior.

At the sensory level, different perceptual modalities provide the navigator with a manifold description of the currently experienced spatial context. The integration of these multimodal signals (that are processed by interrelated brain regions) into a coherent representation is at the core of spatial cognition. A large body of experimental work has been done to elucidate the neural mechanisms subserving the establishment and maintenance of spatial representations in animals and humans. The next section reviews some experimental findings issued from this research, and focuses on those that concern the interrelation between different sensory modalities.

At the action selection level, determining and maintaining a trajectory from one place to another (i.e., navigating, according to Gallistel, 1990) involves multiple concurrent processes, and requires the ability of the subject to adapt its goal-directed strategy to the complexity of the task. The process of dynamically weighing the behavioral contribution of distinct navigation policies depends on contextual variables, like the

Spatial Cognition, Spatial Perception: Mapping the Self and Space, ed. Francine L. Dolins and Robert W. Mitchell. Published by Cambridge University Press.

available sensory inputs and their relative importance, and the motivational state of the animal. The third section focuses on this issue and reviews some experimental findings concerning the cooperative-competitive interaction of multiple spatial navigation strategies.

Similar to animals, autonomous navigating artifacts (e.g., mobile robots) need to interact with their environment, process multimodal sensory signals, and learn both low-level sensory-motor couplings and more abstract context representations supporting spatial behavior. Therefore, in parallel to experimental neuroscience, a large body of research in autonomous robotics has focused on spatial learning-related issues (e.g., self-localization, space representation and wayfinding techniques). Most of the classical control architectures engineered so far (e.g., Elfes, 1987; Kuipers and Byun, 1991; Thrun, 1998; Arleo *et al.*, 1999) provide task-specific (ad hoc) solutions and are not as general and adaptive as animals' spatial learning systems. Therefore, a novel approach to designing autonomous navigating artifacts is being explored, whose principles take inspiration from known behavioral and neurophysiological mechanisms underlying animals' spatial learning capabilities. This approach, termed neuro-mimetic robotics or simply neuro-robotics, focuses more on adaptiveness and flexibility than on optimality and completeness, and stresses the idea that the agent must acquire its own worldview by means of its experience (Brooks, 1991). The last section of this chapter focuses on neuro-mimetic spatial learning and on the importance of combining multisensory information for robust coding of the spatial contexts experienced by the robots.

ACQUIRING REPRESENTATIONS OF SPACE

The variety of sensory modalities conveying spatial information can be dichotomized into two main categories, namely idiothetic and allothetic cues. Idiothetic stimuli are self-motion related signals and include vestibular (inertial), kinesthetic (e.g., information from muscle and joint receptors), motor command efferent copies, and sensory flow information (e.g., optic field flow signals informing the navigator about its own movements). Allothetic signals provide information about the external environment and include visual (e.g., environmental landmarks), olfactory, auditory and somatosensory (e.g., tactile or texture) cues. Learning spatial memories requires the extraction of coherent information from such a redundant and multidimensional sensory input space. This learning process implies, for instance, maintaining idiothetic and allothetic cues congruent (e.g., minimization of interferences or conflicts) both during the exploration of a novel environment and across subsequent visits to a familiar environment.

A given sensory modality is labeled as allothetic or idiothetic to characterize the *type of information* it conveys. On the other hand, if we want to characterize the *way this information is represented* by the navigator, we need to introduce the concept of reference coordinate system (or simply reference frame). This system defines the framework in which spatial information (e.g., the position of an object) can be represented relative to an origin point. Depending on the anchorage of the origin of the reference coordinate system, the same information can be encoded *egocentrically* or *allocentrically*. If the reference frame is centered on the subject (e.g., on a body part such as the head) the representation is said to be egocentric. If the origin of the framework is a fixed point of the environment (e.g., a corner of the room) the representation is called allocentric. As shown in Figure 13.1a, the same allothetic spatial information (e.g., the position of a visual cue in the environment) can be represented either egocentrically (e.g., relative to the body of the navigator) or allocentrically (e.g., relative the room corner). Egocentric coding is simple to build but it varies as the navigator moves in the environment (because the reference frame translates and rotates as the subject moves). Allocentric coding requires more complex processing (e.g., to relate the visual cue position to the world-centered origin), but it is invariant with respect to the subject's position and orientation in the environment. As shown in Figure 13.1b, idiothetic signals (e.g., vestibular information) can also be employed to describe self-motion either egocentrically or relative to an allocentric reference frame.

One factor of the complex multisensory integration process is that different modalities are encoded within different reference frames, each defined by the spatial distribution of the corresponding sensory receptors. For instance, visual information is represented according to the distribution of photoreceptors on the retina, whereas tactile inputs are represented according to the distribution of somatosensory receptors on the skin surface. Thus, in order to combine multisensory spatial information, a navigating system has to infer a representation that is consistent with all the different modality frameworks (Avillac *et al.*, 2005).

Spatial learning on the basis of allothetic and idiothetic signals

Animals employ both idiothetic and allothetic cues to maintain memory traces of the spatial components (e.g., their body position and

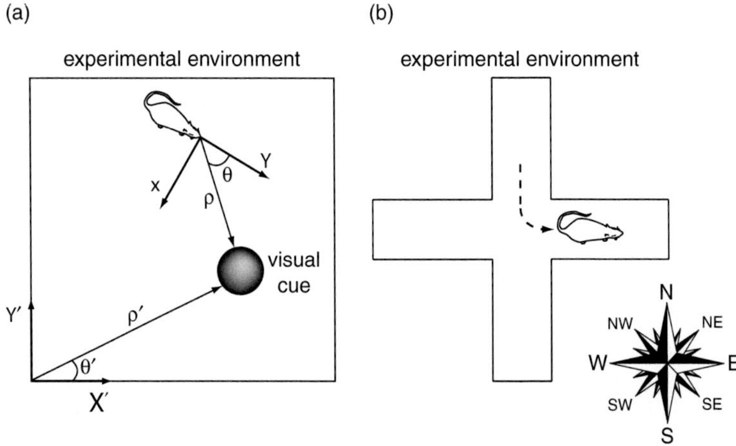

Figure 13.1. Encoding spatial information within a reference coordinate system. **(a)** The circular object provides an *allothetic* (visual) spatial cue to the navigator (rat). The latter can represent the spatial position of the external cue within the *egocentric* reference frame X-Y (centered on its head), that is estimate the distance ρ between its head and the object, as well as the angle θ between its heading and the direction to the object. Alternatively, the rat can encode the same spatial information within the *allocentric* coordinate system X′-Y′ (centered on the bottom-left corner of the experimental environment), that is estimate the distance ρ' and the angle θ'. **(b)** In this example, the navigator can employ *idiothetic* information (e.g., vestibular signals) to represent the change of its motion direction within an *egocentric* reference frame, that is "I turned to my left." Alternatively, it can refer to the *allocentric* directional system based on the geomagnetic north, that is "I turned eastward."

orientation) of experienced events. For instance, they are capable of estimating their current location relative to a starting point (i.e., homing vector) by integrating linear and angular self-motion signals over time (Figures 13.2a and b). This process, termed *path integration* or dead reckoning (Mittelstaedt and Mittelstaedt, 1980; Mittelstaedt and Mittelstaedt, 1982; Etienne *et al.*, 1998a, 1998b; Etienne and Jeffery, 2004; McNaughton *et al.*, 2006), relies upon idiothetic cues like vestibular and kinesthetic signals, motor command efferent copies, and sensory (e.g., optic) flow information. On the other hand, self-localization can occur solely on the basis of allothetic cues like vision, auditory, olfactory and tactile signals. Indeed, locations can be characterized by specific allothetic sensory patterns (e.g., configurations of visual cues), such that memorizing these sensory patterns can enable a subject to recognize familiar places.

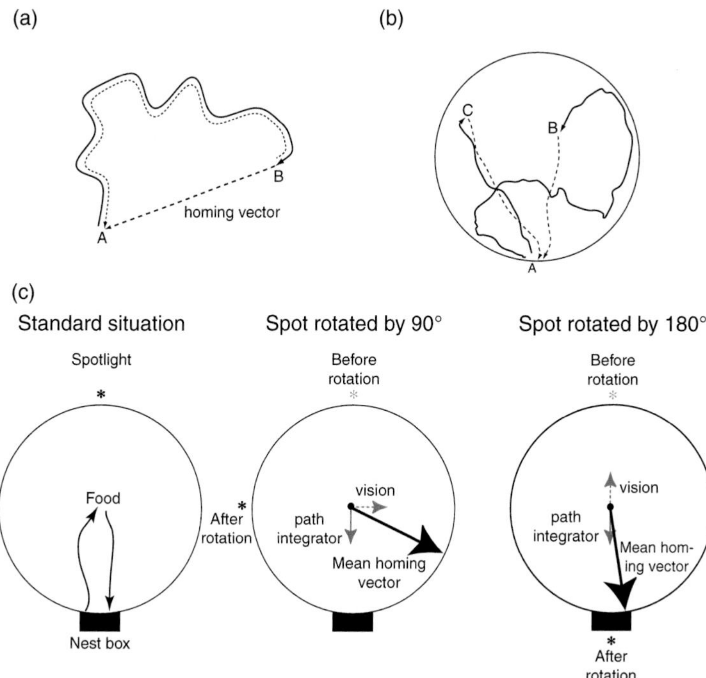

Figure 13.2. Homing behavior based on path integration (PI). **(a)** Difference between path reversal (i.e., inverting the sequence of movements performed from a starting point A to a current location B), and path integration (i.e., integrating translations and rotations over time to generate a homing vector leading the animal directly to the departure point A). The solid line represents the outward journey; the dotted line indicates the return journey based on path reversal; the dashed line is the homing vector obtained by path integration. Adapted from Etienne *et al.* (1998). **(b)** Two examples of homing behavior performed by two hamsters. After having been guided by a bait from the nest location A to points B and C (solid lines), the two animals return home following direct trajectories (dashed lines). The experiment was performed in the dark in a circular arena of 2 m of diameter. Adapted from Etienne *et al.* (1998). **(c)** Hamsters' homing behavior in conflict situations. During training (left), a distal spotlight (asterisk) provides a stable landmark to the animal performing hoarding excursions to a feeder. In probe trials, the spotlight is rotated by either 90° (center) or 180° (right). Animals are guided from the nest to the feeder in darkness conditions, then the spotlight is turned on, which creates a conflict between self-motion (continuous gray arrows) and visual (dashed gray arrows) information. Large arrows indicate the homing vectors followed by the animals and show that in the case of 90° conflicts the visual landmark signal tends to dominate over self-motion, whereas for a 180° mismatch the path integration component becomes predominant. Adapted from Etienne and Jeffery (2004).

Idiothetic and allothetic spatial information have complementary strengths and weaknesses. Since path integration does not depend on external references, it allows a subject to self-localize in an unfamiliar environment from its very first exploring excursion (Griffin and Etienne, 1998). Also, path integration is a basic mechanism suitable for all types of environments (i.e., with or without external cues) and navigators (e.g., agents that can not exploit their interaction with the external world effectively). A limitation of path integration is its vulnerability to cumulative drift over time. Indeed, the idiothetic-based dynamics, consisting of integrating translational and rotational signals over time, is prone to systematic as well as nonsystematic errors that quickly disrupt the position estimate (Mittelstaedt and Mittelstaedt, 1982; Etienne *et al.*, 1998a). This holds for both biological and artificial navigating systems. Allothetic spatial information permits the formation of local[1] sensory views directly suitable for self-localization (McNaughton *et al.*, 1991). Also, if the spatial configuration of the environmental cues (e.g., distal landmark arrays) remains fairly stable over time, the position assessment process is not affected by cumulative errors. However, allothetic (e.g., visual) cues are not always available to the navigator (e.g., in darkness conditions). Additionally, since self-localization based on allothetic cues involves sensory pattern recognition, perceptual aliasing phenomena may occur, that is distinct areas of the environment may be characterized by equivalent local patterns. For instance, visual sensory aliasing can lead to singularities (i.e., ambiguous state representations) in a purely vision-based space coding (Sharp *et al.*, 1990).

Therefore, neither idiothetic nor allothetic cues are sufficient by themselves to establish reliable spatial memories (O'Keefe and Nadel, 1978; Jeffery and O'Keefe, 1999; Redish, 1999). One solution is to combine allothetic and self-motion signals into a unified representation. The combination of allothetic and idiothetic information may yield a mutual benefit in the sense that idiothetic cues may compensate for perceptual aliasing (e.g., discriminate between two locations in a visually symmetrical environment) and, conversely, environmental landmarks may be used to occasionally reset the integrator of self-motion signals. Idiothetic information might provide the spatial framework suitable for "grounding" the knowledge gathered by a navigating animal (McNaughton *et al.*, 1991; Knierim *et al.*, 1995). According to this hypothesis, allothetic local views might be tied onto this framework as the exploration of a novel environment proceeds. But how is this idiothetic–allothetic coupling established and maintained consistent over time? How are conflicts between self-motion and landmark cues solved?

Ethologists have largely investigated the interaction between path integration and landmark cues for spatial navigation (Etienne and Jeffery, 2004). Numerous behavioral studies involve homing tasks in which animals perform hoarding excursions and then return home with the collected food. One method to distinguish the idiothetic and allothetic determinants of the animals' homing behavior consists of setting a conflict between environmental (proximal or distal) and self-motion cues. Then, observing the homing vector makes it possible to assess the relative influence of allothetic and idiothetic information. Etienne *et al.* (1990) have examined the homing behavior of golden hamsters during hoarding trips within a circular open arena (Figure 13.2c). During training, a stable distal spotlight provided a unique visual landmark on an otherwise dark background. Other allothetic cues (e.g., tactile and olfactory stimuli) were masked. In probe trials, hamsters were guided in the dark from the nest (a box located at a fixed peripheral position) toward a feeding location at the center of the arena. During the uptake of food, visual and self-motion information were set in conflict by rotating the spotlight (either by 90 or 180 degrees relative to its standard position) and turning it on. The authors report that animals tended to return home following compromised homing vectors whose visual component dominated over the self-motion component in the case of 90-degree conflicts (Figure 13.2c, center). By contrast, when the divergence between the two types of information was further increased (i.e., 180 degrees) the path integration component became predominant (Figure 13.2c, right). In another series of experiments, Etienne *et al.* (2000) tested the realignment of the path integrator relative to distal landmarks. The arena and the peripheral home base were both rotated before each hoarding excursion. Then, in the darkness, the hamsters were guided from the rotated nest toward a feeding location along a two-leg (L-shaped) journey. Under this condition, the subjects mainly relied on their internally generated homing vector and returned to the new rotated home base. By contrast, if the environmental lights were briefly turned on at the end of the first outward leg and then switched off again, the animals tended to return to the original un-rotated home location, suggesting that a reset of the path integrator had occurred on the basis of the (unchanged) distal visual cues.

The neural bases of spatial learning

In addition to behavioral studies, an extensive body of electrophysiological work has been done to investigate the neural bases of animals' spatial

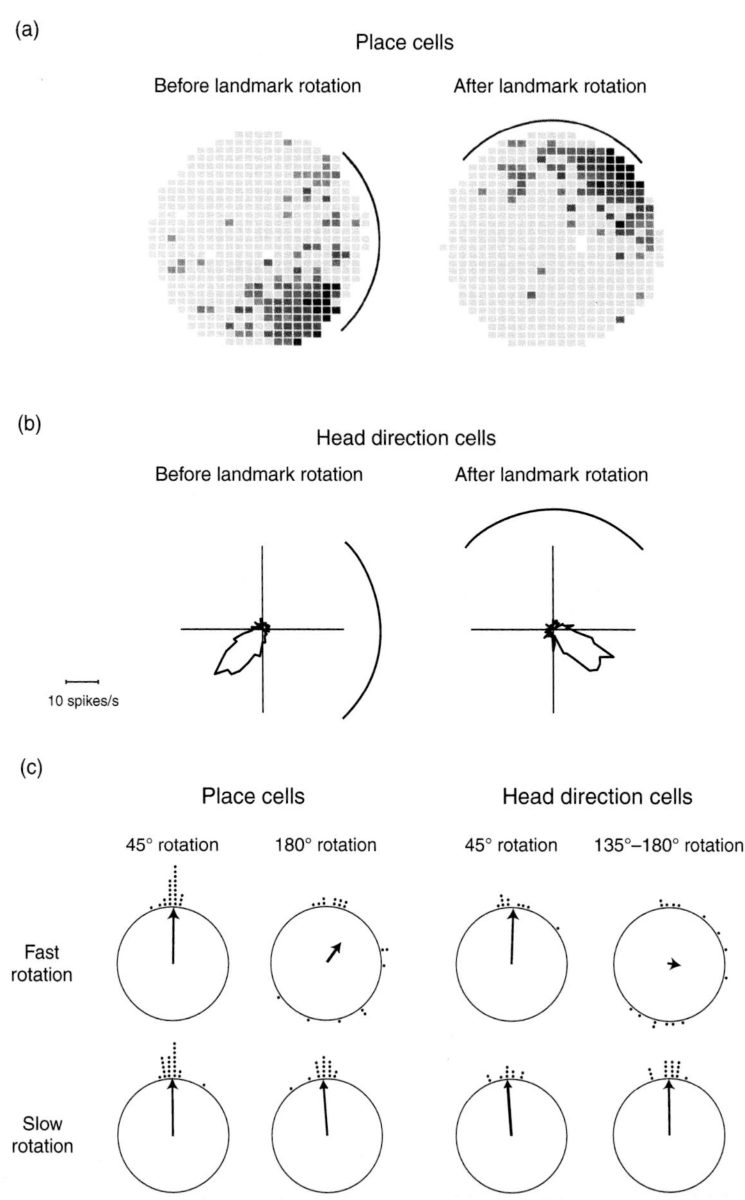

Figure 13.3. (a) Sample of receptive field of a place cell recorded from the rat hippocampus. The plots show the mean discharge of the neuron (blue and yellow denote peak and baseline firing rates, respectively) as a function of the animal position within the environment (a cylindrical arena with a cue card attached to inner wall). The location-selective response of the cell is controlled by the cue card in that rotating the card

learning capabilities. Extracellular single-cell recordings have largely focused on the properties of pyramidal neurons in the hippocampal formation. This limbic region has been thought to mediate spatial memory functions ever since location-sensitive cells (Figure 13.3a) in the hippocampus of freely moving rats were found (O'Keefe and Dostovsky, 1971). These neurons, termed hippocampal place (HP) cells, are likely to provide a spatial representation in allocentric (i.e., world centered) coordinates, thus providing a "cognitive map" to support flexible navigation (O'Keefe and Nadel, 1978). Furthermore, since the spatially selective responses of HP neurons might result from the projection of contextual (relational) memories onto the two-dimensional locomotion space of the animal, a role for the hippocampal formation in a larger class of memories, namely declarative memory, has been suggested (Burgess *et al.*, 2002; Fortin *et al.*, 2002).

The hippocampal formation is well suited for subserving the integration of multimodal spatial information. It receives afferents from numerous subcortical regions (e.g., brainstem, amygdala, septum) via the fornix fiber bundle, and it is the recipient of highly processed sensory-motor signals conveyed by neocortical areas (e.g., the parietal cortex; Burgess *et al.*, 1999) mainly via the entorhinal cortex (Witter, 1993). The entorhinal region plays an important role in mediating the hippocampal–neocortical interactions, and recent electrophysiological findings have brought evidence for a key contribution of this brain structure to the spatial memory function (Fyhn *et al.*, 2004; Hafting *et al.*, 2005; Steffenach *et al.*, 2005; McNaughton *et al.*, 2006; Sargolini *et al.*, 2006).

Caption for Fig. 13.3. (cont.)
by 90° induces an equivalent rotation of the receptive field. Adapted from Muller and Kubie (1987). (b) Sample of tuning curve of a head direction cell recorded in the rat anterodorsal thalamic nucleus. The polar plots indicate that the cell has a unique "preferred" direction and that the response of the cell is controlled by the visual landmark. Data by Arleo and Wiener. (c) Interrelation between visual and self-motion cues in controlling place (left) and head direction (right) cells. Plots indicate the angular deviations of the responses of place and head direction cells relative to the visual landmark in the case of small (45°) and large (135°–180°) conflicts, and for fast and slow induction of the conflict. The angular deviation of 0°, indicating the absolute control of the visual landmark over the cells' response, is plotted at the 12:00 position. Dots indicate individual trials, whereas arrows are averages over all trials. Adapted from Knierim *et al.* (1998). Part (a) is reproduced in the colour plate section.

Binding of multiple spatial representations may occur via correlational learning. According to Hebb (1949), it is now accepted that correlated spiking of pre- and post-synaptic neurons can result in strengthening or weakening of synapses, depending on the temporal order of spiking. The activity-dependent long-term synaptic plasticity in the hippocampus constitutes a neurochemical mechanism suitable for this type of learning (Morris and Frey, 1997). Both pharmacological and genetic approaches have shown that hippocampal NMDA (N-methyl-D-Aspartate) receptors (NMDARs) are required for the induction of hippocampal long-term potentiation (LTP), a temporally correlational learning process that can be understood in terms of Hebbian synaptic modification (Collingridge *et al.*, 1983; Morris *et al.*, 1986; Tsien *et al.*, 1996). NMDAR-mediated plasticity in the recurrent connections of the CA3 hippocampal region is crucial for the rapid encoding of novel experiences (Lee and Kesner, 2002). CA3-NR1-knockout mice are deficient in acquiring novel place/reward information, and CA1 HP cells in these animals are significantly impaired when recorded in a novel environment (Nakazawa *et al.*, 2002, 2003, 2004).

Complementing the allocentric place responses of hippocampal neurons, head direction (HD) cells provide an allocentric representation of the orientation of the animal (see Wiener and Taube, 2005, for a review). The discharge of these neurons is highly correlated with the direction of the head of the animal in the azimuthal plane, regardless of the orientation of the head relative to the body, of the animal's ongoing behavior and of its spatial location. Each HD cell is selective for one specific "preferred" direction (Figure 13.3b), and the preferred directions of a population of HD cells tend to be evenly distributed over 360 degrees. Direction-sensitive neurons have been found in numerous brain regions centered on the limbic system, including postsubiculum (Ranck, 1984; Taube *et al.*, 1990a), anterodorsal thalamic nucleus (Blair and Sharp, 1995; Taube, 1995), lateral mammillary nucleus (Stackman and Taube, 1998), retrosplenial cortex (Chen *et al.*, 1994a), dorsal striatum (Wiener, 1993), and dorsal tegmental nucleus (Sharp *et al.*, 2001). Similar to the HP cell system, the HD circuit receives multimodal afferent information, including angular self-motion signals from the medial vestibular nucleus and visual inputs from neocortical areas (e.g., parietal cortex).

Thus, the discharges of HP and HD cells are determined by the interaction between allothetic and idiothetic cues. Several studies have attempted to identify the nature of the signals relevant for the establishment and maintenance of their firing properties (see Best *et al.*, 2001).

The responses of HP and HD cells are anchored to visual landmarks of the environment (O'Keefe and Conway, 1978; Muller and Kubie, 1987; O'Keefe and Speakman, 1987; Taube *et al.*, 1990b; Bostock *et al.*, 1991; Knierim *et al.*, 1998; Zugaro *et al.*, 2003). A classical experimental apparatus employed to record HP and HD cells consists of a black cylindrical arena in which the rat freely moves while searching for chocolate pellets. The high walls of the cylinder prevent the animal from seeing outside the arena. A large white card, attached to the inner wall of the otherwise black cylinder, is used as a unique salient visual cue. Data show that rotating the white card causes an equal rotation of the receptive fields of HP and HD cells (Figure 13.3a and b). More generally, experimental findings indicate that distal (background) visual cues tend to dominate over proximal (foreground) visual cues in controlling HP (Cressant *et al.*, 1997) and HD cells (Zugaro *et al.*, 2001, 2004). The dominance of background cues may be due to the fact that they provide more stable references than proximal landmarks as the animal moves around. Consistent with this hypothesis, the more stable an animal perceives an allothetic cue to be, the higher its influence upon HP and HD cell dynamics (Biegler and Morris, 1993; Knierim *et al.*, 1995; Jeffery, 1998).

Despite their dependence on exteroceptive signals, both HP and HD cells can maintain stable location and direction tuning for several minutes in the absence of environmental landmarks (Muller and Kubie, 1987; O'Keefe and Speakman, 1987; Quirk *et al.*, 1990; Taube *et al.*, 1990b; Chen *et al.*, 1994b; Markus *et al.*, 1994), which suggests a role for internal movement-related cues. HP and HD cells continue to discharge when the animal moves about in complete darkness (see Wiener and Arleo, 2003, on persistent activity in limbic neurons). Also, the location-selective responses of HP cells can develop in blind animals exploring a novel environment (Hill and Best, 1981). Save *et al.* (1998) studied the HP cell activity in blind rats and found receptive fields and response specifics (e.g., spike parameters) very similar to those recorded from sighted rats. The only major difference concerned the mean peak firing rates that were prominently lower in HP cells from blind animals. Vestibular information seems to be very important for maintaining the selectivity properties of HP and HD cells (Stackman and Taube, 1997). Also, motor signals influence the dynamics of both types of cells, since HP and HD neurons exhibit a dramatic attenuation of their responses if the animal is tightly restrained (Foster *et al.*, 1989; Taube, 1995).

Recently, electrophysiological recordings of HP and HD neurons shed light on the interaction between idiothetic and allothetic cues and their relative importance under different experimental conditions.

Knierim *et al.* (1998) made self-motion and visual cues incongruent (by rotating the animal and a salient familiar landmark relative to each other) and recorded both HP and HD cells before and after the onset of the conflict (Figure 13.3c). For small angular mismatches (45 degrees) between idiothetic and landmark information, the responses of HP and HD cells remained anchored to the visual stimulus. When larger discrepancies (180 degrees) were induced by slow continuous rotations, the landmark still controlled the cell responses. By contrast, for sudden large (180 degree) rotations, either HP and HD cells followed the landmark, or self-motion cues predominated, or a reorganization (remapping) of HP fields occurred. Jeffery and O'Keefe (1999) further examined HP cell responses in the presence of 180-degree conflicts and found that the ability of visual cues to dominate self-motion signals might depend on the "confidence" of the idiothetic information. When animals were prevented from visual update for about three minutes while the conflict was introduced, the visual landmark tended to predominate. Conversely, when animals underwent visual isolation during only thirty seconds, a marked attenuation of the visual control was observed.

Finally, HD cells maintain their directional coding even after the removal of external landmarks, but their preferred directions may drift slowly over time (Taube, 1998). When the visual cue is put back to its standard position, HD cells tend to realign their preferred directions with the external reference (Goodridge *et al.*, 1998). However, this resetting does not always occur during subsequent light–dark–light recording phases (Knierim *et al.*, 1998).

THE COOPERATIVE–COMPETITIVE INTERACTION OF MULTIPLE SPATIAL STRATEGIES

As mentioned in the introduction of this chapter, efficient spatial navigation calls upon the ability of the subject to select the strategy that is most appropriate to the complexity of the task (see Figure 13.4, see also the review by Trullier *et al.*, 1997). For instance, reaching a goal that is either visible or identified by a visible cue (beacon) calls for a simple reactive behavior: orient toward the target (or beacon) and approach it. This procedure, named *target approaching* (or beacon approaching, or cue strategy), requires the acquisition of a single egocentric stimulus–response association and demands limited spatial information processing. If the target is neither directly visible nor identified by a beacon, the subject can learn an ensemble of Pavlovian associations to solve the task, which means that the subject

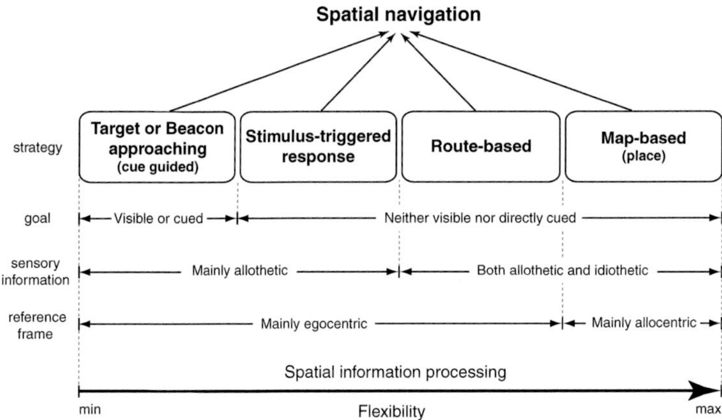

Figure 13.4. Taxonomy of spatial navigation strategies.

employs a *stimulus-triggered response* strategy. This procedure is suitable when the trajectory to a hidden target is identified by an ensemble of choice points where the spatial context (e.g., visual landmarks or geometric configurations) can be associated to specific directions of body movement. However, since the sensory-motor associations of the learned ensemble are treated independently, this strategy does not allow the subject to anticipate subsequent stimuli. One strategy that allows for such anticipation is *route navigation*, in which the subject learns the spatio-temporal relationships among the intermediate events of a sequence. This procedure allows the navigator to predict the next stimulus based on the current stimulus–action association. A route can be thought of as a learned sequence of egocentric stimulus–response–stimulus (S-R-S) associations. Within each S-R-S association, the most prominent element can be either the stimulus itself or the response requirement. In the former case we have "guidance," in the latter "orientation" that may include automated sequences of self-movements that lead to the goal or connect intermediate stimuli. Navigating using such a learned sequence of movements is a skill called "praxis." Tolman (1948) suggested that animals do not solve navigation tasks solely on the basis of sensory-motor associations or fixed motor programs. Rather, they are capable of learning a sort of navigation map encoding the spatio-temporal relationships between their position in the environment, their movements, and the location of rewarding sites. *Map-based navigation*, also referred to as "place" or "locale" navigation (O'Keefe and Nadel, 1978; Redish, 1999), requires complex processing (e.g., allocentric relational learning), but it allows the subject to

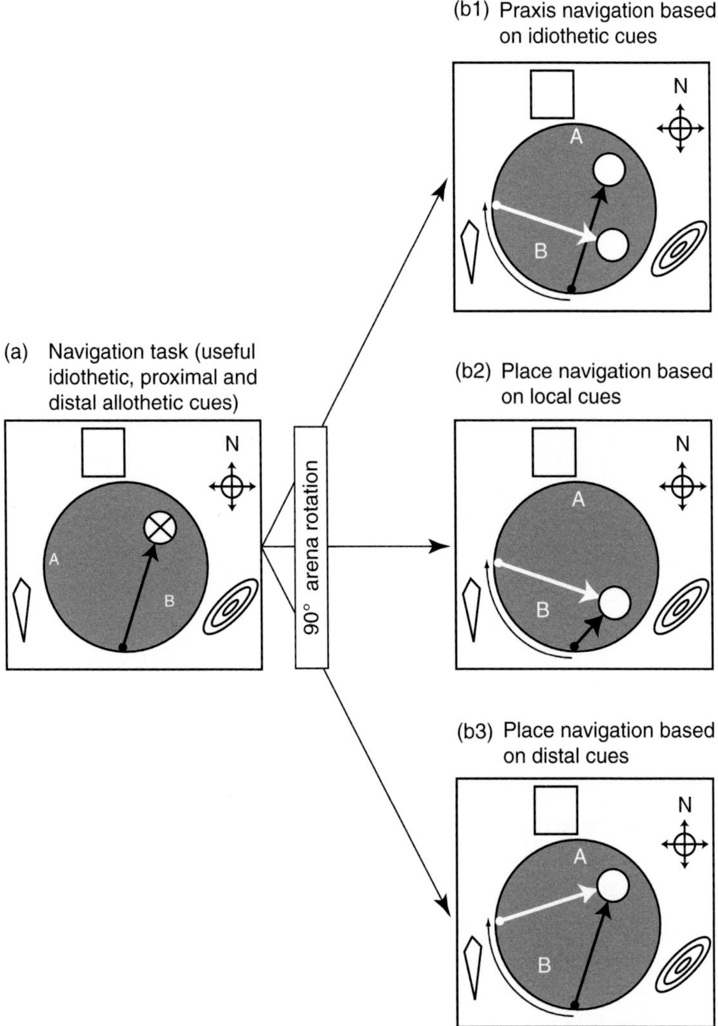

Figure 13.5. Adapted from Bures and Fenton (2000). Different navigation strategies within a circular enclosure in a square room are depicted. The movement trajectory is shown as a vector. Allothetic cues are provided by distal room cues outside the enclosure. Three are shown, along with a compass that indicates the inertial coordinate frame of the earth. There are also arena-bound proximal landmarks such as objects, odors, and textures that can be accessed. They are represented by A and B. **(a)** A rat can learn to go directly from a start point (•) to a goal using distal or proximal allothetic cues or idiothesis. Here the goal is also visible (marked by an X), and thus the animal can also find the target by cued navigation, simply going to the X. **(b)** When the arena is rotated by 90°, it

perform true flexible goal-oriented behavior (e.g., planning new or alternative trajectories, inferring shortcuts, solving multiple-goal tasks).

A coexistence of strategies functionally distinct

The aforementioned navigation strategies are interwoven processes that continuously concur to determine the animal's spatial behavior. The following experiment was designed in order to determine the relative contribution of three of these four strategies in rodents tested in a T-maze paradigm. The animals were first trained to find food located at the end of the right arm of the T. The rewarded goal was distinguished from the unbaited arm by its spatial location (map-based or place strategy), the presence of a tactile cue (cue guided strategy), and by the direction of the body turn required by the animal at the choice point (stimulus-triggered response strategy). Subsequently, probe trials were introduced occasionally in order to characterize the strategy used by each animal in order to solve the task. Normal animals exhibited a strategy distribution of about 30 percent map-based, 25 percent cue and 45 percent response (Barnes *et al.*, 1980). Thus, a task of spatial navigation can be solved on the basis of more than a single strategy.

How are these different strategies of navigation acquired and used? During the navigation process, multisensory information

---

Caption for Fig. 13.5. (cont.)
is possible to detect which navigation strategy is used by the rat. If the rat learned the target position by computing the orientation from the start at the arena periphery (b1), it will look for the goal at different room- and arena-defined places but always according to the vector 60° to the tangent at the start regardless of whether the start is in the room-defined south (black vector) or an arena-defined place 90° away (white vector). This behavior (i.e., navigating using a learned series of movements) is a skill called praxis and is one of the subtypes of the route-based strategies. Navigation learned according to cues on the arena (b2) will lead the rat to the goal between cues A and B regardless of the start position or the placement of the arena cue configuration. Similarly, navigation according to room cues (b3) will lead the rat to search for the goal in the northeast area regardless of the start or the presence of the arena cues. Note that the place navigation according to distal cues has the advantage over the two other strategies to give the correct position of the platform regardless to the starting position (black or white vectors) and despite the 90° arena rotation.

(i.e., allothetic and idiothetic) can be organized in two different reference frames (i.e., allocentric and egocentric) therefore leading to the simultaneous acquisition of different strategies. A rat learning in the Morris water maze to swim to a visible platform that remains at a fixed location will learn at the same time to navigate to the corresponding place (Whishaw and Mittleman, 1986). This capability has been demonstrated in a simple task in which the animal can use idiothetic, proximal and distal allothetic cues to reach a visible platform (Figure 13.5a). After the platform is hidden from view under the water, the rat readily finds the correct position of the platform using either praxis navigation (Figure 13.5b1, black vector) or place navigation according to distal cues (Figure 13.5b3). Animals that previously learned to go to the platform location according to proximal cues were looking for the platform in the wrong place (Figure 13.5b2) therefore showing a hierarchical organization of the different strategies in terms of flexibility. The simultaneous acquisition of these different strategies gives the animal the possibility to adapt its behavior to changes within the environment. For example, when sudden darkness eliminates the access to distant visual cues, the animal can still return to a recently visited part of the environment identifiable by the recalled locomotory movements (Fenton *et al.*, 1998). This finding is in agreement with the Whishaw and Mittleman (1986) experiment, suggesting that even not required strategies can be acquired during a task of spatial navigation (Bures and Fenton, 2000).

The coexistence of different strategies of navigation raises several questions. How does the brain store and coordinate the different strategies in the different brain systems? What are the parameters influencing the use of one strategy over the others?

The shift between strategies as a resultant of multiple parallel memory systems

Multiple and parallel memory systems

A large number of structures are involved in the neuronal network implicated during spatial navigation, including non-exhaustively the hippocampus (O'Keefe and Nadel, 1978), the parahippocampus (Ekstrom *et al.*, 2003), the entorhinal cortex (Hafting *et al.*, 2005; McNaughton *et al.*, 2006), the parietal cortex (Thinus-Blanc, 1996; Burgess *et al.*, 1999), the frontal cortex (Vallar *et al.*, 1999), the striatum (Wiener, 1993; Lavoie and Mizumori, 1994; Devan *et al.*, 1996), and even a new possible player, the cerebellum (Petrosini *et al.*, 1996; Lalonde, 1997; Molinari *et al.*, 1997; Rondi-Reig *et al.*, 1999, 2002).

To detail the contribution of these different structures is beyond the scope of this chapter (see for example Thinus-Blanc, 1996, for a detailed review). The cooperative or competitive interactions of these different systems have been reviewed by White and McDonald (2002). Of particular interest for this chapter is the hippocampus, playing a role in the organization of a spatial representation, and the cerebellum, which seems to be essentially involved in the organization of the motor behavior adapted to the specific context (Thach, 1998). We chose to focus on recent findings and current debated questions concerning these two structures.

Many different studies in humans and animal models have led to a consensus that, among the different structures involved in spatial navigation, the hippocampus supports place learning (O'Keefe and Nadel, 1978; Morris *et al.*, 1982; Eichenbaum *et al.*, 1990; Eichenbaum, 2001). Strong support for the role of the hippocampus in place learning comes from convergent findings of hippocampal lesions, both in humans and in various other mammalian species, and electrophysiological recordings of hippocampal neurons (HP cells, Figure 13.3a).

A current debate concerns the role of the hippocampus in more than only the map-based strategy. It has been proposed that the hippocampus plays a critical role when distinct personal experiences must be encoded in relation to one another and linked temporally (White and McDonald, 2002). Based on the strategy taxonomy defined above (see Figure 13.4), this would lead to the possibility that the hippocampus plays a role in both map-based and route-based strategies. Emerging evidence from recordings of hippocampal neural activity suggests that the hippocampal network encodes episodic memories, i.e., memory requiring spatial or sequential (temporal) organization (Wood *et al.*, 2000, 2001; Lenck-Santini *et al.*, 2001). Nonspatial memory requiring relational and temporal coding has been shown to be dependent on CA1-NMDA receptors (Huerta *et al.*, 2000; Rondi-Reig *et al.*, 2001). This combination of coding and cellular properties suggested that the CA1-NMDA dependent mechanisms could contribute to the construction of a "memory space" composed of multiple episodes spatially or sequentially linked together and therefore sustain multiple strategies of navigation requiring such an organization. Recent findings have demonstrated that mice lacking hippocampal NMDA receptors and presenting a decrease of these receptors in the deep cortical layers are indeed deficient in acquiring the memory of successive stimulus-response-stimulus behaviors requiring the execution of a specific sequence of body rotations associated to an ensemble of choice points (sequence learning) in addition to a deficit in the map-based strategy (Rondi-Reig *et al.*, 2006). This result strengthens the hypothesis of the implication of the hippocampus in a general memory acquisition process, where memories are composed of multiple spatio-temporal events.

Concerning the cerebellum, the question is now to understand which navigation component it is relevant for. Several findings have pointed toward the role of the cerebellum in the procedural part of the navigation process (Petrosini *et al.*, 1996; Lalonde, 1997; Rondi-Reig *et al.*, 1999, 2002). Recent data obtained with subjects asked to navigate through a virtual three-dimensional labyrinth showed a strong activation in the medial temporal area including the parahippocampal region, the hippocampus and the thalamus. The cerebellum was also active in those subjects. The authors proposed that the stronger activation in the thalamic-basal ganglia-cerebellar loop points toward a more automatic support of memory and attentional processes possibly mediating memorization of spatial maps (Jordan *et al.*, 2004). The possible implication of the cerebellum in the organization of the spatial representation per se is matter of strong debate (Rondi-Reig and Burguière, 2005). We recently employed the L7-PKCI transgenic mice model, which presents a specific inactivation of the parallel fiber-Purkinje cell LTD (De Zeeuw *et al.*, 1998), to investigate the potential role of this cellular mechanism during spatial navigation. In order to dissociate the relative importance of the declarative and procedural components of navigation, we adopted two different behavioral paradigms: the Morris water maze (MWM) and a new task called the Starmaze (Rondi-Reig *et al.*, 2006). In both cases, the animal had to find a fixed hidden platform from random departure locations, which requires the declarative capability of learning a spatial representation of the environment. Yet, in contrast to the MWM task, the Starmaze allows the animal to only swim within alleys guiding its movements. This helps to execute goal-directed trajectories effectively, and reduces the procedural demand of the task. Our data bring evidence for a deficit of L7-PKCI mice in the acquisition of an adapted goal-oriented behavior, i.e. in the procedural component of the task. This finding supports the hypothesis that cerebellar LTD may subserve a general sensory-motor adaptation process shared by motor and spatial learning functions (Burguière *et al.*, 2005).

The shift between strategies

Several parameters influence strategy shifts. For example, practice-related changes between strategies have been observed in rodents as well as in humans. In rats, place and stimulus-triggered response strategies involve two different systems of memory including the hippocampus and the striatum (caudate nucleus and putamen), respectively (Morris *et al.*, 1982; McDonald and White, 1993, 1994, 1995). According to McDonald and White (1994) and Packard and McGaugh (1996), rats can reach a target by relying on the contribution of the hippocampal or the striatal neural systems depending on whether the animal is in an early or late phase of training. In the early phase of training, the hippocampus is involved in the rapid acquisition of spatial information, allowing rats to reach a target from any starting position (Morris *et al.*, 1982). The striatum is involved in a slower process that relies on rewarded stimulus–response behavior (Packard and Knowlton, 2002; White and McDonald, 2002).

Similar findings have been found with humans (Iaria *et al.*, 2003). Normal subjects were told to retrieve virtual objects located at the end of virtual arms of a radial maze. These objects were located down a set of stairs and were not visible from the center of the maze. Landscape and trees could be viewed at a distance and used for place navigation. Subjects could also count the visited arms either clockwise or counterclockwise, which corresponds to a more automated strategy. In the early phase of training, one-half of the subjects used spatial landmarks to navigate (place navigation) and these subjects showed increased activation of the right hippocampus. The other half used the counting strategy and showed sustained increased activity within the caudate nucleus during navigation. By the end of the test, 72 percent of the subjects were employing the counting strategy and only 28 percent were using the place strategy, demonstrating a shift from place navigation to a more automated strategy.

However, these results have to be compared to recent data suggesting that the complexity of the task might also account for the strategy used. In mice, using the Starmaze task, which can be solved by using either place navigation or route-based strategies, it has recently been demonstrated that both map-based and route-based strategies involved the hippocampus. Both strategies were used during the entire training period and animals could flexibly switch from one to another (Rondi-Reig *et al.*, 2006). Burgess *et al.* (2002) combined virtual reality and functional imaging and showed that the human

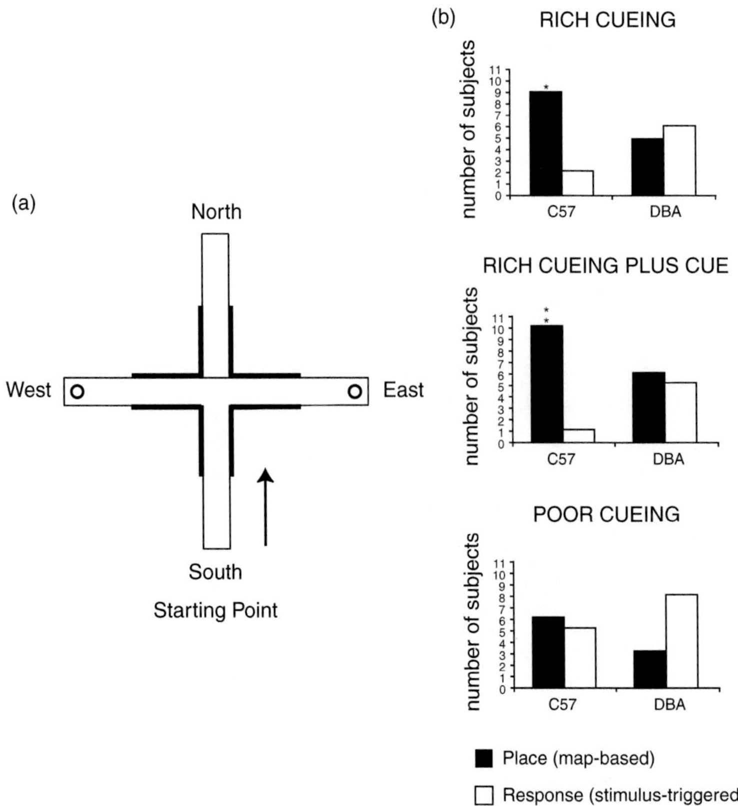

Figure 13.6. From Passino *et al.* (2002). **(a)** Schematic representation of a cross-maze with four identical arms (north, south, east, and west) and transparent Plexiglas high walls fixed on both sides of each arm. During training, access to the north arm was blocked; animals were placed at the starting point of the south arm and allowed to consume the food pellet located at the end of the east arm. During the probe trial, access to the south arm was blocked; animals were released from the north arm, and allowed to choose either the east arm (place learning) or the west arm (response learning). **(b)** Number of C57BL/6 and DBA/2 (two different strains of mice) showing place or response learning under rich, rich plus cue and poor cueing condition. Rich cueing corresponds to a room of 12 m with numerous items. Rich cueing plus cue was similar to the precedent context except that an additional poster was attached to the wall in the direction of the west arm in which the reward was delivered. In the poor cueing condition, all the items were removed. Statistical significant place versus response differences: $p < 0.001$ and $p < 0.05$.

hippocampus is activated when memory for location is required in a complex three-dimensional space, but not within a simple two-dimensional array. Similarly, in young children self-reorientation first occurs solely according to the geometry of the room. Children tend to ignore a large colored cue presented on one wall that under the same conditions enables adults to orient correctly (Hermer and Spelke, 1994). However, this effect is weakened when a larger room is used (Learmonth *et al.*, 2001). In adult mice, the presence of intra-maze and extra-maze cues favors place learning, whereas a poor cueing environment favors stimulus-response learning (Passino *et al.*, 2002; Figure 13.6).

Taken together, these results point toward a crucial role of the complexity of the environment in which the navigation is performed. As suggested by the experiment performed by Iaria *et al.* (2003), the subjects eventually employed the strategy leading to fewer errors during navigation. Therefore, even if multiple strategies are acquired simultaneously, the subject tends to rely on the more advantageous one, that is, the one leading to the best chance to reach the goal without making errors. In simple environments, the behavioral response can be easily automated and therefore relies on a system promoting simple stimulus-response. A complex environment requires more flexible behavior, a process that depends on a brain system including the hippocampal formation.

SPATIAL LEARNING IN NEURO-MIMETIC ROBOTS

The perspective of developing autonomous control systems emulating the spatial navigation capabilities of animals has given rise to a large number of bio-inspired (or neuro-mimetic) robotic models (see Arleo and Gerstner, 2005, for a review). Most of these control architectures rely upon artificial neural networks (ANNs), massively parallel distributed systems suitable for nonsymbolic processing of complex information (Hertz *et al.*, 1991; Haykin, 1994). The elementary constituents of ANNs are formal computing units (artificial neurons), each of which receives and transmits a large number of afferent and efferent connections. The computational power of an ANN derives from the large interconnectivity between its formal neurons. Learning occurs through short- and long-term modification of the strength (or weight) of these connections (modeling synaptic plasticity). Thus, ANNs offer a suitable tool for designing adaptive (experience-dependent) control systems, on the basis of highly simplified models of the anatomo-functional

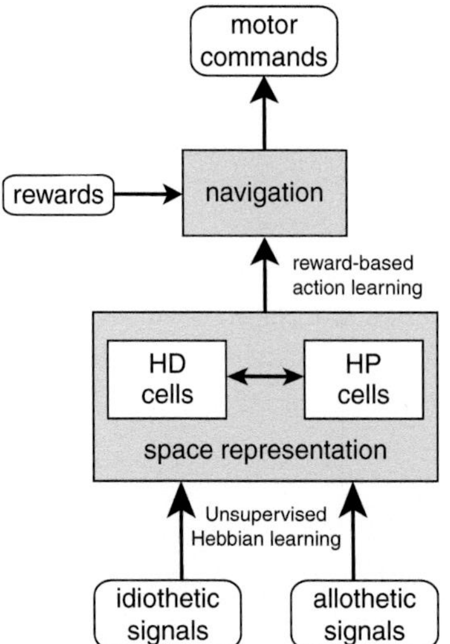

Figure 13.7. Overview of the spatial learning model proposed by Arleo and Gerstner (2000). The system processes idiothetic and allothetic sensory inputs in parallel. The spatial information extracted from these two processing streams is combined by means of LTP-LTD Hebbian learning to generate place and directional coding. Goal-oriented navigation is then achieved by mapping places onto allocentric locomotor actions by means of reward-based learning.

properties of the neural substrates involved in spatial navigation (e.g., hippocampal formation).

Most of the existing models for neuro-robotic spatial learning mainly rely on vision to build space representations (e.g., Schölkopf and Mallot, 1995; Burgess *et al.*, 1997; Trullier and Meyer, 2000; Gaussier *et al.*, 2002). The neuro-mimetic model by Arleo and Gerstner (2000) provides an example of spatial learning system stressing the importance of integrating idiothetic and allothetic cues to establish stable place and head direction coding (Figure 13.7). In the model, the place and direction representations drive a downstream population of action cells mediating motor commands and guide goal-oriented behavior. In particular, reward-based learning is employed to acquire a navigation map allowing the agent to reach hidden goals while avoiding obstacles (Arleo *et al.*, 2004). This section only surveys the place cell learning component of the model and focuses on the interrelation between exteroceptive and self-motion information.

The mobile robot used for the experimental validation of the model is shown in Figure 13.8. Allothetic sensory inputs are provided

Figure 13.8. The mobile Khepera robot (commercialized by K-Team S. A.) used for the experimental validation of the model. A two-dimensional vision system (image resolution: 768 × 576 pixels, view field: 90° pitch, 60° yaw) and eight infrared proximity sensors (detection range: 40 mm) provide the robot with allothetic sensory inputs. Wheel rotation encoders (odometers) estimate linear and angular movements and provide idiothetic signals. See also colour plate section.

by a two-dimensional vision system as well as eight infrared sensors detecting proximal objects (i.e., tactile-like information). Idiothetic signals are provided by wheel rotation encoders that allow the robot to estimate its linear and angular movements (similar to self-motion kinesthetic and vestibular-derived signals). A low-level reactive module takes control whenever the infrared proximity sensors detect an object and endows the robot with obstacle-avoidance capabilities. On the other hand, the spatial learning model is used to develop the high-level controller determining the robot's spatial behavior.

Combining vision and path integration for robust space coding

In the model, space learning occurs via two processing streams that drive two HP cell populations and produce two parallel spatial representations: a place code based on allothetic information (i.e., vision), and a representation obtained by path integration.

Vision-based place coding is a three-step process. First, low-level features are extracted by sampling each image taken by the robot (Figure 13.9a) by means of a family of local visual filters (Figure 13.9b).

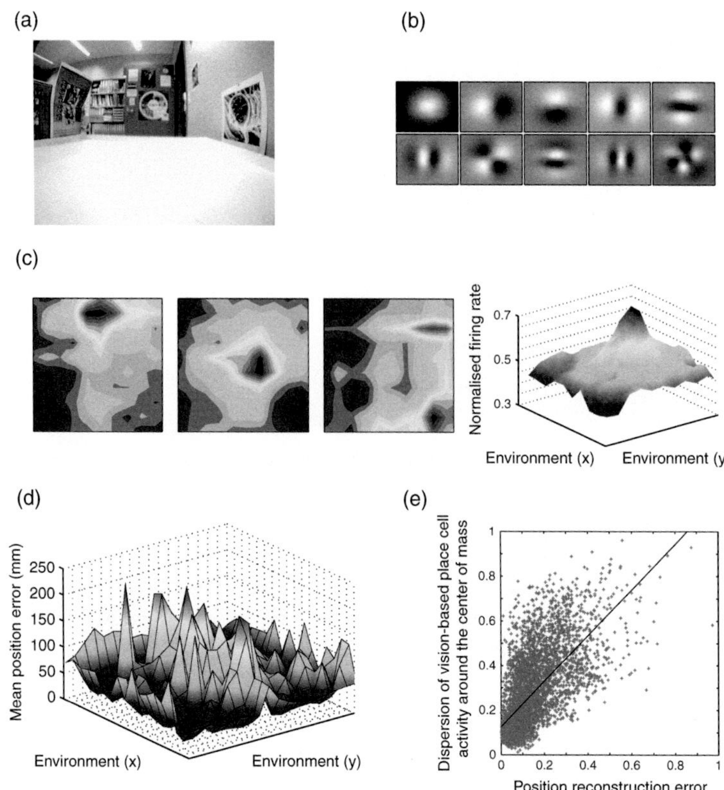

Figure 13.9. Vision-based place representation. (a) A sample image taken by the robot while exploring an open-field square environment. (b) The receptive fields of a set of filters used to sample the images and detect low-level visual features (the ten filters correspond to the first ten principal components, numbered from left to right, top to bottom, obtained by applying the learning algorithm proposed by Sanger 1989). The model has also been tested by employing a set of Gabor filters and a retinotopic sampling method (Arleo *et al.*, 2004). (c) Some samples of vision-based place fields. The squares represent overhead views of the environment. The mean firing rate of each recorded cell is plotted as a function of the locations visited by the robot (red regions denote high activity). Most of the recorded cells showed clean location-correlated firing (e.g., first two plots). However, due to visual aliasing, some cells exhibited multipeak receptive fields (third plot from the left). The three-dimensional diagram suggests that visually driven place cells tend to have rather high baseline firing rates. (d) Accuracy of the vision-based place representation. The diagram has been obtained by separating the environment into an 18×18 grid matrix. First, the robot is let visit the center of each cell of the grid and it has to use its vision-based place map to estimate its position.

Second, the responses of the filters are combined to drive a population of units whose activity becomes correlated to more complex spatial relationships between visual features. We call these units "view cells" because they provide a "neural" encoding of the views perceived by the robot. However, the activity of the view cells is not invariant with respect to the robot's gaze direction and position. Therefore, the third step to achieve vision-based space coding consists of combining multiple gaze-dependent views at each robot's position. This combination produces a local view coding for the spatial relationships between the perceived visual cues and generates allocentric place cell activity (Figure 13.9c).

As discussed in the second section of the chapter, unimodal spatial information is prone to perceptual aliasing and can lead to ambiguous space representations. Indeed, due to visual aliasing, the vision-based HP cells of the model can have multiple subfields and cannot differentiate spatial locations effectively (Figure 13.9c, third plot from the left). As a consequence, the accuracy of the vision-based representation is not uniformly distributed over the surface explored by the robot (Figure 13.9d and e). In the model, path integration is employed to compensate for ambiguities in the visually driven place coding. The robot integrates its linear and angular displacements over

---

Caption for Fig. 13.9. (cont.)

Population vector decoding is employed to map the ensemble place cell activity onto a spatial location (Georgopoulos *et al.*, 1986; Salinas and Abbott, 1994). Second, the position reconstruction error associated to each grid cell is computed by comparing the vision-based position estimate with the location of the center of the grid cell. This process has been iterated n = 10 times (each time corresponds to a different spatial learning session) to calculate the mean position error associated to each grid cell. This mean error function is shown by the three-dimensional diagram. By averaging this error function over the 18 × 18 grid matrix, we obtain a mean position error over the whole environment of about 60 mm. (e) For each position visited by the robot, the reliability of the visual space coding is assessed by measuring the dispersion of the ensemble activity around the center of mass (computed by population coding). The diagram shows the correlation between this dispersion measure and the vision-based position reconstruction error (number of data points: 4600, correlation coefficient: 0.67). The robot utilizes such an on-line reliability criterion to select those local views that are suitable for calibrating its path integrator. Parts (c) and (d) are reproduced in the colour plate section.

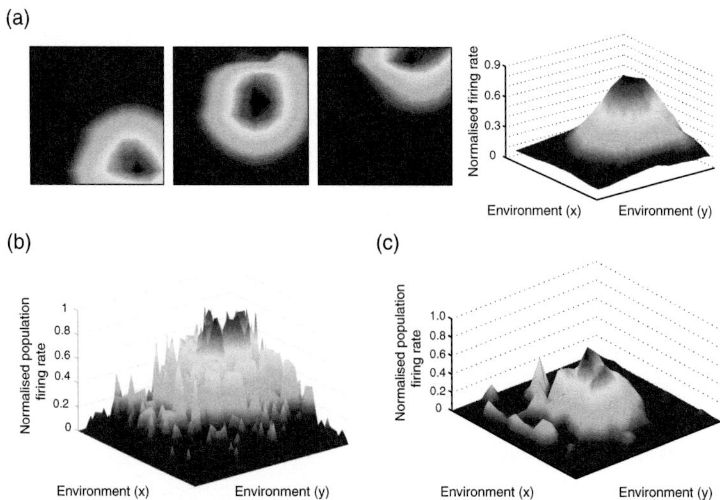

Figure 13.10. Space representation based on multisensory inputs. (a) Examples of place fields obtained by combining vision and path integration. They do not exhibit multiple subfields and they are less noisy than those solely driven by vision (the three dimensional diagram shows very low baseline firing). (b) The robot uses the ensemble place cell activity to self-localize. The diagram shows an example of population activity when the robot is located at the upper-right corner of the arena. (c) In the absence of visual information (e.g., in the dark) place cell firing can be sustained by the input provided by the path integration signal. The figure illustrates the population activity recorded in the dark when the robot is approximately at the center of the arena. See also colour plate section.

time to generate an environment-independent representation of its position relative to a starting point. Such a dead-reckoning mechanism is used to drive a population of HP cells whose activity depends on self-motion signals only (i.e., it provides a space coding based solely on idiothetic information).

The efferents of the two place cell populations (driven by vision and path integration, respectively) converge onto a third downstream network of HP cells. Hebbian learning, inducing both long-term synaptic potentiation (LTP) and depression (LTD), is employed to combine allothetic and idiothetic information based on the agent–environment interaction. This generates a stable space representation consisting of localized place fields similar to those found in hippocampal CA3-CA1 regions (Figure 13.10a). These place fields are less noisy than those solely driven by vision and do not exhibit multi-peak fields, meaning that the system overcomes the sensory aliasing problem of purely vision-based representations.

The goal of the spatial learning process is to generate a large population of overlapping place fields covering the two-dimensional space uniformly and densely. The robot utilizes this ensemble HP cell activity to self-localize (Figure 13.10b). At each time step, a population vector decoding scheme (Georgopoulos *et al.*, 1986; Wilson and McNaughton, 1993) computes the center of mass of the ensemble activity pattern to estimate the robot's current position. Using the population activity, rather than single cell activity, helps in terms of stability and robustness of the self-localization process.

In the model, a place map can emerge and persist even in the absence of visual information (e.g., in darkness conditions). This property is consistent with the experimental observation that hippocampal place fields can arise in darkness (Quirk *et al.*, 1990). Since the activity of the modeled HP cells relies on convergent excitation from both vision and path integration, their mean peak firing rates are lower in visionless conditions than when the robot can use visual spatial cues (Figure 13.10c). The reduced firing activity of the HP units of the model in darkness conditions is in agreement with the experimental findings indicating that HP cells recorded from blind rats exhibit lower discharge frequencies than those observed in sighted animals (Save *et al.*, 1998).

Coherence between allothetic and idiothetic information

Exploring a novel environment

The robot initially explores an unfamiliar environment by relying upon path integration only. As exploration proceeds, local views (encoded by the visually driven HP cells) are coupled (by means of LTP/LTD correlational learning) to the spatial framework provided by the path integrator such that vision and self-motion signals cooperate to form the hippocampal space code (i.e., the CA3-CA1 place representation). However, to maintain this allothetic–idiothetic coupling coherent over time, the robot must prevent the path integrator from accumulating errors. In order to do that, the robot adopts an exploration strategy consisting of looped excursions (i.e., outward and homing journeys) centered at the starting location (Figure 13.11a). During an outward excursion the robot acquires new spatial knowledge and updates its space code. After a while, it starts following its homing

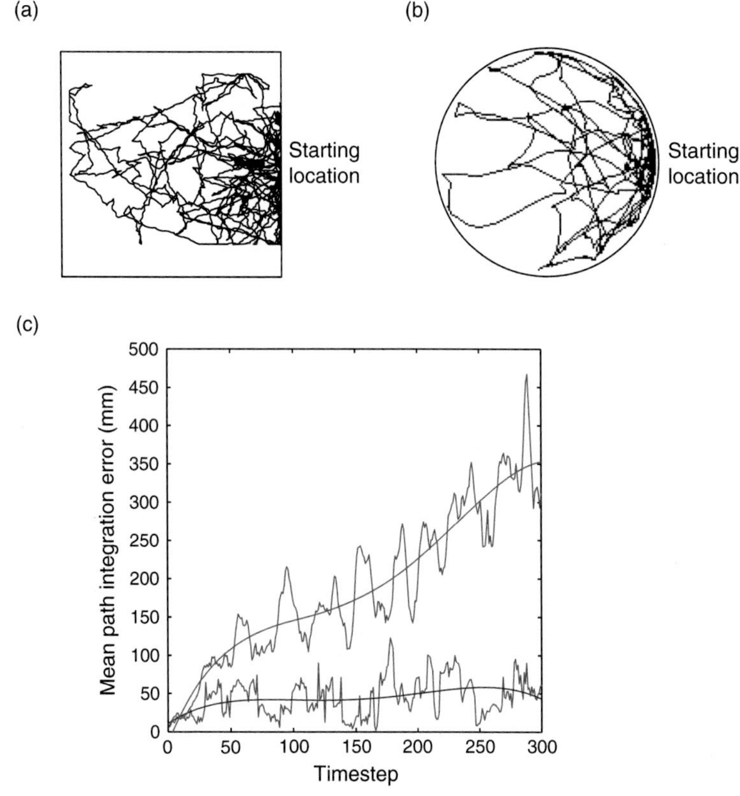

Figure 13.11. Exploratory behavior and path integration calibration. **(a)** To establish a coherent allothetic-idiothetic coupling, the robot starts exploring a novel environment (in this example a square arena) by means of looped excursions centered at the starting location. **(b)** Example of rat's behavior at the beginning of exploration in a novel circular environment (data courtesy C. Brandner, Institute of Psychology, University of Lausanne, Switzerland). **(c)** Uncalibrated (red curves) and calibrated (blue curves) mean path integration error (thin lines are raw data, whereas thick lines are polynomial fittings). At each timestep (*x*-axis) the robot updates its orientation and moves one step further. The difference between the actual position of the robot and the estimate provided by the path integrator is measured at each timestep. The red curves show that this path integration error (averaged over n = 5 trials) tends to grow over time. By contrast, if the system uses the vision-based place representation to calibrate the path integrator occasionally, this error remains bounded over time (blue curves). Part (c) is reproduced in the colour plate section.

vector and as soon as it arrives and recognizes a previously visited location (not necessarily the starting location), it utilizes the vision-based representation to realign the path integrator. Once vision has calibrated the path integrator, a new outward excursion is initiated. By iterating this procedure the robot can keep the dead-reckoning error bounded (Figure 13.11c), and propagate exploration over the entire environment (the probability of calibrating the path integrator at locations other than the starting region increases over time). Behavioral findings concerning the locomotion of rodents exploring novel environments (Drai *et al.*, 2001) show a typical exploratory pattern consisting of looped excursions centered at their home base (Figure 13.11b). The model postulates that maintaining the idiothetic and allothetic signals mutually consistent might be one of the factors determining such a loop-based exploratory behavior.

Importance of landmark stability

The LTP-LTD Hebbian learning used to couple external and internal representations makes stable visual configurations more likely to be correlated to self-motion signals than unstable ones (Figure 13.12a). As a consequence, only those visual configurations that are taken as stable by the robot can influence the dynamics of the space coding process (Arleo and Gerstner, 2001). Stable landmarks can polarize the allocentric space representation across different entries in an environment. This polarization can help the robot to realign the allothetic and idiothetic components of its spatial code and, then, to reactivate a previously learned description of a familiar environment. Failure of such a reactivation process might result in creating a new superfluous representation (Knierim *et al.*, 1998).

In a series of robot experiments conducted by Arleo and Gerstner (2001) the constellation of visual cues was kept stable during spatial learning. Then, the path integrator was reinitialized randomly (simulating a disorientation procedure) and the robot was placed back in the familiar environment. Since the system learned a stable coupling between the idiothetic and allothetic signals, the robot could use the visual information to anchor its allocentric spatial representation, reset its path integrator, and reactivate the previously learned place map (Figure 13.12b, top row). In a second series of experiments, the constellation of visual cues underwent arbitrary rotations during spatial learning. Thus, the Hebbian learning scheme failed to establish stable correlations between idiothetic and allothetic inputs. As a

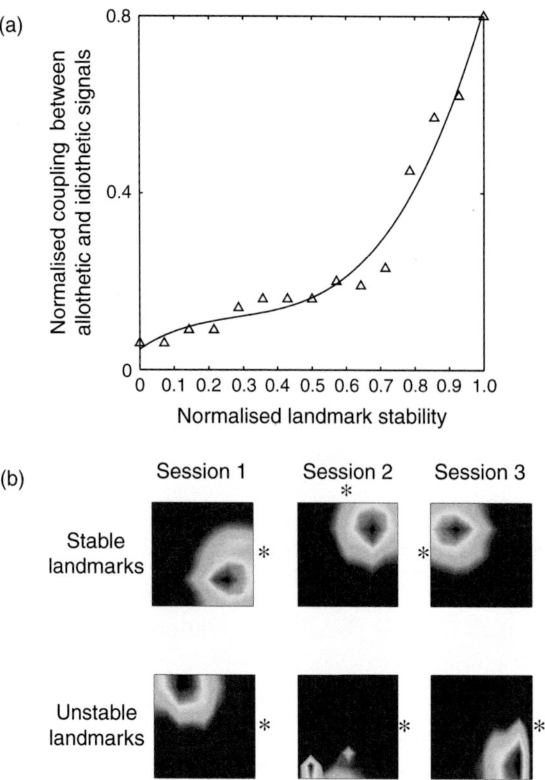

Figure 13.12. Interaction between visual and self-motion signals. **(a)** Due to Hebbian learning, the larger the stability of a visual cue configuration, the strongest its coupling with the path integration-based representation (triangles are sampled data, the curve is a polynomial fitting). **(b)** Intersession responses of one formal place cell after spatial learning. At the beginning of each probe session the robot is disoriented. Top: visual cue configurations that remained stable during spatial learning are able to polarize the place code at the beginning of each probe session. The place cell reorients its receptive field according to the 90° visual cue rotations (the asterisk indicates the centroid of the visual cue configuration). Bottom: unstable visual cues do not allow the disoriented robot to reactivate coherent representations across sessions and remapping occurs. Part (b) is reproduced in the colour plate section.

consequence, when the robot was disoriented and placed back in the explored environment, it was unable to reactivate the learned spatial representation properly and intersession remapping occurred (i.e., HP cell response patterns varied across subsequent visits of the same environment, Figure 13.12b, bottom row). These results are in agreement with those reported by Knierim *et al.* (1995) who recorded HP cells and HD cells from freely moving rats.

Finally, note that since the realigning procedure relies on the allothetic–idiothetic coupling established by the robot via Hebbian learning, impairing this latter mechanism would also lead to unstable intersession representations (i.e., remapping). This result is consistent with experimental findings showing that animals with impaired hippocampal LTP exhibit stable place cell firing patterns within sessions, but unstable mapping between separate runs (Barnes *et al.*, 1997).

Conflict situations

In the above experiments the robot uses external visual fixes to recalibrate an otherwise untrustworthy path integrator (e.g., after disorientation or because of cumulative integration error). Here we consider the situation in which stable allothetic inputs and reliable idiothetic signals provide conflicting spatial information. In the model, the relative importance of coupled external and internal spatial cues is a function of: (1) the degree of confidence of the robot about its self-motion signals; (2) the degree of discrepancy between allothetic and idiothetic spatial information. A series of tests was run inspired by the behavioral experiments by Etienne *et al.* (1990), who studied the homing behavior of hamsters in perceptual conflict situations (see the second section of this chapter for discussion of these experiments). First, we let the robot learn the coupling between a stable visual configuration and its path integrator. Then, during testing, we created both a 90-degree and a 180-degree conflict between external and internal cues and examined the homing behavior of the robot. Results in Figure 13.13a (top row) show that when a 90-degree conflict occurs the visual component tends to influence the robot's homing trajectory more than self-motion signals. By contrast, for 180-degree conflicts (bottom row) the system's response is twofold: if the robot has not been disoriented, then its homing behavior is mainly determined by self-motion signals (bottom row, central plot); on the other hand, if the robot has been disoriented, then it relies on allothetic spatial information even for large discrepancies (i.e., 180 degrees) between allothetic and idiothetic cues (bottom row, right plot). Finally, Figure 13.13b shows the average response of the robot to a 180-degree conflict situation as a function of the degree of confidence about its path integrator. The diagram indicates that as long as self-motion information is given confidence above chance, the robot tends to use it to perform homing behavior. If the confidence falls

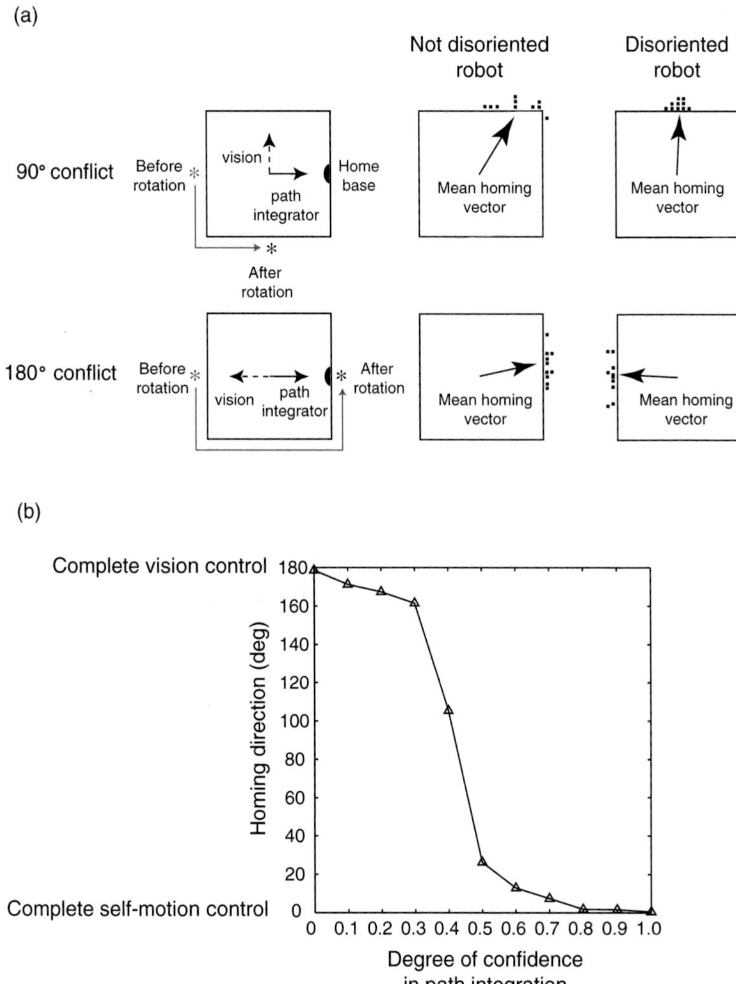

Figure 13.13. Conflict situations between vision and path integration. **(a)** During spatial learning the visual cue configuration is maintained stable. The protocol for the probe trials includes: *(i)* An outward journey during which the robot moves directly from its home base to the center of environment in the dark. *(ii)* A "hoarding" phase during which the robot actively rotates on the spot for a random amount of time (both the amplitude and the sign of the rotation are selected randomly). During hoarding, the visual cue configuration is rotated by either 90° or 180° and the light is switched on. *(iii)* A backward journey during which the robot must compute the homing vector to return home. A conflict occurs between vision and path integration (dashed and continuous arrows,

below chance, then a priority switch occurs and visual information becomes predominant.

CONCLUSIONS

Flexible spatial behavior requires the ability to handle the interaction of multiple parallel processes. At the sensory level, multimodal inputs must be combined to produce a robust description of the spatio-temporal properties of the environment. At the action-selection level, multiple concurrent navigation policies must be dynamically weighted to choose the most adapted to the complexity of the task. Different neural substrates contribute to different spatial processes; elucidating their anatomo-functional interrelations is fundamental for understanding the overall spatial learning mechanism. This chapter has reviewed a series of experimental findings (both behavioral and electrophysiological) that begin to explain the cooperation–competition of the brain areas involved in spatial navigation. It has also addressed the spatial learning issue from the viewpoint of neuro-mimetic robotics. Biologically plausible models and their validation in real experimental conditions help to explore potential connections between findings on the neuronal level (e.g., firing properties of HP and HD cells) and observations on the behavioral level (e.g., an animal's motion trajectory toward a target). Indeed, models permit a scale up to large neural populations, organize them in subsystems, and test hypotheses about their anatomo-functional interactions to produce robust spatial behavior. Such models provide a unique vantage point from which to derive predictions that can then be tested in innovative experiments with animals.

Caption for Fig. 13.13. (cont.)
respectively, in the first column). The thick arrows in the second and third columns indicate the resulting mean homing behavior of the robot averaged over ten trials (black dots). In the case of nondisoriented robot (second column), the familiar visual cues tend to influence the robot's behavior when a 90° conflict occurs. By contrast, the visual control vanishes when the conflict is further increased (i.e., 180°). If the robot is disoriented during the hoarding phase (third column), visual cues predominate for both 90° and 180° conflicts. **(b)** The response of the robot to a 180° conflict depends on its confidence about the path integrator.

NOTE

1. The term "local" indicates "locally invariant," that is constant within a limited region of the environment.

REFERENCES

Arleo, A. and Gerstner, W. (2000). Spatial cognition and neuro-mimetic navigation: a model of hippocampal place cell activity. *Biological Cybernetics*, **83**, 287–299.

Arleo, A. and Gerstner, W. (2001). Spatial orientation in navigating agents: modeling head-direction cells. *Neurocomputing*, **38**–40, 1059–1065.

Arleo, A. and Gerstner, W. (2005). Head direction cells and place cells in models for navigation and robotic applications. In S. I. Wiener and J. S. Taube (Eds.), *Head direction cells and the neural mechanisms of spatial orientation* (pp. 433–457). Cambridge, MA: MIT Press.

Arleo, A., del R. Millán, J. and Floreano, D. (1999). Efficient learning of variable-resolution cognitive maps for autonomous indoor navigation. *IEEE Transactions on Robotics and Automation*, **15**, 990–1000.

Arleo, A., Smeraldi, F. and Gerstner, W. (2004). Cognitive navigation based on nonuniform Gabor space sampling, unsupervised growing networks, and reinforcement learning. *IEEE Trans actions on Neural Networks*, **15**, 639–652.

Avillac, M., Deneve, S., Olivier, E., Pouget, A. and Duhamel, J. R. (2005). Reference frames for representing visual and tactile locations in parietal cortex. *Nature Neuroscience*, **8**, 941–949.

Barnes, C. A., Nadel, L. and Honig, W. K. (1980). Spatial memory deficit in senescent rats. *Canadian Journal of Psychology*, **34**, 29–39.

Barnes, C. A., Suster, M. S., Shen, J. and McNaughton, B. L. (1997). Multistability of cognitive maps in the hippocampus of old rats. *Nature*, **388**, 272–275.

Best, P. J., White, A. M. and Minai, A. (2001). Spatial processing in the brain: the activity of hippocampal place cells. *Annual Review of Neuroscience*, **24**, 459–486.

Biegler, R. and Morris, R. G. (1993). Landmark stability is a prerequisite for spatial but not discrimination learning. *Nature*, **361**, 631–633.

Blair, H. T. and Sharp, P. E. (1995). Anticipatory head direction signals in anterior thalamus: evidence for a thalamocortical circuit that integrates angular head motion to compute head direction. *Journal of Neuroscience*, **15**, 6260–6270.

Bostock, E., Muller, R. U. and Kubie, J. L. (1991). Experience-dependent modifications of hippocampal place cell firing. *Hippocampus*, **1**, 193–205.

Brooks, R. A. (1991). New approaches to robotics. *Science*, **253**, 1227–1232.

Bures, J. and Fenton, A. A. (2000). Neurophysiology of spatial cognition. *News in Physiological Sciences*, **15**, 233–240.

Burgess, N., Jeffery, K. J. and O'Keefe, J. (1999). Integrating hippocampal and parietal functions: a spatial point of view. In *The hippocampal and parietal foundations of spatial cognition* (pp. 3–29). Oxford: Oxford University Press.

Burgess, N., Maguire, E. A. and O'Keefe, J. (2002). The human hippocampus and spatial and episodic memory. *Neuron*, **35**, 625–641.

Burgess, N., Donnett, J. G., Jeffery, K. J. and O'Keefe, J. (1997). Robotic and neuronal simulation of the hippocampus and rat navigation. *Philosophical Transactions of the Royal Society of London. Series B, Biological Sciences*, **352**, 1535–1543.

Burguière, E., Arleo, A., Hojjati, M., Elgersma, Y., De Zeeuw, C. I., Berthoz, A. and Rondi-Reig, L. (2005). Spatial navigation impairment in mice lacking cerebellar LTD: a motor adaptation deficit? *Nature Neuroscience*, **8**, 1292–1294.

Chen, L. L., Lin, L. H., Green, E. J., Barnes, C. A. and McNaughton, B. L. (1994a). Head-direction cells in the rat posterior cortex. I. Anatomical distribution and behavioral modulation. *Experimental Brain Research*, **101**, 8–23.

Chen, L. L., Lin, L. H., Green, E. J., Barnes, C. A. and McNaughton, B. L. (1994b). Head-direction cells in the rat posterior cortex. II. Contribution of visual and ideothetic information to the directional firing. *Experimental Brain Research*, **101**, 24–34.

Collingridge, G. L., Kehl, S. J. and McLennan, H. (1983). Excitatory amino acids in synaptic transmission in the Schaffer collateral-commissural pathway of the rat hippocampus. *Journal of Physiolology*, **334**, 33–46.

Cressant, A., Muller, R. U. and Poucet, B. (1997). Failure of centrally placed objects to control the firing fields of hippocampal place cells. *Journal of Neuroscience*, **17**, 2531–2542.

De Zeeuw, C. I., Hansel, C., Bian, F., Koekkoek, S. K. E., Van Alphen, A. M., Linden, D. J. and Oberdick, J. (1998). Expression of a protein kinase C inhibitor in Purkinje cells blocks cerebellar LTD and adaptation of the vestibulo-ocular reflex. *Neuron*, **20**, 495–508.

Devan, B. D., Goad, E. H. and Petri, H. L. (1996). Dissociation of hippocampal and striatal contributions to spatial navigation in the water maze. *Neurobiology of Learning and Memory*, **66**, 305–323.

Drai, D., Kafkafi, N., Benjamini, Y., Elmer, G. and Golani, I. (2001). Rats and mice share common ethologically relevant parameters of exploratory behavior. *Behavioural Brain Research*, **125**, 133–140.

Eichenbaum, H. (2001). The hippocampus and declarative memory: cognitive mechanisms and neural codes. *Behavioural Brain Research*, **127**, 199–207.

Eichenbaum, H. E., Stewart, C. and Morris, R. G. M. (1990). Hippocampal representation in spatial learning. *Journal of Neuroscience*, **10**, 331–339.

Ekstrom, A. D., Kahana, M. J., Caplan, J. B., Fields, T. A., Isham, E. A., Newman, E. L. and Fried, I. (2003). Cellular networks underlying human spatial navigation. *Nature*, **425**, 184–188.

Elfes, A. (1987). Sonar-based real-world mapping and navigation. *IEEE Journal of Robotics and Automation*, **3**, 249–256.

Etienne, A. S. and Jeffery, K. J. (2004). Path integration in mammals. *Hippocampus*, **14**, 180–192.

Etienne, A. S., Berlie, J., Georgakopoulos, J. and Maurer, R. (1998a). Role of dead reckoning in navigation. In S. Healy (Ed.), *Spatial representation in animals* (pp. 54–68). Oxford: Oxford University Press.

Etienne, A. S., Teroni, E., Hurni, C. and Portenier, V. (1990). The effect of a single light cue on homing behaviour of the golden hamster. *Animal Behaviour*, **39**, 17–41.

Etienne, A. S., Boulens, V., Maurer, R., Rowe, T. and Siegrist, C. (2000). A brief view of known landmarks reorientates path integration in hamsters. *Naturwissenschaften*, **87**, 494–498.

Etienne, A. S., Maurer, R., Berlie, J., Reverdin, B., Rowe, T., Georgakopoulos, J. and Seguinot, V. (1998b). Navigation through vector addition. *Nature*, **396**, 161–164.

Fenton, A. A., Wesierska, M., Kaminsky, Y. and Bures, J. (1998). Both here and there: simultaneous expression of autonomous spatial memories in rats. *Proceedings of the National Academy of Sciences*, **95**, 11493–11498.

Fortin, N. J., Agster, K. L. and Eichenbaum, H. B. (2002). Critical role of the hippocampus in memory for sequences of events. *Nature Neuroscience*, **5**, 458–462.

Foster, T. C., Castro, C. A. and McNaughton, B. L. (1989). Spatial selectivity of rat hippocampal neurons: dependence on preparedness for movement. *Science*, **244**, 1580–1582.

Fyhn, M., Molden, S., Witter, M. P., Moser, E. I. and Moser, M. B. (2004). Spatial representation in the entorhinal cortex. *Science*, **305**, 1258–1264.

Gallistel, C. R. (1990). *The organization of learning*. Cambridge: MIT Press.

Gaussier, P., Revel, A., Banquet, J. P. and Babeau, V. (2002). From view cells and place cells to cognitive map learning: processing stages of the hippocampal system. *Biological Cybernetics*, **86**, 15–28.

Georgopoulos, A. P., Schwartz, A. B. and Kettner, R. E. (1986). Neuronal population coding of movement direction. *Science*, **233**, 1416–1419.

Goodridge, J. P., Dudchenko, P. A., Worboys, K. A., Golob, E. J. and Taube, J. S. (1998). Cue control and head direction cells. *Behavioural Neuroscience*, **112**, 749–761.

Griffin, A. S. and Etienne, A. S. (1998). Updating the path integrator through a visual fix. *Psychobiology*, **26**, 240–248.

Hafting, T., Fyhn, M., Molden, S., Moser, M. B. and Moser, E. I. (2005). Microstructure of a spatial map in the entorhinal cortex. *Nature*, **436**, 801–806.

Haykin, S. (1994). *Neural networks: a comprehensive foundation*. Englewood Cliffs: Macmillan.

Hebb, D. O. (1949). *The organization of behavior*. New York: John Wiley.

Hermer, L. and Spelke, E. S. (1994). A geometric process for spatial reorientation in young children. *Nature*, **370**, 57–59.

Hertz, J., Krogh, A. and Palmer, R. G. (1991). *Introduction to the theory of neural computation*. Redwood City, CA: Addison-Wesley.

Hill, A. J. and Best, P. J. (1981). Effects of deafness and blindness on the spatial correlates of hippocampal unit activity in the rat. *Experimental Neurology*, **74**, 204–217.

Huerta, P. T., Sun, L. D., Wilson, M. A. and Tonegawa, S. (2000). Formation of temporal memory requires NMDA receptors within CA1 pyramidal neurons. *Neuron*, **25**, 473–480.

Iaria, G., Petrides, M., Dagher, A., Pike, B. and Bohbot, V. D. (2003). Cognitive strategies dependent on the hippocampus and caudate nucleus in human navigation: variability and change with practice. *Journal of Neuroscience*, **23**, 5945–5952.

Jeffery, K. J. (1998). Learning of landmark stability and instability by hippocampal place cells. *Neuropharmacology*, **37**, 677–687.

Jeffery, K. J. and O'Keefe, J. M. (1999). Learned interaction of visual and idiothetic cues in the control of place field orientation. *Experimental Brain Research*, **127**, 151–161.

Jordan, K., Schadow, J., Wuestenberg, T., Heinze, H. J. and Jancke, L. (2004). Different cortical activations for subjects using allocentric or egocentric strategies in a virtual navigation task. *NeuroReport*, **15**, 135–140.

Knierim, J. J., Kudrimoti, H. S. and McNaughton, B. L. (1995). Place cells, head direction cells, and the learning of landmark stability. *Journal of Neuroscience*, **15**, 1648–1659.

Knierim, J. J., Kudrimoti, H. S. and McNaughton, B. L. (1998). Interactions between idiothetic cues and external landmarks in the control of place cells and head direction cells. *Journal of Neurophysiology*, **80**, 425–446.

Kuipers, B. J. and Byun, Y. T. (1991). A robot exploration and mapping strategy based on a semantic hierarchy of spatial representations. *Robotics and Autonomous Systems*, **8**, 47–63.

Lalonde, R. (1997). Visuospatial abilities. *International Review of Neurobiology*, **41**, 191–215.

Lavoie, A. M. and Mizumori, S. J. (1994). Spatial, movement- and reward-sensitive discharge by medial ventral striatum neurons of rats. *Brain Research*, **638**, 157–168.

Learmonth, A. E., Newcombe, N. S., Huttenlocher, J., Hermer, L. and Spelke, E. S. (2001). Toddlers' use of metric information and landmarks to reorient: a geometric process for spatial reorientation in young children. *Journal of Experimental Child Psycholology*, **80**, 225–244.

Lee, I. and Kesner, R. P. (2002). Differential contribution of NMDA receptors in hippocampal subregions to spatial working memory. *Nature Neuroscience*, **5**, 162–168.

Lenck-Santini, P. P., Save, E. and Poucet, B. (2001). Place-cell firing does not depend on the direction of turn in a Y-maze alternation task. *European Journal of Neuroscience*, **13**, 1055–1058.

McDonald, R. J. and White, N. M. (1993). A triple dissociation of memory systems: hippocampus, amygdala, and dorsal striatum. *Behavioural Neuroscience*, **107**, 3–22.

McDonald, R. J. and White, N. M. (1994). Parallel information processing in the water maze: evidence for independent memory systems involving dorsal striatum and hippocampus. *Behavioural and Neural Biology*, **61**, 260–270.

McDonald, R. J. and White, N. M. (1995). Hippocampal and nonhippocampal contributions to place learning in rats. *Behavioural Neuroscience*, **109**, 579–593.

McNaughton, B. L., Chen, L. L. and Markus, E. J. (1991). Dead reckoning, landmark learning, and the sense of direction: a neurophysiological and computational hypothesis. *Journal of Cognitive Neuroscience*, **3**, 190–202.

McNaughton, B. L., Battaglia, F. P., Jensen, O., Moser, E. I. and Moser, M. B. (2006). Path integration and the neural basis of the "cognitive map." *Nature Reviews Neuroscience*, **7**, 663–678.

Markus, E. J., Barnes, C. A., McNaughton, B. L., Gladden, V. L. and Skaggs, W. E. (1994). Spatial information content and reliability of hippocampal CA1 neurons: effects of visual input. *Hippocampus*, **4**, 410–421.

Mittelstaedt, H. and Mittelstaedt, M. L. (1982). Homing by path integration. In F. Papi and H. G. Wallraff (Eds.), *Avian navigation* (pp. 290–297). Berlin Heidelberg: Springer.

Mittelstaedt, M. L. and Mittelstaedt, H. (1980). Homing by path integration in the mammal. *Naturewissenschaften*, **67**, 566–567.

Molinari, M., Petrosini, L. and Grammaldo, L. G. (1997). Spatial event processing. *International Review of Neurobiology*, **41**, 217–230.

Morris, R. G. and Frey, U. (1997). Hippocampal synaptic plasticity: role in spatial learning or the automatic recording of attended experience. *Philsosophical Transactions of the Royal Society of London B: Biological Sciences*, **352**, 1489–1503.

Morris, R. G. M., Anderson, E., Lynch, G. and Baudry, M. (1986). Selective impairment of learning and blockade of long-term potentiation by an N-methyl-D-aspartate receptor antagonist, AP5. *Nature*, **319**, 774–776.

Morris, R. G. M., Garrud, P., Rawlins, J. N. P. and O'Keefe, I. (1982). Place navigation impaired in rats with hippocampal lesions. *Nature*, **297**, 681–683.

Muller, R. U. and Kubie, J. L. (1987). The effects of changes in the environment on the spatial firing of hippocampal complex-spike cells. *Journal of Neuroscience*, **7**, 1951–1968.

Nakazawa, K., McHugh, T. J., Wilson, M. A. and Tonegawa, S. (2004). NMDA receptors, place cells and hippocampal spatial memory. *Nature Reviews Neuroscience*, **5**, 361–372.

Nakazawa, K., Sun, L. D., Quirk, M. C., Rondi-Reig, L., Wilson, M. A. and Tonegawa, S. (2003). Hippocampal CA3 NMDA receptors are crucial for memory acquisition of one-time experience. *Neuron*, **38**, 305–315.

Nakazawa, K., Quirk, M. C., Chitwood, R. A., Watanabe, M., Yeckel, M. F., Sun, L. D., Kato, A., Carr, C. A., Johnston, D., Wilson, M. A. and

Tonegawa, S. (2002). Requirement for hippocampal CA3 NMDA receptors in associative memory recall. *Science*, **297**, 211–218.

O'Keefe, J. and Dostovsky, J. (1971). The hippocampus as a spatial map. Preliminary evidence from unit activity in the freely-moving rat. *Brain Research*, **34**, 171–175.

O'Keefe, J. and Nadel, L. (1978). *The hippocampus as a cognitive map*. Oxford: Clarendon Press.

O'Keefe, J. and Conway, D. H. (1978). Hippocampal place units in freely moving rats: why they fire where they fire. *Experimental Brain Research*, **31**, 573–590.

O'Keefe, J. and Speakman, A. (1987). Single unit activity in the rat hippocampus during a spatial memory task. *Experimental Brain Research*, **68**, 1–27.

Packard, M. G. and McGaugh, J. L. (1996). Inactivation of hippocampus or caudate nucleus with lidocaïne differentially affects expression of place and response learning. *Neurobiology of Learning and Memory*, **65**, 65–72.

Packard, M. G. and Knowlton, B. J. (2002). Learning and memory functions of the Basal Ganglia. *Annual Reviews of Neuroscience*, **25**, 563–593.

Passino, E., Middei, S., Restivo, L., Bertaina-Anglade, V. and Ammassari-Teule, M. (2002). Genetic approach to variability of memory systems: analysis of place vs. response learning and fos-related expression in hippocampal and striatal areas of C57BL/6 and DBA/2 mice. *Hippocampus*, **12**, 63–75.

Petrosini, L., Molinari, M. and Dell'Anna, M. E. (1996). Cerebellar contribution to spatial event processing: Morris water maze and T-maze. *European Journal of Neuroscience*, **9**, 1882–1896.

Quirk, G. J., Muller, R. U. and Kubie, J. L. (1990). The firing of hippocampal place cells in the dark depends on the rat's recent experience. *Journal Neuroscience*, **10**, 2008–2017.

Ranck, J. B. J. (1984). Head-direction cells in the deep cell layers of dorsal presubiculum in freely moving rats. *Society for Neuroscience Meeting*, **10**, 599.

Redish, A. D. (1999). *Beyond the cognitive map: from place cells to episodic memory*. London: MIT Press.

Rondi-Reig, L. and Burguière, E. (2005). Is the cerebellum ready for navigation? *Progress in Brain Research*, **148**, 199–212.

Rondi-Reig, L., Caston, J., Delhaye-Bouchaud, N. and Mariani, J. (1999). Cerebellar functions: a behavioral neurogenetic perspective. In B. C. Jones and P. Morméde (Eds.), *Neurobehavioral genetics: methods and applications* (pp. 201–216). Boca Raton: CRC Press.

Rondi-Reig, L., Libbey, M., Eichenbaum, H. and Tonegawa, S. (2001). CA1-specific N-methyl-D-aspartate receptor knockout mice are deficient in solving a nonspatial transverse patterning task. *Proceedings of the National Academy of Sciences*, **98**, 3543–3548.

Rondi-Reig, L., Le Marec, N., Caston, J. and Mariani, J. (2002). The role of climbing and parallel fibers inputs to cerebellar cortex in navigation. *Behavioural Brain Research*, **132**, 11–18.

Rondi-Reig, L., Petit, G., Tobin, C., Tonegawa, S., Mariani, J. and Berthoz, A. (2006). Impaired sequential-egocentric and allocentric memories in forebrain specific NMDA receptor knockout mice during a new task dissociating strategies of navigation. *Journal of Neuroscience*, **26**, 4071–4081.

Salinas, E. and Abbott, L. F. (1994). Vector reconstruction from firing rates. *Journal of Computational Neuroscience*, **1**, 89–107.

Sargolini, F., Fyhn, M., Hafting, T., McNaughton, B. L., Witter, M. P., Moser, M. B. and Moser, E. I. (2006). Conjunctive representation of position, direction, and velocity in entorhinal cortex. *Science*, **312**, 758–762.

Save, E., Cressant, A., Thinus-Blanc, C. and Poucet, B. (1998). Spatial firing of hippocampal place cells in blind rats. *Journal of Neuroscience*, **18**, 1818–1826.
Schölkopf, B. and Mallot, H. A. (1995). View-based cognitive mapping and path planning. *Adaptive Behavior*, **3**, 311–348.
Sharp, P. E., Kubie, J. L. and Muller, R. U. (1990). Firing properties of hippocampal neurons in a visually symmetrical environment: contributions of multiple sensory cues and mnemonic processes. *Journal of Neuroscience*, **10**, 3093–3105.
Sharp, P. E., Tinkelman, A. and Cho, J. (2001). Angular velocity and head direction signals recorded from the dorsal tegmental nucleus of gudden in the rat: implications for path integration in the head direction cell circuit. *Behavioural Neuroscience*, **115**, 571–588.
Stackman, R. W. and Taube, J. S. (1997). Firing properties of head direction cells in the rat anterior thalamic nucleus: dependence on vestibular input. *Journal of Neuroscience*, **17**, 4349–4358.
Stackman, R. W. and Taube, J. S. (1998). Firing properties of rat lateral mammillary single units: head direction, head pitch, and angular head velocity. *Journal of Neuroscience*, **18**, 9020–9037.
Steffenach, H. A., Witter, M., Moser, M. B. and Moser, E. I. (2005). Spatial memory in the rat requires the dorsolateral band of the entorhinal cortex. *Neuron*, **45**, 301–313.
Taube, J. S. (1995). Head direction cells recorded in the anterior thalamic nuclei of freely moving rat. *Journal of Neuroscience*, **15**, 70–86.
Taube, J. S. (1998) Head direction cells and the neurophysiological basis for a sense of direction. *Progress in Neurobiology*, **55**, 225–256.
Taube, J. S., Muller, R. U. and Ranck, J. B. (1990a). Head-direction cells recorded from the postsubiculum in freely moving rats. I. Description and quantitative analysis. *Journal of Neuroscience*, **10**, 420–435.
Taube, J. S., Muller, R. U. and Ranck, J. B., Jr. (1990b). Head-direction cells recorded from the postsubiculum in freely moving rats. II. Effects of environmental manipulations. *Journal of Neuroscience*, **10**, 436–447.
Thach, W. T. (1998). A role for the cerebellum in learning movement coordination. *Neurobiology of Learning and Memory*, **70**, 177–188.
Thinus-Blanc, C. (1996). *Animal spatial cognition*. Singapore: World Scientific.
Thrun, S. (1998). Learning maps for indoor mobile robot navigation. *Artificial Intelligence*, **99**, 21–71.
Tolman, E. C. (1948). Cognitive maps in rats and men. *Psychological review*, **55**, 189–208.
Trullier, O. and Meyer, J. A. (2000). Animat navigation using a cognitive graph. *Biological Cybernetics*, **83**, 271–285.
Trullier, O., Wiener, S. I., Berthoz, A. and Meyer, J. A. (1997). Biologically based artificial navigation systems: review and prospects. *Progress in Neurobiology*, **51**, 483–544.
Tsien, J. Z., Huerta, P. T. and Tonegawa, S. (1996) The essential role of hippocampal CA1 NMDA receptor-dependent synaptic plasticity in spatial memory. *Cell*, **87**, 1327–1338.
Vallar, G., Lobel, E., Galati, G., Berthoz, A., Pizzamiglio, L. and Le Bihan, D. (1999). A fronto-parietal system for computing the egocentric spatial frame of reference in humans. *Experimental Brain Research*, **124**, 281–286.
Whishaw, I. Q. and Mittleman, G. (1986). Visits to starts, routes, and places by rats (*Rattus norvegicus*) in swimming pool navigation tasks. *Journal of Comparative Psychology*, **100**, 422–431.
White, N. M. and McDonald, R. J. (2002). Multiple parallel memory systems in the brain of the rat. *Neurobiology of Learning and Memory*, **77**, 125–184.

Wiener, S. I. (1993). Spatial and behavioral correlates of striatal neurons in rats performing a self-initiated navigation task. *Journal of Neuroscience*, **13**, 3802–3817.

Wiener, S. I. and Arleo, A. (2003). Persistent activity in limbic system neurons: neurophysiological and modeling perspectives. *Journal of Physiology-Paris*, **97**, 547–555.

Wiener, S. I. and Taube, J. S. (Eds.) (2005). *Head direction cells and the neural mechanisms of spatial orientation*. Boston: MIT Press.

Wilson, M. A. and McNaughton, B. L. (1993). Dynamics of the hippocampal ensemble code for space. *Science*, **261**, 1055–1058.

Witter, M. P. (1993). Organization of the entorhinal-hippocampal system: a review of current anatomical data. *Hippocampus*, **3** Spec No: 33–44.

Wood, E. R., Dudchenko, P. A. and Eichenbaum, H. (2001). Cellular correlates of behavior. *International Review of Neurobiology*, **45**, 293–312.

Wood, E. R., Dudchenko, P. A., Robitsek, R. J. and Eichenbaum, H. (2000). Hippocampal neurons encode information about different types of memory episodes occurring in the same location. *Neuron*, **27**, 623–633.

Zugaro, M. B., Arleo, A., Berthoz, A. and Wiener, S. I. (2003). Rapid spatial reorientation and head direction cells. *Journal of Neuroscience*, **23**, 3478–3482.

Zugaro, M. B., Arleo, A., Dejean, C., Burguiere, E., Khamassi, M. and Wiener, S. I. (2004). Rat anterodorsal thalamic head direction neurons depend upon dynamic visual signals to select anchoring landmark cues. *European Journal of Neuroscience*, **20**, 530–536.

Zugaro, M. B., Berthoz, A. and Wiener, S. I. (2001). Background but not foreground, spatial cues are taken as referencesfor head direction responses by rat anterodorsal thalamus neurons. *Journal of Neuroscience*, **21** (RC154) 1–5.

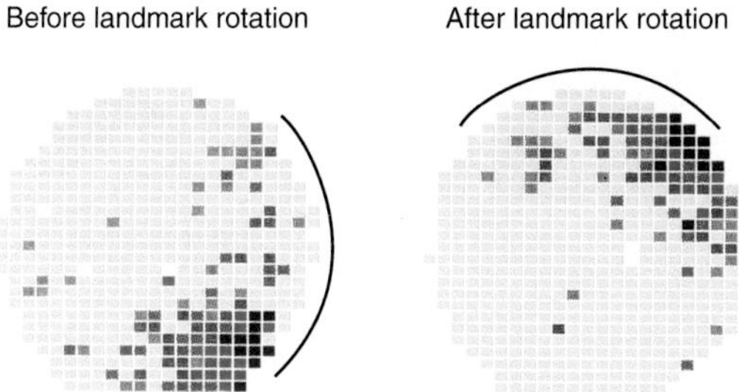

Figure 13.3. (a) Sample of receptive field of a place cell recorded from the rat hippocampus. The plots show the mean discharge of the neuron (blue and yellow denote peak and baseline firing rates, respectively) as a function of the animal position within the environment (a cylindrical arena with a cue card attached to inner wall). The location-selective response of the cell is controlled by the cue card in that rotating the card by 90° induces an equivalent rotation of the receptive field. Adapted from Muller and Kubie (1987).

Figure 13.8. The mobile Khepera robot (commercialized by K-Team S. A.) used for the experimental validation of the model. A two-dimensional vision system (image resolution: 768 × 576 pixels, view field: 90° pitch, 60° yaw) and eight infrared proximity sensors (detection range: 40 mm) provide the robot with allothetic sensory inputs. Wheel rotation encoders (odometers) estimate linear and angular movements and provide idiothetic signals.

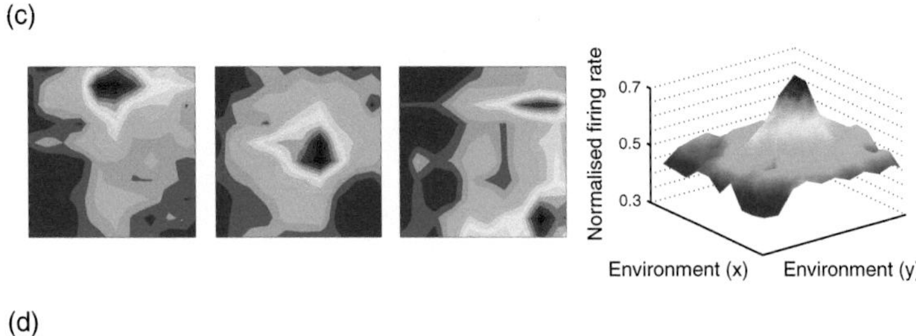

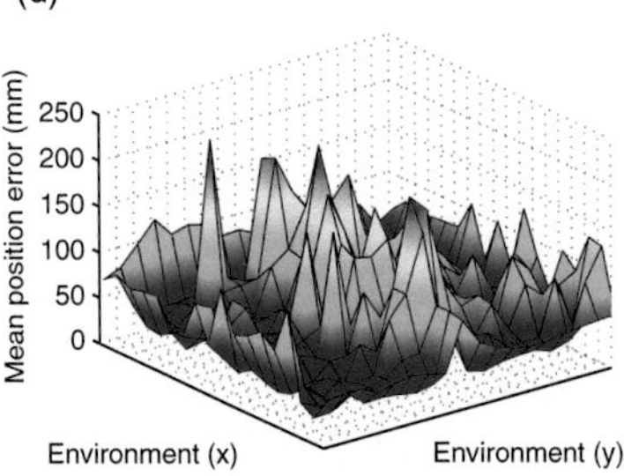

Figure 13.9. (c) Some samples of vision-based place fields. The squares represent overhead views of the environment. The mean firing rate of each recorded cell is plotted as a function of the locations visited by the robot (red regions denote high activity). Most of the recorded cells showed clean location-correlated firing (e.g., first two plots). However, due to visual aliasing, some cells exhibited multipeak receptive fields (third plot from the left). The three-dimensional diagram suggests that visually driven place cells tend to have rather high baseline firing rates. (d) Accuracy of the vision-based place representation. The diagram has been obtained by separating the environment into an 18 × 18 grid matrix. First, the robot is let visit the center of each cell of the grid and it has to use its vision-based place map to estimate its position. Population vector decoding is employed to map the ensemble place cell activity onto a spatial location (Georgopoulos *et al.*, 1986; Salinas and Abbott, 1994). Second, the position reconstruction error associated to each grid cell is computed by comparing the vision-based position estimate with the location of the center of the grid cell. This process has been iterated n = 10 times (each time corresponds to a different spatial learning session) to calculate the mean position error associated to each grid cell. This mean error function is shown by the three-dimensional diagram. By averaging this error function over the 18 × 18 grid matrix, we obtain a mean position error over the whole environment of about 60 mm.

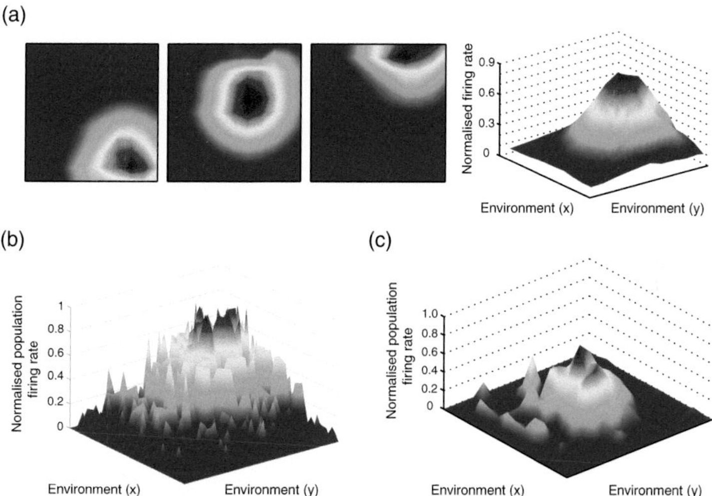

Figure 13.10. Space representation based on multisensory inputs. (a) Examples of place fields obtained by combining vision and path integration. They do not exhibit multiple subfields and they are less noisy than those solely driven by vision (the three dimensional diagram shows very low baseline firing). (b) The robot uses the ensemble place cell activity to self-localize. The diagram shows an example of population activity when the robot is located at the upper-right corner of the arena. (c) In the absence of visual information (e.g., in the dark) place cell firing can be sustained by the input provided by the path integration signal. The figure illustrates the population activity recorded in the dark when the robot is approximately at the center of the arena.

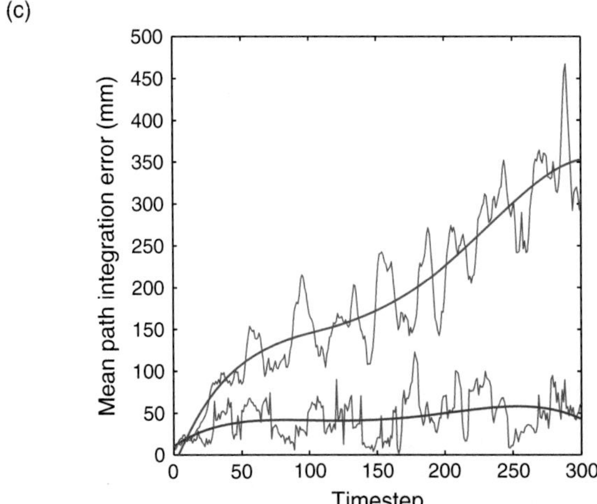

Figure 13.11. (c) Uncalibrated (red curves) and calibrated (blue curves) mean path integration error (thin lines are raw data, whereas thick lines are polynomial fittings). At each timestep (x-axis) the robot updates its orientation and moves one step further. The difference between the actual position of the robot and the estimate provided by the path integrator is measured at each timestep. The red curves show that this path integration error (averaged over n = 5 trials) tends to grow over time. By contrast, if the system uses the vision-based place representation to calibrate the path integrator occasionally, this error remains bounded over time (blue curves).

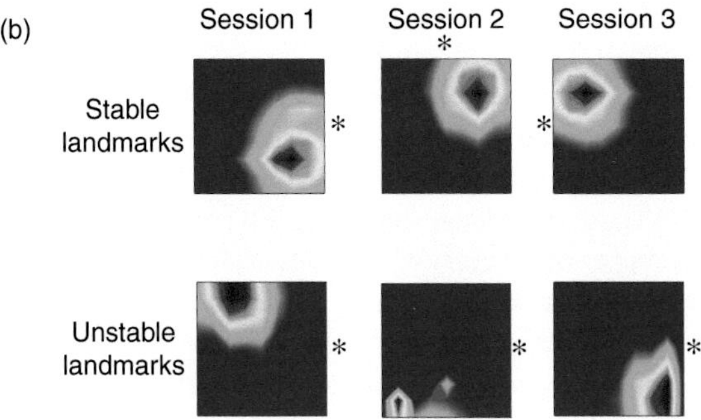

Figure 13.12. (b) Intersession responses of one formal place cell after spatial learning. At the beginning of each probe session the robot is disoriented. Top: visual cue configurations that remained stable during spatial learning are able to polarize the place code at the beginning of each probe session. The place cell reorients its receptive field according to the 90Ê visual cue rotations (the asterisk indicates the centroid of the visual cue configuration). Bottom: unstable visual cues do not allow the disoriented robot to reactivate coherent representations across sessions and remapping occurs.

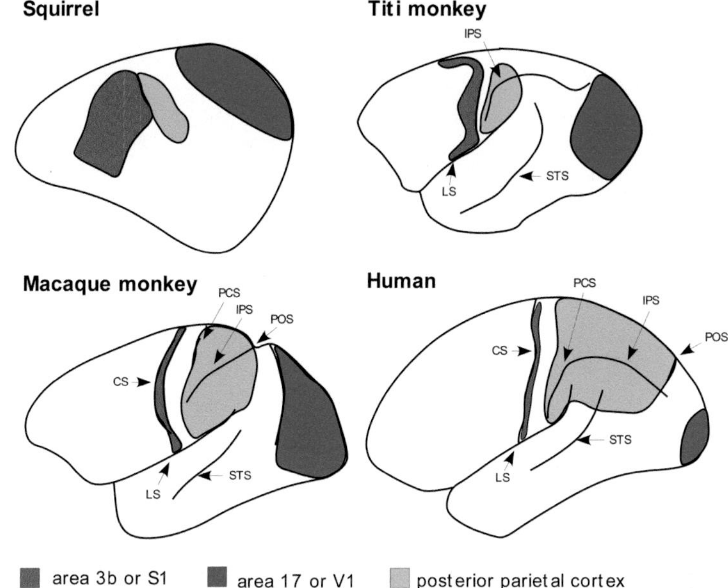

Figure 16.1. The location of association cortex (green) relative to the primary somatosensory area (red) and the primary visual area (blue) in four mammalian species. In mammals with a small neocortex, such as the grey squirrel, primary sensory areas occupy a relatively large portion of the cortical sheet. In mammals with a large neocortex such as monkeys, these primary sensory areas occupy relatively less cortex. In human primates, primary sensory areas occupy a very small portion of the neocortex, while association cortex, including posterior parietal cortex, occupies a large expanse of the cortical sheet. See Table 16.1 for abbreviations.

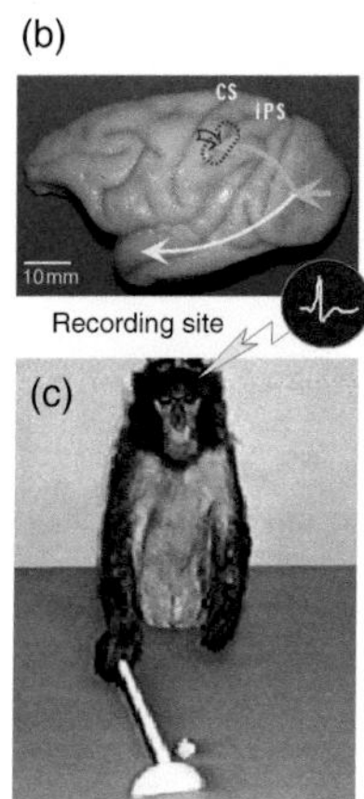

Figure 18.1. (b) recording sites of the intraparietal bimodal neurons (red circle) shown in the lateral surface of the left cerebral cortex. CS, central sulcus. IPS, intraparietal sulcus. Blue arrow, somatosensory information processing pathway. Pink arrows, visual information processing pathway. (c) Neural responses are recorded (yellow arrow) while monkeys retrieve bits of food using a hand-held rake.

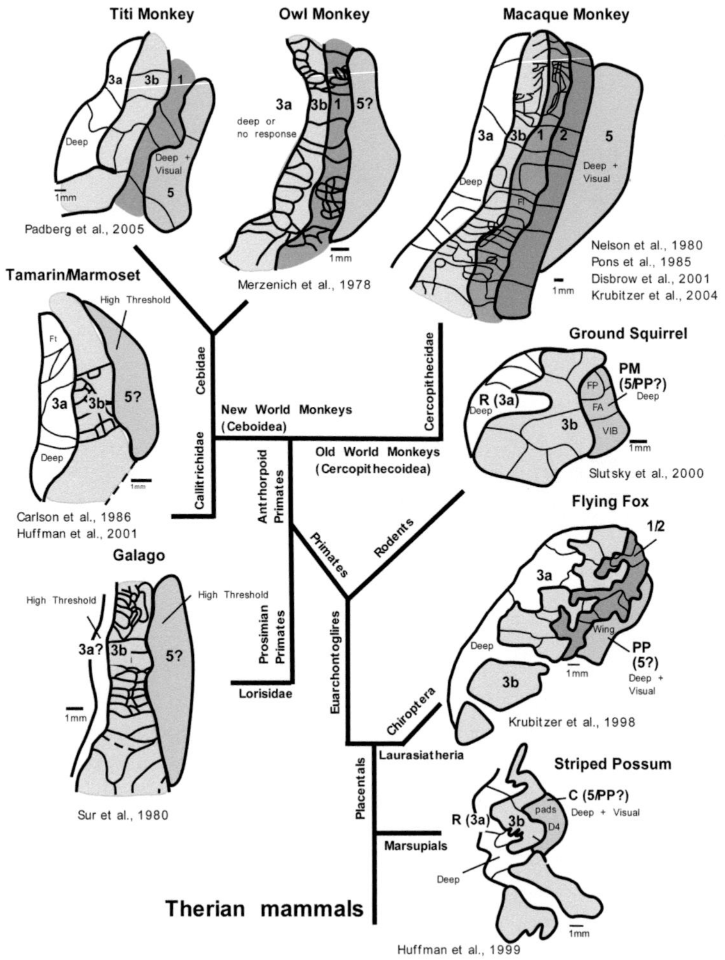

Figure 16.11. See text for caption.

Figure 20.2. “Sacred landscape map, Himalaya” (painting by Shyam Tamang Lama; used by permission of artist).

Part IV Does mapping of the body generate understanding of external space?

MAXINE SHEETS-JOHNSTONE

14

Movement: the generative source of spatial perception and cognition

In their book *Scientific Bases of Human Movement*, in a chapter titled "The Proprioceptors and Their Associated Reflexes," Gowitzke and Milner (1988, p. 256) state: "The voluntary contribution to movement is almost entirely limited to initiation, regulation of speed, force, range, and direction, and termination of the movement." While their point is that voluntary movement is orchestrated by subcortical centers and by individual muscles not normally under voluntary control, the scope and complexity of voluntary kinetic features can hardly be described as *limited* contributions to movement, and in turn passed over so lightly. Speed, force, range and direction are in fact complex dimensions of movement, all movement, whether animate or inanimate. Being core dimensions of animate movement in particular, hence of kinesthesia, they warrant something more than rote specification in a scientific investigation of human movement. The need to examine their centrality to human movement is all the more apparent in the fact that, together, they point toward the kinetic dynamics that inform kinesthetic experience, and hence are necessarily related to spatial perceptions and cognitions. They point toward rather than articulate these dynamics because they are no more than ready everyday labels we apply to movement. In other words, they identify aspects of movement we experience, but movement itself remains unanalyzed and its global kinetic structure remains opaque.

When we examine movement phenomenologically, beginning with our own self-movement and in turn verify our findings in any animate and inanimate movement, we find four kinetic qualities: tensional, linear, areal and projectional qualities constitute kinesthetic experience and are the basis of any movement dynamics (Sheets-Johnstone, 1966, 1999a).[1] They are a finer and more rigorous specification of the four elements identified by Gowitzke and Milner and are

Spatial Cognition, Spatial Perception: Mapping the Self and Space, ed. Francine L. Dolins and Robert W. Mitchell. Published by Cambridge University Press.

directly apparent in any movement we might make. Together, they constitute a movement's particular dynamics: its flaccidity or potency, its contours and paths, its expansiveness or constrictedness, its ballistic, sudden or attenuated character. Looking more closely at each in turn will enable us to see spatial aspects of movement within a global kinetic perspective. On the basis of this initial descriptive analysis, it will be possible to show how movement creates its own space, time and force, and in turn provide substantive empirical grounds for the lead part of Konrad Lorenz's claim that "There is no intuition [*Anschauung*] of space and no intuition of time but there is only one intuition, of movement in space and time."[2] Correlatively, it will be possible to show how the concluding part of Lorenz's claim follows received wisdom in a way that threatens to compromise his lead insight.[3] Moreover on this same basis, it will be possible to show how spatial concepts such as those focused on by Piaget and other psychologists in their studies of infancy are tied to self-movement. Finally, an examination of spatial aspects of movement from a global kinetic perspective will demonstrate how, in meeting our common initial challenge of learning our bodies and learning to move ourselves, we lay the ground for all those further elaborations of space, both kinesthetic and objective, that come to inform our experience.

Tensional quality is perhaps the most immediately evident aspect of self-movement. When we clench our fist, stomp out of the room, wave goodbye, pick up an egg, smile, stretch our arms overhead, or hug someone, we experience a certain kinetic intensity. The intensity may vary as we move through the movement: a hug may start out with a moderate energy and then heighten; a clenched fist may relax and then tense up again; an excited smile may fade and lose its initial vitality. The tensional quality of movement may be and most commonly is a variable within any movement itself. Muscles, after all, are innervated and denervated in the performance of any movement (see Luria, 1973). Whatever its changing character, however, tensional quality is always on the kinesthetic map. We can, in effect, notice its quality in the course of any movement we make, whether in the course of changing a tire, writing our name, executing a tennis serve, making an abdominal incision, or doing a *tour jeté*.

We can be equally aware of the linear quality of our movement, both the linear design of our bodies as we move and the linear pattern created by our movement. In the course of walking, picking something up, swiveling our head, and so on, we bend, extend, curve and/or rotate our bodies in various places. We in fact create ever-changing and often complex linear designs. For example, when we walk, the overall linear design of our body is vertical, but a leg, as it reaches forward, is diagonally aligned, as are our arms at the forwardmost and backwardmost part of their swing. Moreover legs and arms bend and extend at their mid-joints as we walk, and in so doing compound the changing overall linear design by their angularity. In a general sense, the compound linear design created by our body in walking is straight; that is, we create straight rather than rounded bodily lines.

Linear quality is evident in the linear pattern drawn by movement as well as in the changing linear design of the moving body. In other words, the path that any movement traces describes a linear form of some kind: rounded, as when one goes around an obstacle, turns a corner or throws a discus; straight, as when one walks directly toward something or pushes against a door; zig-zag, as in a game of tag or in avoiding a collision with someone, and so on. Moreover multiple patterns may be and often are drawn concurrently by bodily movement. When one jogs, for example, feet draw circles in the process of moving forward; arms draw horizontal lines back and forth; and the body as a whole draws vertical lines up and down as it sequentially takes off and lands with each step.

Linear quality is obviously spatial, but not on that account object-like. On the contrary, in virtue of the bodily movement that creates it, linear quality is ever-changing and cannot be pinned down except in the form of static positionings of the body and imaginary lines drawn by movement. In other words, the linear design of a moving body may be objectified only by bringing movement to a standstill; the linear pattern created by a moving body exists only in the form of imaginary tracings left by movement. Linear designs and patterns are fluidly drawn but distinguishable forms created by any bodily movement. A bodily movement might, for example, have a certain overall linear quality – a dominant aspect such as angularity, straightness or roundedness. As with tensional quality, any time one wishes to pay attention to kinetically etched linear designs and patterns, there they are.

Areal quality, like linear quality, is a compound of design and pattern, the amplitude of any movement having both a bodily and movement component that describes a three-dimensional form or shape. The gamut of the quality in terms of both design and pattern is perhaps most immediately evident in affectively motivated movement, as when one leaps for joy, turns in on oneself in grief, flees in fear, and so on. With respect to areal design, animate bodies may be anywhere along a continuum from expansive to contractive, as the examples suggest. In fact, an animate body continuously changes shape as it moves and moves as it changes shape: in crouching and in ducking, it contracts itself; in stretching and in standing, it expands itself. Because a moving body's body is always spatially articulated in three-dimensional ways, its changing shapes at times create spatially identifiable designs, precisely as in crouching and stretching, ducking and standing. But it is *the bodily movement, not the position arrived at*, that constitutes the areal design of the body: crouching, standing, and so on, are all *kinetic* phenomena. They describe a moving body in the process of contracting and expanding. Areal designs are in effect as fluid as linear designs. Movement is thus not properly defined as a change of position, i.e., as a mere connecting rod from point A to point B; it is a dynamic phenomenon, and as such is properly defined in terms of dynamics, not statically in terms of positions.

Movement itself is similarly a three-dimensional phenomenon describable in terms of its areal pattern: intensive to extensive, according to its amplitude, which, as noted above, is perhaps most readily evident in affective movement. When one flees in fear, the areal pattern of one's movement is extensive; when one turns in on oneself in grief, it is intensive. Similarly, the areal pattern created in shaking someone's hand is intensive in comparison to the extensive pattern created in opening one's arms to embrace someone. In each instance, the space that movement creates – its overall shape – is distinct. Implements may augment the apparent shape of a movement. When one sweeps the floor or swings an ax to chop a log, the implement extends the areal design of the body and thereby extends the areal pattern of the body's movement, making it seem more extensive than it actually is.

All of the above qualities are themselves qualified by projectional quality, the manner in which one moves, the way in which energy is released. In quite general terms, one can move in a sustained manner – as in stretching or in an evenly paced walk; in an abrupt manner – as in slamming a door or grabbing one's hat; in a ballistic manner – as in skipping or throwing a ball; or in a collapsing manner – as in falling down or keeling over. The qualitative possibilities are often compounded in everyday movement as well as in sport activities. Knocking on a door may combine subtle shadings of sustained and abrupt or even ballistic movement; a basketball shot may do the same. The manner in which one moves is obviously a facet of voluntary movement. One can reach out for something suddenly or with a sustained effort; one can jump abruptly or in a bouncing manner, rebounding upward with each landing; and so on. In short, one can control not only the intensity and the linear and areal character of one's movement, but the way in which one moves. In effect, the qualitative aspects of movement are all created with movement itself. Indeed, all qualities of movement are internally bound to one another and mutually influential, which is to say that no quality exists apart from the unique *Gestalt* that any particular movement presents. The linear quality of any movement, for example, does not exist apart from the tensional quality subtending the linear design and pattern, the areal quality created in creating the linear design and pattern, or the manner in which the linear design and pattern are projected. The movement - whatever it might be - creates its own dynamics on the basis of its qualitative relationships. The qualities are thus basic to understandings of voluntary movement. A consideration of kinetic degrees of freedom will allow us to spell out the voluntary in voluntary movement more closely.

The fact that movement creates its own space-time-force, and hence its own particular dynamics, is implicit in the notion of kinetic degrees of freedom.[4] Physiologist Nicholas Bernstein's conception of the degrees of freedom problem in human movement highlights - one might even say, dramatically highlights - the open dynamic character of animate movement. However much Bernstein "was committed to achieving a mechanistic understanding of human motor control" (Hinton, 1984, p. 424), and however much "muscles are smoothly regulated by reflex mechanisms" in the execution of movement (Gowitzke and Milner, 1988, p. 193), voluntary animate movement is by definition initiated, carried through, and terminated at will, and is thus experientially replete with degrees of freedom. Its dynamic may be changed midcourse, for example, as when one begins to stand up and then sits down again. Whatever the movement, its kinetic dynamics may be analyzed and its degrees of freedom shown in greater detail through analysis. A leg lift, for example, might be forceful (tensional quality), straight-legged and forwardly directed (linear design of the body and linear pattern of the movement), barely elevated above the floor (areal design of the body and areal pattern of the movement), and abrupt (projectional quality). Alternatively - and antithetically stated to indicate the continuum between extremes and the range of possible variations - the lift might be weak, bent-legged and diagonally directed, elevated high off the floor, and sustained - or the lift might be moderately strong, bent-legged and diagonally directed, barely elevated off the floor, and abrupt - or it might be performed in myriad other ways reflecting different qualitative combinations. Further still, of course, the overall specific qualitative structure of the lift might be ineffable. Language, after all, is not experience and can at times fail to provide a ready means of transliteration (Sheets-Johnstone, 1999b, 2002; see also Stern, 1985, pp. 181–182; 1990). Indeed, in a fully literal sense, we may find that – to borrow an observation of philosopher Edmund Husserl (on the nature of the temporally constitutive flux of consciousness) – "For all this, names are lacking" (Husserl, 1964, p. 100), and in effect be at a loss for words.[5] As might be evident from the above descriptions, our vocabulary for movement is comparatively limited. Where are the words that would describe, for example, gradient degrees of tension between a weak and moderate tensional quality, or between a moderate and strong tensional quality (or, in the language of Gowitzke and Milner, describe gradient degrees of "force")? When the overall quality of movement is described globally from the perspective of its kinetic/affective dynamics, the movement vocabulary is richer and can enhance the qualitative description – "hesitantly," "apprehensively," or "warily," for example, exemplify the greater richness and can serve to inform qualitative analyses. However ineffable the qualities might be,[6] we experience them and are thereby aware of the specific qualitative character of our movement, its surgings, fadings, expansions, contractions, intensities, attenuations, and so on. In short, through kinesthesia, we are or can be aware of the kinetic dynamics we are creating in the very act of creating them and can alter those dynamics at any time, changing any quality at will and thereby the overall quality of our movement. By this very token, i.e., because we experience our own movement, we can at any time investigate how movement creates a spatio-temporal-energic dynamic, and, in effect, determine what is actually there in the perception of self-movement.

For example, and as might now be evident, what Gowitzke and Milner identify as speed is actually a combination of tensional and projectional qualities. Fast does not necessarily mean greater tension but only an accelerated projection of force. A fast movement may, in effect, have either a lightly buoyant or a strongly urgent quality. Speed similarly depends on both qualities when a question of slowness. Slow does not necessarily mean limp or weak. Pushing a heavy piece of furniture into place, for example, one moves slowly but with great force. It is not only with respect to speed that a finer analysis of movement is instructive. Deeper understandings of direction and range are equally possible in a phenomenological account. Indeed, the compound and complex nature of the spatial qualities of movement – linear quality and areal quality – otherwise remain unelucidated. In short, and as should be apparent, the qualities of movement are a more fundamental and precise way of identifying and describing the dynamic constituents of animate movement in general and voluntary animate movement in particular. They in turn offer fine-grained insights into the perceptual and cognitional aspects of kinesthesia because they elucidate at a deeper level the dynamic nature of animate movement; that is, they attest to the fact that movement creates its own space, time, and force and thereby a unique dynamic. This insight in particular clearly calls received wisdom into question.

The commonplace and unexamined idea that movement takes place *in* space and *in* time is in need of clarification. Space is commonly regarded a container and time a matter of measurable durations within that container.[7] In the course of everyday life, both are commonly taken for granted. They define heres and theres, nows and thens, locations and punctualities that are objectively tethered in the sense of objects in space – trees, houses, offices, cars, clouds, stars, and so on – and objects that measure time – clocks and calendars. Movement in everyday life is in consequence conceived precisely as occurring *in* space and *in* time. Being thus conceived, movement is caught up in the objective constituents of everyday space-time realities and the daily doings that go with them: alarm clocks, bus schedules, freeways, meetings, lunch rooms, doctor appointments, and so on. The result is that the qualitative realities of movement itself are occluded. As the above phenomenological analysis shows, any movement is the creation of a thoroughly unique spatio-temporal-energic dynamic that is anchored in movement itself and that flows forth with its own particular qualitative structure. Our concern in what follows is to bring the qualitative structure of that dynamic to light in a way that demonstrates how spatial perceptions and cognitions are rooted in movement from the very beginning, but not movement *in* space and *in* time.

How we build our knowledge of space may be described in basically two ways: from the viewpoint of an adult and that of a maturing newborn. The former perspective commonly takes for granted knowledge gleaned from the latter perspective, in part because movement is itself commonly taken for granted by adults, in part because, as adults, we are already experienced in the ways of the world and take "the world" for granted. As adults, we nevertheless have the possibility of experiencing fundamental aspects of space, that is, fundamental aspects of what we already know as "space" simply by paying attention to our experience of movement. However untrained we might be in such an endeavor, by paying close attention, we have the possibility of experiencing a diversity of spatial qualities, including ones thus far unmentioned – for example, a resistant space as we walk into a strong wind or shove a heavy box across a floor; an angular space when we feel ourselves cutting sharp corners; a yielding space in the course of running across an open meadow; a circular space in molding our arms to pick up a large bowl. Whatever the spatial quality, we experience it kinesthetically. As two of the examples make evident, an "exterior" as well as "interior" space of movement may be brought to life through kinesthesia; space "out there" may have a specific resonant character – resistant or yielding, for example – in virtue of our movement.

We can shift attention from a felt dynamics to a felt exterior space any time we wish. The shift is not unlike shifting attention from the touching to the touched, an experience that has received sizably more attention (see especially Merleau-Ponty, 1968) than its barely recognized kinesthetic analogue. But we can also, and most often do, shift attention in such a way that we turn spatial qualities of our movement into objective features of "space." Objectifying our movement in this way puts us – along with our movement – directly and immediately *in* space: spatial qualities of our moving body and of movement itself become spatial quantities specified or specifiable as measured or measurable distances and as particular directional coordinates. While certain aspects of the idea that movement takes place in space and in time were clarified earlier, a further clarification is apposite here with respect to an objectifying shift in attention. The shift brings with it a subtle but distinguishable shift in the conception of space that categorically changes its nature: the three-dimensional expanse co-extensive with our movement becomes a three-dimensional container *in which* we move. We become precisely objects *in* space, objects among the many natural and cultural objects that surround us. Correlatively, "the world" becomes a synonym for space, the space in which all objects are located.[8] As adults, we of course navigate our way in space deftly and efficiently. We are familiar with it and know what to expect. We are familiar with it and know what to expect, however, only because we have built up a reservoir of first-hand spatial knowledge from the time we were born. In a word, we all learned our bodies and learned to move ourselves many years ago in the course of infancy and early childhood (Sheets-Johnstone, 1999a, chapter V).

If we ask specifically what corporeal–kinetic knowledge we glean as maturing infants, we readily see that, in the beginning, movement is not a pre-given program of proficiencies and capacities, but something we must actively learn – precisely by moving ourselves.[9] Kinesthesia – the experience of self-movement – is the ground on which we do so. In reaching and kicking, we discover particular kinetic possibilities of our bodies and correlative spatio-temporal-energic dynamics in the process. In each instance, our movement has a particular flow, the dynamics of which are kinesthetically felt. When we learn to turn over, we experience a spatio-temporal dynamics quite different from reaching and kicking, a kinesthetically-felt coordinative dynamics that grounds our capacity ultimately to turn over any time we wish.[10] When we learn to walk, we learn a complex and challenging coordination dynamics that is, again, kinesthetically felt. Indeed, kinesthesia is an ever-present modality whether one is an infant or an adult. We cannot in fact close off kinesthesia in the way we can readily close our eyes and turn away from the visual, close our mouths, pinch our nose, clamp our hands over our ears, and similarly turn away from other sensory modalities.

While negligibly treated in physiology and psychology textbooks, kinesthesia is central to animate life, a fact dramatically illustrated both by the loss of the modality (Cole, 1991; Gallagher and Cole, 1998) and by its neuro-embryology: kinesthesia and tactility are the first sensory systems to develop. In brief, kinesthesia is the gateway to those coordinative dynamics that make the world familiar to us and allow us to know what to expect (Sheets-Johnstone, 1999a, 2003; see also Kelso, 1995). Famed neurologist Aleksandr Romanovich Luria implicitly indicated as much when, in describing "the working brain," he spoke of kinesthetic as well as kinetic melodies (Luria, 1973) and of "integral kinaesthetic structures" (ibid., p. 176). His insights into "complex sequential movement" show that a close study of kinesthesia is essential reading for those studying animate movement.

Given our common heritage of learning our bodies and learning to move ourselves, taking heed of the perspective of a maturing newborn is mandatory. The necessity is aptly highlighted by Lorenz's claim that we do not intuit space and time but only "movement in space and time." The previous analyses and clarifications support and provide strong empirical justification for the lead part of Lorenz's claim: we intuit space and time not directly but by way of movement. They support and justify only the lead part because subsequently, the notions of space and time – *the very concepts of space and time* – are erroneously if unwittingly conceived as containers and taken conceptually for granted in the process. Indeed, where do the concepts come from? If we do not intuit space and time directly, as Lorenz asserts, but derive them from self-movement, as Lorenz asserts and the foregoing analyses have shown, then clearly, the assertion that "there is only one intuition, of movement in space and time" is unsustainable. A container notion of space and time – something movement is *in* – objectifies the spatio-temporal experience of self-movement and reifies what Lorenz has earlier denied, namely, the direct experience of space and time: that which movement is *in*. In short, to intuit movement *in* space and time is to have a concept of space and time apart from, and in advance of movement. From the viewpoint of a maturing newborn, the having of such a concept – or intuition – is absurd. Space and time are not ready-mades, either as observational items *or* as concepts. In consequence, they cannot be taken for granted conceptually any more than they can be intuitively.

To say that there is no space and no time without movement is not a hypothesis or a logical or grammatical deduction but an experiential datum, that is, an empirically resonant fact of life. What is requisite to an understanding of the fact is not only a phenomenological analysis of movement but a fine and thoroughgoing diachronic examination of kinesthetic experience, an examination that not only presupposes nothing but holds the experience of self-movement up to the light of its original history. In what follows, the examination will narrow to spatial aspects of kinesthesia alone.

Spatial perceptions and cognitions are intimately tied to movement and constituted from the ground up by the infant itself, without a manual or instruction from anyone. We see this process clearly if at times implicitly in descriptive accounts of developing infant/child spatial awarenesses. Psychologists describe the fascination of infants and young children with *insideness*, for example, with being *in* or *inside*, or with *putting inside*. Piaget, for instance, documents Lucienne watching her hand as she repeatedly opens and closes it (Piaget, 1968, e.g., ob. 60, p. 90; ob. 67, p. 96). He does not mention *insideness* or *insides* in relation to the movement pattern, but clearly, watching one's hand alternately closing and opening alternately hides and discloses *insides*. T. G. R. Bower, corroborating other observations of Piaget, writes: "Piaget's son was surely typical in finding the relation 'inside' fascinating ... One of my own daughters spent the best part of one night placing small objects in my hand, closing my hand on them, moving my hand to a new location, and then opening it up to see if the object were still there. This kept her happy and busy till nearly 4 a.m." (Bower, 1974, p. 238).

As with *insides*, putting one thing inside another has an attraction for infants and small children. Although the attraction is explained in terms of learning the rule that "two objects can be in the same place provided one is inside the other" (Bower, 1974), such an explanation falls short of its mission: it says nothing of the *natural penchant* of infants and young children to put one thing inside another and nothing of their elemental fascination with *insideness*. Piaget's descriptive account of a sixteen-month-old child trying to open a matchbox testifies dramatically to the oversight. The child, Piaget says, looks "very attentively" at the slit into which one slides one's finger to open the matchbox, then "opens and closes [her] mouth several times, at first weakly, then wider and wider!" (Piaget, 1968, obs. 180, p. 294, my translation). To judge from his punctuation, Piaget is astonished by the child's corporeally rooted spatial intention and understanding. He does not interpret his observation along the lines of a developing tactile–kinesthetic body, however, because an experientially resonant thinking body does not enter into his theory of the development of human intelligence. On the contrary, he explains the child's lingual movement as a *faute de mieux*: "Lacking the power to think in words or clear visual images," Piaget writes, "the infant uses, as 'signifier' or symbol, a simple motor indication" (ibid.). The problem with such an explanation is readily apparent: the tactile–kinesthetic analogy between lingual gesture and manual gesture is far too powerful to be reduced to "a simple motor indication" (see Sheets-Johnstone, 1990, p. 238). Most significantly too, the explanation suffers from an adultist bias; that is, movement is regarded prelinguistic when in actuality language is – and should be regarded – *post-kinetic* (Sheets-Johnstone, 1999a).

The same adultist bias is apparent in studies of language acquisition showing the prepositional primacy of the word *in* as both locative state and locative act, *in* being the first spatial concept understood by an infant/child perceptually as signifying a certain locational relationship and behaviorally as signifying a certain locational act or acts (Clark, 1973, 1979; Cook, 1978; see also Grieve *et al.*, 1977). What requires explanation in this context is, again, the primacy of *in* in an experiential sense. If children as early as one-and-a-half years have a conceptual mastery of *in* as locative state and locative act, then surely their life experiences must be taken into account; that is, their conceptual mastery is less plausibly explained by rules, motor programs, and the like (as per Huttenlocher *et al.*, 1983, p. 211), than by the fact that their experiences of *in*, *inside*, *being inside*, and *putting inside* have been reiterated many times over every day of their lives in such acts as sucking, eating, defecating, urinating, being held in the arms of others, being put into a crib, grasping something in their hand, putting a foot into a shoe, an arm into a sleeve, a thumb into a mouth, and so on.

Research studies of *insideness*, *in*, and so on, are of particular interest for what they say and what they do not say about kinesthesia and movement. Careful experiential reflection on these studies shows that we put the world together in a spatial sense through movement and do so from the very beginning of our lives. Spatial concepts are born in kinesthesia. Accordingly, the constitution of space begins not with adult thoughts about space but in infant experience. A consideration of perceptions and cognitions of *near* and *far* will further exemplify this claim.

The concept of "distance" is commonly taken for granted and in turn far less researched by infant/child psychologists than other spatial concepts such as inside and outside, open and closed, appearing and disappearing, under and over. Yet both the perception and conception of *near* and *far* are part of our everyday lives: we reach for things that are reachable; we walk to something not quite within reach; we drive or fly to a place that is distant; and so on. Even brief reflection on such everyday facts of life readily points to the certainty that *near* and *far* are basically facts of *bodily* life: they are rooted in bodily experience, specifically experiences of one's tactile-kinesthetic body. A summary phenomenological sketch will suffice to show the soundness of the corporeal linkage and the perceptual and conceptual lines of its development.

Studies of infants watching objects appear and disappear show that in the beginning what occupies an infant's attention is solely what is present. They show that an infant later keeps track of objects appearing and disappearing. Studies of infants show too that in the beginning, infants put whatever is present at hand – grasped objects as well as their own thumbs and fingers – into their mouths. They show furthermore that infants make inchoate reaching movements toward objects they see, movements that over time become refined into coordinated movements toward things within reach. They show further that infants point and often make a sound of some kind when they want something that is out of reach. All such studies take for granted the tactile-kinesthetic body that is the center of an infant's world, the primary sensory base on which infants experience and explore what is about them. In finer terms, contact – touching or not touching something – grounds an infant's developing spatial knowledge. *Near* and *far* are thus tethered to its tactile-kinesthetic body. In Husserlian terms, they are tied to the "zero-point" of all possible orientations; that is, everything anyone might experience is "'there' – with the exception of … the Body, which is always 'here'" (Husserl, 1989, p. 166) – or as

Husserl later says, "I do not have the possibility of distancing myself from my Body" (ibid., p. 167).

The zero-point or "hereness" of an infant is, one might say, a concentrated zero-point or "hereness;" a limited range and repertoire of movement restrict a freely changing hereness. In effect, what is near is *something present and immediately touchable, something present which the infant can move toward and touch*; what is far is *something present that is beyond its range of movement or movement possibilities, hence beyond touch.* The perception of something as near or far and the concept of near and far are thus originally tactile-kinesthetic perceptions and concepts. Correlatively, foundational meanings of *near* and *far* are nonlinguistic corporeal meanings: is X within reach, or is X not within reach? Visual percepts and concepts of near and far develop on the basis of these original tactile-kinesthetic perceptions and concepts.[11] The original tactile-kinesthetic meanings of *near* and *far* are indeed the basis on which not only later visual percepts and concepts develop but more complex tactile-kinesthetic meanings arise. In other words, original meanings are kinesthetically as well as objectively elaborated in the course of development. For example, the tactile-kinesthetic experiences of near and far come to be fleshed out along the lines of ease, fatigue or effort – along the lines of a specific felt flow of movement, its strains and tensions or lack thereof in pursuit of some aim. They come to be objectively elaborated when distance becomes a measured or measurable quantity, a specific space to be traversed in a physico-mathematical sense – so many doors to pass on the way to a meeting, so many blocks to the store, so many miles to the airport, and so on. The objective sense of near and far has its roots in an objective body, a body no longer exclusively experienced as the "zero-point" of orientation but as an item in the prevailing landscape, a body that, passing a certain number of doors, walking a set number of blocks, or driving a set number of miles, is experienced as an object *in* space and moving *in* space.

Experiencing one's body in this way, one experiences not movement, i.e., not a kinesthetic dynamics, but oneself as an object in motion (Sheets-Johnstone, 1979, 1999a); the kinesthetic dynamics of self-movement are swallowed up in the objectification of both body and space. The dynamics, however, remain the *sine qua non* of the objectification; that is, experiences of a kinetic dynamics precede in both a chronological and logical sense the experience of oneself as an object in motion. Indeed, the possibility of the latter experience rests on and could not arise without the former experiences. When we learn our bodies and learn to move ourselves, we do so not analytically as objects in motion or objects in space, but dynamically as animate forms. Studies of infant development implicitly document this fact in their descriptions of movements such as reaching, turning over, sitting, standing, and so on (e.g., Bower, 1979, 1982; Thelen and Smith, 1994). In short, the experience of ourselves as objects in motion, objects in space, is possible only on the basis of our having learned our bodies and learned to move ourselves, and these learnings are anchored in the fundamental kinesthetic dynamics that form the basis of our everyday adult capacities to move effectively and efficiently in the world.

An important conceptual aspect of the distinction between experiencing the kinesthetic dynamics of self-movement and experiencing oneself as an object in motion – an aspect that ultimately warrants clarification of the relationship between a fascination with *insides* and the idea of being or moving *in* space – turns on the concept of space as an expanse and as a container. The conceptual difference is implicit in descriptive studies of infancy, which show that an infant's world stretches out in the beginning to whatever is present. Space is not objectified by the infant any more than its body is. The expanse an infant perceives defines whatever is *beyond its hereness*. The expanse has no terminable boundaries. What is *there*, in other words, is neither perceived nor conceived as being *in* space, but is simply *there*, an unbounded presence. What is near and what is far are thus in the beginning markers of a spatially open distance between a *here* and a *there*, in a tactile–kinesthetic sense, an expanse uncontained within or by a larger universal or worldly space – even the space of a room or a house.

The moment one speaks of being "in space," one has already objectified it by conceiving it a place filled with objects, a populated container whose population also includes oneself. No longer is it conceived as experienced originally, i.e., in and through movement of one's own voluminously-felt body that in moving, moves *through* – not *in* – space, precisely *through an expanse* – not *in a container*. That movement creates its own space testifies to the experiential distinction. When we come into the world and attend to the task of learning our bodies and learning to move ourselves, we are first and foremost attentive to movement (Bloom, 1993; Bower, 1971, 1974; Bruner, 1990). Through that attentiveness, we continually shape our movement spatially to the intentional urgings that prompt us to move. In a very real sense, we play with movement, discovering kinesthetic awarenesses and possibilities in the course of moving; over time, we hone our movement to better effect – changing our orientation, for example, or the range of our movement. Our focus of attention is not on ourselves as objects in motion, but forefronts the spatiality of movement itself, what it affords and does not afford with respect to touching things that are near, grasping them, pulling them toward us, crawling toward those that are distant, pointing toward them, and so on. This experiential space, or better this *tactile-kinesthetic spatiality*, has nothing to do with measured or measurable distances but is a dimension of movement itself. Space is precisely *not* a container in which movement takes place but a dynamic tactile-kinesthetically charged expanse.

An infant's fascination with *insides* and with putting one thing *inside* another testifies to this elemental tactile-kinesthetic spatiality as does the primary preposition *in*. An adultist bias easily deflects us from a recognition of this spatiality, precipitating us toward container interpretations of infant perceptions and conceptions of space. Yet clearly, an infant does not have a sense of its mouth as a container; it has a sense of its mouth as a complex tactile/gustatory center of activity and center of movement in the form of a tongue. Similarly, opening and closing its hand does not make and unmake a container, but makes an inside appear and disappear. Similarly again, putting one thing inside another is *putting one thing inside* another, like putting food or a thumb in one's mouth; the inside is not a container but a spatial aspect of what, for the infant, is a complex sequential activity. The concept of *in* operative in the notion of containment – and of "being *in* space" – is foreign to this original spatiality. Only as the tactile-kinesthetically charged expanse becomes perceived as bounded – a becoming undoubtedly helped along by language, which emphasizes early on the naming of containers such as cups, bottles, rooms, houses, cars, and so on – does a young child begin to perceive and conceive her/himself as being in space and to perceive and conceive the world of objects about her/him as being similarly in space.

In sum, we build our perceptions and conceptions of space originally in the process of moving ourselves, in tactile-kinesthetic experiences that in fact go back to prenatal life where the movement that takes a thumb to a mouth originates (Furuhjelm *et al.*, 1966). When we learn our bodies and learn to move ourselves, we are kinesthetically attuned to a kinetic dynamics and our concepts of space are grounded in that dynamics. On the basis of our original learnings, we develop more complex notions of space, later coming to perceive and conceive ourselves as spatially bounded bodies and objects in motion. Our early experiences are the foundation of this transition to objectification.

NOTES

1. I have reverted to the original term I used to describe the third quality. In *The Primacy of Movement* (Sheets-Johnstone, 1999a), I changed "areal" to "amplitudinal" to avoid any suggestion of a two-dimensional space, but have since found the original a far more useful and precise description. I might note too that, according to people expert in Labanotation (e.g. Sally Ann Ness, personal communication), the four phenomenologically derived kinetic qualities accord in basic ways with Labanotation, a system of movement analysis that, as one researcher states, "provides a way of perceiving, describing, and analyzing movement, as well as a way of notating it" (Youngerman, 1984, p. 109). Finally, it might be noted that some ethologists have profitably used a movement notation system (the Eshkol-Wachmann system) in their investigations of various mammalian behaviors (see Sheets-Johnstone, 1983 for a discussion and references).
2. Lorenz, 13 March 1980. Dr Anton Fuërlinger, medical doctor and long-time researcher in evolutionary biology, met with other researchers, including Konrad Lorenz, each Thursday afternoon in Altenberg, Germany in 1980 in a seminar devoted to evolutionary issues. Dr Fuërlinger taped the sessions and kindly shared the relevant tape with me.

   Elaborating on the tape, Dr Fuërlinger writes, "What movement really is we experience 'a priori' in mother's womb from conception on and after birth every time our whole body is moved actively or passively. Only humans succeeded in decomposing the unique quality or 'primacy' of movement (Sheets-Johnstone, 1999[a]) into dimensions, usually four, three for space and one for time. In this way of reduction descriptions became possible, but physics, ethology, psychology or the neurosciences can never replace or model the inner experience" (personal communication).
3. Given the informal nature of the seminars in which Lorenz participated and spoke, the concluding part of his claim was undoubtedly made without attention to the import and implications of everyday ways of speaking of space and time.
4. We might note that "kinetic degrees of freedom" contrasts markedly with the idea that features of voluntary movement are *limited*: as noted earlier, "initiation, regulation of speed, force, range, and direction, and termination" are sizable as well as elemental dimensions of animate movement.
5. Infant psychiatrist Daniel Stern makes a related observation in introducing the new descriptive term "vitality affects:" "It is necessary," he says, "because many qualities of feeling that occur do not fit into our existing lexicon or taxonomy of affects" (Stern, 1985, p. 54).
6. One could, of course, make up a word, but making up a word misses the point: qualitative aspects are not easily packaged. Not only are they fleeting and evanescent, but they have complex, subtle and intricate shadings. The degrees of freedom problem in human movement attests to the linguistic impasse at the same time it attests to the qualitative realities of actual experience.
7. For just such a ready-made container concept of space, see Johnson, 1987. For a critique, see Sheets-Johnstone, 1999b.
8. Husserl speaks explicitly of "a *common time-form*" with respect to "the constitution of *a world and a world time*" (Husserl, 1973, p. 128). We humans inhabit a "common space-form" as well, as Husserl implicitly indicates in distinguishing between individual "surrounding worlds" and "a single

Objective world, which is *common* to them" (p. 140). The point is that we can elucidate the common space-form phenomenologically no less than we can elucidate the common time-form. The common space-form is constituted in and through self-movement, that is, in and through our capacity to learn our bodies and to move ourselves.

9. Dynamic systems theorist J. A. Scott Kelso's concept of "intrinsic dynamics," which infant/child psychologists Esther Thelen and Linda Smith utilize in their analyses of the development of movement proficiencies and capacities, is significant in this respect. The concept is furthermore akin to what I have termed "primal animation." See Kelso, 1995, Thelen and Smith, 1994, and Sheets-Johnstone 1999a.
10. For perspicuous and finely-detailed analyses of coordinative dynamics, see Kelso, 1995.
11. They do so in a way similar to the way in which the perception and conception of visually drawn straight lines derive from the perception and conception of tactilely felt straight edges (see Sheets-Johnstone, 1990, Chapter I).

REFERENCES

Bloom, L. (1993). *The transition from infancy to language: acquiring the power of expression*. New York: Cambridge University Press.

Bower, T. G. R. (1971). The object in the world of the infant. *Scientific American* **225**(4), 30–38.

Bower, T. G. R. (1974). *Development in infancy*. San Francisco: W. H. Freeman.

Bower, T. G. R. (1979). *Human development*. San Francisco: W. H. Freeman.

Bower, T. G. R. (1982). *Development in infancy* (2nd edn). San Francisco: W. H. Freeman.

Bruner, J. (1990). *Acts of meaning*. Cambridge, MA: Harvard University Press.

Clark, E. V. (1973). Non-linguistic strategies and the acquisition of word meanings. *Cognition* **2**, 161–182.

Clark, E. V. (1979). Building a vocabulary: words for objects, actions and relations. In P. Fletcher and M. Garman (Eds.), *Language acquisition* (pp. 149-160). Cambridge: Cambridge University Press.

Cole, J. (1991). *Pride and a daily marathon*. London: Duckworth.

Cook, N. (1978). In, on, and under revisited again. In *Papers and Reports on Child Language Development* 15 (pp. 38–45). Stanford, CA: Stanford University Press.

Furuhjelm, M. Ingelman-Sundberg, A. and Wirsén, C. (1966). *A child is born*. New York: Dell Publishing.

Gallagher, S. and Cole, J. (1998). Body image and body schema in a deafferented subject. In D. Welton (Ed.), *Body and flesh* (pp. 131–147) Oxford: Blackwell.

Gowitzke, B. A. and Milner, M. (1988). *Scientific bases of human movement* (3rd edn.). Baltimore: Williams and Wilkins.

Grieve, R., Hoogenraad, R. and Murray, D. (1977). On the young child's use of lexis and syntax in understanding locative instructions. *Cognition*, **5**, 235–250.

Hinton, G. (1984). Some computational solutions to Bernstein's problems. In H. T. A. Whiting (Ed.), *Human motor actions: Bernstein reassessed* Amsterdam: Elsevier Science Publishers.

Husserl, E. (1964). *The phenomenology of internal time consciousness*, trans. James S. Churchill, M. Heidegger (Ed.). Bloomington: Indiana University Press.

Husserl, E. (1973). *Cartesian meditations*, trans. Dorion Cairns. The Hague: Martinus Nijhoff.

Husserl, E. (1989). *Ideas pertaining to a pure phenomenology and to a phenomenological philosophy*, trans. R. Rojcewiczs and A. Schuwer. Dordrecht: Kluwer Academic.

Huttenlocher, J., Smiley, P. and Ratner, H. (1983). What do word meanings reveal about conceptual development? In T. B. Seiler and W. Wannemacher (Eds.), *Concept development and the development of word meanings* (pp. 210–233). Berlin: Springer-Verlag.

Johnson, M. (1987). *The body in the mind*. Chicago: University of Chicago Press.

Kelso, J. A. S. (1995). *Dynamic patterns*. Cambridge, MA: Bradford Books/MIT Press.

Lorenz, K. (1980). Altenberg Seminar tapes.

Luria, A. R. (1973). *The working brain*, trans. B. High. New York: Harper & Row.

Merleau-Ponty, M. (1968). *The visible and the invisible*, trans. Alphonso Lingis, Claude Lefort (Ed.). Evanston: Northwestern University Press.

Piaget, J. (1968). *La naissance de l'intelligence chez l'enfant* (?6th edn.). Neuchatel: Delachaux et Niestlé.

Sheets-Johnstone, M. (1966). *The phenomenology of dance*. Madison: University of Wisconsin Press. Second editions: London: Dance Books Ltd, 1979; New York: Arno Press, 1980.

Sheets-Johnstone, M. (1979). On movement and objects in motion: the phenomenology of the visible in dance. *Journal of Aesthetic Education*, **13** (2), 33–46.

Sheets-Johnstone, M. (1983). Evolutionary residues and uniqueness in human movement. *Evolutionary Theory*, **6**, 205–209.

Sheets-Johnstone, M. (1990). *The roots of thinking*. Philadelphia: Temple University Press.

Sheets-Johnstone, M. (1999a). *The primacy of movement*. Amsterdam/Philadelphia: John Benjamins.

Sheets-Johnstone, M. (1999b). Sensory-kinetic understandings of language: an inquiry into origins. *Evolution of Communication*, **32**, 149–183.

Sheets-Johnstone, M. (2002). Descriptive foundations, keynote address, American Society for Literature and the Environment, Flagstaff, Arizona; subsequently published in *Interdisciplinary Studies in Literature and Environment*, **9**(1), winter 2002: 165–179.

Sheets-Johnstone, M. (2003). Kinesthetic memory. *Theoria et Historia Scientiarum*, **7**(1), 69–92.

Stern, D. N. (1985). *The interpersonal world of the infant*. New York: Basic Books.

Stern, D. N. (1990). *Diary of a baby*. New York: Basic Books.

Thelen, E. and Smith, L. B. (1994). *A dynamic systems approach to the development of cognition and action*. Cambridge, MA: Bradford Books/MIT Press.

Youngerman, S. (1984). Movement notation systems as conceptual frameworks: the Laban System. In M. Sheets-Johnstone (Ed.), *Illuminating Dance: Philosophical Explorations* (pp. 101–112). Lewisburg, PA: Bucknell University Press.

ROBERT W. MITCHELL

15

Understanding the body: spatial perception and spatial cognition

A topic of concern in the study of the development of spatial cognition is how the spatiality present in perception is related to cognition about space. To highlight how cognition might arise from perception, I focus on a phenomenon that appears to be present in both incarnations: kinesthetic-visual matching (KVM). Many theorists believe that cognitions derive (somehow) developmentally from perceptions for all perceptual systems (see Finke, 1980, 1986; Mitchell, 1994a; Piaget, 1945/1962, 1954; Reisberg and Heuer, 2005; Ryland, 1909). It may be that perception implicitly includes understanding, and cannot occur without it (Noë, 2004). In that case, the question becomes how perception develops into other forms of cognition.

KVM is the recognition and/or creation of spatial similarity (or identity) between kinesthetic (proprioceptive, somasthethic) and visual experiences. In essence, matchers experience what they feel of their body - the position of their body parts and their movements (including their timing) - as isomorphic with what they see of it. How the body is isomorphic is the correspondence problem delineated by Nehaniv and Dautenhahn (2002) - what corresponds to what in the isomorphism. Correspondence is handily assured in that the spatiality of the kinesthetic sense of the body is normally and biologically coherent with the spatiality of the visual experience of the body. (For discussion of the development of ideas about kinesthesis from what was termed the "muscle sense," and its relation to other terms such as proprioception and somasthesis, see Bastian, 1902, pp. 540–544, 691–700; Boring, 1942; Finger, 1994, p. 204; Mitchell, 2007; Scheerer, 1987.) It is the nature of kinesthesis to match the body's form when static or in motion, and vision shares the same skill at matching forms (whether bodily or not). KVM as a causal factor in self-recognition and

Spatial Cognition, Spatial Perception: Mapping the Self and Space, ed. Francine L. Dolins and Robert W. Mitchell. Published by Cambridge University Press.

imitation has been "discovered" by multiple authors since 1898 (for review, see Mitchell, 1997a, 2007), and was again recently "discovered" by still more authors (e.g., Bischof-Köhler, 1994, p. 353; Lingis, 2007; Miyazaki and Hiraki, 2006; Rochat and Striano, 2002; Suddendorf and Whiten, 2001, pp. 632, 635; van den Bos and Jeannerod, 2002). (See also discussion of ideo-motor action in Prinz, 1987.) KVM as discussed here assumes either the visual-kinesthetic matching or the visuomotor-kinesthetic matching (or both) suggested by Hecht *et al.* (2001, p. 12), in that the motor representations in the latter involve "expected kinesthetic afferents." Visual observation of actions induces "specific innervation in the corresponding muscles" (Grèzes *et al.*, 1999, p. 1875) that may or may not be experienced kinesthetically. Kinesthesis apparently includes both efferent and afferent information about one's own movement (Tsakiris and Haggard, 2005).

There are two relevant perceptual experiences subsumed under kinesthesis (also called "proprioception" – Boring, 1942, p. 535). One is the feeling of the spatial outline of one's body, one's sense of "the *somatic periphery* ... the peripheral body surface" (von Fieandt, 1966, p. 327). (This perception is sometimes called "somesthesis" [e.g., Funk and Brugger, 2002].) The other is the feeling of movement of one's body and awareness of the spatial arrangement of or relations among the body parts. Under normal circumstances, these two perceptual experiences are not differentiated in experience, and are usually experienced as one. However, the somatic periphery can be experienced when muscles are static, but the feeling of movement usually requires movement. ("By sensing the effort made by those muscles necessary to bring a limb to a given position or to hold it there, we gradually come to be aware continuously of the position of our limbs. If, for instance, the hands be held behind the back, so that they cannot be seen, and we allow somebody else to place them in positions where they are not touching each other, we still know at every moment just where they are" [Weber, 1846/1996, p. 194]; "Though his eyes be shut, a man knows about the position and posture of his body, and whether its parts are moving or exerting effort ... If this be kinesthesis, then no philosopher can ever have doubted its existence" [Boring, 1942, p. 524].) Kinesthesis is either identical to or part of the body schema. According to Schwoebel *et al.* (2004, p. 285), the body schema "provides a dynamic mapping of the current positions of body parts relative to one another," just as kinesthesis does. However, the body schema is not a perception, but a representation, albeit an "on-line" one, that incorporates multiple inputs: "the 'body schema' has been characterized as an on-line representation of the current configuration of the body, derived from numerous motor and sensory inputs (e.g., proprioceptive, vestibular, tactile, visual, efference copy), that is involved in the production of action" (p. 285). (See discussion of different notions of the body schema in Chapters 17 and 18.)

Some have argued (Cratty and Sams, 1968; Robertson, 1896, p. 116) that kinesthesis creates the expectation of external as well as bodily space: blind children purportedly extrapolate from their understanding of their body's spatiality to create their understanding of external space. However, such an extrapolation is unsupported by evidence (Morsley *et al.*, 1991), though clearly some aspects of space incorporate both bodily and external space (e.g., right vs. left).

Kinesthesis may seem a somewhat odd perception, in that experience of it can seem to be lost when one engages in habitual action. However, what we lose is not kinesthesis per se, but our attention to it. We become inattentive to the kinesthetic perception of our movement (e.g., James, 1890, pp. 496–497). Similarly, while driving on a well-known route but thinking of other things, we continue to see our surroundings even though we may remain unaware – are not attending to the fact – that sight is occurring. In such dissociation, it is only when something unexpected or jarring occurs that we are pulled back into attentively experiencing our sight as such.

I present two modes of KVM in this chapter – one perceptual, the other conceptual – and offer an account of how the former becomes the latter.

PERCEPTUAL KVM: WHEN KINESTHESIS IS HERE, WHERE I AM

There is an inherent match between our visual and kinesthetic experiences of our body. Kinesthetic experience is our body being felt by us without touch. It provides us with the location and extent of our body's parts. Vision, from a different reference frame, tells us the location and extent of parts of our body under particular conditions – i.e., when we can see them. What we see is (usually) matched by what we feel, likely because we (and other animals as well) have a program that produces the feel of our bodily shape attached to our body and allows us to match this feel to other perceptual experiences of our body shape (Melzack, 1989). Indeed, given that the body grows and changes in form, it would be best to have a program that takes the actual location of our body's parts into account in formulating our perceptual experience of it (see, for example, Kinsbourne, 1995, pp. 216–217; Melzack, 1989). Without an actual body part, as in cases of congenital or later amputation, the program still produces the feel of the body, usually in the shape of a normal body (Brugger *et al.*, 2000; Melzack, 1989; von Fieandt, 1966). As Erasmus Darwin (1796, p. 20) described it, "After [an] amputation ... the patient has complained of a sensation of pain in the foot or finger that was cut off ... the pain or sensation ... was at the same time accompanied with a *visible* idea of the shape and place, and with a *tangible* idea of the solidity of the affected limb" (emphasis added).

To show KVM at the perceptual level, consider the case of a congenitally legless and armless woman. This woman experienced phantom limbs almost always, except when visual experiences made their felt existence impossible: "Awareness of her phantom limbs is transiently disrupted only when some object or person invades their felt position or when she sees herself in a mirror" (Brugger *et al.*, 2000, p. 6168). Vision maps the kinesthetic experience to the body, so that the nonvisual but felt body parts of the amputee lose their existence when vision shows them to be absent. People with one amputated arm, upon seeing through a mirror the reflection of their real arm in the place where their amputated arm would be, are able to experience the movements of both arms (real and phantom) when asked to move both arms simultaneously (Ramachandran, 1998, p. 1621; Ramachandran and Blakesee, 1998, p. 47). KVM is also present for normal people who experience their kinesthetic feelings in misplaced appendages. When someone's arm is placed just below a platform, out of sight, and a rubber arm is visibly placed on the platform above the real arm, if the rubber arm and the person's arm are lightly touched at the same time and on the same area of the arm, people will feel that the touch occurs where they see the rubber arm being touched (Botvinick and Cohen, 1998; Graziano and Botvinick, 2002). The same effect with a shoe or a table can make people feel that their arm has become the shoe or table (Ramachandran, 1998, p. 1622), suggesting that the kinesthetic experience can map to almost any potential shape. Apparently, implausible arm positions – fake arm position 90 degrees off from where the real arm is – show no kinesthetic mapping to the fake arm (see Austen *et al.*, 2001, 2004; Farné *et al.*, 2000; Pavani *et al.*, 2000). Similarly, five-month-old infants show some experience of perceptual KVM when shown visual images of hands – one of their own hidden hand as it explored a hidden object, the other the hand of another infant exploring the same object – that retained the left–right spatial axis egocentrically (i.e., from the point of view of the child), but not when shown the same images with the left–right spatial axis reversed (Schmuckler, 1996). The evidence for perceptual KVM is that infants preferred to look more to the other infant's hand than to their own with the egocentric displays, but showed no preference with the reversed images.

KVMs also occur when people wear prisms that distort their vision, such that they have kinesthetic experiences of their arm where they see it, rather than where it is (Cohen, 1967; Harris, 1965; Kornheiser, 1976). In these studies, humans wear prisms which distort the visual field (Harris, 1965). When they see their arm dislocated through a prism, people actually have kinesthetic feeling in a new location (where they *see* their arm); that is, they maintain the match between kinesthesis and vision even when they know that the match is nonsensical. When subjects looking through a prism *continuously* observe their displaced hand pointing toward various locations over several trials (such that the hand becomes adapted to the deviation) and are then asked after the prism is removed to point to an object with each hand separately, they tend to misjudge their point with their adapted hand (the one they used while the prism was on), but to point accurately with their nonadapted hand (if they had misjudged with this hand, it would be indicative of transfer of adaptation). However, when subjects are allowed to observe only their *completed* point over several trials (called terminal exposure) and are then asked after the prism is removed to point to an object with each hand separately, they tend to misjudge their points with either hand, indicating transfer of the pointing error to the nonadapted hand (Cohen, 1967; Kornheiser, 1976, p. 793).

The explanation for these differences has to do with which sensory modality subjects take as veridical (see Mitchell, 1997a): if the visual modality is taken as veridical (as it would be in the continuous observation task, as vision is the most salient perception), subjects map their kinesthetic sensations to the visual (and therefore do not transfer to the other hand); however, if the kinesthetic modality is taken as veridical (as it would be in the terminal exposure task, as kinesthesis is the only perception the subject experiences for most of the task), subjects map the visual sensations to the felt kinesthetic sensations (and therefore transfer to the other nonadapted hand: kinesthetic perceptions are apparently not as easily separable into parts as visual perceptions are). KVM is rarely brought to one's attention, as the visual and kinesthetic images of one's self are so finely attuned.

Researchers describe these re-placements of kinesthetic sensations into the visually observed (and tactually felt, in the case of the rubber arm) location as an attempt to maintain consistency among perceptual modalities, in which more dominant perceptual modalities hold sway (see discussion in Graziano and Botvinick, 2002; Mitchell, 1997a, 2002a; Ramachandran and Blakesee, 1998). What is normally experienced as a correspondence or match between actual phenomena (seen and felt body parts) becomes lost, and instead a correspondence is created between similar phenomena (seen object that is not one's own body part to which feeling becomes attached). What is interesting is that kinesthesis is set up to solve the correspondence problem by matching itself to the visual body. This built-in directedness to match kinesthesis to the visual body may be malleable because the visual body changes developmentally, so that kinesthesis must be capable of matching what is a moving target – the developing body (Melzack, 1989; see discussion in Mitchell, 2002b). Kinesthesis is a normal part of mammalian experience (Sheets-Johnstone, 1999; also see Chapter 14, this volume), and the spontaneous matching between kinesthesis and vision in relation to one's own body is likely a constant among sighted mammals and other vertebrates. Indeed, monkeys can use (and apparently thus recognize) the image of their own arm on a concurrent video-image when using a tool (Iriki *et al.*, 2001; also see Chapter 18, this volume), suggesting at least a perceptual matching between kinesthesis and vision.

Matchings occur across other modalities as well, such as visual-tactile matching (di Pellegrino *et al.*, 1997; Pavani *et al.*, 2000). However, matchings without vision, such as tactile–kinesthetic matching in blind children, occur less spontaneously than matchings with vision, requiring representation (memory) and attention to create an overview or "image" of a body (Millar, 1994). Disparate tactual perceptions must be consolidated into a representation comparable to the "impression of the overall form" that vision immediately provides (Kinsbourne and Lempert, 1980, p. 36).

CONCEPTUAL KVM: WHEN KINESTHESIS CONNECTS WITH SOMETHING VISUALLY OVER THERE

Whereas perceptual KVM seems common among mammals, conceptual KVM appears to be present only among a select group of mammals: apes, humans, some cetaceans, and elephants (see Mitchell, 2002b, for overview; Plotnik *et al.*, 2006); and perhaps one bird, the magpie, which apparently shows both contingency testing and passing the mark test

(Prior *et al.*, 2008). The evidence for conceptual KVM is an animal's ability to match between its kinesthetic experience of its own body and visual images of its own body (as in a mirror) or another's body (as in imitation). This evidence shows that the animal is not just extending the kinesthetic to the visual experiences of its physical body, but beyond to "another" body. Specifically, the evidence includes self-recognition, a generalized ability for bodily (including facial) imitation, recognizing that your bodily actions are being imitated, and pretending to be someone you have observed by enacting its behaviors (which requires re-enacting kinesthetically a visually experienced bodily action of another). So far, only some apes and cetaceans studied have shown generalized imitation, self-recognition and/or pretending to be another individual, and one chimpanzee recognized being imitated, though most humans are believed to show all four pieces of evidence (see Mitchell, 2007). Specific skills at KVM may be present in the different evidence: skill at synchronic (relatively simultaneous) bodily imitation may be necessary for self-recognition, and deferred bodily imitation seems necessary for pretending to be another. I have argued extensively for the involvement of KVM in self-recognition, generalized bodily imitation, recognition of being imitated, and pretending to be another (e.g., Mitchell, 1993a, 1993b, 1997a, 1997b, 2007). The general idea is that, if I am to do what another does, or recognize my own image in a mirror or in another's actions, I must be able to match my kinesthetic experiences (which contain information about my form and the temporal order of my actions) to my visual experiences of a similar form and temporal order. Consistent with this idea, human infants exhibit imitation of another's actions that they cannot see themselves perform (e.g., putting their hand on their head) at sixteen to eighteen months of age (Jones, 2007), the same time they show self-recognition (Courage *et al.*, 2004). (See Jones, 2007, concerning why neonatal "imitation" is not problematic for this view.) Simply matching visual experiences of another to visual experiences of myself does not appear to be enough, as most mammals have visual–visual matching skills but fail to show evidence of any of these activities (Mitchell, 1993a; Mitchell and Anderson, 1993). Interestingly, something comparable to kinesthetic–visual matching can be created in robots, and allows for self-recognition (Gold and Scassellati, 2007).

KVM is also implied in the development of mental imagery for use in producing or recognizing some actions (Féry, 2003; Jeannerod, 1994; Mitchell, 1994b; Neuper *et al.*, 2005), though whether KVM is present at the beginning (planning) stage or after motor imagery is well learned (or both) is unclear (see Johnson, 2000). Note, however, that how perception is employed in spatial imagery is uncertain, as some spatial imagery, including that about the body, can be represented amodally, i.e., without specifically visual or kinesthetic (or other) content (Reisberg and Heuer, 2005).

Kinesthesis can be entrained to make matching to a visually experienced novel body position easier (Casile and Giese, 2006). This finding was discovered using images of people walking in a dark space in which only their joints are lighted – so-called point-light displays. People normally walk with a phase difference of about 180 degrees between their arms, and between their legs, but can be taught to walk with other phase differences. If people are simultaneously shown two point-light displays that are either the same or differ in phase (in this study, the phase differences were 180, 225 or 270 degrees), they are best (about 85 percent correct) at detecting when the two images are the same for normal walking (180 degrees), and are somewhat better at detecting when the two images are the same for the phase difference of 225 degrees (about 70 percent correct) than for the phase difference of 270 degrees (63 percent correct). However, if these same people are trained (while blindfolded) to walk with a phase difference of 270 degrees, they become better able to detect when the phase differences are the same for the two images with a 270-degree phase difference (now 72 percent correct), but show no increase in ability to detect for the other two phase differences (compared to their ability before this training). Apparently the kinesthetic experience of walking at a phase difference of 270 degrees assisted in matching this form of walking in two visual displays. Note that people can detect their own point-light display, further supporting an ability for KVM as people rarely (if ever) see themselves walking (Beardsworth and Buckner, 1981). In addition, people are able to detect known individuals from their point-light displays, even from various perspectives and under various transformations in (e.g.) the speed, size and shape of walking humans (Troje *et al.*, 2005).

Conceptual KVM coordinates perspectives of one's body and images of it. In a study by van den Bos and Jeannerod (2002), participants had to determine which of two hands was their own. Each participant sat opposite a confederate and observed for two seconds, via a screen, two gloved right hands with palms down – one hand was his/her own, and the other (opposite hand) was the confederate's. Both hands were placed under a mirror, above which was a screen which presented the hand images to the participant, but which shielded the participant from seeing the actual hands. A program transformed the mirror image of the two hands into a visual display on the screen, manipulating their placement in several ways: the image was not rotated (such that the hand image above the participant was his/her own), was rotated 180 degrees (such that the hand image above the participant was the confederate's), or was rotated 90 degrees to the left or to the right. (Participants were not informed which of these rotations was present, though certainly they recognized after a few trials that all were possible.) In addition, the participant was instructed to move his/her thumb or index finger, or make no movement. Simultaneously, when the participant moved, the confederate was instructed to move either his/her thumb or index finger; when the participant was instructed not to move, the confederate was also instructed not to move. When there was movement, half the time the two movements were congruent (e.g., participant moves thumb as confederate moves thumb), and half they were incongruent (e.g., participant moves thumb as confederate moves index finger). After the two seconds of observation, the screen went dark, and an arrow pointed to the position of one of the two hands. Participants had to say "yes" when they believed the hand pointed to was their own, and "no" when they believed it was not their hand. Clearly KVM was the main information available to participants, but the 90 and 180 degree misalignments are likely to induce uncertainty (and thus error) when the thumb and finger actions of the participant and confederate are congruent. (The authors acknowledge that slight variations in thumb and finger actions of participant and confederate might assist in using KVM to detect which is the participant's hand.) Predictably, participants were most accurate (almost perfect performance) when their movements and those of the confederate were incongruent, and worst when neither moved (34 percent). Overall, in all same-movement and no-movement conditions in all four rotations, participants were more likely to inaccurately say "yes" when the hand was the confederate's than to inaccurately say "no" when the hand was their own (i.e., they tended to over-attribute the hand to themselves). Better than chance responding in correctly attributing the hand to self or other in most of the same-movement and no-movement conditions seems likely to derive from differences in size and shape of the visual display of the gloved hands, and closer similarity in contiguity (contingency) between the felt action and the visual display of the participant's action than the confederate's (i.e., KVM). Without differences in the size and shape of gloved hands, one would expect in the no-movement condition that selections would be at chance levels. Apparently, the size and shape of the gloved hands were distinctive in the no-movement condition: the average number of errors when the hands were not rotated (only about 23 percent) and when the hands were rotated 90 degrees in either direction (about 31 percent for both) were smaller than the average when the hands were rotated 180 degrees (about 52 percent). This last percentage suggests chance performance when the hands were rotated 180 degrees, but in this condition participants were more likely to recognize their own hand (about 58 percent of the time) than to recognize the confederate's hand (about 40 percent). (Or, put inversely, participants were more likely to err by saying that the hand was their own when it was the confederate's hand – just over 60 percent – than by saying that the hand was not their own when it was – about 42 percent.) Thus, it would appear that participants tended to view the 180-degree transformation as if it were not transformed at all, and thus viewed the hand in the same position as their hand as their own, applying perceptual KVM.

Elaborating upon this experiment, Tsakiris and Haggard (2005) examined whether active and passive movements influenced self-recognition of hand movements. In this study, participants saw an index finger, either their own or a confederate's, of a right hand move on a screen. Prior to seeing this finger movement, the participant either performed an action with the left hand that made the right index finger move, or did not perform an action (in which case, the confederate performed the action that made the participant's finger move). In all cases, the participant's right index finger moved. The question asked was whether participants could better detect their own finger movement (and better detect that the confederate's finger movement was not their own) when they caused the finger movement by pressing the lever compared to when they had not. Participants were excellent (better than 90 percent accurate) in detecting their own hand whether it was self-generated or externally generated (though slightly more accurate when it was self-generated). By contrast, they were not as good at detecting when the hand was not their own: when viewing the confederate's hand, they were more accurate in stating that the hand was not their own (62 percent of the time) when the movement was self-generated than when it was not (only 45 percent). Conversely stated, when viewing the confederate's hand, participants falsely stated that the hand was their own only 38 percent of the time when their own left hand produced an action, but did so 55 percent when their left hand performed no action (and the hand movement was caused by the confederate). Clearly, doing an action that was paired with the passive finger movement influenced participants to see the finger movement as their own. From the point of view of KVM, what seems noteworthy here is that participants matched extraordinarily well when the felt movement matched the visual image of their own hand, regardless of how the movement was initiated.

The suggested explanation for the difference in the success (or error) rate when the participant viewed the confederate's hand is that, when the movement was self-generated, the temporal dis-contiguity between the left hand action and the right index finger movement provided additional information that the index finger movement was not the participant's own, compared to when no left hand action occurred but the right index finger moved and was felt more frequently to match the visible display of the confederate's finger movement. Thus, without any disconfirmatory evidence in the form of temporal contiguity of action, participants had to rely on the match between the passive feeling of their index finger moving and the intended-to-be-similar visual movement of the confederate's finger movement, which was not a reliable feature.

Self-recognition is usually discerned in several ways by a person's response to a mirror. Initially, a child (or other organism) may engage in contingency testing in relation to the mirror image. Later, the child may explore parts of its body that are indiscernible without the mirror (self-exploration). To make sure that children see the mirror-image as themselves, researchers mark the child's face and see if the child locates the mark on the face (rather than in the mirror) – the so-called "mark test." Given that children do not know what their face looks like without the mirror, it would appear that KVM is necessary to pass the mark test. Children appear able to match kinesthetic and visual images of the self in a mark test even if the visual image is delayed (on videotape) by between one and two seconds (Miyazaki and Hiraki, 2006). However, it could very well be the case that infants could recognize other parts of their body in the mirror through visual–visual matching of the mirror-image of the body part and the actual body part (Mitchell, 1993a, p. 309), and I suggested (p. 298) that this be tested:

> An interesting extrapolation of the ... technique of placing dye on an animal's forehead and then allowing the animal to observe its image in a mirror would be to put dye on a 5-month-old infant's leg following recognition of proprioceptive-visual contingency and then, without letting the infant see its leg, to observe whether the infant reached for its own leg. If the infant reached for its leg, this behavior would suggest that

> the infant recognized parts of its body as its own in a mirror-image prior to recognizing its full body as its own.

Just such a test was enacted by Nielsen *et al.* (2006), although they used a sticker (rather than dye) and tested eighteen- and twenty-four-month-old children (rather than five-month-old children) to determine if recognition of one's own (unseen) leg in a mirror co-occurred with recognition of one's own face in a mirror – it did. (It would, of course, be helpful to know whether mirror-leg-recognition occurs prior to age eighteen months and in nonhuman animals.) In fact, the authors tested to see whether it was KVM (what they call "proprioceptive-visual matching") that accounted for this co-occurrence of facial and leg mirror-self-recognition by seeing if children recognized their own legs when they were surreptitiously covered with a novel material, and when they were covered while the child watched. The children mostly failed when the leg covering was surreptitious, but were usually successful when they observed the legs being covered. In addition, the frequency of these same children who passed the facial mark test was comparable to those who passed the leg mark test. From these findings, the authors argue that the KVM model is discounted because children could solve both facial and leg mark tests with the same frequency, but there are a few problems with their interpretation. First, the co-occurrence of two skills does not entail that they are based on the same skills, especially when leg recognition may begin developmentally earlier than face recognition. However, the authors assume that identical skills are used in facial self-recognition and leg self-recognition: "Theories that have emphasized the special status of the face in explaining task performance ... do not explain why toddlers perform equivalently when the mark in on a ... usually directly observable body part: their legs. We can hence reject these proposals" (Nielsen *et al.*, 2006, p. 182). But there is a clear difference: children see their legs and the clothes they wear, but they do not see their own face, and cannot know what their own face *looks* like until they self-recognize (Mitchell, 1993a). Indeed, it seems clear from the authors' own data that knowing what you look like is essential for leg self-recognition. In essence, to know it is their leg is to know what they are wearing. Imagine if the previously described studies by van den Bos and Jeannerod (2002) and Tsakiris and Haggard (2005) had been done with the participant having a red bow on his/her finger, and the confederate not having one. I would expect close to 100 percent accuracy in selecting the finger with a bow on it as one's own, no matter what the conditions. Would this suggest that participants do not have KVM or that they do not use it when the bow is not present? I think not. (Oddly, the authors reject the theories presented in Mitchell [1993a] because they purportedly "emphasize the special status of the face" as "a body part that is not directly observable" [p. 182], but in fact one of the theories [on p. 308] focuses specifically on how recognizing any body part of oneself in a mirror could lead to whole body self-recognition.)

A second problem, closely related to the first, is that children are quite good at visual-visual matching well before they pass the facial mark test: for example, they are able to recognize legs and hands as such (Mitchell, 1993a). Thus, it is likely that they know what their leg should look like: that is, if non-color-blind children are wearing green trousers, they would be unlikely to suspect that a pair of legs in red trousers is their own. Consequently, when they observed a marked leg in the mirror that showed different trousers than they knew that they were wearing, it is not surprising that they usually failed to respond as if it were their own leg. Although KVM might be used by some children to detect their leg in these circumstances (as Nielsen *et al.* [2006] acknowledge on p. 183), it is not necessary to use KVM when it is clear, from knowledge of what they are wearing, that the leg they observe in the mirror is their own (when it looks like what they know they are wearing), or is unlikely to be their own (when it does not look like what they know they are wearing). Why would or should children suspect that legs sporting clothes they are not wearing are their legs? Indeed, a further test must be done to see what children do when they see their leg itself, unclothed, in the mirror (assuming it has no distinguishing features); in that case, KVM would more likely be necessary, as legs per se are not particularly distinguishable visually (see Wuilleman and Richardson, 1982) the way that clothing is. Strangely, the authors claim that their results "substantiate the view that, at least by 24 months of age, toddlers pass the mark test because they know what they look like" (p. 183), yet fail to provide any means by which these toddlers *could* know what they look like when it comes to their face (other than KVM). Although the authors acknowledge that infants have some "mental self-image" of their appearance (p. 183), something that almost no one doubts (see Mitchell, 1993a, 1993b, 1997a), they never specify what that mental representation consists of. Since the mental self-image of their face cannot be a visual image, it seems likely to be based on KVM.

A third problem is that, in fact, children are not showing own leg recognition, but rather clothing-on-one's-leg recognition. A similar experiment examining for self-recognition using the face would require that the child be wearing a mask (which the child knows the look of), and then be shown either its own face in a mask or another child's face in another mask. Just as this experiment would not show self-recognition of one's own face but rather recognition of what covers one's face, so the study by Nielsen *et al.* does not show own leg recognition but rather recognition of what covers one's own leg.

HOW PERCEPTUAL KVM BECOMES CONCEPTUAL: CONNECTING HERE WITH OVER THERE

Conceptual KVM (as I imagine it) requires some kinesthetic involvement in visual experiences of bodies, including one's own and those of others. If perceptual KVM is the groundwork for conceptual KVM, how does conceptual KVM develop?

I have two hypotheses to offer. One derives from data about how we decide which of our hands, right or left, is presented in an image of a hand. The other derives from work concerning our understanding of mirrors.

If people are shown an image of a disembodied hand in diverse orientations and are asked to decide if this hand "over there" is a right or left hand, they usually imagine moving their appendage and/or body (as if kinesthetically) until it matches the hand in order to make or confirm their decision. This imagined appendage or imagined body movement is called "motor imagery." (With images of objects that are not body parts, people usually visually imagine transforming the image of the object to align – or not – with another image, rather than imagining their body moving. But of course either method, or variants of them, can be used with images of body parts or non-body parts.) When people engage in motor imagery, they often experience kinesthesis in the body part they imagine moving (Parsons and Fox, 1998; Sekiyama, 1982). Just as visual imagery seems to operate using brain processes peculiar to vision, and auditory imagery similarly depends on audition, motor imagery depends on motor movement and its kinesthetic accompaniments, in that the time required to make the imagined and real movements are correlated (Parsons, 1994). (Some auditory imagery, such as inner speech, may depend on motor movements – see Reisberg, 1992.) Although some argue that motor imagery and motor preparation (or the intention to move) are identical and that motor imagery is kinesthetic imagery (Jeannerod, 1994), in fact much of what is called motor imagery seems more like visual imagery – one imagines a visual image, or something that is vaguely like a visual image, of one's body or body part moving through space to match the image of the body part (Mitchell, 1994b). This "visual" imagery seems tied to possible kinesthetic experiences (i.e., actions), such that one knows how to (and can) do what is visually imagined. Visual images of one's actions can derive from experienced actions or from planning itself (Johnson, 2000), although certainly one can act without experiencing a mental image of oneself acting (Vogt, 1996).

If one wishes to connect the matching of hand images to one's own hand to the matching that occurs during mirror-self-recognition or imitation, a problem arises: in a mirror image of one's hand (and in mirroring imitation, which is apparently the usual form – see Schofield, 1976), the kinesthetic image would have to invert in some way to match the visual image. (Note that this problem does not occur with video-self-recognition, or with people's imitations in which they use their own right hand to imitate actions of my right hand.) Thus, if my right hand is going to match the image of my right hand in the mirror by kinesthetic extension of my right hand, there is no easy way to have my kinesthetic image "attach" to the mirrored hand. The kinesthetic image would have essentially to reverse – the felt back of my hand would "attach" to the unseen back of the mirrored hand, and the felt front of my hand would "attach" to the front of my hand visible in the mirror. This seems an overly cumbersome method of matching kinesthesis and vision.

Instead, what may happen is comparable to what happens in the fake arm studies – we see (and consequently kinesthetically feel) "our" hand over there, at least in our initial forays with mirrors. This account is also problematic, however, because the felt and seen hands are disjointed – our thumb would be where our pinky finger is, and vice versa. Thus, it would appear that some additional process is involved in recognizing one's own mirror-image (and in mirror imitating another's actions). We may have to assume that, when people recognize their image, they can recognize the same object from different perspectives or reference frames. When looking in the mirror, I can see that the two hands are the same (visually and kinesthetically) because I include the reference frame of the mirror. I acknowledge that THAT (in the mirror) looks like what THIS (my body) feels like from a different perspective. In other words, I am able to recognize that my hand in one (canonical) position is the same as another hand from a different (non-canonical, inverted) perspective. I know enough about objects like my hands and body to know that they have a particular topography, and I can use that knowledge as part of what allows me to match myself and my mirror image.

Thus, an additional process may simply be an understanding of mirrors (Mitchell, 1993a). To understand how understanding of mirrors might be necessary, the reader will benefit from a description of a series of studies of visuo-tactile cross-modal matching.

In these studies, a participant holds a small cube in each hand, say between the thumb and the index finger. The cube can vibrate and/or light up a small light on top (by the index finger) or on the bottom (by the thumb). The participant is asked to indicate where (top or bottom) the cube is vibrating, while a light is simultaneously flashing either on the top or bottom. (Participants are told to ignore the light.) When the flash and vibration are at the same place (i.e., congruent), people are better at detecting where the vibration is than when the flash and vibration are at different places (i.e., incongruent). Specifically, people take longer to designate where the vibration is occurring (top or bottom), and make more errors, in the incongruent than in the congruent contexts. This difference in the reaction times and error rates for incongruent and congruent contexts is called the cross-modal congruency effect. This effect supports the idea that people use the location of the visual (flash) information to make decisions about location of the tactile (vibration) information. Tactile perception is, thus, influenced by visual information (Spence *et al.*, 2004). The cross-modal congruency effect is larger the closer visual flash and the tactile vibration are; if the flash is distant from the vibration, the effect is much less strong. In addition, if the participant's hands each holding a cube are covered in rubber gloves and hidden, and a pair of rubber gloves each (apparently) holding a cube are situated above where the participant's hands actually are (and the participant knows they are there), the cross-modal congruency effect increases (compared to the normal situation), and the participant feels his or her hands to be where the gloves are (as in perceptual KVM).

These facts become important in interpreting an elaboration of the usual study. This elaboration used a mirror to show participants their otherwise hidden hands holding the cubes. If you look into a mirror, the image appears to be "over there," a distance approximately double the distance you are from the mirror. Using this information as well as the cross-modal congruency effect paradigm, researchers had participants hold a cube in each hand (itself covered with a glove), and hid them from direct view. However, participants were told that they would either see their own cubes and hands in a mirror, or they would see cubes (and gloves filled to look like hands) in the distance, such that the retinal image of these distant objects was the same size as that for the images in the mirror. In these contrasting contexts, the cross-modal congruency effect was larger when the cubes and gloved hands were visible in the mirror rather than viewed in the distance (Maravita *et al.*, 2002). The authors took this finding to indicate that images in the mirror were accepted as being in peri-personal space (i.e., the area around the body), rather than in distant space (see also Maravita *et al.*, 2000). Although this would seem to answer, to some degree, how it is that we can match the kinesthetic feelings in our body to the visual image in the mirror – we simply know that the visual objects in the mirror are here, rather than over there – but actually the problem of how we know this still remains, because this understanding is not immediately given in our perception. Children prior to eighteen months of age appear to have no problem recognizing that things *other* than themselves in mirror images are here (rather than over there). It is only around eighteen months of age that they discover that they themselves are like these other objects. I called the understanding of mirror images prior to self-recognition the "understanding of mirror-correspondence" (Mitchell, 1993a, 2007), and argued that it was part of the explanation for mirror-self-recognition. In understanding mirror-correspondence, a perceiver knows that mirrors reflect accurate and contingent images of objects (other than the perceiver) in front of him or her.

What Maravita *et al.* (2002) discovered is the thing that needs to be explained: *how is it that* mirror images come to be located in peripersonal space? From all that I've written above, it would appear that recognition that what is seen is what is felt – kinesthetic–visual matching – is the basis for that recognition. Once KVM is established, the understanding of mirror-correspondence comes to include one's own body as another object that can be localized via the mirror. It would appear that simply knowing that the visual hand or body is in its gross shape (though not in its perspective) "identical" to the kinesthetic hand or body may be enough to match the two, without any concern with coordinating the perspectives toward the two (as discussed above). On the other hand, it may be that what allows for KVM to be understood is the ability to take on an alternate perspective. After all, what Schmuckler's (1996) research above showed was that young infants, prior to mirror-self-recognition via standard methods, seem to have some awareness of self in a video-image only when the image is shown from an obvious perspective, but not from an inverted one – which seems to be the perspective seen in a mirror image. Thus, it would appear that applying understanding of mirror-correspondence in relation to one's own body is dependent upon understanding an additional perspective (or reference frame) of the body, which can be derived from, or can be implicit in, KVM. Similar processes occur with tool use, when distant objects become encoded as in peri-personal space once a tool is used toward those objects (Maravita and Driver, 2004; see Chapter 18, this volume). Indeed, simple reaching by monkeys appears to require a mechanism in the posterior parietal cortex (PPC) coordinating visual and kinesthetic information about the relationship between the monkey and the object reached for into a reference frame in peri-personal space: "The PPC could transform visual target locations from retinal coordinates to hand-centered coordinates by combining sensory signals in a serial manner to yield a body-centered representation of target location" (Buneo *et al.*, 2002, p. 632).

THE PARIETAL REGION AND KVM

As a final note, I want to emphasize that recent work supports my contention (Mitchell, 1997b, 2007) that KVM is "located in" (created by) the parietal region. Although clearly the parietal region does not act independently of other regions of the brain (Fogassi and Gallese, 2004; Jackson and Husain, 1996; Mitchell, 2007; Rizzolatti *et al.*, 2001), it appears that the bulk of the processing and matching of kinesthetic and visual information occurs there (see Andersen, 1995; Andersen and Buneo, 2002; Buneo *et al.*, 2002; Calvo-Merino *et al.*, 2005; Daprati *et al.*, 2000; Dawson *et al.*, 1985; di Pellegrino and Frassinetti, 2000; Grèzes *et al.*, 1999; Iriki *et al.*, 2001; Mitchell, 2007; Mühlau *et al.*, 2005; Sirigu *et al.*, 1999; Tsakiris and Haggard, 2005). KVM may be localized differentially in the two hemispheres, with recognition of another's imitation of oneself represented in the right parietal cortex and imitation of another's actions by the self in the left parietal region (Decety *et al.*, 2002; Funk and Brugger, 2002; Mühlau *et al.*, 2005). Difficulties of brain-injured people in pointing to their own body parts but not to someone else's, or vice versa, suggest that the left superior parietal lobule processes somatosensory representation of the body, whereas the left inferior parietal lobule processes visuospatial representations of the body (Felician *et al.*, 2003; see also Felician *et al.*, 2004). In addition, although some research suggests that the specific brain areas involved in imitation of meaningful and meaningless actions (Grèzes *et al.*, 1999), or in recognizing well-rehearsed and unknown actions (Calvo-Merino *et al.*, 2005), are different, in fact all such actions have parietal involvement. (There may be more parietal involvement with meaningless than meaningful imitation, and more with well-rehearsed than unknown actions, but the parietal region is engaged nonetheless.) As is evident from its evolutionary history (see Chapter 16, this volume), the parietal region is quite diverse in its functions. Overall, functional considerations – i.e., what is actually required to fulfill the experimenters' requests, which vary across studies – may influence how much parietal involvement is necessary for any given task.

REFERENCES

Andersen, R. A. (1995). Encoding of intention and spatial location in the posterior parietal cortex. *Cerebral Cortex*, **5**, 457–469.

Andersen, R. A. and Buneo, C. A. (2002). Intentional maps in posterior parietal cortex. *Annual Review of Neuroscience*, **25**, 189–220.

Austen, E. L., Soto-Faraco, S., Enns, J. T. and Kingstone, A. (2004). Mislocalizations of touch to a fake hand. *Cognitive, Affective, & Behavioral Neuroscience*, **4**, 170–181.

Austen, E. L., Soto-Faraco, S., Pinel, J. P. J. and Kingstone, A. F. (2001). Virtual body effect: factors influencing visual-tactile integration. *Abstracts of the Psychonomic Society*, **6**, 2.

Bastian, H. C. (1902). *The brain as the organ of mind.* London: Kegan Paul, Trench, Trübner & Co.

Beardsworth, T. and Buckner, T. (1981). The ability to recognize oneself from a video recording of one's movements without seeing one's body. *Bulletin of the Psychonomic Society*, **18**, 19–22.

Bischof-Köhler, D. (1994). Selbstobjektivierung und fremdbezogene Emotionen. Identifikation des eigenen Spiegelbildes, Empathie und prosoziales Verhalten im 2. Lebensjahr [Self-objectification and other-oriented emotions: self-recognition, empathy, and prosocial behaviour in the second year]. *Zeitschrift für Psychologie*, **202**, 349–377.

Boring, E. G. (1942). *Sensation and perception in the history of experimental psychology.* New York: D. Appleton-Century Co.

Botvinick, M. and Cohen, J. (1998). Rubber hands "feel" touch that eyes see. *Nature*, **39**, 756.

Brugger, P., Kollias, S. S., Müri, R. M., Crelier, G., Hepp-Reymond, M.-C. and Regard, M. (2000). Beyond re-membering: phantom sensations of congenitally absent limbs. *Proceedings of the National Academy of Sciences USA*, **97**, 6167–6172.

Buneo, C. A., Jarvis, M. R., Batista, A. P. and Andersen, R. A. (2002). Direct visuomotor transformations for reaching. *Nature*, **416**, 632–636.

Calvo-Merino, B., Glaser, D. E., Grèzes, J., Passingham, R. E. and Haggard, P. (2005). Action observation and acquired motor skills: an fMRI study with expert dancers. *Cerebral Cortex*, **15**, 1243–1249.

Casile, A. and Giese, M. A. (2006). Nonvisual motor training influences biological motion perception. *Current Biology*, **16**, 69–74.

Cohen, M. M. (1967). Continuous versus terminal visual feedback in prism aftereffects. *Perceptual and Motor Skills*, **24**, 1295–1302.

Courage, M. L., Edison, S. C. and Howe, M. L. (2004). Variability in the early development of visual self-recognition. *Infant Behavior and Development*, **27**, 509–532.

Cratty, B. J. and Sams, T. A. (1968). *The body-image of blind children*. New York: American Foundation for the Blind.

Daprati, E., Sirigu, A., Pradat-Diehl, P., Franck, N. and Jeannerod, M. (2000). Recognition of self-produced movement in a case of severe neglect. *Neurocase*, **6**, 477–486.

Darwin, E. (1796). *Zoonomia, or the laws of organic life* (vol. 1). New York: T. & J. Swords.

Dawson, G., Warrenburg, S. and Fuller, P. (1985). Left hemisphere specialization for facial and manual imitation. *Psychophysiology*, **22**, 237–243.

Decety, J., Chaminade, T., Grèzes, J. and Meltzoff, A. N. (2002). A PET exploration of the neural mechanisms involved in reciprocal imitation. *NeuroImage*, **15**, 265–272.

di Pellegrino, G. and Frassinetti, F. (2000). Direct evidence from parietal extinction of enhancement of visual attention near a visible hand. *Current Biology*, **10**, 1475–1477.

di Pellegrino, G., Làdavas, E. and Farné, A. (1997). Seeing where your hands are. *Nature*, **388** (6644), 730.

Farné, A., Pavani, F., Meneghello, F. and Làvadas, E. (2000). Left tactile extinction following visual stimulation of a rubber hand. *Brain*, **123**, 2350–2360.

Felician, O., Ceccaldi, M., Didic, M., Thinus-Blanc, C. and Poncet, M. (2003). Pointing to body parts: a double dissociation study. *Neuropsychologia*, **41**, 1307–1316.

Felician, O., Romaiguère, P., Anton, J.-L., Nazarian, B., Roth, M., Poncet, M. and Roll, J.-P. (2004). The role of human left superior parietal lobule in body part localization. *Annals of Neurology*, **55**, 749–751.

Féry, Y.-A. (2003). Differentiating visual and kinesthetic imagery in mental practice. *Canadian Journal of Experimental Psychology*, **57**, 1–10.

Finger, S. (1994). *Origins of neuroscience: a history of explorations into brain function*. Oxford: Oxford University Press.

Finke, R. A. (1980). Levels of equivalence in imagery and perception. *Psychological Review*, **87**, 113–132.

Finke, R. A. (1986). Mental imagery and the visual system. *Scientific American*, **254**, 88–95.

Fogassi, L. and Gallese, V. (2004). Action as the binding key to multisensory integration. In G. A. Calvert, C. Spence and B. E. Stein (Eds.), *The handbook of multisensory processes* (pp. 425–441). Cambridge, MA: MIT Press.

Funk, M. and Brugger, P. (2002). Visual recognition of hands by persons born with only one hand. *Cortex*, **38**, 860–863.

Gold, K. and Scassellati, B. (2007). A Bayesian robot that distinguishes "self" from "other." *Proceedings of the 29th Annual Meeting of the Cognitive Science Society (CogSci 2007)*, Nashville, Tennessee. Available online at www.cs.yale.edu/homes/scaz/papers/Gold-CogSci-07.pdf, **31**.

Graziano, M. and Botvinick, M. (2002). How the brain represents the body: insights from neurophysiology and psychology. In W. Prinz and B. Hommel (Eds.), *Common mechanisms in perception and action: attention and performance XIX* (pp. 136–157). Oxford: Oxford University Press.

Grèzes, J., Costes, N. and Decety, J. (1999). The effects of learning and intention on the neural network involved in the perception of meaningless actions. *Brain*, **122**, 1875–1887.

Harris, C. S. (1965). Perceptual adaptation to inverted, reversed, and displaced vision. *Psychological Review*, **72**, 419–444.

Hecht, H., Vogt, S. and Prinz, W. (2001). Motor learning enhances perceptual judgment: a case for action-perception transfer. *Psychological Research*, **65**, 3–14.

Iriki, A., Tanaka, M., Obayashi, S. and Iwamura, Y. (2001). Self-images in the video monitor coded by monkey intraparietal neurons. *Neuroscience Research*, **40**, 163–173.

Jackson, S. R. and Husain, M. (1996). Visuomotor functions of the lateral premotor cortex. *Current Opinion in Neurobiology*, **6**, 788–795.

James, W. (1890). *Psychology* (vol. II). New York: Holt.

Jeannerod, M. (1994). The representing brain: neural correlates of motor intention and imagery. *Behavioral and Brain Sciences*, **17**, 187–246.

Johnson, S. H. (2000). Thinking ahead: the case for motor imagery in prospective judgements of prehension. *Cognition*, **74**, 33–70.

Jones, S. S. (2007). Imitation in infancy: the development of mimicry. *Psychological Science*, **18**, 593–599.

Kinsbourne, M. (1995). Awareness of one's own body: an attentional theory of its nature, development, and brain basis. In J. L. Bermúdez, A. Marcel and N. Eilan (Eds.), *The body and the self* (pp. 205–223). Cambridge, MA: MIT Press.

Kinsbourne, M. and Lempert, H. (1980). Human figure representation by blind children. *The Journal of General Psychology*, **102**, 33–37.

Kornheiser, A. S. (1976). Adaptation for laterally displaced vision: a review. *Psychological Bulletin*, **83**, 783–816.

Lingis, A. (2007). Understanding avian intelligence. In L. Simmons and P. Anderson (Eds.), *Knowing animals* (pp. 43–56). Leiden: Brill.

Maravita, A. and Driver, J. (2004). Cross-modal integration and spatial attention in relation to tool use and mirror use: representing and extending multisensory space near the hand. In G. A. Calvert, C. Spence and B. E. Stein (Eds.), *The handbook of multisensory processes* (pp. 819–835). Cambridge, MA: MIT Press.

Maravita, A., Spence, C., Clarke, K., Husain, M. and Driver, J. (2000). Vision and touch through the looking glass in a case of crossmodal extinction. *NeuroReport*, **11**, 3521–3526.

Maravita, A., Spence, C., Sergent, C. and Driver, J. (2002). Seeing your own touched hands in a mirror modulates cross-modal interactions. *Psychological Science*, **13**, 350–356.

Melzack, R. (1989). Phantom limbs, the self, and the brain. *Canadian Psychology*, **30**, 1–16.

Millar, S. (1994). *Understanding and representing space: theory and evidence from studies with blind and sighted children*. Oxford: Clarendon Press.

Mitchell, R. W. (1993a). Mental models of mirror-self-recognition: two theories. *New Ideas in Psychology*, **11**, 295–325.

Mitchell, R. W. (1993b). Recognizing one's self in a mirror? A reply to Gallup and Povinelli, De Lannoy, Anderson, and Byrne. *New Ideas in Psychology*, **11**, 351–377.

Mitchell, R. W. (1994a). The evolution of primate cognition: simulation, self-knowledge, and knowledge of other minds. In D. Quiatt and J. Itani (Eds.), *Hominid culture in primate perspective* (pp. 177–232). Boulder: University Press of Colorado.

Mitchell, R. W. (1994b). Are motor images based on kinesthetic-visual matching? *Behavioral and Brain Sciences*, **17**, 214–215.

Mitchell, R.W. (1997a). A comparison of the self-awareness and kinesthetic-visual matching theories of self-recognition: autistic children and others. *Annals of the New York Academy of Sciences*, **818**, 39–62.

Mitchell, R.W. (1997b). Kinesthetic-visual matching and the self-concept as explanations of mirror-self-recognition. *Journal for the Theory of Social Behavior*, **27**, 101–123.

Mitchell, R. W. (2002a). Imitation as a perceptual process. In K. Dautenhahn and C.L. Nehaniv (Eds.), *Imitation in animals and artifacts* (pp. 441–469). Cambridge, MA: MIT Press.

Mitchell, R.W. (2002b). Subjectivity and self-recognition in animals. In M.R. Leary and J. Tangney (Eds.), *Handbook of self and identity* (pp. 567–593). New York: Guilford Press.

Mitchell, R.W. (2007). Mirrors and matchings: imitation from the perspective of mirror-self-recognition, and why the parietal region is involved in both. In K. Dautenhahn and C.L. Nehaniv (Eds.), *Imitation and social learning in robots, humans and animals* (pp. 103–130). Cambridge: Cambridge University Press.

Mitchell, R. W. and Anderson, J. R. (1993). Discrimination learning of scratching, but failure to obtain imitation and self-recognition in a long-tailed macaque. *Primates*, **34**, 301–309.

Miyazaki, M. and Hiraki, K. (2006). Delayed intermodal contingency affects young children's recognition of their current self. *Child Development*, **77**, 736–750.

Morsley, K., Spencer, C. and Baybutt, K. (1991). Is there any relationship between a child's body image and spatial skills? *British Journal of Visual Impairment*, **9**, 41–43.

Mühlau, M., Hermsdörfer, J., Goldenberg, G., Wohlschläger, A.M., Castrop, F., Stahl, R., Röttinger, M., Erhard, P., Haslinger, B., Ceballos-Baumann, A.O., Conrad, B. and Boecker, H. (2005). Left inferior parietal dominance in gesture imitation: an FMRI study. *Neuropsychologia*, **43**, 1086–1098.

Nehaniv, C.L. and Dautenhahn, K. (2002). The correspondence problem. In K. Dautenhahn and C.L. Nehaniv (Eds.), *Imitation in animals and artifacts* (pp. 41–61). Cambridge, MA: MIT Press.

Neuper, C., Scherer, R., Reiner, M. and Pfurtscheller, G. (2005). Imagery of motor actions: differential effects of kinesthetic and visual-motor mode of imagery in single-trail EEG. *Cognitive Brain Research*, **25**, 668–677.

Nielsen, M., Suddendorf, T. and Slaughter, V. (2006). Mirror self-recognition beyond the face. *Child Development*, **77**, 176–185.

Noë, A. (2004). *Action in perception*. Cambridge, MA: MIT Press.

Parsons, L.M. (1994). Temporal and kinematic properties of motor behavior reflected in mentally simulated action. *Journal of Experimental Psychology: Human Perception and Performance*, **20**, 709–730.

Parsons, L.M. and Fox, P.T. (1998). The neural basis of implicit movements used in recognising hand shape. *Cognitive Neuropsychology*, **15**, 583–615.

Pavani, F., Spence, C. and Driver, J. (2000). Visual capture of touch: out-of-the-body experiences with rubber gloves. *Psychological Science*, **11**, 353–359.

Piaget, J. (1945/1962). *Play, dreams, and imitation in childhood*. New York: W.W. Norton.

Piaget, J. (1954). *The construction of reality in the child*. New York: Basic Books.

Plotnik, J. M., de Waal, F. B. M. and Reiss, D. (2006). Self-recognition in an Asian elephant. *Proceedings of the National Academy of Sciences*, **103** (45), 17053–17057.

Prinz, W. (1987). Ideo-motor action. In H. Heuer and A. F. Sanders (Eds.), *Perspectives on perception and action* (pp. 47–76). Hillsdale: Erlbaum.

Prior, H., Schwarz, A. and Güntürkün, O. (2008). Mirror-induced behavior in the magpie (*Pica pica*): evidence of self-recognition. *PLoS Biology*, **6**, 1642–1650.

Ramachandran, V. S. (1998). The perception of phantom limbs. *Brain*, **121**, 1603–1630.

Ramachandran, V. S. and Blakesee, S. (1998). *Phantoms in the brain*. New York: HarperCollins.

Reisberg, D. (Ed.) (1992). *Auditory imagery*. Hillsdale: Erlbaum.

Reisberg, D. and Heuer, F. (2005). Visuospatial images. In P. Shah and A. Miyake (Eds.), *The Cambridge handbook of visuospatial thinking* (pp. 35–80). New York: Cambridge University Press.

Rizzolatti, G., Fogassi, L. and Gallese, V. (2001). Neurophysiological mechanisms underlying the understanding and imitation of action. *Nature Reviews*, **2**, 661–670.

Robertson, G. C. (1896). *Elements of psychology*. New York: Charles Scribner's Sons.

Rochat, P. and Striano, T. (2002). Who's in the mirror? Self-other discrimination in specular images by four-and nine-month-old infants. *Child Development*, **73**, 35–46.

Ryland, F. (1909). *Thought and feeling*. London: Hodder and Stoughton.

Scheerer, E. (1987). Muscle sense and innervation feelings: a chapter in the history of perception and action. In H. Heuer and A. F. Sanders (Eds.), *Perspectives on perception and action* (pp. 171–194). Hillsdale: Erlbaum.

Schmuckler, M. A. (1996). Visual-proprioceptive intermodal perception in infancy. *Infant Behavior and Development*, **19**, 221–232.

Schofield, W. N. (1976). Hand movements which cross the body midline: findings relating age differences to handedness. *Perceptual and Motor Skills*, **42**, 643–646.

Schwoebel, J., Buxbaum, L. J. and Coslett, H. B. (2004). Representation of the human body in the production and imitation of complex movements. *Cognitive Neuropsychology*, **21**, 285–298.

Sekiyama, K. (1982). Kinesthetic aspects of mental representations in the identification of left and right hands. *Perception and Psychophysics*, **32**, 89–95.

Sheets-Johnstone, M. (1999). *The primacy of movement*. Amsterdam: John Benjamins Publishing Company.

Sirigu, A., Daprati, E., Pradat-Diehl, P., Franck, N. and Jeannerod, M. (1999). Perception of self-generated movement following left parietal lesion. *Brain*, **122**, 1867–1874.

Spence, C., Pavani, F., Maravita, A. and Holmes, N. (2004). Multisensory contributions to the 3-D representation of visuotactile peripersonal space in humans: evidence from the crossmodal congruency task. *Journal of Physiology – Paris*, **98**, 171–189.

Suddendorf, T. and Whiten, A. (2001). Mental evolution and development: evidence for secondary representation in children, great apes and other animals. *Psychological Bulletin*, **127**, 629–650.

Troje, N. F., Westhoof, C. and Lavrov, M. (2005). Person identification from biological motion: effects of structural and kinematic cues. *Perception and Psychophysics*, **67**, 667–675.

Tsakiris, M. and Haggard, P. (2005). Experimenting with the acting self. *Cognitive Neuropsychology*, **22**, 387–407.

Van den Bos, E. and Jeannerod, E. (2002). Sense of body and sense of action both contribute to self-recognition. *Cognition*, **85**, 177–187.

Vogt, S. (1996). Imagery and perception-action mediation in imitative actions. *Cognitive Brain Research*, **3**, 79–86.

von Fieandt, K. (1966). *The world of perception*. Homewood, IL: Dorsey Press.

Weber, E.H. (1846/1996). The sense of touch and common sensibility [Trans. of *Tastsinn und Gemeingefühl*, by D.J. Murray]. In *E.H. Weber on the tactile senses* (2nd edn). Hove: Erlbaum (UK) Taylor & Francis.

Wuilleman, D. and Richardson, B. (1982). On the failure to recognize the back of one's own hand. *Perception*, **11**, 53–55.

LEAH KRUBITZER AND ELIZABETH DISBROW

16

The evolution of parietal areas involved in hand use in primates

Anthropocentric /, anθrəpə'sɛntrik / *a.* M19. [f. prec. + CENTRIC] Centering in humans; regarding humanity as the central fact of the universe.

(*The New Shorter Oxford English Dictionary*)

As a beginning for a review on areas of the primate brain involved in generating frames of reference and ultimately a sense of self, the definition of anthropocentrism may seem misplaced. Yet anthropocentrism is, paradoxically, central to this scientific endeavor: our fascination with ourselves is simultaneously the driving force behind our desire to understand biological organisms and the activities they generate (including anthropocentrism itself) and probably the largest stumbling block to achieve an objective understanding. That we are anthropocentric is without question. Indeed, when not actually pursuing activities necessary for our literal survival, we spend the majority of our time pursuing an understanding of ourselves. We strive to understand our uniqueness, our past, our future, how we behave together, how we behave individually, how we think, how we expend our resources, and how to determine if there is any other sentient being "out there" like us. These pursuits encompass a variety of disciplines such as psychology, sociology, philosophy, anthropology, history and economics, to name a few. All of these disciplines are based on the premise that humans are radically different from other animals – that we abide by a different set of rules, that we are unconstrained by evolution, and that we have miraculously developed emergent properties, such as a mind, intelligence, language, and even a soul. While we poke fun at this egocentric drive that all of us possess and the anthropocentric nature of our institutions, we must concede that this magnificent sense of self is biologically driven. Specifically, there are regions of the neocortex

Spatial Cognition, Spatial Perception: Mapping the Self and Space, ed. Francine L. Dolins and Robert W. Mitchell. Published by Cambridge University Press.

that are involved in generating a sense of our own body with respect to the world around us. These regions allow us to distinguish ourselves from external animate and inanimate objects, and to interact with them via highly specialized morphological tools, such as our hands.

Despite this natural preoccupation with ourselves, and our need to understand how complexity has arisen in our species, many authors believe that it is more informative, when trying to appreciate our unique attributes with respect to other species, to examine similarities across groups, rather than trying to explain this uniqueness or variability in isolation. For example, are there similar principles of organization for all mammal brains? Once this question has been answered, then one can examine the departures from the common plan of organization to determine how these departures or modifications are achieved, and ultimately how they generate variable behavior like that described above for humans. A good illustration of this approach, and the focus of this review, is provided by an examination of the somatosensory areas of anterior and posterior parietal cortex in both primate and nonprimate mammals. These regions appear to have expanded in humans, and are proposed to be associated with a number of sophisticated behaviors.

SENSORY VERSUS ASSOCIATION CORTEX

Traditionally the mammalian neocortex has been divided into three broad categories: motor cortex, sensory cortex (including primary and second sensory fields such as S1, S2, V1, V2, A1 and R; see Table 16.1), and association cortex. This broad classification became popular in the middle part of the twentieth century, at which time electrophysiological mapping studies demonstrated that all mammals possessed motor cortex and primary and second sensory cortices, and that primates in particular had a great deal of association cortex (Woolsey and Fairman, 1946; Woolsey, 1958; see Kaas and Collins, 2004 for review). Association cortex, as defined by modern textbooks, includes temporal, prefrontal and posterior parietal cortex, and is hypothesized to mediate complex behaviors such as perception, attention, cognition, and other high level mental functions (see Saper *et al.*, 2000). The argument for defining cortex as "association" is somewhat circular and is based on the premise, which emerged from earlier mapping studies of Woolsey and colleagues, that the amount of cortex that could not be defined as unimodal sensory cortex in primates was relatively large compared to other mammals (Figure 16.1). Since primates were thought to be

Table 16.1. *List of abbreviations.*

Cortical fields	
A1	primary auditory area
AIP	anterior intraparietal area
C1	caudal region 1
C2	caudal region 2
DSG	dysgranular zone
LIP	lateral intraparietal area
M1	primary motor area
MIP	medial intraparietal area
PM	parietal medial area
PP	posterior parietal cortex
PR	rostroventral parietal area
PV	parietal ventral area
R	rostral area
R1	rostral region 1
R2	rostral region 2
S1	primary somatosensory area
S2	secondary somatosensory area
S3	third somatosensory area
SMA	supplementary motor area
UZ	unresponsive zone
V1	primary visual area
VIP	ventral intraparietal area
VPZ	ventral posterior nucleus, recipient zone
Sulci	
CS	central sulcus
IPS	intraparietal sulcus
LS	lateral sulcus
PCS	post central sulcus
POS	parietal occipital sulcus
STS	superior temporal sulcus
Thalamic nuclei	
CL	central lateral nucleus
MG	medial geniculate nucleus
Pa	anterior pulvinar
PO	posterior nucleus
VL	ventral lateral nucleus
VP	ventral posterior nucleus
Vpi	ventral posterior nucleus, inferior division

Table 16.1. (*cont.*)

VPl	ventral posterior nucleus, lateral division
VPm	ventral posterior nucleus, medial division
VPs	ventral posterior nucleus, superior division
Body parts	
ch	chin
ck	cheek
dig	digits
D1	digit one
D2	digit two
D3	digit three
D4	digit four
D5	digit five
F or ft	foot
fa	forearm
fl	forelimb
fp	forepaw
gen	genitals
hl	hindlimb
hp	hindpaw
j	jaw
l	lip
nb	nail bed
ne	neck
sh	shoulder
sn	snout
t	toes
te	teeth
to	tongue
tr	trunk
ut	upper trunk
vib	vibrissae
w	wing
web	finger web
wr	wrist
Neuroanatomical directions	
dor	dorsal
prox	proximal

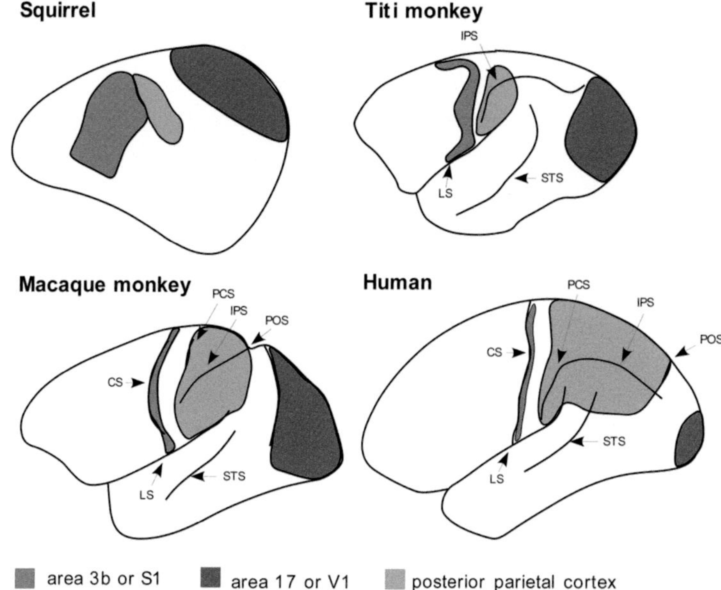

Figure 16.1. The location of association cortex (green) relative to the primary somatosensory area (red) and the primary visual area (blue) in four mammalian species. In mammals with a small neocortex, such as the grey squirrel, primary sensory areas occupy a relatively large portion of the cortical sheet. In mammals with a large neocortex such as monkeys, these primary sensory areas occupy relatively less cortex. In human primates, primary sensory areas occupy a very small portion of the neocortex, while association cortex, including posterior parietal cortex, occupies a large expanse of the cortical sheet. See Table 16.1 for abbreviations. See also colour plate section.

more cognitively complex than other mammals, this expanded cortex became associated with higher mental processes and was considered a primate phenomenon.

Unfortunately, the early mapping studies that provided support for these ideas were hampered by technical difficulties such as suboptimal anesthetics, recording methods, and stimulation parameters, all of which made it difficult to elicit responses from neurons in cortex other than primary and secondary sensory fields. Despite these limitations, these early studies fostered several ideas regarding the human brain that still persist today. Probably the most noteworthy are that primary fields are evolutionarily older, and that association cortex is a new evolutionary phenomenon found mainly in human and nonhuman primates. Thus, association cortex is proposed to be the hallmark of human brain evolution.

These ideas were firmly entrenched for most of the twentieth century. Indeed, a number of psychologists still hold this view, and it is currently circulated as the reigning model in most popular textbooks (e.g., Carlson, 1998; Saper *et al.*, 2000). However, work in the early 1970s by Allman and Kaas (1971, 1974, 1975, 1976) upended these traditional views by demonstrating in nonhuman primates that much of extrastriate cortex that was considered association cortex actually contained a number of unimodal visual fields (also see Felleman and Van Essen, 1991). Somewhat later, work in both somatosensory (Merzenich *et al.*, 1978; Robinson and Burton, 1980a, 1980b; Pons *et al.*, 1985; Krubitzer and Kaas, 1990; Krubitzer *et al.*, 1995a) and auditory cortex (e.g., Imig *et al.*, 1977; Hackett *et al.*, 1998) demonstrated that parietal cortex and portions of temporal cortex that were thought to be association regions were occupied by somatosensory and auditory cortical fields respectively (see Kaas and Collins, 2004 for review).

Despite the ever-dwindling classical association cortex in primate brains (including human primates), there are still a few strongholds that seem indomitable, namely prefrontal and posterior parietal cortex. Work on prefrontal cortex in humans demonstrates that it is involved in higher order cognitive processes such as the ability to detect and respond to novel events, to discriminate internally motivated versus externally driven models of the world, and to "extract oneself from the present and fluidly move forward and backward in time" (p. 1319; Knight and Grabowecky, 2000 for review). However, this region of cortex also has a number of visuomotor, olfactory and limbic functions.

In humans, posterior parietal cortex is considered to be involved in coding the spatial location of objects within a particular frame of reference, both egocentric and extracentric (e.g., Mishkin *et al.*, 1983; Behrmann, 2000; Robertson and Rafal, 2000 for review). In nonhuman primates, posterior parietal cortex is divided into a number of cortical areas that are thought to be involved in visuospatial processing related to limb and hand use such as monitoring limb location during visually guided reaching and grasping, converting sensory locations into motor coordinates for intentional movement, and perceiving the movements of the body in extra-personal space (Andersen *et al.*, 1997; Snyder *et al.*, 1997; Wise *et al.*, 1997; Debowy *et al.*, 2001; Andersen and Buneo, 2002). Thus, much of the region traditionally defined as posterior parietal association cortex has actually evolved in primates as a consequence of, and for the generation of specialized hand use, rather than for general higher mental functions.

The following review will focus on several fields in anterior parietal cortex (areas 3a, 3b, 1 and 2) involved in processing somatic inputs from the skin, muscles and joints, as well as one area of posterior parietal cortex (area 5) involved in manual dexterity, bilateral coordination of the hands, intentional reaching and grasping, and interhemispheric transfer of information. We contend that our sense of self, or an internal representation of our body with respect to the external world, is a concrete characteristic common to all mammals. This internal representation is generated by interactions between visual, proprioceptive and vestibular systems, all of which are intricately intertwined with the motor system. Such an interaction allows an individual to interface an internal representation of self with the external world via a particular morphological structure, such as a hand, thereby generating a sensory-motor feedback loop that allows one to distinguish self from non-self.

ORGANIZATION OF ANTERIOR PARIETAL CORTEX IN PRIMATES (AREAS 3B, 3A, 1 AND 2)

Somatosensory cortex in primates is divided into three major divisions: anterior parietal cortex, posterior parietal cortex, and cortex of the lateral sulcus (see Kaas and Pons, 1988 for review). Each of these major divisions contains several cortical areas. Anterior parietal cortex includes the primary somatosensory area, S1, which corresponds to area 3b in primates (see Kaas, 1983 for review), area 3a, area 1 and area 2. Posterior parietal cortex has been subdivided differently by different investigators, but most would agree that at least one of these regions, area 5, processes somatic inputs. Finally, like posterior parietal cortex, cortex in the lateral sulcus has been subdivided differently by different investigators. Recently our laboratory and others have examined this region of cortex using multiunit electrophysiological recording techniques and neuroanatomical tracing methods combined with architectonic analysis, and we have divided this cortex into several fields including the second somatosensory area, S2, the parietal ventral area, PV, the ventral somatosensory area, VS, and the rostroventral parietal area, PR (e.g., Krubitzer and Kaas, 1990; Krubitzer *et al.*, 1995a; Disbrow *et al.*, 2002, 2003; Wu and Kaas, 2003).

In this chapter we focus only on somatosensory areas of the anterior and posterior parietal cortex. While several anterior parietal fields such as areas 3a, 3b, 1 and 2 are architectonically distinct in some primates, architectonic comparisons across species are difficult to make for fields other than areas 3b and 3a. Ideally, the consideration of homology should be based on a number of criteria such as architecture, electrophysiological recording data, cortical and subcortical connections, and lesions to cortical fields and resultant behavioral deficits (Kaas, 1982). Indeed, determining homology based on architecture alone, in the absence of corroborative electrophysiological data, leads to too many inaccuracies in subdividing the neocortex and erroneous conclusions regarding the evolution of cortical fields. For this reason, the following review focuses primarily on electrophysiological recording studies, studies of connections and studies of lesions, rather than studies which only use architecture to subdivide the cortex.

AREA 3B OR S1

Functional organization

The topographic organization of the primary somatosensory area, S1 or area 3b, has been described in a variety of primates including Old World macaque monkeys (Nelson *et al.*, 1980); New World monkeys such as owl (Merzenich *et al.*, 1978), squirrel (Sur *et al.*, 1982), cebus (Felleman *et al.*, 1983), spider (Pubols and Pubols, 1971), tit (Padberg *et al.*, 2005), tamarin (Carlson *et al.*, 1986), and marmoset monkeys (Krubitzer and Kaas, 1990); prosimian galagos (Sur *et al.*, 1980); and humans (e.g., Penfield and Rasmussen, 1968; Woolsey *et al.*, 1979; Fox *et al.*, 1987; Moore *et al.*, 2000; Hlushchuk *et al.*, 2003; Blankenburg *et al.*, 2003 for review). In all primates investigated, area 3b forms a systematic representation of the contralateral body surface with the tail, genitals and feet represented most medially, followed by the representations of the hindlimb, trunk, forelimb, hand, face and oral structures in a mediolateral progression (Figure 16.2).

Single unit studies and optical imaging studies indicate that neurons in area 3b have small receptive fields compared to other anterior parietal fields (e.g., Gardner, 1988), are rapidly or slowly adapting to cutaneous stimulation (e.g., Sur *et al.* 1984; Chen *et al.*, 2001), and respond to high frequency stimulation (e.g., Lebedev and Nelson, 1996), pressure and flutter (Chen *et al.*, 2001). In awake, behaving monkeys, firing rates are modulated by contact with an object (DeBowy *et al.*, 2001), and neurons respond prior to wrist movements (Nelson *et al.*, 1991). It is proposed that area 3b is involved in texture and form

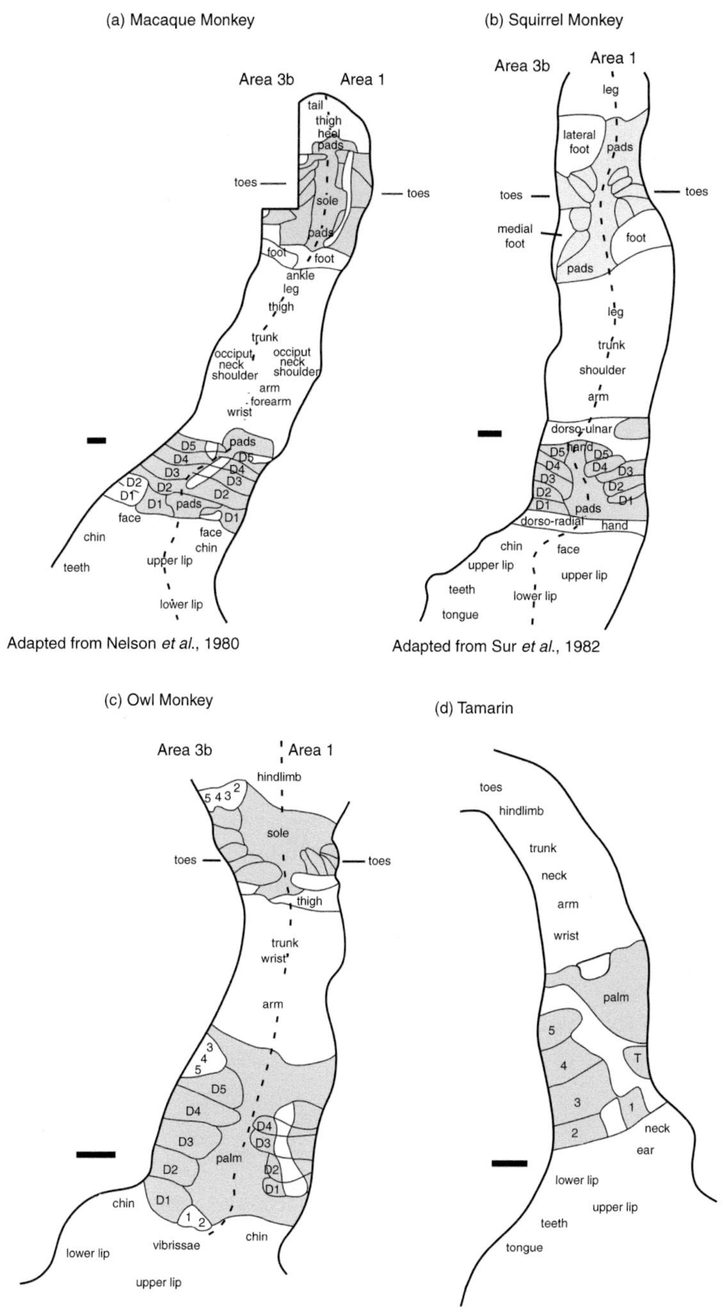

Figure 16.2. The topographic organization of the primary somatosensory area (S1 or 3b) and area 1 in New and Old World monkeys. [cont. over]

discrimination (Johnson and Lamb, 1981; Chapman and Ageranioti-Belanger, 1991; Sinclair and Burton, 1991; Ageranioti-Belanger and Chapman, 1992; Tremblay *et al.*, 1996; Jiang *et al.*, 1997; Dicarlo *et al.*, 1998; see Johnson and Yoshioka, 2002, for review) as well as topographic tactile learning (e.g., Romo *et al.*, 1998, 2000; Diamond *et al.*, 2002 for review). While most researchers have examined the neural response to stimulation of the hand or have designed behavioral tasks associated with the hand, one group of studies examined the properties of neurons in area 3b in response to tongue and face movements (Lin *et al.*, 1994a, 1994b). These investigators found that neurons in area 3b (and area 1) increase their firing rate for preferred tongue directions, and suggest that neurons in 3b and 1 provide sensory inputs to M1 necessary for generating coordinated tongue and facial movements. Such a homologous processing network in primates could form the basis, in a modified form, for the highly derived articulation behaviors that have evolved in humans.

Studies in which area 3b has been lesioned in monkeys are consistent with single unit studies. For example, lesions in area 3b result in an inability to make discriminations of roughness, hardness and angle of an object (Randolph and Semmes, 1974). Other studies demonstrate that lesions in area 3b result in an inability to make static tactile discriminations (detection of edges), as well as tactile discriminations that require movement, such as discriminating a many-sided object (Schwartz, 1983). While the studies of LaMotte and Mountcastle (1979) support these findings, the lesions in their study encompassed the entire postcentral gyrus making it difficult to relate a particular deficit with a specific cortical area.

A primary somatosensory area has also been identified in a wide range of nonprimate mammals including monotremes (e.g., Krubitzer *et al.*, 1995b), marsupials (e.g., Pubols *et al.*, 1976; Beck *et al.*, 1996; Huffman *et al.*, 1999; Frost *et al.*, 2000) and eutherians (e.g., Felleman

---

Caption for Fig. 16.2. (cont.)
Both areas 3b and 1 contain a systematic representation of the contralateral body surface and form mirror representations of each other. In some primates (A-C), area 1 is very well developed; while in other primates (D) area 1 has not been identified. We propose that area 1 co-evolved with a well-developed, glabrous hand in some lineages. This is supported by the fact that tamarins have a more primitive hand with claws, compared to the other primates shown.

et al., 1983; Chapin and Lin, 1984; Ledoux *et al.*, 1987; Catania *et al.*, 1993; see Kaas, 1983; Johnson, 1990; Krubitzer, 1995 for review). Like area 3b in primates, S1 or area 3b in other mammals is topographically organized with the foot represented most medially, followed by the representation of the trunk, forelimb, forepaw and face in a mediolateral progression (Figure 16.3). While single units have been recorded in the barrel field of some rodents, a full description of work in barrel cortex is beyond the scope of this chapter. However, Diamond *et al.* (2002) propose that S1 in whisking rodents may be involved in topographic learning via the whiskers, as is the hand representation in area 3b in monkeys. For our purpose it is important to note that not only is the gross topographic organization of S1 similar in all species examined, but also the types of systems level modifications to S1, or the rules of change, are consistent across species.

Probably the most salient modification in S1 across species is the amount of neocortex that represents body parts associated with specialized use, which we term behaviorally relevant body surfaces. For example, animals such as the duck-billed platypus have evolved electrosensory receptors that run in parallel strips along the bill (Figure 16.4). Together with mechanosensory receptors, platypus use electrosensory receptors almost exclusively for prey capture, navigation and most other activities (Scheich *et al.*, 1986; Manger and Pettigrew, 1995; see Krubitzer, 1998 for review). The importance of this morphological structure is reflected in the neocortical organization in that a large percentage of S1, and indeed most of the neocortex is devoted to the representation of the bill (Krubitzer *et al.*, 1995b). Other examples of cortical magnification of specialized morphology (Figure 16.3) include the nose representation of the star-nosed mole (Catania *et al.*, 1993), the digit 4 (D4) representation of the striped possum (Huffman *et al.*, 1999), the D1 and tongue representation of the flying fox (e.g., Calford *et al.*, 1985; Krubitzer *et al.*, 1998), and the hand representation of the raccoon (Welker and Seidenstein, 1959; Johnson *et al.*, 1982).

Most primates, including humans, have a large representation of the glabrous hand in area 3b that reflects the special use of the hand in discrimination of texture and form (see above), and a larger representation of oral structures (Figure 16.4; e.g., Cusick *et al.*, 1986), possibly involved in discriminations necessary for omnivorous feeding and for the articulation of species specific sounds. Similarly, some primates have a relative enlargement of other body parts that are related to specialized use. For example, cebus and spider monkeys have an

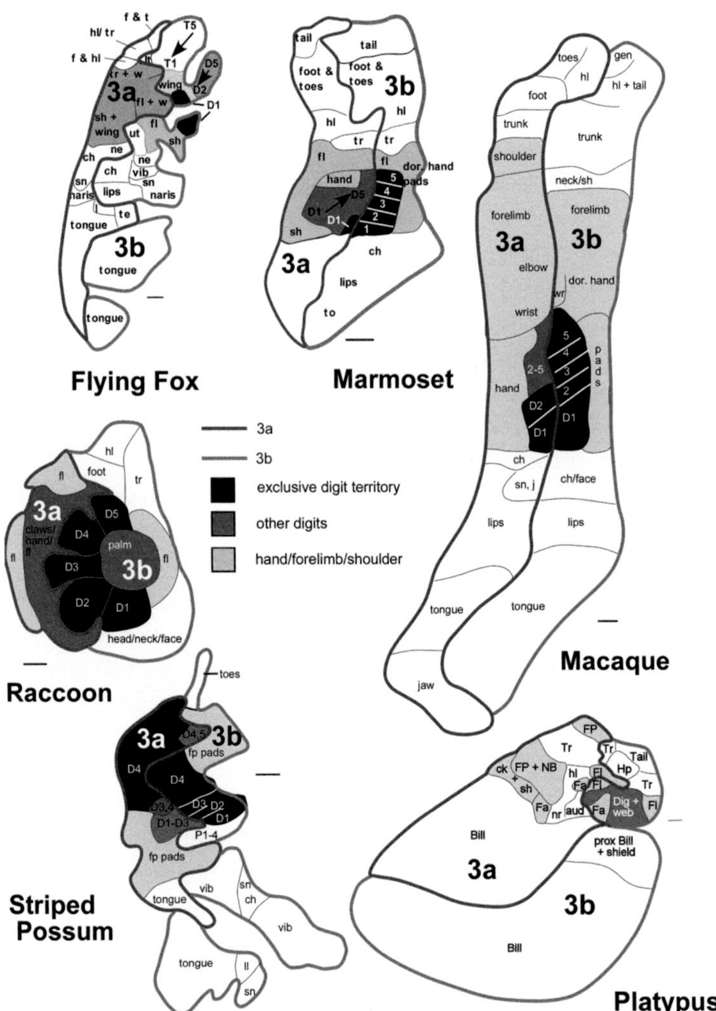

Figure 16.3. The topographic organization of area 3b (red) and 3a (blue) in a variety of mammals including primates (macaques and marmosets), megachiropteran bats (flying fox), carnivores (raccoon), marsupials (striped possum) and monotremes (duck-billed platypus). Although the region of cortex immediately rostral to area 3b is termed differently in different mammals, there are a number of features of organization that are similar across groups of mammals including neural responsiveness to stimulation of deep receptors, dense connections with motor and posterior parietal cortical areas, and a large magnification of behaviorally relevant body parts. For these reasons, we believe that area 3a is a homologous cortical field in mammals. The exaggerated representation of specialized peripheral sensory receptor arrays in area 3a argues that the evolution of this field is highly dependent on the use of a particular body part and suggests that the motor system plays a crucial role in the development of area 3a. This figure was adapted from Krubitzer *et al.*, 2004.

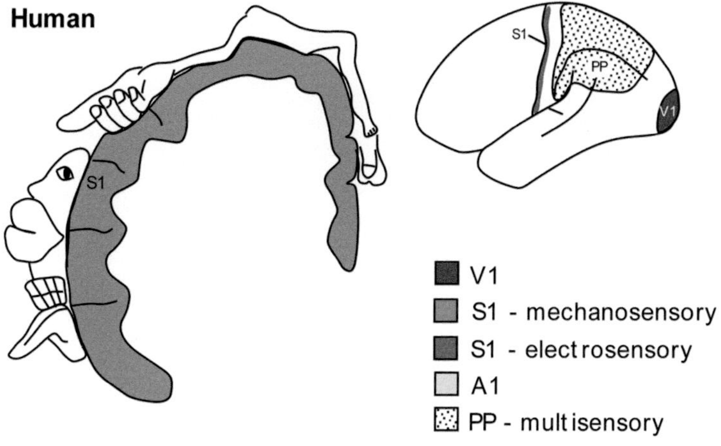

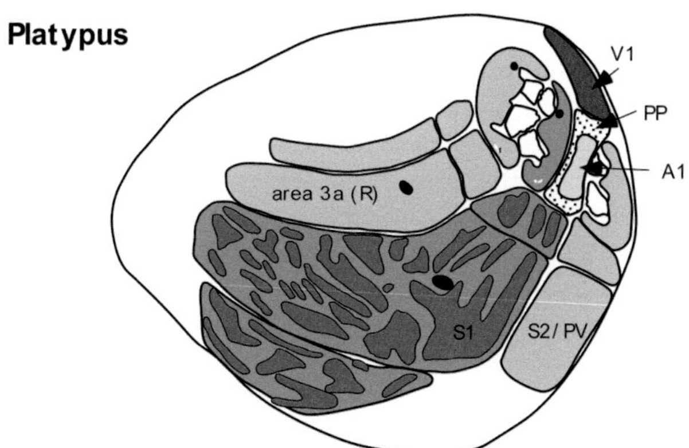

Figure 16.4. A schematic rendition that demonstrates the cortical magnification of behaviorally relevant body part representations in area 3b across different mammals. In humans, the hand, lips, and oral structure representations are magnified, while in platypus, the bill representation is magnified. Although such magnification occurs for different body part representations, the rules of modification of S1 are similar in both groups of mammals despite approximately 200 million years of independent evolution. This observation indicates that S1 is highly constrained in evolution. The schematic of the human is adapted from Penfield and Rasmussen, 1968, and the schematic of the platypus is adapted from Krubitzer *et al.*, 1995b.

enlarged representation of their prehensile tail (Pubols and Pubols, 1971; Felleman *et al.*, 1983;). While this magnification of S1 in different species is related to innervation density at the periphery (e.g., Lee and Woolsey, 1975; Catania and Kaas, 1997), recent evidence also indicates

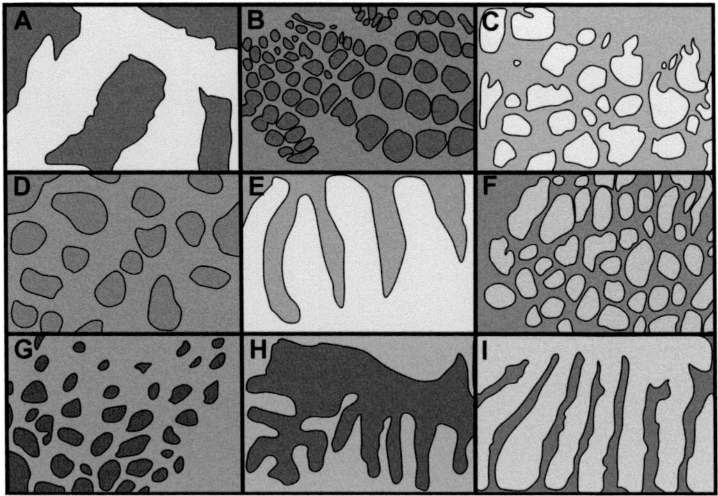

Figure 16.5. A schematic illustrating a second type of modification, modular organization, that has been observed in slices of flattened sensory cortex in a variety of different mammals. A, myelin bands in area 18 of squirrel monkeys; B, the barrel field in S1 of rats; C, modules in the insula of dolphin neocortex; D, entorhinal clusters in macaque monkeys; E, ocular dominance columns in monkeys; F, entorhinal clusters in humans; G, the barrel field in S1 of brush-tailed possums; H, mechanosensory and electrosensory bands in S1 of duck-billed platypuses; I, nose follicle representation of S1 in star-nosed moles. This figure was adapted from Manger *et al.*, 1998.

that the use, or the unique pattern of activity from peripheral sensory receptor arrays, also contributes to this cortical magnification (Recanzone *et al.*, 1992b; Xerri *et al.*, 1996; Catania and Kaas, 1997). It has been proposed that this magnification generates the high sensory spatial resolution necessary for texture and form discriminations made with specialized structure (see Johnson and Yoshioka, 2002 for review).

A second type of consistent modification that has evolved in a number of different lineages is the segregation of inputs from similar body parts into distinct isomorphs, or modules, defined electrophysiologically or histochemically (Figure 16.5). For example, in primates, rapidly adapting and slowly adapting inputs are segregated into separate clusters or bands within S1 (e.g., Sur *et al.*, 1984; Chen *et al.*, 2001), and the face representation is divided into myelin light and dense zones that segregate different portions of the face and oral structures (Krubitzer and Kaas, 1990; Jain *et al.*, 2001). In a number of rodents, inputs from the vibrissae are segregated into cortical barrels (Woolsey and Van der Loos, 1970), in the star-nosed mole, nose follicles form isomorphs (Catania *et al.*, 1993), and in the duck-billed platypus, electrosensory and mechanosensory inputs are segregated into interdigitating bands (Krubitzer *et al.*, 1995b). The evolutionary and developmental significance of modules has been discussed previously (Krubitzer, 1995; Krubitzer and Kahn, 2003). In brief, we hypothesize that modules represent a stage in the evolution of a cortical field, and that selection for shorter connection length and/or increased speed of transmission has led to the aggregation of modules in different cortical fields and different species over time. If this is the case, then it suggests that a cortical field is actually a pattern of activation (connectivity) from a variety of sources including the thalamus and ipsilateral and contralateral hemisphere, that shifts and redistributes across the cortical sheet within an individual lifetime (development) and across species over time (evolution).

Connections

The ipsilateral cortical connections of area 3b have been described in macaque monkeys (e.g., Jones *et al.*, 1978; Pearson and Powell, 1985; Shanks *et al.*, 1985b; Juliano *et al.*, 1990; Darian-Smith *et al.*, 1993; Burton and Fabri, 1995; Burton *et al.*, 1995), New World titi monkeys (Coq *et al.*, 2004; Padberg *et al.*, 2005), marmosets (Krubitzer and Kaas, 1990) and prosimian galagos (Wu and Kaas, 2003). In these primates, restricted injections in area 3b result in a relatively tight distribution of connections predominantly with adjacent somatosensory cortical fields including areas 3a, cortex immediately caudal to area 3b (areas 1 and 2 in macaque monkeys, area 1/2 in New World monkeys and galagos), S2 (and PV where described), and primary motor cortex (Figure 16.6). In primates, thalamic connections of area 3b are predominantly from the ventral posterior nucleus, both VPm and VPl (e.g., Jones *et al.*, 1979; Nelson and Kaas, 1981; Mayner and Kaas, 1986; Darian-Smith *et al.*, 1990; Krubitzer and Kaas, 1992; Rausell and Jones, 1995; Coq *et al.*, 2004; Padberg and Krubitzer, 2006). However, sparse projections from VPi, VPs and Pa have also been observed (e.g., Cusick and Gould, 1990).

Both the cortical and subcortical connections of area 3b (S1) have been described in a number of mammals such as rodents (e.g., Wise and Jones, 1976; Akers and Killackey, 1978; Krubitzer *et al.*, 1986; Chapin *et al.*, 1987; Krubitzer and Kaas, 1987; Koralek *et al.*, 1990; Fabri and

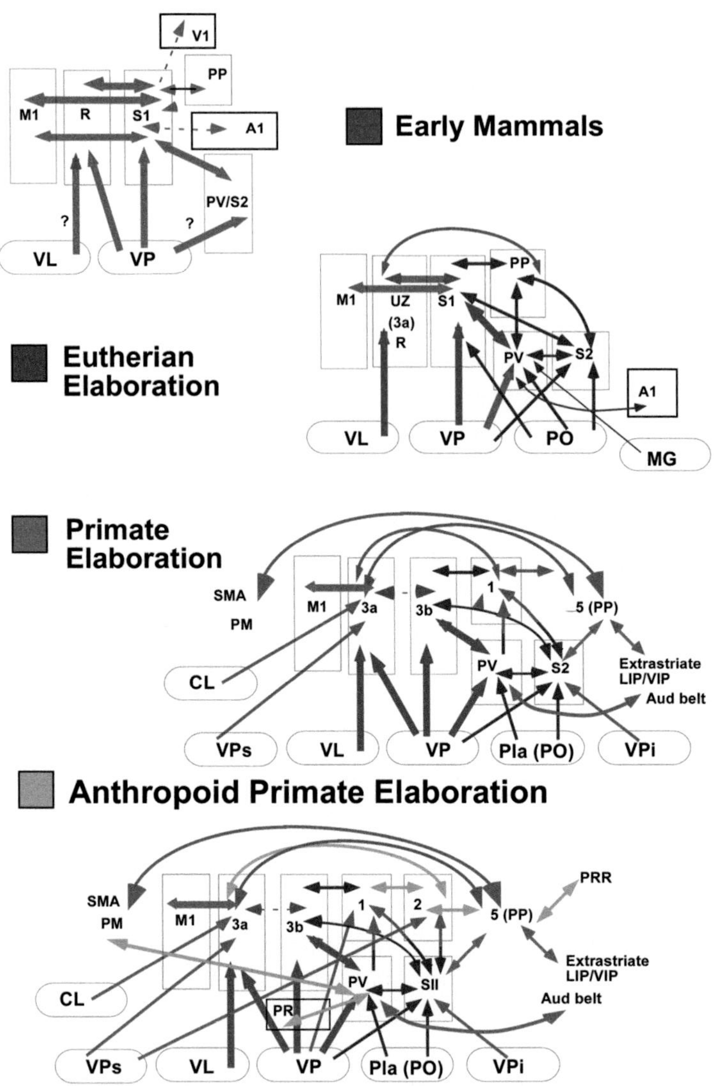

Figure 16.6. An overview of the cortical and thalamocortical connections of anterior parietal cortical fields, lateral sulcus fields, and posterior parietal area 5 in mammals. This figure is not an exhaustive overview of the connections of particular fields in different mammals. Rather, this figure depicts that particular patterns of connections can be identified in all groups of mammals due to retention from a common ancestor (grey arrows). Such connections are homologous. Additional connections arose later in mammalian evolution and have been added to this retained network in eutherians

Burton, 1991; Paperna and Malach, 1991) carnivores (Alloway and Burton, 1985; Barbaresi *et al.*, 1987; Herron and Johnson, 1987), and marsupials (Beck *et al.*, 1996; Elston and Manger, 1999; see Johnson, 1990 for review). As in primates, S1 is densely connected with area 3a (R, UZ and dysgranular cortex, see below), M1 and S2/PV. The major thalamic projection to S1 is from VPl and VPm, and connections are also observed with the posterior nucleus, PO.

The total pattern of callosal connections of areas 3b, 1, 2 and 5 indicates that the hand representation of area 3b is almost completely acallosal in primates (e.g., Pandya and Vignolo, 1968; Killackey *et al.*, 1983; Shanks *et al.*, 1985a, 1985b). Surprisingly, specific callosal connections of individual anterior parietal fields have only been described for a few species of primates (e.g., Manzoni *et al.*, 1986; Krubitzer and Kaas, 1990; Padberg *et al.*, 2005). In macaque monkeys it was found that while the hand representation of area 3b does not project to the hand representation of area 3b of the opposite hemisphere, it does project to S2 (e.g., Manzoni *et al.*, 1986). In marmosets, electrophysiologically defined body part representations in area 3b are differentially interconnected across hemispheres. Myelin light portions of area 3b appear to be strongly interconnected, while myelin dense portions of area 3b are acallosal. Furthermore, as described in early studies, the hand representation of area 3b appears to be mostly acallosal. Area 3b is also interconnected with S2 and PV of the opposite hemisphere. Similar findings were reported in titi monkeys in which the hand representation was acallosal, and myelin light regions formed callosal zones in area 3b (Padberg *et al.*, 2005).

Callosal connections have been described for area 3b or S1 in rats (e.g., Wise and Jones, 1976; Akers and Killackey, 1978), mice (White and DeAmicis, 1977), cats (e.g., Ebner and Myers, 1965; Caminiti *et al.*, 1979; McKenna *et al.*, 1981), raccoons (Ebner and Myers, 1965; Herron and Johnson, 1987), flying foxes (Krubitzer *et al.*, 1998), rabbits (Ledoux *et al.*,

---

Caption for Fig. 16.6. (cont.)
mammals (blue arrows). In different lineages, such as primates, these networks were further elaborated with the addition of new cortical areas, modules to existing cortical areas, and the addition of new thalamic nuclei. Such modifications to the neocortex make determining homologous areas and connection patterns difficult. Furthermore, these types of anatomical modifications suggest that homologous cortical fields may not be analogous (see Figure 16.8). See also colour plate section.

1987) and tree shrews (Cusick *et al.*, 1985; Weller *et al.*, 1987; see Krubitzer *et al.*, 1998 for review). As in primates, there is a heterogeneity of callosal connectivity of different body parts that appear to be associated with use. For example, the representations of specialized body parts in area 3b are acallosal, and zones such as the unresponsive zone in squirrels, the dysgranular zone (septa) in rats, the heterogeneous zones of raccoons, and the myelin light area 1/2 interdigitations in flying foxes are rich in callosal connections, as is the unmyelinated region in monkeys. Thus, studies of callosal connectivity for all mammals investigated report that S1 contains callosal zones related to different body part representations often associated with distinct myelin light zones (Figure 16.7; see Krubitzer *et al.*, 1998 for review). Body part representations associated with specialized use such as the barrels in rodents, lateral face representations in rabbits, and the hand in primates are devoid of callosal connections. Previous studies suggest that callosal free zones are the regions of cortex where thalamocortical afferents terminate (e.g., Wise and Jones, 1976; Gould and Kaas, 1981; Krubitzer and Kaas, 1990; Krubitzer *et al.*, 1998). It appears that selection for specialized peripheral morphology in the form of receptor dense regions like finger tips, also operates on cortical representations to both conserve discrete receptive fields of neurons created by thalamocortical inputs and preserve neuroanatomical interactions between neighboring specialized skin regions. This conservation of thalamocortical inputs carrying information from specialized skin regions, and of short intrinsic cortical connections within the representation of these specialized skin regions, results in cortical zones or islands that are uninterrupted by callosal afferents (Herron and Johnson, 1987; Ledoux *et al.*, 1987; Krubitzer *et al.*, 1998). We further suggest that in area 3b this type of connection labyrinth may be necessary to maintain the integrity of sensory discriminations derived from inputs from behaviorally significant body parts, and as such constrains the evolution of S1.

Taken together, the data illustrate that there is a common organization of S1 across species, and that modifications to this field take a similar form reflected as a magnification of the representation of specialized peripheral sensory morphology. How this magnification is related to specialized behavior is not clear, and why this mode of cortical specialization has emerged (compared to any number of potential types of modification) is not known. However, Kaas (1997) suggests that it is a by-product of development that has been functionally optimized. Regardless of the functional significance of this

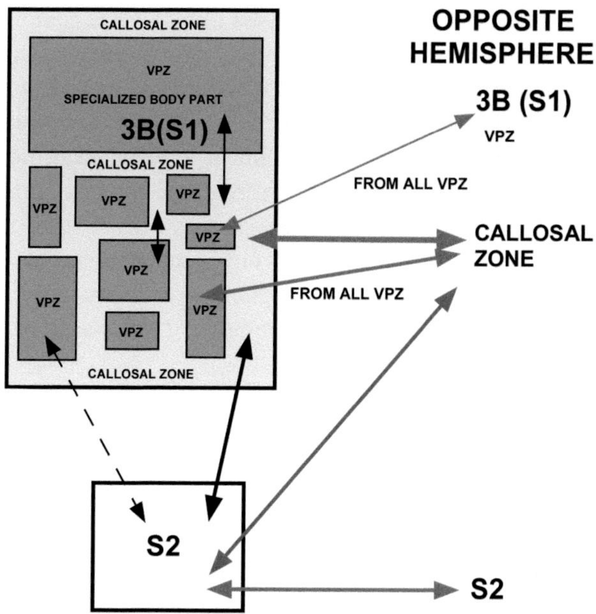

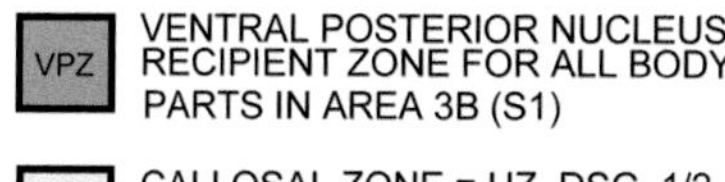

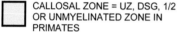

Figure 16.7. An illustration of the callosal connections of the primary somatosensory area (3b or S1) that depicts some common features of connectivity across all mammals. This figure demonstrates that area 3b appears to be modularly organized based on several important principles. First, S1 is broken into thalamocortical recipient zones (VPZ) and callosal recipient zones. Thalamocortical recipient zones of behaviorally relevant body parts (e.g., hand, bill, nose, wing) are acallosal. Some of the VPZs have sparse callosal connections with VPZs in the opposite hemisphere and more dense connections to callosal zones in the opposite hemisphere. Finally, information from 3b can also reach the opposite hemisphere indirectly via S2. We and others propose that this type of organization evolved to maintain close intracortical connections between neurons in the same VPZ, uninterrupted by connections from the opposite hemisphere. This type of organization would allow for the maintenance of small receptive fields in neurons over a large region of cortex, and would allow the integrity of sensory discriminations derived from inputs from behaviorally significant body parts to be maintained.

magnification, comparative surveys demonstrate that the neocortex must be highly constrained since only a few modifications to S1, other than cortical magnification, have emerged. In addition to cortical magnification, the cortex has been modified to segregate subclasses of input, such as the segregation of rapidly adapting and slowly adapting inputs as well as the callosal and acallosal zones described above. Again, it is not understood why this feature has emerged independently in a number of different lineages, but it has been proposed that it reflects developmental contingencies and is therefore highly constrained (e.g., Krubitzer, 1995; Kaas, 1997; Krubitzer and Kahn, 2003).

As with cortical organization, connections of S1 also have similar patterns across mammals. This homologous network may in part subserve similar functions, but it is likely that with the addition of new fields and connections (as in primates), and a re-weighting of existing synaptic interactions, new functions or at least a species-specific refinement of particular behaviors is likely to have emerged (Figures 16.6 and 16.8). This observation, of course, implies that even for primary fields, direct extrapolation between nonhuman and human primates regarding function is problematic.

AREA 3A

Functional organization

A representation of deep receptors of the contralateral body, termed area 3a, is located immediately rostral to area 3b. Area 3a was originally described as an architectonically distinct cortical field in humans by Vogt and Vogt (1919), but has since been described in a variety of mammals. The functional organization of area 3a in primates has been described for marmosets (Huffman and Krubitzer, 2001a) and macaque monkeys (Krubitzer *et al.*, 2004) in multiunit recording studies, and recently in humans using modern imaging techniques (Moore *et al.*, 2000). These studies demonstrate that area 3a contains a topographically organized representation of deep receptors and musculature of the contralateral body that parallels that of area 3b (Figure 16.3), and ultimately receives input from proprioceptors of the muscles and joints (Phillips *et al.*, 1971; Schwarz *et al.*, 1973; Heath *et al.*, 1975; Hore *et al.*, 1976; see Tanji and Wise, 1981 for review). Single unit studies in awake monkeys report that neurons in area 3a modulate activity prior to wrist flexion and extension (Nelson, 1987) and are active under a

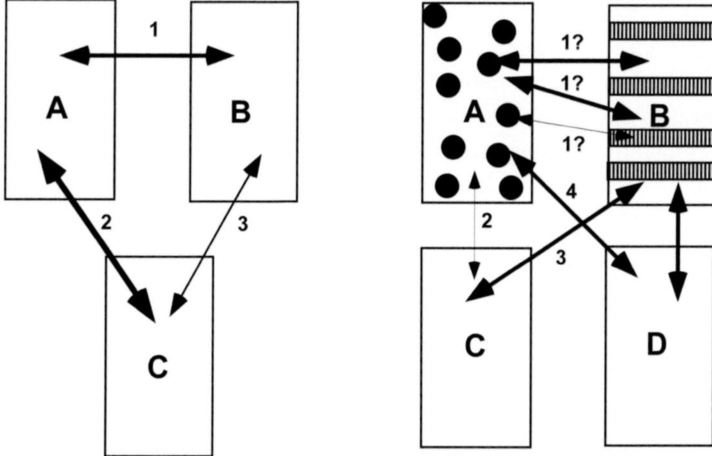

Figure 16.8. A theoretical rendition of the evolution of cortical fields and their connections. This figure illustrates the difficulties inherent in assigning homology across groups of mammals. It also illustrates the problems with making direct extrapolations regarding analogy of cortical fields, and their circuitry, between any groups of mammals (such as macaque monkeys and humans). Hypothetical cortical fields include field A, B and C with connection patterns 1, 2 and 3 (left). When comparing cortical organization and connections across groups to determine homology, several problems arise. First, new cortical fields have been added in some groups (D). Second, modules may have been added to existing cortical fields (A and B). Third, the density of connections between fields may have changed (connection 2). Fourth, existing connections have been modified by the generation of modules (connection 1). Finally, new connections (4) have developed between retained fields (A) and new fields (D). All of these events are complicated by the fact that connection patterns are often used to determine homology between fields. Thus, even when one can convincingly determine homology of cortical fields in different groups of animals, it is unlikely that homologous fields are strictly analogous.

variety of joint movements (Gardner, 1988). Neural activity in area 3a increases with maintained limb position and is modulated with the velocity of movement toward a limb position, as well as the ultimate position of the limb (Tanji, 1975; Wise and Tanji, 1981). Furthermore, lateral portions of area 3a (termed 3aV) contain neurons that respond to vestibular stimulation (Guldin *et al.*, 1992). Thus, area 3a in primates is a proprioceptive field that is closely involved in the kinetics of movement. This contention is supported by microstimulation studies in monkeys which demonstrate that body movements can be elicited by stimulating area 3a with very small currents (e.g., Stepniewska *et al.*, 1993).

Studies in other mammals have identified a region of cortex just rostral to area 3b in which neurons respond to stimulation of deep receptors. In carnivores such as cats (e.g., Oscarsson and Rosen, 1963, 1966; Oscarsson *et al.*, 1966; Landgren and Silfvenius, 1969; Zarzecki *et al.*, 1978; Felleman *et al.*, 1983), raccoons (Johnson *et al.*, 1982; Feldman and Johnson, 1988) and ferrets (Hunt *et al.*, 2000) a region just rostral to area 3b exhibits a number of characteristics of area 3a in monkeys, such as neural response to stimulation of deep receptors, architectonic appearance, and gross topographic organization (Figure 16.3). This field has been termed area 3a in cats and ferrets and kinesthetic cortex in raccoons. In other mammals such as marsupials (see Huffman *et al.*, 1999), rodents (see Slutsky *et al.*, 2000 for review), insectivores (Krubitzer *et al.*, 1997) and monotremes (Krubitzer *et al.*, 1995b), these features are observed in a field just rostral to S1, termed the rostral field, R (Figure 16.6). In all of these mammals, this cortical area immediately rostral to S1 and caudal to area 4 contains a gross mediolateral organization that mirrors that of area 3b, although there is a great deal of variability across groups (see below). Thus, all mammals have a cortical field located just rostral to area 3b that receives inputs from deep receptors of the contralateral body and is distinguished from motor cortex (area 4) based on a number of criteria including functional organization, neural stimulus preference, architectonic distinctiveness, and when known cortical and subcortical connectivity. We propose that area 3a is homologous in all mammals.

As demonstrated for area 3b, there are several features of area 3a organization that have been consistently modified in different lineages, and these modifications appear to be more related to the use of behaviorally relevant body parts than in area 3b. For instance, when the topographic organization of area 3b is compared with that of area 3a in macaque monkeys, area 3b is observed to be more topographically precise, particularly for representations of the digits where every digit is represented in an exclusive cortical zone. This aspect of representation in area 3b has been observed in every primate examined regardless of the use of the hand or whether the primate in question has an opposable thumb. In contrast, in area 3a in macaque monkeys' exclusive cortical territory is devoted to the representation of D1 and D2, while digits three through five are represented together. In area 3a in marmosets, little if any cortical territory is devoted to the exclusive representation of a single digit. Digit representation in area 3a in the two species appears to parallel differences in use (Figure 16.3). Macaque monkeys have opposable thumbs and are highly skilled "graspers" who employ two general techniques. The first technique is to oppose D1 to D2 (precision grip), and the second technique is to oppose all four remaining digits to the palm (power grip; Welles, 1976; Roy *et al.*, 2000). Marmoset monkeys on the other hand, generally employ only a power grip, and their cortical representation in area 3a reflects this behavior. These observations suggest that in primates area 3a emerges in development and evolution as a result of the actual use of the body part and, therefore, the motor system must play an important role in its construction.

Studies of the organization of area 3a in a variety of other species who use the hand quite differently than primates support this contention (Figure 16.3). For instance, the flying fox has a highly derived hand in which the digits have evolved membranes that span between them. This digit/membrane configuration functions as a whole unit, the wing, which is adapted for flight. Consequently, all digits form a single representation in area 3a, the wing, and there is no cortical territory that exclusively represents any one digit (Krubitzer *et al.*, 1998), although such exclusive digit representation is present in area 3b in these mammals (Calford *et al.*, 1985; Krubitzer and Calford, 1992). Another interesting example of use-dependent organization of area 3a is the marsupial striped possum. This animal has a specialized D4 that it uses almost entirely to capture insects (Van Dyck, 1983). While exclusive cortical territory for all of the digits is observed in area 3b (with a magnification of D4), only D4 is represented in area 3a. Differences in use between members of the same species can produce a good deal of variability in the topographic organization of area 3a (e.g., Recanzone *et al.*, 1992b), and this variability is even more pronounced across species due to specialized motor sequences or exaggerated behaviors that co-evolved with specialized peripheral morphology and associated sensory receptor arrays (see Krubitzer *et al.*, 2004).

Connections

There are several studies of the connections of area 3a in primates including macaque monkeys (e.g., Jones *et al.*, 1978; Darian-Smith *et al.*, 1993), squirrel monkeys (Guldin *et al.*, 1992) and marmosets (Huffman and Krubitzer, 2001a). These studies demonstrate that area 3a is densely connected with M1, SMA, area 2 (or cortex caudal to area

3b), S2/PV, cingulate cortex, insular cortex and posterior parietal cortex (area 5; Figure 16.6). The surprising result from these previous studies is that area 3a has more numerous connections with motor and posterior parietal areas of the neocortex than with traditionally defined somatosensory areas. Studies of thalamic connections of area 3a in Old World (Jones *et al.*, 1979; Darian Smith *et al.*, 1990) and New World monkeys (Akbarian *et al.*, 1992; Huffman and Krubitzer 2001b) indicate that area 3a receives input from somatic nuclei such as the ventral posterior superior nucleus (VPs, VPLc of Jones *et al.* 1979; Friedman and Jones, 1981 and Darian-Smith *et al.*, 1993) and the anterior pulvinar (Pla), some of which (e.g., VPi, VPs and VLc) are also associated with vestibular processing (e.g., Lang *et al.*, 1979). Area 3a also receives input from nuclei associated with the motor system such as the ventral lateral (VL), ventral anterior (VA) and central lateral (CL) nucleus of the thalamus.

The cortical and subcortical connections of area 3a, or cortex in the location of area 3a, has only been described in a few mammals such as squirrels (Gould *et al.*, 1989; dysgranular UZ), flying foxes (Krubitzer *et al.*, 1998), cats (e.g., Avendano *et al.*, 1992) and briefly for ferrets (e.g., Hunt *et al.*, 2000). While there is variability in the patterns of cortical connections of area 3a across animals, there are consistent patterns of connections with motor cortex, areas of the lateral sulcus such as S2/PV, and with cortex immediately caudal to S1 (Figure 16.6; termed area 5, PP, 1/2 and C in different animals). Thalamic connections of the UZ/R region of squirrels (area 3a) were from nuclei situated rostral, dorsal and caudal to VP, and from the central medial nucleus (Gould *et al.*, 1989).

As with 3b, in most studies, the total pattern of callosal connections of anterior parietal fields 3a, 3b, 1 and 2 has been examined rather than the specific connections of area 3a (e.g., Jones and Powell, 1969a; Killackey *et al.*, 1983; Shanks *et al.*, 1985a). These studies suggest that like area 3b, the hand representation of area 3a is acallosal. There are only a few studies of callosal connections of area 3a in primates and other mammals. In marmosets, specific connections of the hindlimb and forelimb representation in area 3a have been described (e.g., Huffman and Krubitzer, 2001a). Interestingly, at least for the marmoset, the hand representation of area 3a does have callosal connections with the hand representation of area 3a in the opposite hemisphere. Furthermore, area 3a is callosally connected to areas M1 and SMA as well. Similar findings were demonstrated for the wing representation of the flying fox (Krubitzer *et al.*, 1998).

Taken together, the data indicate that area 3a is a proprioceptive area that integrates somatic and vestibular inputs with the motor system. This cortical field is involved in the kinetics of movement (determining the load and force), in maintaining posture and limb position, and in regulating the velocity of limb movement.

AREA 1

Functional organization

A third somatosensory cortical field just caudal to area 3b, termed area 1, has been described in macaque monkeys and three species of New World monkeys including owl, squirrel and cebus monkeys (Merzenich *et al.*, 1978; Nelson *et al.*, 1980; Sur *et al.*, 1982; Felleman *et al.*, 1983, respectively). Recent evidence indicates that area 1 is present in humans as well (e.g., Krause *et al.*, 2001; Blankenburg *et al.*, 2003). In some primates, such as macaques, squirrel monkeys and owl monkeys, area 1 forms a mirror reversal representation of area 3b, and contains a precise, topographically organized representation of the contralateral body surface, much like that of area 3b (Figure 16.2). As in area 3b, there is a magnification of the glabrous hand and oral structures, and receptive fields for neurons are small, and limited to single digits. In other primates, only a rudimentary area 1 has been observed (e.g., titi monkeys, Coq *et al.*, 2004; Padberg *et al.*, 2005) or is absent under similar recording conditions (e.g., tamarins, Carlson *et al.*, 1986; marmosets, Krubitzer and Kaas, 1990; prosimian galagos, Sur *et al.*, 1980). Although we and others have termed cortex caudal to area 3b "area 1/2" in previous studies in marmoset monkeys (e.g., Krubitzer and Kaas, 1990) and galagos (Wu and Kaas, 2003), this terminology was based solely on position with respect to area 3b, rather than any solid electrophysiological data. Indeed, in these species, cortex caudal to area 3b contains neurons that are either unresponsive to any type of sensory stimulation under the anesthetic conditions utilized, or are inconsistently driven by stimulation of deep receptors or high-threshold cutaneous receptors of the contralateral body (Figure 16.11). Thus, area 1 is well developed in only a few primates including Old World macaque monkeys and a few New World monkeys. Given that these species have a well-developed glabrous hand compared to the clawed New World marmoset and tamarin monkeys, we hypothesize that this field coevolved with the highly specialized glabrous hand and is associated with sophisticated hand use.

Single unit studies demonstrate that like neurons in area 3b, neurons in area 1 are modulated by contact with an object (e.g., Debowy *et al.*, 2001), are selective for motion across the skin, and are involved in edge orientation (e.g., Gardner, 1988). Furthermore, neurons in area 1 respond to noxious stimulation (Kenshalo *et al.*, 2000). Lesions that are restricted to area 1 in primates demonstrate that animals are unable to discriminate between hard versus soft and rough versus smooth objects (Randolph and Semmes, 1974; Carlson, 1981).

There is very limited evidence for an area 1 in other mammals. While investigators have subdivided cat cortex into architectonic areas 3a, 3b, 1 and 2 (e.g., Hassler and Muhs-Clement, 1964), electrophysiological recording experiments in which this cortex was densely surveyed demonstrate that only a single representation of the contralateral body surface (area 3b or S1) resides in anterior parietal cortex of the cat (e.g., Felleman *et al.*, 1983). Although cortex caudolateral to area 3b has been electrophysiologically explored and termed S3 by Garraghty *et al.*, (1987), these investigators propose that S3 may be homologous to S2 or PV in other mammals, rather than area 1. The data in ferrets are equivocal. Some laboratories report multiple cutaneous representations in anterior parietal cortex (termed C1, C2, R1 and R2; e.g., Leclerc *et al.*, 1993), while others report that neurons in cortex caudal to area 3b respond to stimulation of deep receptors (Hunt *et al.*, 2000). In the former study, only the snout representation was explored. It is not clear if indeed four separate cutaneous representations exist in ferrets since the mapping data were limited; or if they do exist, which of the four corresponds to area 3b and area 1. Finally, a field termed area 1/2 has been described for the flying fox (e.g., Krubitzer and Calford, 1992). This region of cortex interdigitates with area 3b and contains neurons that respond to cutaneous stimulation of the contralateral body surface as well as islands of neurons that respond to stimulation of deep receptors. Like area 1 in primates, area 1/2 in the flying fox contains a complete representation of the contralateral body surface.

In all other mammals investigated, cortex caudal to area 3b contains neurons that respond to stimulation of deep receptors (Figure 16.11). For example, in rodents (Slutsky *et al.*, 2000), insectivores (Krubitzer *et al.*, 1997), monotremes (Krubitzer *et al.*, 1995b) and marsupials (Beck *et al.*, 1996; Huffman *et al.*, 1999), cortex immediately caudal to 3b contains neurons that respond to stimulation of deep receptors of the contralateral body, which are often multimodal (e.g., short-tailed possums and platypus). This field has been called the caudal field (C), the caudal somatosensory area (SC), the parietal medial field (PM), PPC, or area 1/2 in different mammals, and has been proposed to correspond to posterior parietal cortex including area 5 or a combination of areas 1 and 2 (Reep *et al.*, 1994; Krubitzer *et al.*, 1997, 1998; Huffman *et al.*, 1999; Slutsky *et al.*, 2000 for review). Interestingly, studies in which the behavioral effects of lesions in PPC were examined in rats report that these animals have deficits in spatial attention and navigation (Kolb *et al.*, 1994). Similar types of deficits are observed when posterior parietal cortex (rather than area 1) is lesioned in primates (see below).

It seems unlikely that cortex immediately caudal to area 3b in nonprimate mammals is homologous to area 1 in primates for three reasons. The first is that there are only two nonprimate mammals in which neurons in cortex that adjoins area 3b at its caudal boundary respond to cutaneous stimulation, and this cortex has only been thoroughly explored in one of these species (e.g., flying fox). In all other mammals, cortex immediately caudal to 3b contains neurons that respond to stimulation of deep receptors, that are often multimodal. Second, the location of an area 1 described in these animals as just caudal to S1 is not a conclusive indicator of homology because in small-brained mammals such as tenrecs and marsupial northern quolls, cortex immediately caudal to area 3b is also immediately rostral to V1 or V2 (Figure 16.9). Therefore, assigning homology based on position alone is problematic. Finally, prosimians, marmosets and tamarins do not appear to have an area 1, which suggests that this field arose later in primate evolution and is therefore a purely primate phenomenon, with the cutaneous field immediately caudal to area 3b arising independently in flying foxes (and possibly ferrets, if area 1 does indeed exist in these animals), and serving different, although overlapping functions in these species (see below).

Connections

Highly restricted injections into electrophysiologically identified portions of area 1 have only been made in macaque monkeys (Pons and Kaas, 1986; Burton and Fabri, 1995; Burton *et al.*, 1995) and titi monkeys (Coq *et al.*, 2004; Padberg *et al.*, 2005). Connections of area 1 in macaque monkeys are more broadly distributed than those in area 3b and are observed with areas 3b, 2, S2/PV, 5 AIP/7b and sparsely with areas 3a, M1 and frontal cortex. Previous reports on connections of the architectonically defined area 1 with local parietal cortical areas support these electrophysiological studies (e.g., Jones *et al.*, 1978; Vogt and Pandya,

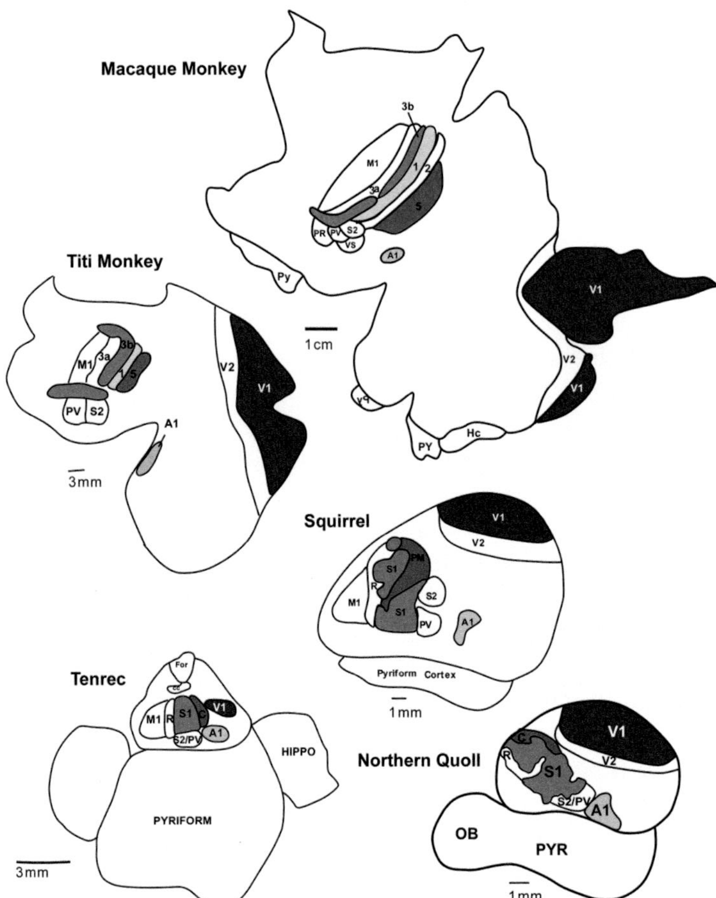

Figure 16.9. The organization of neocortex in several different mammals. This figure demonstrates that assigning homology of a cortical field based on the location of a particular field relative to another field is problematic. For instance, in mammals with large brains, such as macaque monkeys, area 1 is immediately caudal to area 3b, area 5 is close to the caudal boundary of area 1, and both are very far removed from V1 and V2. In mammals such as squirrels, cortex immediately caudal to area 3b, PM, is relatively close to V2 compared to macaque monkeys. In tenrecs, cortex immediately caudal to area 3b (S1) is also immediately adjacent to V1. If one were to assign homology based solely on location, this cortex in tenrecs could as easily be termed area 1 as it could be termed V2.

1978; Pearson and Powell, 1985; Shanks *et al.*, 1985b). In macaque monkeys, thalamocortical connections from electrophysiologically identified locations in area 1 indicate that like area 3b, area 1 receives the majority of its inputs from VP proper (e.g., Nelson and Kaas, 1981; Pons and Kaas, 1985). However, in New World titi monkeys, thalamocortical connections of area 1 are also known from VL, Pa and VPs (Padberg and Krubitzer, 2006).

Examination of callosal connections of areas 3a, 3b, 1 and 2 indicates that the hand representation of area 1 is acallosal (e.g., Pandya and Vignolo, 1968; Killackey *et al.*, 1983; Shanks *et al.*, 1985a; Conti *et al.*, 1986). Recent work in titi monkeys in our laboratory in which the injections in area 1 were made under electrophysiological guidance demonstrates sparse callosal connections for the hand representation in area 1 (Padberg and Krubitzer, 2006). In this primate, area 1 was most densely interconnected with areas 5 and AIP/7 of the opposite hemisphere.

Connections, lesions and single unit studies in macaque monkeys indicate that areas 3b and 1 in primates are involved in integrating local inputs from restricted portions of the glabrous hand necessary for fine tactile discriminations such as ascertaining object texture and form (Randolph and Semmes, 1974; LaMotte and Mountcastle, 1979; Carlson, 1981; Sinclair and Burton, 1991; Ageranioti-Belanger and Chapman, 1992; Jiang *et al.*, 1997). We propose that area 1 is a recently evolved field restricted to primates, although some species (e.g., flying foxes) have independently evolved a cutaneous representation caudal to area 3b. The location of this field in flying foxes is similar to that of area 1 in primates, but comparative data and examination of behavior in this species indicate that it is unlikely that this area is homologous, or strictly analogous to area 1 in primates since flying foxes do not use their wings for making form and texture discriminations of objects. It is possible that this area arose with the modification of the distal forelimb for flight, and is involved in making fine tactile discriminations regarding small changes in air pressure and velocity across the wing during flight.

AREA 2

Functional organization

The functional organization of area 2 has only been investigated in one species of nonhuman primate, the macaque monkey (e.g., Pons *et al.*, 1985; Toda and Taoka, 2001, 2002). In macaque monkeys,

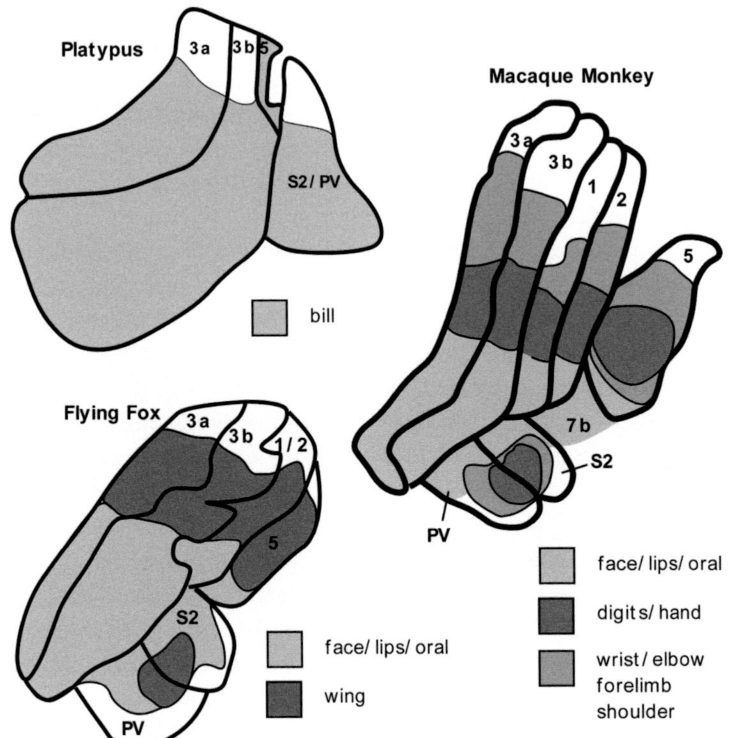

Figure 16.10. This figure demonstrates that parietal cortex obeys similar rules of construction regardless of which body part has been specialized. For example, the platypus has only three somatosensory areas and, we propose, a posterior parietal area (area 5). The three somatosensory areas are dominated by the representation of the bill, and area 5 exclusively represents the bill. In other groups, more cortical fields have been added between areas 3b and 5, such as areas 1 and 2 in macaque monkeys. These fields contain an enlarged representation of the hand and oral structures, as do 3b and 3a. Like the bill representation of area 5 in the platypus, area 5 in primates almost exclusively represents the hand. Even in species with a derived hand, such as the wing of the bat, similar principles of expansion and magnification are upheld.

neurons in area 2 contains a complete representation of deep receptors of the contralateral body, and the gross mediolateral topography is much like that described for areas 3b and 1, although the somatotopic organization is not as precise. The representation of the hand and forelimb in area 2 is highly magnified, more so than in areas 3a, 3b and 1 (Figure 16.10).

Studies of response properties of neurons in area 2 (e.g., Hyvärinen and Poranen, 1978; Taoka *et al.*, 1998, 2000; Iwamura *et al.*, 2002) report that neurons here respond well to stimulation of deep receptors, although in some portions of area 2 neurons also responded to cutaneous stimulation (e.g., Pons *et al.*, 1985; Ageranioti-Belanger and Chapman, 1992), and may participate with areas 3b and 1 in texture discriminations (e.g., Salimi *et al.*, 1999b). Receptive fields for neurons in area 2 are relatively large (sometimes bilateral) when compared to areas 3b and 1 (e.g., Taoka *et al.*, 2000; Iwamura *et al.*, 2002). In awake behaving monkeys, neurons in area 2 respond to both passive and active flexion of joints (Wolpaw, 1980; Gardner, 1988) and are facilitated or inhibited during grasping (Debowy *et al.*, 2001). Furthermore, neuronal burst duration is correlated with arm movement duration (Burbaud *et al.*, 1991), suggesting that area 2 directly participates in the online maintenance of movement. There is limited evidence for an area 2 in humans, and the organization and the subclass of receptors represented appears to be similar to that of macaque monkeys (e.g., Moore *et al.*, 2000). Lesions in macaque monkeys that include area 2 affect the animal's ability to discriminate object shape, size and curvature (Randolph and Semmes, 1974; Carlson, 1981). Furthermore, the animal is unable to make discriminations that require active exploration with the hands (Schwartz, 1983).

Cortex immediately caudal to area 1 has been explored in a limited fashion in New World monkeys. Neurons in this region have been reported to be unresponsive to somatic stimulation, responsive only at the caudal border of area 1, or sometimes responsive to stimulation of deep receptors (Figure 16.11; e.g., Merzenich *et al.*, 1978; Carlson *et al.*, 1986; Huffman and Krubitzer, 2001a). Thus, unlike macaque monkeys it appears that New World monkeys do not possess an area 2. As noted above, cortex caudal to area 3b in nonprimate mammals contains neurons that respond to stimulation of deep receptors. While these regions could correspond to area 2 in these mammals and New World monkeys, it is also possible that they correspond to area 5. We suggest that this is the case based on several lines of evidence including the organization and magnification of representations in these fields, location relative to visual and somatosensory cortex, and the presence of neurons that also respond to stimulation of other sensory modalities (see below), but are often hard to drive under any given anesthetic condition (unlike neurons in area 2).

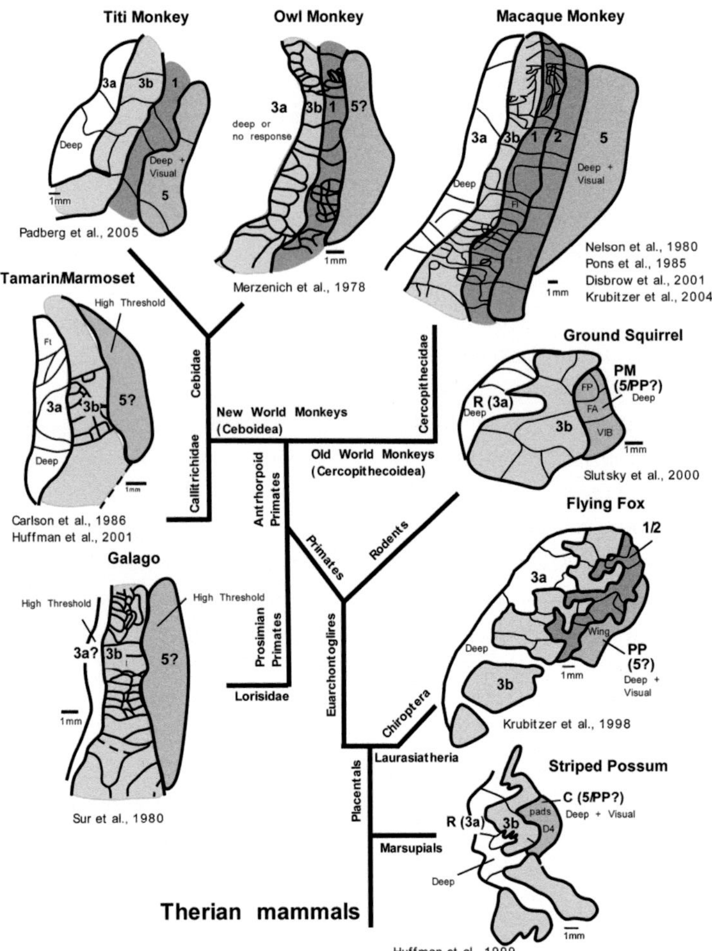

Figure 16.11. A simplified cladogram depicting the phylogenetic relationship of primates and other mammals, and the organization of anterior and posterior parietal cortex in several species. Comparative data from these and other mammals indicate that early therian mammals possessed a primary somatosensory area (3b or S1, light gray), a rostral field (3a or R, white), and a caudal area (5 or PP, green). Some species, such as simian primates, have evolved a low threshold cutaneous representation just caudal to 3b, termed area 1 (blue). Since area 1 has not been identified in other mammals, or even in prosimian primates, it is likely that area 1 evolved after the simian and prosimian divergence. Area 2 (orange) has only been identified in macaque monkeys. Comparisons across mammals indicate that areas 3a, 3b, and 5 are evolutionarily old fields, and that new, unimodal somatosensory fields such as areas 1 and 2 evolved later in some lineages, and are interspersed between existing

Connections

Studies of connections of area 2 in which injections were placed under electrophysiological guidance indicate that this field is connected with other somatosensory cortical areas such as 3b, 1, 3a, and S2, as well as with M1 and area 5 (Pons and Kaas, 1985). Earlier studies of the connections of architectonically defined area 2 with local parietal cortical areas support these later studies (e.g., Jones *et al.*, 1978). Area 2 receives thalamic input predominantly from VPs (VPLc of Friedman and Jones, 1981) and the anterior pulvinar (PA; Pons and Kaas, 1985). The callosal connections of area 2 have only been described collectively with other anterior parietal fields. Unlike areas 3b and 1, the hand representation of area 2 does have callosal connections (see Manzoni, 1997). Taken together, all data indicate that area 2 is a proprioceptive area that is involved in the discrimination of shape and in the online maintenance of hand and forelimb movements necessary for reaching and grasping and is only present in Old World monkeys and humans (and likely great apes).

POSTERIOR PARIETAL AREA 5 IN PRIMATES

Functional organization

Area 5 was first described as a very large cortical field that occupied the entire rostral bank of the IPS and much of the caudal post-central gyrus (e.g., Brodmann, 1909). However, modern electrophysiological and anatomical studies indicate that area 5 is much smaller and resides in the middle of the rostral bank of the IPS and folds around the sulcal crown to spread onto the adjacent postcentral gyrus (e.g., Pons *et al.*, 1985; Iwamura, 2000 for review). Most data pertaining to posterior parietal cortex, and to area 5 in particular, have been collected in macaque monkeys and recently in titi monkeys (Figure 16.11), with limited information available in humans (see Culham and Kanwisher, 2001). In all of these primates, area 5 is dominated by the

---

Caption for Fig. 16.11. (cont.)
fields (i.e., are not added hierarchically). We propose that areas 1 and 2 evolved with the modified morphology of the hand in anthropoid primates, and that older, retained field, such as area 5, were modified both functionally and connectionally for sophisticated hand use. Phylogenetic relationships come from Murphy *et al.*, 2001 and Eisenberg, 1981. Cortical organization of different species depicted here is taken from studies listed below each species. See also colour plate section.

representation of the hand and forelimb (Figure 16.10); neurons in area 5 have contralateral, ipsilateral and bilateral receptive fields (particularly on the hand and forelimb), and most neurons respond to stimulation of deep receptors of the skin and joints (e.g., Sakata *et al.*, 1973; Mountcastle *et al.*, 1975; Taoka *et al.* 2000; Iwamura *et al.*, 1994, 2002; Padberg *et al.*, 2005; see Iwamura, 2000 for review). Single unit studies in awake, behaving macaque monkeys indicate that area 5 is involved in coordinating or programming intention of movement (Snyder *et al.*, 1997; Debowy *et al.*, 2001), in preshaping the hand before grasping an object (e.g., Debowy *et al.*, 2001), and in generating body- or shoulder-centered, rather than eye-centered coordinates for reaching (Ferraina and Bianchi, 1994; Lacquaniti *et al.*, 1995; see Wise *et al.*, 1997 for review). Furthermore, area 5 appears to be involved in the kinematics (e.g., spatiotemporal coordinates) rather than the kinetics (e.g., load and force of muscle) of reaching (see Kalaska, 1996; Wise *et al.*, 1997 for review). The proposition that area 5 is involved in the kinematics of movement is in part supported by studies which demonstrate that a large proportion of neurons are active prior to an arm movement, that these neurons are direction-selective for arm movement, and that activity is dependent on the behavioral context in which the movement will occur (Burbaud *et al.*, 1991).

Recent studies indicate that neurons in area 5 are involved in generating an internal representation of the body that can be modified by experience (Graziano *et al.*, 2000). For example, neurons in area 5 can change their receptive field size and location when a limb used to perform a task is artificially extended with a tool (Iriki *et al.*, 1996). Furthermore, Iriki and colleagues (2001) demonstrated that monkeys can be trained to recognize an image in a video monitor as part of their own body, and that neurons in the IPS, in the location of what we consider to be area 5, change their visual receptive fields to incorporate changes in hand location and size, as viewed on the video monitor, into an internal frame of reference. In these studies, the position and the size of the visual receptive field was modified with respect to modifications of the image in the monitor (e.g., expansions, contractions and displacement of the hand). These investigators speculate that a symbolic representation of the self in humans has a precursor in monkeys that is latent and can be activated with training. We would argue that these monkeys always possess an internal frame of reference, much like that in humans, and that the experimenters devised a clever and objective way in which to evaluate this internal representation in a nonhuman mammal. We would further argue that such a representation is a critical characteristic of most living things, and constitutes a sense of self. Thus, this "sense of self" is present in all mammals and the region of cortex we propose to be involved in generating this behavioral characteristic (area 5 or PP) is homologous across groups of mammals.

Our recent studies in New World titi monkeys, as well as preliminary findings in macaque monkeys, indicate that many neurons in area 5 can be driven by visual stimulation (Disbrow *et al.*, 2001; Padberg *et al.*, 2005; Figure 16.11). While responsiveness to visual stimulation has not been described in previous studies for area 5, the studies often did not test whether or not neurons would be responsive to visual stimulation, or may have subdivided the cortex in a different fashion than other laboratories. This issue of whether neurons in area 5 respond to visual stimulation is important for our discussion of the evolution of posterior parietal cortex in mammals since, as noted above, cortex immediately caudal to 3b in many primitive mammals contains neurons that respond to both somatic and visual stimulation.

There are several studies in which lesions, that include area 5, were made in posterior parietal cortex in monkeys. For example, studies in which unilateral lesions included areas 5, MIP and 7b demonstrated that animals had deficits in coordinating arm velocity with hand velocity, that the postural relationship between the arm and wrist was disrupted, and that there were disruptions in coordinating the hand in shoulder-centered space (Rushworth *et al.*, 1997). However, these lesions did not affect the range or velocity of movements or the hand's trajectory. In a related investigation in which areas 5, 7b and MIP (which may have also included V6) were bilaterally ablated, the monkey had deficits in reaching to the same target under different starting positions (Rushworth *et al.*, 1998). Finally, a study in green monkeys in which bilateral ablations were made to area V6A (but may have incorporated portions of area 5) demonstrated that monkeys showed a reluctance to move, and had deficits in reaching, grasping and wrist orientation (Battaglini *et al.*, 2002). While previous work on the effects of posterior parietal lesions on behavior are difficult to interpret because the size of the lesion was so large (e.g., Ettlinger and Kalsbeck, 1962; Hartje and Ettlinger, 1973; Brown *et al.*, 1983), they do indicate that lesions to posterior parietal cortex result in non-visually guided reaching deficits but spare roughness discrimination abilities (Brown *et al.*, 1983; Murray and Mishkin, 1984).

There are a number of studies in humans on the deficits that occur with insults to posterior parietal cortex. These studies indicate that the most severe deficit, termed spatial hemi-neglect, is in coding spatial location of objects within a particular frame of reference (see Robertson and Rafal, 2000; Behrmann, 2000 for review). While the data suggest that there are several egocentric frames of reference, including that of the forelimb or shoulder, there are little data on areas of posterior parietal cortex that contribute to these frames of reference. Furthermore, in these studies, the lesions are extremely large and encompass a number of cortical areas, and it is difficult to interpret which area is associated with which aspect of the deficit.

Connections

There are only a few studies of connections of area 5 in the macaque monkey, and those that exist are limited in scope. For example, only one study used electrophysiological guidance to place injections in area 5, and in this study, only one injection in one animal was performed, and the injection spread into area 2 (Pons and Kaas, 1986). Furthermore, only local connections, or connections in neighboring fields, were examined. This previous investigation demonstrated connections with areas 1, 7b, S2, M1 and premotor cortex, which is a subset of the connections we describe for area 5 in titi monkeys (Padberg *et al.*, 2005). Early studies of connections of architectonically defined area 5 to local parietal cortical areas support these recent findings in macaques (e.g., Jones and Powell, 1969b; Jones *et al.*, 1978; Pandya and Seltzer, 1982). In titi monkeys, area 5 connections are widespread compared to anterior parietal fields, and some of the strongest connections of area 5 are with motor and premotor cortex, the supplementary motor area, S2, PV, extrastriate visual areas, 7b, and cingulate cortex (Figure 11.6; Padberg *et al.*, 2005). Studies of vestibular processing in human and nonhuman primates indicate that cortex at the juncture of areas 5 and 7b projects to vestibular brainstem nuclei, contains neurons that respond to optokinetic and vestibular stimulation, and is interconnected with other areas of the neocortex that process vestibular information (Akbarian *et al.*, 1988; Guldin *et al.*, 1992; Akbarian *et al.*, 1993, 1994; Brandt and Dieterich, 1999; Lobel *et al.*, 1999; see Guldin and Grüsser, 1998 for review).

Most studies that examine callosal connectivity have studied total patterns of connections of large regions of cortex (e.g., Karol and Pandya, 1971; Killackey *et al.*, 1983), or connections of several fields grouped together such as 3a, 3b, 1, 2 and 5 collectively (e.g., Jones and Powell, 1969a; Boyd *et al.*, 1971; Jones *et al.*, 1975; Jones *et al.*, 1979; Shanks *et al.*, 1985a). A consistent observation is that area 5 receives callosal inputs throughout the field (i.e., including the hand representation). There is one study in which the connections of cortex in the location of area 5 were examined, although the location of the injection site was not verified electrophysiologically (Caminiti and Sbriccoli, 1985). In this study in macaque monkeys, label in the hemisphere contralateral to that injected was observed throughout area 5, the supplementary motor area, 7b, and on the dorsal bank of the lateral sulcus (in the S2/PV region). Our recent studies in titi monkeys indicate that the hand representation of area 5 has dense callosal connections with the contralateral area 5 in the expected location of the hand representation (Padberg *et al.*, 2005). Area 5 in titi monkeys is also connected with areas 7b/AIP, S2/PV, motor, premotor and cingulate cortex, a finding similar to that of Caminiti and Sbriccoli (1985). Thus, area 5 is one of the few somatosensory cortical areas involved in integrating inputs between the hands and in the interhemispheric transfer of information necessary for limb and hand coordination.

This latter notion is supported by studies of the behavioral consequences of lesions to the posterior portions of the corpus callosum through which axons from posterior parietal cortex travel (Seltzer and Pandya, 1983). For example, after lesions to the posterior portion of the corpus callosum, monkeys have deficits in intermanual transfer of information about shape, roughness or size of an object (Manzoni *et al.*, 1984; Hunter *et al.*, 1976; Myers and Ebner, 1976). However, if animals were allowed to view the manual task to be transferred, thus supplying visual input regarding the task to both hemispheres, there was no noticeable deficit in information transfer across hemispheres (Kohn and Meyers, 1969). Studies of tactile and tactuomotor transfer in humans who have undergone complete section of the corpus callosum, or partial sections of the posterior portion of the corpus callosum indicate a number of abnormalities associated with transferring manual information regarding one hand to the opposite hemisphere. For example, complete section of the corpus callosum (Geffen *et al.*, 1985) or a section in the portion of the callosum through which axons connecting the posterior parietal cortex of each hemisphere travel (Geffen *et al.*, 1985; Risse *et al.*, 1989), result in an inability to perform cross-localization, intermanual tasks (Geffen *et al.*, 1985; Lassonde *et al.*, 1986; Risse *et al.*, 1989). Other studies demonstrate that with sections of the posterior corpus callosum, individuals cannot perform posture matching tasks (kinesthesis) in which they are required to match the position of one forelimb and hand with the opposite forelimb and hand in the absence of visual guidance (Risse *et al.*, 1989). Finally, humans with sections of the corpus callosum performed poorly on the transfer of information regarding object shape, and the magnitude of the effect was dependent on the difficulty of the task (Lassonde *et al.*, 1986). The studies described above in both human and nonhuman primates indicate that the connections between the posterior parietal cortex of each hemisphere, including area 5, transfer crucial information necessary for intermanual tactile learning and coordination of the hands.

Taken together, electrophysiological, connection and lesion studies in nonhuman primates indicate that area 5 is involved in generating an egocentric, shoulder-centered frame of reference necessary for object exploration with the hands. Related abilities such as kinesthetic–visual matching, which may require an understanding of both object permanence and body part objectification, are proposed to be generated by regions in posterior parietal cortex (possibly area 5) in humans as well (see Mitchell, 2005 for review). This field also is involved in coordinating both hands to accomplish visually guided and perhaps nonvisually guided tasks and for interhemispheric transfer of information between the hands. While area 5 alone may not be responsible for generating an internal representation of the self, the activity of neurons in area 5 is coincident with tasks that require the animal to have such an internal representation, and receptive fields for neurons in area 5 are modified in a manner consistent with alterations in the internal frame of reference.

IS AREA 5 HOMOLOGOUS ACROSS MAMMALS?

Throughout the text we intimated that cortex caudal to area 3b or S1 in nonprimate mammals is homologous to area 5. Although traditionally cortex immediately caudal to 3b is considered to be area 1 in nonprimate mammals such as cats, the predominant evidence for this traditional view is that this presumptive area is immediately caudal to area 3b. Furthermore, there is a tendency to make the cat somatosensory system both homologous and analogous to the primate somatosensory system, despite the fact that cats have a derived forepaw that is not used in a fashion analogous to that of the hand of primates. Indeed, Johnson (1990) states in his comparative analysis of mammal somatosensory cortex:

> Doubtless due to the great number of investigations rather than the specializations of the species, no less than five distinct somatic sensory areas have been identified and formally numbered in cat neocortex. (p. 399)

It should be noted that in a number of species in which cortex caudal to area 3b has been mapped, this cortex is rostrally adjacent to V2 (which would make it V3 if one were to assume homology based solely on relative location; Figure 16.9). As noted above, cortex immediately caudal to 3b, and immediately rostral to V1 or V2 has been explored using electrophysiological recording techniques in a variety of mammals and has been termed C, PP, PM or SC. Like area 5 in primates, there is an extreme magnification of particular body parts, such that a complete representation of the body surface is often not present (Figures 16.10 and 16.11). For example, in murine rodents, this cortex is dominated by the representation of the vibrissae and in squirrels by the representation of the forepaw. In marsupials such as the striped possum, with its specialized fourth digit, this cortex represents only D4. Also like area 5 and unlike areas 1 and 2, neurons in this cortex often respond to visual stimulation as well as to stimulation of deep receptors. Species such as the flying fox have a cortical field immediately caudal to 3b in which neurons respond to cutaneous and deep stimulation, termed area 1/2 (see above). Just caudal to this region, a cortical field in which neurons respond to the stimulation of deep receptors as well as to visual stimulation has been described and termed the posterior parietal cortex or PP (Krubitzer and Calford, 1992). While studies of connections of this caudal region of cortex in a variety of mammals would allow us to more accurately infer homology, we believe that most evidence supports the hypothesis that this field (termed C, SC, PM and PP) is more like area 5 in primates than like area 1 or area 2.

THE EVOLUTION OF ANTERIOR AND POSTERIOR PARIETAL CORTEX

All mammals examined have a primary somatosensory area, S1 or 3b, in which the organization clearly reflects specializations of peripheral morphology, innervation density of peripheral receptors, and use. Furthermore, there appears to be a basic pattern of interconnections with several cortical and subcortical areas, suggesting that there is a network that has been inherited from a common ancestor, but modified in different lineages with the evolution of peripheral morphology, as well as with the addition of new cortical fields and thalamic nuclei to the network (Figures 16.6 and 16.8). While one would like to speculate as to the function of S1, most single unit recording studies in awake, behaving animals and/or lesions of S1 have been done in primates, namely macaque monkeys. Thus, these data must be interpreted with caution when discussing other species since homologous fields need not be analogous. In primates, area 3b appears to be involved in texture and form discrimination, and such discriminations are made with the glabrous hands. Although the representation of the oral structures in area 3b of primates is greatly magnified, suggesting some behavioral specialization related to the mouth, most studies only examine response to stimulation of the hand, or the activity of neurons in area 3b during discriminations made with the hand. If S1 serves as a general-purpose processor of cutaneous inputs necessary for fine discriminations using a morphologically distinct and specialized structure for a particular animal, then one would expect different magnifications of different body parts for different animals. One would also expect that the greatest variability within a species would be for the specialized body part that is used extensively, such as the hand (e.g., Merzenich *et al.*, 1987) and lips (Cusick *et al.*, 1986) of primates, versus a less actively used body part such as the trunk, which appears to be the case.

Like area 3b, area 3a appears to be part of a common plan of cortical organization in all mammals (Figure 16.11). Area 3a is a proprioceptive area involved in integrating somatic and vestibular inputs (at least in primates) with the motor system to generate specialized behavior that allows the animal to maximally interface receptor-dense morphological structures (e.g., hand, nose and bill) with an external object or animal to be explored. Such a sensorimotor interface likely contributes to an enlarged representation of a specialized peripheral structure not only in area 3a, but also in all anterior parietal fields. While the body part in question may be different for different animals, the same rules of modification are implemented in all mammals.

The presence of an area 1 or a rudimentary area 1 in several New World primates and macaque monkeys, and what appears to be the absence of area 1 in marmosets, tamarins, prosimian galagos, as well as other mammals indicates that this cortical field arose later in primate evolution (Figure 16.11). We believe that the most parsimonious explanation for the observations made about anterior parietal cortex in a variety of primates is that in early primates, areas 3b and 3a were present. A rudimentary area 1 arose after the simian and prosimian divergence, and this was lost or greatly reduced in some lineages (e.g., Callitricidae), retained in a primitive form in some lineages (e.g., titi monkeys), or became well developed in some lineages (e.g., squirrel monkeys and macaque monkeys), possibly with the evolution of the hand and consequent tactile abilities associated with hand use. It should be noted that the two species of New World monkey (tamarins and marmosets) that do not possess an area 1 have a modified hand with claws used to a large extent for climbing, and to a lesser extent for tactile discrimination.

The functional organization of area 2 has only been investigated in one species of nonhuman primates, the macaque monkey (Pons *et al.*, 1985). While cortex immediately caudal to area 1 has been explored in New World monkeys and prosimian galagos, neurons in this region have been reported to be unresponsive to somatic stimulation, responsive only at the caudal border of area 1, or sometimes responsive to the stimulation of deep receptors, features generally associated with area 5 in macaque monkeys. Thus, if one relies on electrophysiological mapping data, it appears that New World monkeys do not possess an area 2 (e.g., Padberg *et al.*, 2005; Figure 16.11). It is tempting to postulate that area 2 arose or co-evolved with the emergence of an opposable thumb, and is related to the behaviors associated with using a variety of grips for tactile exploration and identification.

The presence of an area 5 in both New World and Old World monkeys, and a rudimentary form of area 5 (posterior parietal cortex) in most mammals studied suggests that this posterior parietal field arose early in evolution and has been retained in most or all mammals (Figure 16.11). While area 5 may be a homologous cortical area in all mammals, the addition of new areas, such as 1 and 2, and new connections likely promotes new functions of this cortical field in primates (Figures 16.6 and 16.8). For example, PM in squirrels and PP in flying foxes may be homologous to area 5 in primates, but not strictly analogous. Much like the magnification of behaviorally relevant body parts in area 3b, in area 5 these representations and associated functions are magnified to the extreme in particular lineages. An important point of these observations in primate and nonprimate mammals is that cortical fields are not added in a functional hierarchy in evolution, but rather are interspersed between existing fields. Indeed, we believe that the data indicate that areas 3a and 3b and 5 are evolutionarily old fields and that areas 1 and 2 are recent primate phenomenon, likely associated with sophisticated hand use.

With respect to the ideas put forward in the beginning of this chapter regarding the evolution of association cortex, we propose a modified scenario based on data in a variety of other mammals. In primates, unimodal somatosensory cortex has expanded with the addition of areas 1 and 2. Posterior parietal area 5 in primates, although homologous to area 5 in other mammals, has undergone a number of changes including a magnification of the hand and forelimb representation, the preponderance of neurons active under different reaching and grasping paradigms, and the broad distribution of ipsilateral and contralateral connections of the hand and limb representation with proprioceptive, limbic and motor cortex (Figure 16.6). All of these features are coincident with the evolution of the hand and opposable thumb in a number of primates, as well as with a larger repertoire of grips and manual and bimanual hand configurations (Napier, 1960, 1962; Welles, 1976). A reasonable hypothesis is that anterior parietal cortex in primates has expanded as a result of sophisticated hand use, which distinguishes primates, and humans in particular, from other mammals. An evolutionary old cortical area, area 5, has been modified due to the addition of these new anterior parietal fields, and the evolution of highly derived manual behavior. Thus, modification of homologous cortical fields (area 5), and the addition of new unimodal cortical fields (e.g., areas 1 and 2) devoted to hand use is one of the hallmarks of human brain evolution.

The evolution of sophisticated, visually guided, hand use and the addition of anterior parietal fields and the elaboration of posterior parietal cortical areas associated with this behavior may ultimately have led to the emergence of a more refined and species-specific internal representation of self, and an increased number of permutations of how this internal representation can interact with objects in extrapersonal space via the hands. It should be noted that this species-specific internal representation is not a static, enigmatic property that emerged in anthropoid primates alone, but is a dynamic sensorimotor loop that all mammals possess in a derived form based on their morphological distinctions and distribution of sensory receptors. Studies of connections as well as electrophysiological recording data indicate that the motor system is a critical component for distinguishing self from non-self, an attribute traditionally delegated solely to association cortex. It follows then that any discussion of an internal representation of self and how an individual distinguishes itself from non-self should incorporate the motor system in this more than human phenomenon.

GLOSSARY

Acallosal: a neural pathway that does not project to the contralateral hemisphere via the corpus callosum. An acallosal or callosal free zones refers to regions of the cortex that does not receive inputs via the corpus callosum. Such a region can be determined in studies of callosal connections by noting an absence of callosally projecting axon terminals in an otherwise terminal rich area.

Analogous: a structure that has the same function. An analogous structure need not be homologous.

Callosal: a neural pathway that projects to the contralateral hemisphere via the corpus callosum. A callosal zone refers to a region of the cortex that receives inputs via the corpus callosum.

Caudal: anatomical location – toward the tail.

Contralateral: a relative term referring to the opposite side of the brain or body.

Deep receptors: peripheral receptor in the skin, muscles or joints.

Extrastriate: visual cortex that does not include V1 (striate cortex).

Glabrous: regions of the skin lacking hairs, e.g., the pads of hands and feet.

Homologous: a structure that is inherited from a common ancestor. A homologous structure is not always analogous.

Interdigitate: to interweave two or more structures.

Lateral: anatomical location – away from the midline of the brain or body

Medial: anatomical location – toward the midline of the brain or body.

Myelin: a material composed of lipids and lipoproteins that surrounds certain axons and functions as an electrical insulator. Myelin stains are often used to demarcate cortical areas and thalamic nuclei.

Optokinetic: movement of the eyes when a moving visual stimulus is viewed.

Rapidly adapting receptor: a mechanoreceptor that responds to the initial presentation of a stimulus, but stops responding or reduces its firing rate throughout the presentation of a sustained stimulus.

Rostral: anatomical location – toward the rostrum or snout.

S1 or 3b: traditionally S1 in primates was considered to incorporate areas 3a, 3b, 1 and 2. While some investigators still treat these four separate areas as S1, it has been clearly demonstrated that only area 3b should be considered as S1. While 3b in nonprimate mammals is often referred to as S1, it should be noted that it is homologous only to area 3b in primates, not areas 3a, 3b, 1 and 2.

Slowly adapting receptor: a mechanoreceptor that responds throughout the presentation of a sustained stimulus.

Vibrissae: whiskers and/or stiff hair located on the face of an animal that often functions as a tactile organ.

Whisking: sweeping movement of the vibrissae during tactile exploration of the environment.

ACKNOWLEDGMENTS

This work was supported by an NINDS grant (R01-NS035103-11) and McDonnell Foundation grant to Leah Krubitzer.

This chapter is reproduced from J. Kass and E. Gardner (Eds.) (2008), *The Senses: A Comprehensive Reference*, vol. 6, *Somatosensation* (pp. 183–214). London: Elsevier. With permission from Elsevier.

REFERENCES

Ageranioti-Belanger, S. A. and Chapman, C. E. (1992). Discharge properties of neurones in the hand area of primary somatosensory cortex in monkeys in relation to the performance of an active tactile discrimination task: II. Area 2 as compared to areas 3b and 1. *Experimental Brain Research*, **91**, 207–228.

Akbarian, S., Grusser, O. J. and Guldin, W. O. (1992). Thalamic connections of the vestibular cortical fields in the squirrel monkeys (*Saimiri sciureus*). *Journal of Comparative Neurology*, **326**, 423–441.

Akbarian, S., Grusser, O. J. and Guldin, W. O. (1993). Corticofugal projections to the vestibular nuclei in squirrel monkeys: further evidence of multiple cortical vestibular fields. *Journal of Comparative Neurology*, **332**, 89–104.

Akbarian, S., Grusser, O.-J. and Guldin, W. O. (1994). Corticofugal connections between the cerebral cortex and brainstem vestibular nuclei in the macaque monkey. *Journal of Comparative Neurology*, **339**, 421–437.

Akbarian, S., Berndl, K., Grusser, O.-J., Guldin, W., Pause, M. and Schreiter, U. (1988). Responses of single neurons in the parietoinsular vestibular cortex of primates. *Annals of the New York Academy of Sciences*, **545**, 187–202.

Akers, R. M. and Killackey, H. P. (1978). Organization of corticocortical connections in the parietal cortex of the rat. *Journal of Comparative Neurology*, **181**, 513–538.

Allman, J. M. and Kaas, J. H. (1971). A representation of the visual field in the caudal third of the middle temporal gyrus of the owl monkey (*Aotus trivirgatus*). *Brain Research*, **31**, 85–105.

Allman, J. M. and Kaas, J. H. (1974). A crescent-shaped cortical visual area surrounding the middle temporal area (MT) in the owl monkey (*Aotus trivirgatus*). *Brain Research*, **81**, 199–213.

Allman, J. M. and Kaas, J. H. (1975). The dorsomedial cortical visual area: a third tier area in the occipital lobe of the owl monkey (*Aotus trivirgatus*). *Brain Research*, **100**, 473–487.

Allman, J. H. and Kaas, J. H. (1976). Representation of the visual field on the medial wall of the occipital-parietal cortex in the owl monkey. *Science*, **191**, 572–576.

Alloway, K. D. and Burton, H. (1985). Homotypical ipsilateral cortical projections between somatosensory areas I and II in the cat. *Neuroscience*, **14**, 15–35.

Andersen, R. A. and Buneo, C. A. (2002). Intentional maps in posterior parietal cortex. *Annual Review of Neuroscience*, **25**, 189–220.

Andersen, R. A., Snyder, L. H., Bradley, D. C. and Xing, J. (1997). Multimodal representation of space in the posterior parietal cortex and its use in planning movements. *Annual Review of Neuroscience*, **20**, 303–330.

Avendano, C., Isla, A. J. and Rausell, E. (1992). Area 3a in the cat: II. Projections to the motor cortex and their relations to corticocortical connections. *Journal of Comparative Neurology*, **321**, 373–386.

Barbaresi, P., Fabri, M., Conti, F. and Manzoni, T. (1987). D-[^{3}H]Aspartate retrograde labelling of callosal and association neurones of somatosensory areas I and II of cats. *Journal of Comparative Neurology*, **263**, 159–178.

Battaglini, P. P., Muzur, A., Galletti, C., Skrap, M., Brovelli, A. and Fattori, P. (2002). Effects of lesions in area V65 in monkeys. *Experimental Brain Research*, **144**, 419–422.

Beck, P. D., Pospichal, M. W. and Kaas, J. H. (1996). Topography, architecture, and connections of somatosensory cortex in opossums: evidence for five somatosensory areas. *Journal of Comparative Neurology*, **366**, 109–133.

Behrmann, M. (2000). Spatial reference frames and hemispatial neglect. In M. Gazzaniga (Ed.), *The new cognitive neurosciences* (pp. 651–666). Cambridge, MA: MIT Press.

Blankenburg, F., Ruben, J., Meyer, R., Schwiemann, J. and Villringer, A. (2003). Evidence for a rostral-to-caudal somatotopic organization in human primary somatosensory cortex with mirror-reversal in areas 3b and 1. *Cerebral Cortex*, **13**, 987–993.

Boyd, E. H., Pandya, D. N. and Bignall, K. E. (1971). Homotopic and nonhomotopic interhemispheric cortical projections in the squirrel monkey. *Experimental Neurology*, **32**, 256–274.

Brandt, T. and Dieterich, M. (1999). The vestibular cortex, its locations, functions, and disorders. *Annals of the New York Academy of Sciences*, **871**, 293–312.

Brodmann, K. (1909). *Vergleichende Lokalisationsiehre der Grosshirnrinde in ihren Prinzipien Dargestellt auf Grund des Zellenbaues*. Leipzig: Barth.

Brown, J. V., Ettlinger, G. and Garcha, H. S. (1983). Visually guided reaching and tactile discrimination in the monkey: the effects of removals of parietal cortex soon after birth. *Brain Research*, **267**, 67–79.

Burbaud, P., Doegle, C., Gross, C. and Bioulac, B. (1991). A quantitative study of neuronal discharge in areas 5, 2 and 4 of the monkey during fast arm movements. *Journal of Neurophysiology*, **66**, 429–443.

Burton, H. and Fabri, M. (1995). Ipsilateral intracortical connections of physiologically defined cutaneous representations in areas 3b and 1 of macaque monkeys: projections in the vicinity of the central sulcus. *Journal of Comparative Neurology*, **355**, 508–538.

Burton, H., Fabri, M. and Alloway, K. (1995). Cortical areas within the lateral sulcus connected to cutaneous representations in areas 3b and 1: a revised interpretation of the second somatosensory area in macaque monkeys. *Journal of Comparative Neurology*, **355**, 539–562.

Calford, M. B., Graydon, M. L., Huerta, M. F., Kaas, J. H. and Pettigrew, J. D. (1985). A variant of the mammalian somatotopic map in the bat. *Nature*, **313**, 477–479.

Caminiti, R. and Sbriccoli, A. (1985). The collosal system of the superior parietal lobule in the monkey. *Journal of Comparative Neurology*, **2237**, 85–99.

Caminiti, R., Innocenti, G. M. and Manzoni, T. (1979). The anatomical substrate of callosal messages from SI and SII in the cat. *Brain Research*, **35**, 295–314.

Carlson, M. (1981). Characteristics of sensory deficits following lesions of Brodmann's areas 1 and 2 in the postcentral gyrus of *Macaca mulatta*. *Brain Research*, **204**, 424–430.

Carlson, M., Huerta, M. F., Cusick, C. G. and Kaas, J. H. (1986). Studies on the evolution of multiple somatosensory representations in primates: the organization of anterior parietal cortex in the New World Callitrichid, *Saguinus*. *Journal of Comparative Neurology*, **246**, 409–426.

Carlson, N. R. (1998). *Physiology of behavior* (6th edn). Needham Heights, MA: Allyn & Bacon.

Catania, K. C. and Kaas, J. H. (1997). Somatosensory fovea in the star-nosed mole: behavioral use of the star in relation to innervation patterns and cortical representation. *Journal of Comparative Neurology*, **387**, 215–233.

Catania, K. C., Northcutt, R. G., Kaas, J. H. and Beck, P. D. (1993). Nose stars and brain stripes. *Nature*, **364**, 493.

Chapin, J. K. and Lin, C.-S. (1984). Mapping the body representation in the SI cortex of anesthetized and awake rats. *Journal of Comparative Neurology*, **229**, 199–213.

Chapin, J. K., Sadeq, M. and Guise, J. L. U. (1987). Corticocortical connections within the primary somatosensory cortex of the rat. *Journal of Comparative Neurology*, **263**, 326–346.

Chapman, C. E. and Ageranioti-Belanger, S. (1991). Discharge properties of neurones in the hand area of primary somatosensory cortex in monkeys in relation to the performance of an active tactile discrimination task: I. Areas 3b and 1. *Experimental Brain Research*, **87**, 319–339.

Chen, L., Friedman, R., Ramsden, B., LaMotte, R. and Roe, A. (2001). Fine-scale organization of S1 (area 3b) in the squirrel monkey revealed with intrinsic optical imaging. *Journal of Neurophysiology*, **86**, 3011–3029.

Conti, F., Fabri, M. and Manzoni, T. (1986). Bilateral receptive fields and callosal connectivity of the body midline representation in the first somatosensory area of primates. *Somatosensory Research*, **3**, 273–289.

Coq, J. O., Qi, H., Collins, C. E. and Kaas, J. H. (2004). Anatomical and functional organization of somatosensory areas of the lateral fissure of the new world titi monkey *(Callicebus moloch)*. *Journal of Comparative Neurology*, **476**, 363–387.

Culham, J. C. and Kanwisher, N. G. (2001). Neuroimaging of cognitive functions in human parietal cortex. *Current Opinion in Neurobiology*, **11**, 157–163.

Cusick, C. G. and Gould, H. J. I. (1990). Connections between area 3b of the somatosenosry cortex and subdivisions of the ventroposterior nuclear complex and the anterior pulvinar nucleus in squirrel monkeys. *Journal of Comparative Neurology*, **292**, 83–102.

Cusick, C. G., MacAvoy, M. G. and Kaas, J. H. (1985). Interhemispheric connections of cortical sensory areas in tree shrews. *Journal of Comparative Neurology*, **235**, 111–128.

Cusick, C. G., Wall, J. and Kaas, J. (1986). Representations of the face, teeth and oral cavity in areas 3b and 1 of somatosensory cortex in squirrel monkeys. *Brain Research*, **370**, 359–364.

Darian-Smith, C., Darian-Smith, I. and Cheema, S. S. (1990). Thalamic projections to sensorimotor cortex in the macaque monkey: use of multiple retrograde fluorescent tracers. *Journal of Comparative Neurology*, **299**, 17–46.

Darian-Smith, C., Darian-Smith, I., Burman, K. and Ratcliffe, N. (1993). Ipsilateral cortical projections to areas 3a, 3b, and 4 in the macaque monkey. *Journal of Comparative Neurology*, **335**, 200–213.

Debowy, D. J., Ghosh, S., Ro, J. Y. and Gardner, E. P. (2001). Comparison on neuronal firing rates in somatosensory and posterior parietal cortex during prehension. *Experimental Brain Research*, **137**, 269–291.

Diamond, M., Harris, J. and Petersen, R. (2002). Sensory learning and the brain's body map. In R. Nelson (Ed.), *The somatosensory system: deciphering the brain's own body image* (pp. 183–195). Boca Raton: CRC Press.

Dicarlo, J., Johnson, J. K. and Hsiao, S. (1998). Structure of receptive fields in area 3b of primary somatosensory cortex in the alert monkey. *The Journal of Neuroscience*, **18**, 2626–2645.

Disbrow, E., Litinas, E., Recanzone, G., Padberg, J. P. and Krubitzer, L. A. (2003). Cortical connections of the parietal central area and the second somatosensory area in macaque monkeys. *Journal of Comparative Neurology*, **462**, 382–399.

Disbrow, E., Litinas, E., Recanzone, G., Slutsky, D. and Krubitzer, L. A. (2002). Thalamocortical connections of the parietal ventral area (PV) and the second somatosensory area (S2) in macaque monkeys. *Thalamus and Related Systems*, **1**, 289–302.

Disbrow, E. A., Murray, S. O., Roberts, T. P., Litinas, E. D. and Krubitzer, L. A. (2001). Sensory integration in human posterior parietal area 5. *Society for Neuroscience Abstract*, **27**, 511–526.

Ebner, F. F. and Myers, R. E. (1965). Distribution of corpus callosum and anterior commissure in cat and raccoon. *Journal of Comparative Neurology*, **124**, 353–356.

Eisenberg, J. H. (1981). *The mammalian radiations: an analysis of trends in evolution, adaptation and behavior*. Chicago: University of Chicago Press.

Elston, G. N. and Manger, P. R. (1999). The organization and connections of somatosensory cortex in the brush-tailed possum *(Trichosurus vulpecula)*: evidence for multiple, topographically organized and interconnected representations in an Australian marsupial. *Somatosensory and Motor Research*, **16**, 312–337.

Ettlinger, G. and Kalsbeck, J. E. (1962). Changes in tactile discrimination in visual reaching after successive and simultaneous bilateral posterior parietal ablations in the monkey. *Journal of Neurology, Neurosurgery and Psychiatry*, **25**, 256–268.

Fabri, M. and Burton, H. (1991). Topography of connections between primary somatosensory cortex and posterior complex in rat: a multiple fluorescent tracer study. *Brain Research*, **538**, 351–357.

Feldman, S. H. and Johnson, J. I. (1988). Kinesthetic cortical area anterior to primary somatic sensory cortex in the raccoon *(Procyon lotor)*. *Journal of Comparative Neurology*, **277**, 80–95.

Felleman, D. J. and Van Essen, D. C. (1991). Distributed hierarchical processing in primate cerebral cortex. *Cerebral Cortex*, **1**, 1–47.

Felleman, D. J., Wall, J. T., Cusick, C. G. and Kaas, J. H. (1983). The representation of the body surface in S-I of cats. *The Journal of Neuroscience*, **3**, 1648–1669.

Ferraina, S. and Bianchi, L. (1994). Posterior parietal cortex: functional properties of neurons in area 5 during an instructed-delay reaching task within different parts of space. *Experimental Brain Research*, **99**, 175–178.

Fox, P. T., Burton, H. and Raichle, M. E. (1987). Mapping human somatosensory cortex with positron emission tomography. *Journal of Neurosurgery*, **67**, 34–43.

Friedman, D. and Jones, E. (1981). Thalamic input to areas 3a and 2 in monkeys. *Journal of Neurophysiology*, **45**, 59–85.

Frost, S. B., Milliken, G. W., Plautz, E. J., Masterton, R. B. and Nudo, R. J. (2000). Somatosensory and motor representations in cerebral cortex of a primitive mammal (*Monodelphis domestica*): a window into the early evolution of sensorimotor cortex. *Journal of Comparative Neurology*, **421**, 29–51.

Gardner, E. (1988). Somatosensory cortical mechanisms of feature detection in tactile and kinesthetic discrimination. *Candian Journal of Physiology and Pharmacology*, **66**, 439–454.

Garraghty, P. E., Pons, T. P., Huerta, M. F. and Kaas, J. H. (1987). Somatotopic organization of the third somatosensory area (SIII) in cats. *Somatosensory Research*, **4**, 333–357.

Geffen, G., Nilsson, J. and Quinn, K. (1985). The effects of lesions of the corpus callosum on finger localization. *Neuropsychologia*, **23**, 497–514.

Gould, H. J. I. and Kaas, J. H. (1981). The distribution of commissural terminations in somatosensory areas I and II of the grey squirrel. *Journal of Comparative Neurology*, **196**, 489–504.

Gould, H. J., III, Whitworth, R. H., Jr. and LeDoux, M. S. (1989). Thalamic and extrathalamic connections of the dysgranular unresponsive zone in the grey squirrel (*Sciurus carolinensis*). *Journal of Comparative Neurology*, **287**, 38–63.

Graziano, M. S. A., Cooke, D. F. and Taylor, C. S. R. (2000). Coding the location of the arm by sight. *Science*, **290**, 1782–1786.

Guldin, W. and Grusser, O.-J. (1998). Is there a vestibular cortex? *Trends in Neuroscience*, **21**, 254–259.

Guldin, W. O., Akbarian, S. and Grusser, O. J. (1992). Cortico-cortical connections and cytoarchitectonics of the primate vestibular cortex: a study in squirrel monkeys (*Saimiri sciureus*). *Journal of Comparative Neurology*, **326**, 375–401.

Hackett, T. A., Stepniewska, I. and Kaas, J. H. (1998). Subdivisions of auditory cortex and ipsilateral cortical connections of the parabelt auditory cortex in macaque monkeys. *Journal of Comparative Neurology*, **394**, 475–495.

Hartje, W. and Ettlinger, G. (1973). Reaching in light and dark after unilateral posterior parietal ablations in the monkey. *Cortex*, **9**, 346–354.

Hassler, R. and Muhs-Clement, K. (1964). Architectonic construction of the sensomotor and parietal cortex in the cat. *Journal für Hirnforschung*, **20**, 377–420.

Heath, C. J., Hore, J. and Philips, C. G. (1975). Inputs from low threshold muscle and cutaneous affernts of hand and forearm to areas 3a and 3b of baboon's cerebral cortex. *Journal of Physiology (London)*, **257**, 199–227.

Herron, P. and Johnson, J. I. (1987). Oragnization of intracortical and commissural connections in somatosensory cortical areas I and II in the raccoon. *Journal of Comparative Neurology*, **257**, 359–371.

Hlushchuk, Y., Forss, N. and Hari, R. (2003). Distal-to-proximal representation of volar index finger in human area 3b. *NeuroImage*, **21**, 696–700.

Hore, J., Preston, J. B. and Cheney, P. D. (1976). Responses of cortical neurons (areas 3a and 4) to ramp stretch of hindlimb muscles in the baboon. *Journal of Neurophysiology*, **39**, 484–500.

Huffman, K. J. and Krubitzer, L. (2001a). Area 3a: topographic organization and cortical connections in marmoset monkeys. *Cerebral Cortex*, **11**, 849–867.

Huffman, K. J. and Krubitzer, L. A. (2001b). Thalamo-cortical connections of areas 3a and M1 in marmoset monkeys. *Journal of Comparative Neurology*, **435**, 291–310.

Huffman, K., Nelson, J., Clarey, J. and Krubitzer, L. (1999). The organization of somatosensory cortex in three species of marsupials, *Dasyurus hallucatus, Dactylopsila trivirgata*, and *Monodelphis domestica*: neural correlates of morphological specializations. *Journal of Comparative Neurology*, **403**, 5–32.

Hunt, D. L., Slutsky, D. A. and Krubitzer, L. A. (2000). The organization of somatosensory cortex in the ferret. *Society for Neuroscience Abstract*, **26**, 243.13.

Hunter, M., Maccabe, J. J. and Ettlinger, G. (1976). Intermanual transfer of tactile training in the monkey: the effects of bilateral parieto-prestriate ablations. *Neuropsychologia*, **14**, 385–389.

Hyvärinen, J. and Poranen, A. (1978). Receptive field integration and submodality convergence in the hand area of the post-central gyrus of the alert monkey. *Journal of Physiology*, **283**, 539–556.

Imig, T. J., Ruggero, M. A., Kitzes, L. M., Javel, E. and Brugge, J. F. (1977). Organization of auditory cortex in the owl monkey (*Aotus trivirgatus*). *Journal of Comparative Neurology*, **171**, 111–128.

Iriki, A., Tanaka, M. and Iwamura, Y. (1996). Coding of modified body schema during tool use by macaque postcentral neurons. *NeuroReport*, **7**, 2325–2330.

Iriki, A., Tanaka, M., Obayashi, S. and Iwamura, Y. (2001). Self-images in the video monitor coded by monkey intraparietal neurons. *Journal of Neuroscience Research*, **40**, 163–173.

Iwamura, Y. (2000). Bilateral receptive field neurons and callosal connections in the somatosensory cortex. *Philosophical Transactions of the Royal Society of London Series B: Biological Sciences*, **355**, 267–273.

Iwamura, Y., Iriki, A. and Tanaka, M. (1994). Bilateral hand representation in the postcentral somatosensory cortex. *Nature*, **369**, 554–556.

Iwamura, Y., Tanaka, M., Iriki, A., Taoka, M. and Toda, T. (2002). Processing of tactile and kinesthetic signals from bilateral sides of the body in the postcentral gyrus of awake monkeys. *Behavioral Brain Research*, **135**, 185–190.

Jain, N., Qi, H.-X., Catania, K. and Kaas, J. (2001). Anatomic correlates of the face and oral cavity representations in the somatosensory cortical area 3b of monkeys. *Journal of Comparative Neurology*, **429**, 455–468.

Jiang, W., Tremblay, F. and Chapman, C. E. (1997). Neuronal encoding of texture changes in the primary and the secondary somatosensory cortical areas of monkeys during passive texture discrimination. *Journal of Neurophysiology*, **77**, 1656–1662.

Johnson, J. I. (1990). Comparative development of somatic sensory cortex. In E. G. Jones and A. Peters (Eds.), *Cerebral cortex* (pp. 335–449). New York: Plenum.

Johnson, J. I., Ostapoff, E.-M. and Warach, S. (1982). The anterior border zones of primary somatic sensory (SI) neocortex and their relation to cerebral convolutions, shown by micromapping of peripheral projections to the region of the fourth forepaw digit representation in raccoons. *Neuroscience*, **7**, 915–936.

Johnson, K. and Yoshioka, T. (2002). Neural mechanisms of tactile form and texture perception. In R. Nelson (Ed.), *The somatosensory system: deciphering the brain's own body image* (pp. 73–101). Boca Raton: CRC Press.

Johnson, K. O. and Lamb, G. D. (1981). Neural mechanisms of spatial tactile discrimination: neural patterns evoked by Braille-like dot patterns in the monkey. *The Journal of Physiology*, **310**, 117–144.

Jones, E. G. and Powell, T. P. S. (1969a). Connexions of the somatic sensory cortex of the rhesus monkey II contralateral cortical connexions. *Brain*, **92**, 717–730.

Jones, E. G. and Powell, T. P. S. (1969b). Connexions of the somatic sensory cortex of the rhesus monkey: I. Ipsilateral cortical connections. *Brain*, **92**, 477–502.

Jones, E. G., Burton, H. and Porter, R. (1975). Commissural and cortico-cortical "columns" in the somatic sensory cortex of primates. *Science*, **190**, 572–574.

Jones, E. G., Coulter, J. D. and Hendry, S. H. C. (1978). Intracortical connectivity of architectonic fields in the somatic sensory, motor and parietal cortex of monkeys. *Journal of Comparative Neurology*, **181**, 291–348.

Jones, E. G., Wise, S. P. and Coulter, J. C. (1979). Differential thalamic relationships of sensory-motor and parietal cortical fields in monkeys. *Journal of Comparative Neurology*, **183**, 833–882.

Juliano, S., Friedman, D. and Eslin, D. (1990). Corticocortical connections predict patches of stimulus-evoked metabolic activity in monkey somatosenosry cortex. *Journal of Comparative Neurology*, **298**, 23–39.

Kaas, J. H. (1982). The segregation of function in the nervous system: why do the sensory systems have so many subdivisions? *Contributions in Sensory Physiology*, **7**, 201–240.

Kaas, J. H. (1983). What, if anything, is SI? Organization of first somatosensory area of cortex. *Physiological Reviews*, **63**, 206–230.

Kaas, J. H. (1997). Topographic maps are fundamental to sensory processing. *Brain Research Bulletin*, **44** (2), 107–112.

Kaas, J. H. and Collins, C. E. (2004). The resurrection of multimodal cortex in primates: connection patterns that integrate modalities. In G. Calvert, C. Spence and B. E. Stein (Eds.), *Handbook of multisensory processing* (pp. 285–294). Cambridge: MIT Press.

Kaas, J. H. and Pons, T. P. (1988). The somatosensory system of primates. *Comparative Primate Biology*, **4**, 421–468.

Kalaska, J. F. (1996). Parietal cortex area 5 and visuomotor behavior. *Canadian Journal of Physiology and Pharmacology*, **74**, 483–498.

Karol, E. A. and Pandya, D. N. (1971). The distribution of the corpus callosum in the rhesus monkey. *Brain*, **94**, 471–786.

Kenshalo, D., Iwata, K., Sholas, M. and Thomas, D. (2000). Response properties and organization of nociceptive neurons in area 1 of monkey primary somatosensory cortex. *Journal of Neurophysiology*, **84**, 719–729.

Killackey, H. P., Gould, H. J. I., Cusick, C. G., Pons, T. P. and Kaas, J. H. (1983). The relation of corpus callosum connections to architectonic fields and body surface maps in sensory motor cortex of new and old world monkeys. *Journal of Comparative Neurology*, **219**, 384–419.

Knight, R. T. and Grabowecky, M. (2000). Prefrontal cortex, time and consciousness. In M. Gazzaniga (Ed.), *The new cognitive neurosciences* (pp. 1319–1330). Cambridge, MA: MIT Press.

Kohn, B. and Myers, R. E. (1969). Visual information and intermanual transfer of latch box problem solving in monkeys with commissures sectioned. *Experimental Neurology*, **23**, 303–309.

Kolb, B., Buhrmann, K., McDonald, R. and Sutherland, R. J. (1994). Dissociation of the medial prefrontal, posterior parietal, and posterior temporal cortex for spatial navigation and recognition memory in the rat. *Cerebral Cortex*, **4**, 664–680.

Koralek, K. A., Olavarria, J. and Killackey, H. P. (1990). Areal and laminar organization of corticocortical projections in rat somatosensory cortex. *The Journal for Comparative Neurology*, **299**, 133–150.

Krause, R., Kurth, R., Ruben, J., Schwiemann, J., Villringer, K., Deuchert, M., Moosmann, M., Brandt, S., Wolf, K., Curio, G. and Villringer, A. (2001). Representational overlap of adjacent fingers in multiple areas of human primary somatosensory cortex depends on electrical stimulus intensity: an fMRI study. *Brain Research*, **899**, 36–46.

Krubitzer, L. (1995). The organization of neocortex in mammals: are species differences really so different? *Trends in Neuroscience*, **18**, 408–417.

Krubitzer, L. (1998). What can monotremes tell us about brain evolution? *Philosophical Transactions of the Royal Society of London Series B: Biological Sciences*, **353**, 1127–1146.

Krubitzer, L. and Calford, M. B. (1992). Five topographically organized fields in the somatosensory cortex of the flying fox: microelectrode maps, myeloarchitecture, and cortical modules. *Journal of Comparative Neurology*, **317**, 1–30.

Krubitzer, L. and Kaas, J. (1987). Thalamic connections of three representations of the body surface in somatosensory cortex of grey squirrels. *Journal of Comparative Neurology*, **265**, 549–580.

Krubitzer, L. and Kaas, J. (1990). The organization and connections of somatosensory cortex in marmosets. *The Journal of Neuroscience*, **10**, 952–974.

Krubitzer, L. and Kaas, J. (1992). The somatosensory thalamus of monkeys: cortical connections and a redefinition of nuclei in marmosets. *Journal of Comparative Neurology*, **319**(1), 123–140.

Krubitzer, L. and Kahn, D. (2003). Nature vs. nurture: an old idea with a new twist. *Progressive Neurobiology*, **70**, 33–52.

Krubitzer, L., Künzle, H. and Kaas, J. (1997). Organization of sensory cortex in a madagascan insectivore, the tenrec (*Echinops telfairi*). *Journal of Comparative Neurology*, **379**, 399–414.

Krubitzer, L., Sesma, M. A. and Kaas, J. (1986). Microelectrode maps, myeloarchitecture, and cortical connections of three somatotopically organized representations of the body surface in the parietal cortex of squirrels. *Journal of Comparative Neurology*, **250**, 403–430.

Krubitzer, L., Clarey, J., Tweedale, R. and Calford, M. (1998). Interhemispheric connections of somatosensory cortex in the flying fox. *Journal of Comparative Neurology*, **402** (4), 538–539.

Krubitzer, L., Huffman, K. J., Disbrow, E. and Recanzone, G. (2004). Organization of area 3a in macaque monkeys: contributions to the cortical phenotype. *Journal of Comparative Neurology*, **471**, 97–111.

Krubitzer, L., Manger, P., Pettigrew, J. and Calford, M. (1995b). The organization of somatosensory cortex in monotremes: in search of the prototypical plan. *Journal of Comparative Neurology*, **351**, 261–306.

Krubitzer, L., Clarey, J., Tweedale, R., Elston, G. and Calford, M. (1995a). A redefinition of somatosensory areas in the lateral sulcus of macaque monkeys. *Journal of Neuroscience*, **15**, 3821–3839.

Lacquaniti, F., Guigon, E., Bianchi, L., Ferraina, S. and Caminiti, R. (1995). Representing spatial information for limb movement: the role of area 5 in monkey. *Cerebral Cortex*, **5**, 391–409.

LaMotte, R. H. and Acuna, C. (1978). Defects in accuracy of reaching after removal of posterior parietal cortex in monkeys. *Brain Research*, **139**, 309–326.

LaMotte, R. H. and Mountcastle, V. B. (1979). Disorders in somesthesis following lesions in parietal lobe. *Journal of Neurophysiology*, **42**, 400–419.

Landgren, S. and Silfvenius, H. (1969). Projection to cerebral cortex of group I muscle afferents from the cat's hind limb. *Journal of Physiology*, **200**, 353–372.

Lang, W., Büttner-Ennever, J. A. and Büttner, U. (1979). Vestibular projections to the monkey thalamus: an autoradiographic study. *Brain Research*, **177**, 3–17.

Lassonde, M., Sauerwein, H., Geoffroy, G. and DèCarie, M. (1986). Effects of early and late transection of the corpus callosum in children. *Brain*, **109**, 953–967.

Lebedev, M. and Nelson, R. (1996). High-frequency vibratory sensitive neurons in monkey primary somatosensory cortex: entrained and non-entrained responses to vibration during the performance of vibratory-cued hand movements. *Experimental Brain Research*, **111**, 313–325.

Leclerc, S. S., Rice, F. L., Dykes, R. W., Pourmoghadam, K. and Gomez, C. M. (1993). Electrophysiological examination of the representation of the face in the suprasylvan gyrus of the ferret: a correlative study with cytoarchitecture. *Somatosensory and Motor Research*, **10**, 133–159.

Ledoux, M. S., Whitworth, R. H. and Gould, H. J. I. (1987). Interhemispheric connections of the somatosensory cortex in the rabbit. *Journal of Comparative Neurology*, **258**, 145–157.

Lee, K. and Woolsey, T. (1975). A proportional relationship between peripheral innervation density and cortical neuron number in the somatosensory system of the mouse. *Brain Research*, **99**, 349–353.

Lin, L.-D., Murray, G. and Sessle, B. (1994a). Functional properties of single neurons in the primate face primary somatosensory cortex: I. Relations with trained orofacial motor behaviors. *Journal of Neurophysiology*, **71**, 2377–2390.

Lin, L.-D., Murray, G. and Sessle, B. (1994b). Functional properties of single neurons in the primate face primary somatosensory cortex: II. Relations with different directions of trained tongue protrusion. *Journal of Neurophysiology*, **71**, 2391–2400.

Lobel, E., Kleine, J., Leroy-Willig, A., Van de Moortele, P.-F., Le Bihan, D., Grusser, O.-J. and Berthoz, A. (1999). Cortical areas activated by bilateral galvanic vestibular stimulation. *Annals of the New York Academy of Sciences*, **871**, 313–323.

Manger, P. and Pettigrew, J. D. (1995). Electroreception and feeding behaviour of the platypus. *Philosophical Transactions of the Royal Society of London Series B: Biological Sciences*, **347**, 359–381.

Manger, P., Sum, M., Szymanski, M., Ridyway, S. and Krubitzer, L. (1998). Modular subdivisions of dolphin anterior ensular cortex: does evolutionary history repeat itself? *Journal of Cognitive Neuroscience*, **10** (2), 153–166.

Manzoni, T. (1997). The callosal connections of the hierarchically organized somatosensory areas of primates. *Journal of Neurosurgical Sciences*, **41**, 1–13.

Manzoni, T., Barbaresi, P. and Conti, F. (1984). Callosal mechanism for the hemispheric transfer of hand somatosensory information in the monkey. *Behavioural and Brain Research*, **11**, 155–170.

Manzoni, T., Conti, F. and Fabri, M. (1986). Callosal projections from area SII to SI in monkeys: anatomical organization and comparison with association projections. *Journal of Comparative Neurology*, **252**, 245–263.

Mayner, L. and Kaas, J. (1986). Thalamic projections from electrophysiologically defined sites of body surface representations in areas 3b and 1 of somatosensory cortex of cebus monkeys. *Somatosensory Research*, **4**, 13–29.

McKenna, T. M., Whitsel, B. L., Dreyer, D. A. and Metz, C. B. (1981). Organization of cat anterior parietal cortex: relations among cytoarchitecture, single neuron functional properties, and interhemispheric connectivity. *Journal of Neurophysiology*, **45**, 667–697.

Merzenich, M. M., Kaas, J. H., Sur, M. and Lin, C.-S. (1978). Double representation of the body surface within cytoarchitectonic areas 3b and 1 in "SI" in the owl monkey *(Aotus trivirgatus)*. *Journal of Comparative Neurology*, **181**, 41–74.

Merzenich, M. M., Nelson, R. J., Kaas, J. H., Stryker, M. P., Jenkins, W. M., Zook, J. M., Cynader, M. S. and Schoppmann, A. (1987). Variability in hand surface representations in area 3b and 1 in adult owl and squirrel monkeys. *Journal of Comparative Neurology*, **258**, 281–296.

Mishkin, M., Ungerleider, L. G. and Macko, K. A. (1983). Object vision and spatial vision: two cortical pathways. *Trends in Neuroscience*, **6**, 414–417.

Mitchell, R. W. (2005). Mirrors and matchings: imitation from the perspective of mirror-self recognition, and why the parietal region is involved in both. In

K. Dautenhahn and C. L. Nehaniv (Eds.), *Imitation and social learning in robots, humans and animals* (pp. 103–130). Cambridge: Cambridge University Press.

Moore, C., Stern, C., Corkin, S., Fischl, B., Gray, A., Rosen, B. and Dale, A. (2000). Segregation of somatosensory activation in the human rolandic cortex using fMRI. *Journal of Neurophysiology*, **84**, 558–569.

Mountcastle, V. B., Lynch, J. C., Georgopoulos, A., Sakata, H. and Acuña, C. (1975). Posterior parietal association cortex of the monkey: command functions for operations within extrapersonal space. *Journal of Neurophysiology*, **38**, 871–908.

Murphy, W. J., Eizirik, E., O'Brein, S. J., Madsen, O., Scally, M., Douady, C. J., Telling, E., Ryder, O. A., Stanhope, M. J., de Jong, W. W. and Springer, M. S. (2001). Resolution of the early placental mammal radiation using Bayesian phylogenetics. *Science*, **294**, 2348–2351.

Murray, E. A. and Mishkin, M. (1984). Relative contributions of SII and area 5 to tactile discrimination in monkeys. *Behavioral Brain Research*, **11**, 67–83.

Myers, R. E. and Ebner, F. F. (1976). Localization of function in corpus callosum: tactual information transmission in *Macaca mulatta*. *Brain Research*, **103**, 455–462.

Napier, J. (1962). The evolution of the hand. *Scientific American*, **207**, 56–61.

Napier, J. R. (1960). Studies of the hands of living primates. *Proceedings from the Zoological Society of London*, **134**, 647–657.

Nelson, R. (1987). Activity of monkey primary somatosensory cortical neurons changes prior to active movement. *Brain Research*, **406**, 402–407.

Nelson, R. J. and Kaas, J. H. (1981). Connections of the ventroposterior nucleus of the thalamus with the body surface representations in cortical areas 3b and 1 of the cynomologus macaque, *Macaca fascicularis*. *Journal of Comparative Neurology*, **199**, 29–64.

Nelson, R., Smith, B. and Douglas, V. (1991). Relationship between sensory responsiveness and premovement activity of quickly adapting neurons in areas 3b and 1 of monkey primary somatosensory cortex. *Experimental Brain Research*, **84**, 75–90.

Nelson, R. J., Sur, M., Felleman, D. J. and Kaas, J. H. (1980). Representations of the body surface in postcentral parietal cortex of *Macaca fascicularis*. *Journal of Comparative Neurology*, **192**, 611–643.

Oscarsson, O. and Rosen, I. (1963). Projection to cerebral cortex of large muscle-spindle afferents in forelimb nerves of the cat. *Journal of Physiology*, **169**, 924–945.

Oscarsson, O. and Rosen, I. (1966). Short-latency projections to the cat's cerebral cortex from skin and muscle afferents in the contralateral forelimb. *Journal of Physiology*, **182**, 164–184.

Oscarsson, O., Rosen, I. and Sulg, I. (1966). Organization of neurones in the cat cerebral cortex that are influenced from group I muscle afferents. *Journal of Physiology*, **183**, 189–210.

Padberg, J. and Krubitzer, L. (2006). Thalamocortical connections of anterior and posterior parietal cortical areas in New World tit monkeys. *Journal of Comparative Neurology*, **497**, 416–435.

Padberg, J. P., Disbrow, E. and Krubitzer, L. (2005). The organization and connections of anterior and posterior parietal cortex in titi monkeys: do New World monkeys have an area 2? *Cereberal Cortex*, **15**, 1938–1963.

Pandya, D. N. and Seltzer, B. (1982). Intrinsic connections and architectonics of posterior parietal cortex in the rhesus monkey. *Journal of Comparative Neurology*, **204**, 196–210.

Pandya, D. N. and Vignolo, L. A. (1968). Interhemispheric neocortical projections of somatosensory areas I and II in the rhesus monkey. *Brain Research*, **7**, 300–303.

Paperna, T. and Malach, R. (1991). Patterns of sensory intermodality relationships in the cerebral cortex of the rat. *Journal of Comparative Neurology*, **308**, 432–456.

Pearson, R. C. A. and Powell, T. P. S. (1985). The projection of the primary somatic sensory cortex upon area 5 in the monkey. *Brain Research Review*, **9**, 89–107.

Penfield, W. and Rasmussen, T. (1968). Secondary sensory and motor representation. In *Cerebral cortex of man: a clinical study of localization of function* (pp. 109–134). New York: Hafner Publishing Company.

Phillips, C. B., Powell, T. P. S. and Wiesandanger, M. (1971). Projections from low threshold muscle afferents of hand and forearm to area 3a of baboon's cortex. *Journal of Physiology*, **217**, 419–446.

Pons, T. P. and Kaas, J. H. (1985). Connections of area 2 of somatosensory cortex with the anterior pulvinar and subdivisions of the ventroposterior complex in macaque monkeys. *Journal of Comparative Neurology*, **240**, 16–36.

Pons, T. P. and Kaas, J. H. (1986). Corticocortical connections of area 2 of somatosensory cortex in macaque monkeys: a correlative anatomical and electrophysiological study. *Journal of Comparative Neurology*, **248**, 313–335.

Pons, T. P., Garraghty, P. E., Cusick, C. G. and Kaas, J. H. (1985). The somatotopic organization of area 2 in macaque monkeys. *Journal of Comparative Neurology*, **241**, 445–466.

Pubols, B. and Pubols, L. (1971). Somatotopic organization of spider monkey somatic sensory cerebral cortex. *Journal of Comparative Neurology*, **141**, 63–76.

Pubols, B. H., Pubols, L. M., DePette, D. J. and Sheely, J. C. (1976). Opossum somatic sensory cortex: a microelectrode mapping study. *Journal of Comparative Neurology*, **165**, 229–246.

Randolph, M. and Semmes, J. (1974). Behavioral consequences of selective subtotal ablations in the postcentral gyrus of *Macaca mulatta*. *Brain Research*, **70**, 55–70.

Rausell, E. and Jones, E. G. (1995). Extent of intracortical arborization of thalamocortical axons as a determinant of representational plasticity in monkey somatic sensory cortex. *Journal of Neuroscience*, **15**, 4270–4288.

Recanzone, G. H., Merzenich, M. M. and Jenkins, W. M. (1992a). Frequency discrimination training engaging a restricted skin surface results in an emergence of a cutaneous response zone in cortical area 3a. *Journal of Neurophysiology*, **67**, 1057–1070.

Recanzone, G. H., Merzenich, M. M., Jenkins, W. M., Grajski, K. A. and Dinse, H. R. (1992b). Topographic reorganization of the hand representation in cortical area 3b of owl monkeys trained in a frequency discrimination task. *Journal of Neurophysiology*, **67**, 1031–1056.

Reep, R. L., Chandler, H. C., King, V. and Corwin, J. V. (1994). Rat posterior parietal cortex: topography of corticocortical and thalamic connections. *Experimental Brain Research*, **100**, 67–84.

Risse, G. L., Gates, J., Lund, G., Maxwell, R. and Rubens, A. (1989). Interhemispheric transfer in patients with incomplete section of the corpus callosum. *Archives of Neurology*, **46**, 437–443.

Robertson, L. C. and Rafal, R. (2000). Disorders of visual attention. In M. Gazzaniga (Ed.), *The new cognitive neurosciences* (pp. 633–649). Cambridge, MA: MIT Press

Robinson, C. J. and Burton, H. (1980a). Organization of somatosensory receptive fields in cortical areas 7b, retroinsula, postauditory, and granular insula of *M. fascicularis*. *Journal of Comparative Neurology*, **192**, 69–92.

Robinson, C. J. and Burton, H. (1980b). Somatotopographic organization in the second somatosensory area of *M. fascicularis*. *Journal of Comparative Neurology*, **192**, 43–67.

Romo, R., Hernandez, A., Zainos, A. and Salinas, E. (1998). Somatosensory discrimination based on cortical microstimulation. *Nature*, **392**, 387.

Romo, R., Hernandez, A., Zainos, A., Brody, C. and Lemus, L. (2000). Sensing without touching: psychophysical performance based on cortical microstimulation. *Neuron*, **26**, 273.

Rosa, M., Gattas, R. and Soares, J. (1991). A quantitative analysis of cytochrome oxidase-rich patches in the primary visual cortex of cebus monkeys: topographic distribution of effects of late monocular enucleation. *Experimental Brain Research*, **84**, 195–209.

Roy, A., Paulignan, Y., Farne, A., Jouffrais, C. and Boussaoud, D. (2000). Hand kinematics during reaching and grasping in macaque monkey. *Behavioral Brain Research*, **117**, 75–82.

Rushworth, M. F. S., Johansen-Berg, H. and Young, S. A. (1998). Parietal cortex and spatial-postural transformation during arm movements. *Journal of Neurophysiology*, **79**, 478–482.

Rushworth, M. F. S., Nixon, P. D. and Passingham, R. E. (1997). Parietal cortex and movement. *Experimental Brain Research*, **117**, 311–323.

Sakata, H., Takaoka, Y., Kawarasaki, A. and Shibutani, H. (1973). Somatosensory properties of neurons in the superior parietal cortex (area 5) of the rhesus monkey. *Brain Research*, **64**, 85–102.

Salimi, I., Brochier, T. and Smith, A. (1999a). Neuronal activity in somatosensory cortex of monkeys using a precision grip: II. Responses to object texture and weights. *Journal of Neurophysiology*, **81**, 835–844.

Salimi, I., Brochier, T. and Smith, A. M. (1999b). Neuronal activity in somatosensory cortex of monkeys using a precision grip. I. Receptive fields and discharge patterns. *J Neurophysiol*, **81** (2), 825–834.

Saper, C. B., Iversen, S. and Frackowiak, R. (2000). Integration of sensory and motor function: the association areas of the cerebral cortex and the cognitive capabilities of the brain. In E. R. Kandel, J. H. Schwartz and R. M. Jessel (Eds.). *Principles of neural science* (pp. 349–380). New York: McGraw-Hill.

Scheich, H., Langner, G., Tidemann, C., Coles, R. B. and Guppy, A. (1986). Electroreception and electrolocation in platypus. *Nature*, **319**, 401–402.

Schwartz, A. (1983). Functional relationship between somatosensory cortex and specialized afferent pathways in the monkey. *Experimental Neurology*, **79**, 316–328.

Schwarz, D. W., Deeck, L. and Fredrickson, J. M. (1973). Cortical projections of group I muscle afferents to areas 2, 3a, and the vestibular field in the rhesus monkey. *Experimental Brain Research*, **17**, 516–526.

Seltzer, B. and Pandya, D. N. (1983). The distribution of posterior parietal fibers in the corpus callosum of the rhesus monkey. *Experimental Brain Research*, **49**, 147–150.

Shanks, M. F., Pearson, R. C. A. and Powell, T. P. S. (1985a). The callosal connexions of the primary somatic sensory cortex in the monkey. *Brain Research Review*, **9**, 43–65.

Shanks, M. F., Pearson, R. C. A. and Powell, T. P. S. (1985b). The ipsilateral cortico-cortical connexions between the cytoarchitectonic subdivisions of the primary somatic sensory cortex in the monkey. *Brain Research Review*, **9**, 67–88.

Sinclair, R. J. and Burton, H. (1991). Neuronal activity in the primary somatosensory cortex in monkeys (*Macaca mulatta*) during active touch of textured surface gratings: responses to groove width, applied force, and velocity of motion. *Journal of Neurophysiology*, **66**, 153–169.

Slutsky, D.A., Manger, P.R. and Krubitzer, L. (2000). Multiple somatosensory areas in the anterior parietal cortex of the California ground squirrel (*Spermophilus beecheyii*). *Journal of Comparative Neurology*, **416**, 521–539.

Snyder, L.H., Batista, A.P. and Andersen, R.A. (1997). Coding of intention in the posterior parietal cortex. *Nature*, **386**, 167–170.

Stepniewska, I., Preuss, T.M. and Kaas, J.H. (1993). Architectonics, somatotopic organization, and ipsilateral cortical connections of the primary motor area (MI) of owl monkeys. *Journal of Comparative Neurology*, **330**, 238–271.

Sur, M., Nelson, R.J. and Kaas, J.H. (1980). Representation of the body surface in somatic koniocortex in the prosimian *Galago*. *Journal of Comparative Neurology*, **189**, 381–402.

Sur, M., Nelson, R.J. and Kaas, J.H. (1982). Representations of the body surface in cortical areas 3b and 1 of squirrel monkeys: comparisons with other primates. *Journal of Comparative Neurology*, **211**, 177–192.

Sur, M., Wall, R.J. and Kaas, J.H. (1984). Modular distribution of neurons with slowly adapting and rapidly adapting responses in area 3b of somatosensory cortex in monkeys. *Journal of Neurophysiology*, **51**, 724–744.

Tanji, J. (1975). Activity of neurons in cortical area 3a during maintenance of steady postures by the monkey. *Brain Research*, **88**, 549–553.

Tanji, J. and Wise, S. (1981). Submodality distribution in sensorimotor cortex of the unanesthetized monkey. *Journal of Neurophysiology*, **45**, 467–481.

Taoka, M., Toda, T. and Iwamura, Y. (1998). Representation of the midline trunk, bilateral arms, and shoulders in the monkey postcentral somatosensory cortex. *Experimental Brain Research*, **123**, 315–322.

Taoka, M., Toda, T., Iriki, A., Tanaka, M. and Iwamura, Y. (2000). Bilateral receptive field neurons in the hindlimb region of the postcentral somatosensory cortex in awake macaque monkeys. *Experimental Brain Research*, **134**, 139–146.

Toda, T. and Taoka, M. (2001). The complexity of receptive fields of periodontal mechanoreceptive neurons in the postcentral area 2 of conscious macaque monkey brains. *Archives of Oral Biology*, **46**, 1079–1084.

Toda, T. and Taoka, M. (2002). Integration of the upper and lower lips in the postcentral area 2 of conscious macaque monkeys (*Macaca fuscata*). *Archives of Oral Biology*, **47**, 449–456.

Tremblay, F., Ageranioti-Belanger, S.A. and Chapman, C.E. (1996). Cortical mechanisms underlying tactile discrimination in the monkey: I. Role of primary somatosensory cortex in passive texture discrimination. *Journal of Neurophysiology*, **76**, 3382–3403.

Van Dyck, S. (1983). *The complete book of Australian mammals*. Sydney: Angus and Robertson.

Vogt, B.A. and Pandya, D.N. (1978). Cortico-cortical connections of somatic sensory cortex (areas 3, 1 and 2) in the Rhesus monkey. *Journal of Comparative Neurology*, **177**, 179–192.

Vogt, C. and Vogt, O. (1919). Allgemeinere ergelnisse unserer hirnforschung. *Journal of Psychology (Leipzig)*, **25**, 279–462.

Welker, W.I. and Seidenstein, S. (1959). Somatic sensory representation in the cerebral cortex of the raccoon (*Procyon lotor*). *Journal of Comparative Neurology*, **111**, 469–501.

Weller, R.E., Sur, M. and Kaas, J.H. (1987). Callosal and ipsilateral cortical connections of the body surface representations in SI and SII of tree shrews. *Somatosensory Research*, **5**, 107–133.

Welles, J. F. (1976). A comparative study of manual prehension in anthropoids. *Saugetierkundliche Mittielungen*, **24**, 26–38.

White, E. L. and DeAmicis, R. A. (1977). Afferent and efferent projections of the region in mouse SmI cortex which contains the posteromedial barrel subfield. *Journal of Comparative Neurology*, **175**, 455–481.

Wise, S. P. and Jones, E. G. (1976). The organization and postnatal development of the commissural projection of the rat somatic sensory cortex. *Journal of Comparative Neurology*, **168**, 313–344.

Wise, S. and Tanji, J. (1981). Neuronal responses in sensorimotor cortex to ramp displacements and maintained positions imposed on hindlimb of the unanesthetized monkey. *Journal of Neurophysiology*, **45**, 482–500.

Wise, S. P., Boussaoud, D., Johnson, P. B. and Caminiti, R. (1997). Premotor and parietal cortex: corticocortical connectivity and combinatorial computations. *Annual Review of Neuroscience*, **20**, 25–42.

Wolpaw, J. (1980). Correlations between task-related activity and responses to perturbation in primate sensorimotor cortex. *Journal of Neurophysiology*, **44**, 1122–1138.

Woolsey, C. N. (1958). *Organization of somatic sensory and motor areas of the cerebral cortex*. Madison: University of Wisconsin Press.

Woolsey, C. N. and Fairman, D. (1946). Contralateral, ipsilateral, and bilateral representation of cutaneous receptors in somatic areas I and II of the cerebral cortex of pig, sheep, and other mammals. *Surgery*, **19**, 684–702.

Woolsey, C. N., Erickson, T. and Gilson, W. (1979). Localization in somatic sensory and motor areas of human cerebral cortex as determined by direct recording of evoked potentials and electrical stimulation. *The Journal of Neurosurgery*, **51**, 476–506.

Woolsey, T. A. and Van der Loos, H. (1970). The structural organization of layer IV in the somatosensory region (SI) of the mouse cerebral cortex: the description of a cortical field composed of discrete cytoarchitectonic units. *Brain Research*, **17**, 205–242.

Wu, C.-H., and Kaas, J. (2003). Somatosensory cortex of prosimian galagos: Physiological recording, cytoarchitecture, and corticocortical connections of anterior parietal cortex and cortex of the lateral sulcus. *Journal of Comparative Neurology*, **457**, 263–292.

Xerri, C., Coq, J. O., Merzenich, M. M. and Jenkins, W. M. (1996). Experience-induced plasticity of cutaneous maps in the primary somatosensory cortex of adult monkeys and rats. *Journal of Physiology (Paris)*, **90**, 277–287.

Xerri, C., Merzenich, M., Jenkins, W. M., and Santucci, S. (1999). Representational plasticity in cortical area 3a paralleling tactual-motor skill acquisition in adult monkeys. *Cerebral Cortex*, **9**, 1047–3211.

Zarzecki, P., Shinoda, Y. and Asanuma, H. (1978). Projection from area 3a to the motor cortex by neurons activated from group I muscle afferents. *Experimental Brain Research*, **33**, 269–282.

SARAH H. CREEM-REGEHR

17

Body mapping and spatial transformations

INTRODUCTION

The human ability to imagine spatial transformations is important for accomplishing many daily goals such as action planning, object recognition, spatial navigation and problem solving. One specific component of spatial cognition that is likely to play a role in these goals is an individual's representation of his or her own body. A history of examples of neuropsychological disorders of body knowledge beginning at the turn of the twentieth century (Head and Holmes, 1911–1912; Munk, 1890) supports the claim that spatial body-representations are a separable component from other types of spatial knowledge, linked to processing of the posterior parietal cortex. Patients with selective lesions to primarily posterior parietal cortical regions have been described with deficits in representing their own or others' body parts, deficits in spatial representation with respect to one side of the body, or left–right orientation problems.

Over recent years it has become clear that representations of one's body directly influence spatial cognitive processing in normal individuals. In fact, a research area labeled *embodied cognition* has emerged and become quickly accepted, suggesting that cognition is best understood in the context of an active, physical body (Wilson, 2002). The goal of the present chapter is to examine the role of body-specific representation in one category of spatial tasks: mental spatial transformations. In the early 1970s, Roger Shepard and colleagues (Cooper and Shepard, 1973; Shepard and Metzler, 1971) first demonstrated that the mental rotation of an object to determine shape congruency was analogous to physical rotation of an object in space. Although the seminal claim of the spatial rather than propositional nature of mental imagery was controversial, variations of experiments involving differing mental imagery tasks and subsequent functional

Spatial Cognition, Spatial Perception: Mapping the Self and Space, ed. Francine L. Dolins and Robert W. Mitchell. Published by Cambridge University Press.

neuroimaging research led to the presently believed claim of a functional analogue between mental and physical rotation of objects.

But what about bodies? Research has demonstrated that the human ability to mentally rotate non-biological objects extends to decisions about bodies and body parts. Physical movement of the limbs results in proprioceptive and efferent feedback that provides information to an individual about body-position, known as the body schema. What is the role of this physical representation of one's own body and body-movement in the process of mental transformations of the self?

Answers to this question about body-specific representations and spatial cognition in normal individuals have emerged from both cognitive and neuroscience approaches to research paradigms. Cooper and Shepard (1975) first extended their object mental rotation studies to decisions about the handedness of rotated pictures of hands, suggesting little difference in performance from their results with two-dimensional and three-dimensional shapes. Parsons (1987a, 1987b, 1994) conducted an extensive study of the human ability to mentally transform bodies, hands and feet, with a series of experimental investigations suggesting that the task recruited a process of imagining one's own body to match the orientation of the visually presented stimulus. Subsequently, the growth of functional neuroimaging paradigms using positron emission tomography (PET) and functional magnetic resonance imaging (fMRI) has led to further investigation of the behavioral and neural mechanisms of body and body-part transformations (Bonda *et al.*, 1996, 1995; Kosslyn *et al.*, 1998; Parsons *et al.*, 1995; Vingerhoots *et al.*, 2002; Wraga *et al.*, 2003; Zacks *et al.*, 2002b; Zacks *et al.*, 1999). In addition to mental rotation tasks, other types of cognitive paradigms also have tested the nature of body-specific transformations such as decisions about body motion (Shiffrar and Freyd, 1990), body-position (Reed and Farah, 1995) and visual perspective taking (Creem *et al.*, 2001b; Wraga, 2003; Wraga *et al.*, 2000; Zacks *et al.*, 2003). Finally, recent related neuroimaging efforts have investigated the role of the self in observation, imitation and imagination of actions (Buccino *et al.*, 2001; Grézes and Decety, 2001; Grèzes *et al.*, 2003)

Taken together, the growth of cognitive neuroscience research concerning the self and mental spatial transformations warrants a review of recent findings and conclusions. This chapter will define recent concepts of body schema and embodied cognition and test the hypothesis that some imagined spatial transformations recruit unique body-mapping processes.

BODY SCHEMA AND ITS RELATION TO "EMBODIED COGNITION"

The body schema is defined as knowledge of the spatial relations among the parts of the body that can be used to represent oneself as well as others (Reed, 2002; Reed and Farah, 1995). Reed (2002) has differentiated between the *body schema*, defined as long-term representations of body-part relations and knowledge of body function, and the *body percept*, a more immediate and dynamic representation of current body position. However, others have used the body schema itself to refer to on-line, dynamic representations of the body in space (Buxbaum *et al.*, 2000; Coslett, 1998; Gallagher, 1995) differing from the *body image*, defined as conscious representation and knowledge of the body (Coslett, 1998). In the present chapter, the term body schema will be used to incorporate both knowledge of body-part relations and immediate or dynamic representations of body position.

The idea that cognitive processes are best understood in the context of a physical body interacting with an environment has become an emerging viewpoint referred to as *embodied cognition* (Barsalou *et al.*, 2003; Clark, 1998). As Wilson (2002) has recently discussed, the term "embodied cognition" involves diverse meanings that vary in the extent of their validity and usefulness. Among them are the views that cognition is situated, time-pressured, and ultimately for action. A dominant view in embodied cognition is that cognition involves on-line interaction with an environment in which processing is modified by incoming information and outgoing motor responses (situated) and that cognition is "off-loaded" onto the environment (actions are performed in place of mental storage or manipulation). These claims seem to be especially valid for spatial cognition, falling somewhat short in some other aspects of cognitive processes such as nonspatial decision-making or reasoning. However, most relevant to the present discussion is an *off-line* aspect of embodied cognition. In this view, mechanisms that may have evolved for overt action serve to facilitate more abstract cognitive thought. For example, body-based representations such as motor programs can be used to simulate actions in imagined external situations and provide effective strategies for solving complex cognitive problems. The present focus on body schema and spatial transformations directly examines this view. In addition to its role in spatial imagery, sensorimotor simulation has a likely role in other areas of cognition such as working, episodic and implicit memory and reasoning and problem-solving (Wilson, 2002).

COGNITIVE BODY MAPPING

The ability to mentally simulate events allows humans to extend inferences beyond what is visually perceived. In the context of spatial transformations, this enables tasks such as object and scene recognition and spatial localization. Much of the early work on mental rotation focused on the human ability to make a decision about the congruency of one rotated object with respect to another. Shepard and colleagues (Cooper and Shepard, 1973; Shepard and Metzler, 1971) found that the time required to make a decision about the similarity of the structure of two rotated objects was a function of the angular disparity between the two objects. This monotonic rotation function was upheld for rotations in the picture plane and in depth. Since these revolutionary studies, the processes involved in mental rotation have been examined in numerous ways. Much mental rotation work has focused on 2D or 3D objects (Farah and Hammond, 1988; Hinton and Parsons, 1981; Just and Carpenter, 1985; Pani and Dupree, 1994; Shepard and Metzler, 1971; Tarr, 1995; Tarr and Pinker, 1989; Wexler *et al.*, 1998), but other related research has involved imagined transformations of hands (Gerardin *et al.*, 2000; Kosslyn *et al.*, 1998; Parsons, 1987a, 1994; Parsons *et al.*, 1995) and bodies (Parsons, 1987b; Zacks *et al.*, 2002b; Zacks *et al.*, 1999) using both cognitive and neuroimaging approaches.

Mental transformations of bodies and body-parts may serve to facilitate planning of actions, predicting or understanding other's behavior, or other complex tasks of spatial reasoning. Parsons' (1987a, 1987b, 1994) work with imagined spatial transformations of biological objects such as hands, feet and bodies set the stage for the claim that representations of one's own body are involved in this spatial cognitive task. Parsons (1987a) used a left–right judgment task in which participants viewed individual pictures of left and right hands and feet rotated in the picture plane and judged whether the image presented was a left or right limb. The results indicated that the response time to make a left–right decision about the hands or feet (given no explicit instructions on a strategy to use) was highly correlated with the time required to imagine a limb movement (without the left–right decision). Both types of judgments were also highly correlated with participant's ratings of the awkwardness of moving into a given limb orientation. Later work by Parsons (1994) further investigated the claim that biomechanical representations of hand movement were used in the left–right decision task. Parsons showed that the time required to *physically* move to a given hand position was highly correlated with the time to imagine moving the hand and the time required for the left–right hand decision. Together, these results support the claim that the body schema is used in cognitive decisions about body-relevant stimulus orientation.

Other investigations have involved whole-body transformations again suggesting a special role of body-based transformations. Parsons (1987b) presented visual pictures of a human body with an outstretched hand and required a decision about whether the hand belonged to the left or right side of the body. The results indicated that time to make a left–right decision varied as a function of the orientation difference between the observer and visual body, and differed significantly for different planes of rotation. Parsons suggested that observers imagine their own body rotating to match the visually presented body in order to make a decision. An additional task that explicitly asked observers to imagine a body rotation (without a left–right decision) showed highly correlated response time functions with the left-right task, supporting this claim. Zacks and colleagues (Zacks *et al.*, 1999, 2002a) later compared two types of spatial judgments in the context of visually presented bodies with the goal of comparing cognitive and neural mechanisms in object- versus egocentric perspective-transformations. Using pictures of bodies with an outstretched hand, they required observers to make a same-different judgment (e.g., are the two figures presented the same or mirror image?) or a left–right judgment (e.g., is the figure's extended hand a left or right hand?). They found very different response time functions for the two types of tasks. The same-different task showed a monotonic increase in response time with increasing orientation disparity, as in earlier object-based mental rotation tasks. In contrast, the left–right task indicated a flat response time function, likely a result of the ability to perform an egocentric perspective transformation of the body.

A related paradigm has compared object- and perspective-based transformations in the context of spatially updating external objects. Several studies have directly compared both cognitive and neural mechanisms involved in mental transformations of *objects* versus *egocentric* (or *viewer*) *perspective* (Amorim and Stucchi, 1997; Carpenter and Proffitt, 2001; Creem *et al.*, 2001b; Creem-Regehr, 2003; Presson, 1982; Tversky *et al.*, 1999; van Lier, 2003; Wraga *et al.*, 2000; Zacks *et al.*, 2002a, 2003). When participants are given a spatial updating task to name an object in a given location after a specified imagined transformation, a large systematic performance advantage has been found for *viewer* versus *object* or *array* transformations. For example, Wraga *et al.*'s

(2000) experiments, extending a paradigm of Presson (1982), required participants to perform imagined viewer and array rotations with eyes closed after memorizing the positions of four objects in an array. Participants were presented with a degree of rotation and a position (e.g., 90 degrees, what's on the left?). In the viewer task, they imagined the rotation of their own perspective, and named the object that fell in the given position after the transformation. In the array task, they imagined the rotation of the array itself. Response time and accuracy functions differed for the two types of rotations, indicating superior performance for the viewer transformation. This viewer advantage is consistent with the notion that humans have evolved as organisms moving relative to their environment.

To further define the apparent advantage for imagined self-rotations, Creem *et al.* (2001) examined the role of physical constraints in imagined transformations. Whereas Parsons (1994) had demonstrated a tight link between biomechanical plausibility of hand movements and imagined hand-rotation or left–right hand decisions, Creem *et al.* (2001) found that participants could perform "physically impossible" viewer rotations that defied the physics of gravity, given an imagined transverse rotation (rotating one's principal axis) with respect to the environment. For example, in one experiment, using the paradigm of Wraga *et al.* (2000) participants stood facing an array of objects presented on the wall and imagined rotating around the objects (as if walking on the wall) or imagined the rotation of the objects themselves. Again the viewer transformation was superior to the array transformation. Creem *et al.* (2001) posited that critical to the self-rotation advantage was the geometrical relation between the imagined environmental and body position rather than the physical constraint of gravity. To determine this, they enacted a final experiment that required a coronal rotation of the viewer with respect to the objects (as if performing a cartwheel flip), and the advantage over array rotation was lost. Thus, efficient transformations of the egocentric reference frame relied on body–environment relations that allowed for rotation of the observer's principal axis. Although this paradigm has generally indicated an advantage for imagined self-rotation, Creem *et al.*'s study introduces the additional factor of the relevance of ecologically valid self-transformations.

Creem *et al.*'s (2001) findings introduce an interesting distinction between body perspective transformations and body-part transformations. Whereas imagined whole-body movement may be less tied to one's own physical constraints, performance on tasks involving positions of body-parts supports the notion that mental representations of bodies are constrained by the plausibility of biomechanical movement. In addition to the hand/foot rotation studies described above, other cognitive spatial tasks involving body-parts conform to physical properties of movement. For example, in a task assessing perception of apparent motion by Shiffrar and Freyd (1990), participants were more likely to perceive trajectories that did not violate laws of physical biological motion. In another study, De'Sperati and Stucchi (1997) created a task that required observers to determine whether a rotating screwdriver was "screwing" or "unscrewing" by imagining their hand grasping the tool. They found that orientations of the screwdriver that were awkward for a right-handed grasp led to longer response times for the screw/unscrew decision, again supporting a direct relationship between the biomechanics of movement and mental spatial representation. Reed and Farah (1995) distinguished between the role of the body schema in perception of others' body positions compared to other complex object forms. They used a dual-task paradigm in which the primary task was to determine whether two models' body positions separated by a five-second delay were the same or different. The secondary task required the subject to physically move his or her arms or legs in an unconstrained but non-repetitive way. Reed and Farah found that participant movement, specific to the limb being moved, facilitated detection of an arm- or leg-position difference in the two models. For example, errors were lower for detecting an arm-position change when subjects moved their arms. They concluded that humans use their own representations of body-positions to make decisions about others' body positions suggesting a connection between perception of proprioceptive information about oneself and cognitive representations of others.

Given the potential differences between the role of body schema in body-part and body transformations, Creem-Regehr *et al.* (2002) conducted a study to directly compare hand and perspective rotations in the context of real-world spatial updating. They varied both the nature of the object and the potential for employing an egocentric strategy during imagined object-rotation. Using an object and viewer rotation paradigm based on Wraga *et al.* (2000), they manipulated the object-rotation task in an attempt to reduce the distinction between the two types of transformations. They asked whether the object task could be facilitated by a more egocentric rotation strategy involving embodiment of the object, predicting that presenting a *hand* as the object to rotate would lead to performance that would begin to resemble the viewer rotation task. Participants imagined a block or an artificial hand rotating in the object task, and they imagined themselves rotating around the block/hand in the viewer task. Response time and accuracy were measured as in previous experiments. Creem *et al.* (2001a) found that the nature of the object interacted with body-posture and instructions to embody the object in influencing object-rotation performance. Implementing a hand as the object but constraining the observer's own hand posture to a position behind the back, led to little difference in object-rotation performance compared a control object-rotation task using a four-colored block. However, when the participants' hands were placed in their lap, a change in hand posture that would be more likely to afford physical hand rotation, there was a trend in improvement in hand-rotation performance. In a final experiment, explicit instructions to treat the hand as a part of one's own body significantly improved performance on the hand-rotation task.

These results have implications for the processes involved in different imagined spatial transformations and the flexibility of selecting frames of reference. As referred to above, Zacks and colleagues demonstrated that mental transformations with respect to visually presented body-figures may recruit either object-based or egocentric transformations leading to very different response time functions, depending on the task instructions (Zacks *et al.*, 2002a). The findings of Creem-Regehr *et al.* (2002) are consistent with these studies, in that they suggest an apparent flexibility of humans to represent images with respect to their own egocentric reference frame. In addition, the results suggest both similarities and differences between imagined rotations of one's body and body-parts. Although there was an effect of "embodiment" on the object-task, an overall advantage for the viewer rotation remained (even when the task involved a self-rotation around one's own hand, a biomechanically impossible task). Hand and body transformation tasks may both recruit processes that manipulate the egocentric reference frame, but these mechanisms are likely to differ as well. The importance of biomechanical plausibility during mental representation of body-parts has been demonstrated by several groups (Parsons, 1994; Shiffrar and Freyd, 1990). However, this direct correspondence between mental and physical movement in hand rotation is not evidenced in the flexible and less constrained nature of viewer transformations.

NEURAL BODY MAPPING

The role of body mapping in cognitive spatial transformations is further supported by research from neurology, neuropsychology and neuroimaging, showing neural evidence of the recruitment of one's body schema in motor imagery tasks. In recent years, there has been a burst of findings using functional neuroimaging methods suggesting shared neural motor representations among mental simulation, observation and execution of actions (Grèzes and Decety, 2001).

Imagined actions used in neuroimaging research may be categorized into two broad types of tasks: *explicit* goal-directed actions and *implicit* motor imagery often involving spatial decisions that recruit mental body transformations. Goal-directed actions involving imagined grasping, imagined joystick control and imagined hand/finger-movement tasks have indicated activity in the supplementary motor area (SMA), anterior cingulate, lateral premotor cortex, inferior frontal gyrus, posterior parietal cortex, and cerebellum (Decety *et al.*, 1994; Gerardin *et al.*, 2000; Grafton *et al.*, 1996). Some have also found dorsal prefrontal cortex, basal ganglia (Gerardin *et al.*, 2000) and primary motor activation (Porro *et al.*, 1996). A second category of imagined action tasks includes *implicit* motor imagery, in which an observer is asked to make a spatial decision, and in doing so, recruits motor processing. These types of tasks typically involve handedness or same/different decisions about visually presented hands or objects (Kosslyn *et al.*, 1998, 2001; Parsons *et al.*, 1995; Vingerhoots *et al.*, 2002; Wraga *et al.*, 2003; Zacks *et al.*, 2002b). These tasks have led to similar regions of activation as explicit goal-directed tasks, e.g., posterior parietal cortex, posterior temporal cortex, premotor and some primary motor cortex and cerebellar activation.

Beyond these generalizations, more specific neural distinctions have been found in implicit spatial imagery tasks based on strategies and spatial frames of reference used in the transformation. Parsons *et al.* (1995) and Bonda *et al.* (1995) conducted early PET studies on imagined transformations of hands requiring a left–right decision about visually presented drawings of hands. They found similar patterns of activation falling in motor processing regions, including the bilateral superior parietal lobule extending on the left to the intraparietal sulcus, premotor cortex and SMA, suggesting that observers map their own body-part representation to match the visually presented body-part. Given evidence for the recruitment of motor processing in mental transformations tied to body-part position, the question arises of whether these same regions are active for mental rotation of non-biological objects. Mixed results to this question led Kosslyn *et al.* (2001) and Wraga *et al.* (2003) to suggest that differences in strategies used to mentally rotate objects may recruit egocentric body representations to a greater or lesser extent. For example, Kosslyn *et al.* (2001) presented Shepard and Metzler (1971) block-like figures in a mental rotation task in a PET study. However, prior to the scanning session, subjects were given one of two strategies to visualize the rotation of the objects and they were instructed to imagine the particular strategy during the testing session. In the external action strategy, subjects viewed the rotation of the figure by an external motor. In the internal action strategy, subjects were given the experience of rotating the figure with their own hands. The results indicated a clear difference between the two strategy conditions. When an internal strategy of rotating the object with one's own hand was used, activation was found in primary motor cortex as well as premotor cortex and the superior parietal lobule. The external condition showed activation in similar premotor and parietal regions, but not in primary motor cortex. These findings are suggestive of both the flexibility in body strategies used to mentally rotate objects, and the tightly linked association between representations for mental body transformations and physical motor movement.

Given the evidence that explicit strategies may influence the extent of motor representations involved in mental spatial transformations, Wraga *et al.* (2003) asked the question of whether motor strategies may be implicitly transferred from one task to another. More specifically, if participants performed a hand rotation task first that was predicted to recruit an egocentric body-mapping strategy, would this strategy then transfer to the object task? They included both object (cube figures) and hand stimuli, presenting pairs of images at different orientations and required a decision about whether the two images were the same or mirror-images. The manipulation varied the order of the two tasks. One group of subjects performed the hand task before the object task and a second group performed two sessions of the object task. The results indicated greater motor activation in the object task when it followed the hand task than when it followed the first object task. This activation focused on the bilateral premotor cortex and the junction between the premotor and primary motor cortex in the left hemisphere, supporting the notion that a representation based in one's own body movement may be used implicitly in an object rotation task, given a prior context of this strategy.

Neuropsychological research further supports the notion that one's body schema influences body-part mental transformations. Coslett (1998) tested performance of patients with right hemisphere lesions (three with spatial neglect and three without) and patients with left-hemisphere damage (without neglect) on a version of a left–right hand decision task based on Parsons (1994). Coslett hypothesized that if neglect involves a disorder of body schema, then right-hemisphere damaged patients should show a deficit in making decisions about left, but not right hands. This prediction was upheld, as the three patients diagnosed with neglect (but not the other right or left hemisphere lesion patients) showed impaired performance on identifying pictures of hands contralateral to their lesion. In further support of spatial representations of the body schema as being distinct from other types of spatial processing, Buxbaum *et al.* (2000) tested a patient with apraxia who made errors in pantomiming and on a hand-rotation task, but who was able to perform extrinsic egocentric tasks such as reaching and grasping as accurately as healthy subjects.

The neuroimaging and neuropsychology results described above strongly suggest shared motor processing regions in overt motor movement and implicit motor imagery tasks such as hand, foot and some object rotation. A novel study by Downs (2002) investigating cortical reorganization of phantom limb patients further examined this claim by asking the extent to which motor and cognitive representations regarding the body are interdependent. The phantom limb phenomenon is described as sensory or motor experience of a missing limb and is often discussed with respect to ideas of the body schema (Downs, 2002; Melzack, 1990). Using fMRI, Downs (2002) tested the claim that (1) cortical reorganization occurs after limb amputation; and (2) the same reorganization is evident in cognitive spatial transformation tasks. First, individuals who had undergone a leg amputation and matched controls were tested on the localization of activity associated with overt foot flexion (of the intact limb for the patients) and the two groups showed no differences in activation in motor cortical areas. Second, it was shown that activation resulting from "overt" foot flexion of patients' phantom limb was shifted 0.5 cm in the primary and supplementary motor areas, indicating the presence of cortical reorganization of the phantom compared to the intact leg. Third, an implicit motor imagery task in which a left–right decision about visually presented images of feet was required, showed cortical changes consistent with the foot flexion task. Downs (2002) concluded that the neural systems involved in overt movement are not dissociable from those involved in motor imagery. These results further support the notion that cognitive representations for movement closely parallel physical representations for movement, consistent with a view that cognition might be an extension of more immediate perceptual or motor processing mechanisms.

In a related study involving patients with only one hand, Funk and Brugger (2002) asked whether performance on a hand discrimination task would be influenced by biomechanical constraints if the participants had never experienced these constraints. One of the findings of Parsons (1994) was a medial to lateral gradient, in which rotations of the hand toward the body midline are faster than those away from the body midline. The authors found that individuals with congenitally absent left or right hands (none of which had phantom limb sensations) overall showed slower response times to stimuli corresponding to their missing hand. The difference in performance associated with hand stimuli that could or could not map to an individual's body supports the contribution of the body schema in spatial representation. More specifically, those with an absent left hand showed a normal medial-lateral gradient for right-hand stimuli only.

Extending beyond body-*part* transformations, a number of recent studies have investigated the neural substrates of imagined whole-body and perspective rotations, contributing to our understanding of the role of the body schema in spatial transformations (Creem *et al.*, 2001; Zacks *et al.*, 1999, 2002b, 2003). For example, Zacks *et al.* (2002b) conducted an fMRI study based on earlier cognitive paradigms involving two distinct decisions with respect to visually presented bodies, a same–different decision between two bodies (object-based transformation) or a left–right decision about an outstretched arm (egocentric transformation). Right hemisphere dominance in the parietal-occipital cortex was found for the object-based task relative to the egocentric perspective transformation task. The posterior parietal cortex has been a dominant neural area recruited in imagined body and perspective tasks with healthy, neurologically intact participants. A recent patient study further supports the role of parietal cortex in body-schema and spatial representation. Zacks *et al.* (2004) used fMRI to test a patient who underwent surgical resection of the right inferior and superior parietal cortex on the body spatial transformation tasks used in Zacks *et al.* (2002b). The patient's behavioral performance was quantitatively similar in all tasks to a healthy control group, although response times and errors were greater. Interestingly, the patient showed reduced activation in right parietal regions overlapping his lesion (particularly for the same–different task) and increased activation in the left parietal cortex for all tasks. As in Downs (2002), these findings suggest cortical reorganization of motor regions, in which the left hemisphere took over some of the functioning of the impaired right parietal cortex.

The nature of body perspective transformations has also been investigated with neuroimaging using spatial updating paradigms as described in the section above (Creem *et al.*, 2001b; Wraga *et al.*, 2000). The spatial updating paradigm differs from the perspective transformation tasks involving visual bodies because an array of objects is presented (without a body), and the observer is explicitly told to imagine either an object or viewer transformation in order to name an external object in a given position. As described earlier, a viewer rotation advantage results from this task, given that the question requires a holistic rotation of the array. Creem *et al.* (2001a) extended this paradigm to an fMRI task examining only the viewer rotation component of the task, finding brain activation in the posterior parietal cortex, lateralized to the left hemisphere. Zacks *et al.* (2003) recently compared object to viewer rotations using a similar spatial updating task. They found clear distinctions in activation between the two tasks. The right intraparietal sulcus showed larger activation in the object versus the viewer rotation, and a region of the left superior temporal sulcus showed more activation in the viewer versus the object rotation. Although neither of the comparisons resulted in additional activation in "motor" regions described earlier of the SMA, premotor and posterior parietal cortex, activation in the superior temporal sulcus (STS) in the viewer tasks is consistent with recent studies examining biological motion (Grèzes *et al.*, 2001) and mechanisms involved in representations of the self and imitation (Decety *et al.*, 2002).

CONCLUSIONS

Together, cognitive and neural investigations of spatial tasks involving body and body-part transformations have shown that mental representations of visually presented objects are likely tied to physical body representations or the body schema. Several studies have demonstrated that humans imagine transformations of their own body-part or body to make left–right decisions about visually presented stimuli. The posterior parietal cortex as well as "motor" regions of the frontal lobe, including the premotor, supplementary motor and primary cortex, have been consistently recruited among many spatial transformation tasks. However, differences exist among self-transformation tasks as well. Whereas body-part transformations appear constrained by biomechanical plausibility of movement, body and perspective transformations appear more physically flexible but constrained by experience. Furthermore, neuroimaging studies of perspective rotation tasks have not demonstrated the activation in premotor and primary cortex seen in some of the limb rotation tasks.

One way of conceptualizing the extent to which the body schema is involved in spatial transformations is to ask whether the transformation requires *extrinsic* coding of an external object/location *or intrinsic coding* of the positions of the body relative to the body itself. Extrinsic coding of objects relative to oneself after an imagined transformation of one's body is seen in spatial updating paradigms. The intrinsic system, providing dynamic information about the position of body parts seems most relevant to body-part transformations, most likely involving input from proprioceptive and vestibular systems. This distinction may partly explain the neural differences seen in body and body-part rotation tasks, with body-part tasks being more tied to activation in regions of the motor cortex than body tasks. Work presently underway in our laboratory is investigating this question by directly comparing perspective and hand transformations using both cognitive and functional neuroimaging approaches (Creem-Regehr *et al.*, 2007).

Spatial representation involves a wide range of functions that allow for people to competently accomplish everyday goals. Spatial transformations are an important component of this processing, facilitating the perception of a stable and constant environment. Given the human ability to make spatial orientation decisions about bodies and body-parts, as well as to update object locations from new perspectives, this chapter explored the notion of body-specific representations in spatial transformations. Many spatial cognitive tasks do not involve overt action and, thus, need not necessarily follow implicit constraints of acting observers. However, the evidence of body mapping during imagined transformations suggests an intimate relation between perceptual-motor representations and cognitive processing. Future research and theory should allow for serious consideration of the links between perception (including visual, proprioceptive and motoric information) and cognition in the context of spatial representation.

REFERENCES

Amorim, M. and Stucchi, N. (1997). Viewer- and object-centered mental explorations of an imagined environment are not equivalent. *Cognitive Brain Research*, **5**, 229–239.

Barsalou, L. W., Simmons, W. K., Barbey, A. K. and Wilson, C. D. (2003). Grounding conceptual knowledge in modality-specific systems. *Trends in Cognitive Sciences*, **7**, 84–91.

Bonda, E., Frey, S. and Petrides, M. (1996). Evidence for a dorso-medial parietal system involved in mental transformations of the body. *Journal of Neurophysiology*, **76** (3), 2042–2048.

Bonda, E., Petrides, M., Frey, S. and Evans, A. (1995). Neural correlates of mental transformations of the body-in-space. *Proceedings of the National Academy of Sciences*, USA, **92**, 11180–11184.

Buccino, G., Binkofski, F., Fink, G. R., Fadiga, L., Fogassi, L., Gallese, V., Seitz, R. J., Zilles, K., Rizzolatti, G. and Freund, H. J. (2001). Action observation activates premotor and parietal areas in a somatotopic manner: an fMRI study. *European Journal of Neuroscience*, **13**, 400–404.

Buxbaum, L. J., Giovannetti, T. and Libon, T. (2000). The role of the dynamic body schema in praxis: evidence from primary progressive apraxia. *Brain and Cognition*, **44**, 166–191.

Carpenter, M. and Proffitt, D. R. (2001). Comparing viewer and array mental rotations in different planes. *Memory & Cognition*, **21**, 441–448.

Clark, A. (1998). Embodied, situated, and distributed cognition. In W. Bechtel and G. Graham (Eds.), *A companion to cognitive science* (pp. 506–517). Malden, MA: Blackwell.

Cooper, L. A. and Shepard, R. N. (1973). The time required to prepare for a rotated stimulus. *Memory and Cognition*, **1** (3), 246–250.

Cooper, L. A. and Shepard, R. N. (1975). Mental transformations in the identification of left and right hands. *Journal of Experimental Psychology: Human Perception & Performance*, **104**, 48–56.

Coslett, H. B. (1998). Evidence for a disturbance of the body schema in neglect. *Brain and Cognition*, **37**, 527–544.

Creem, S. H., Wraga, M. and Proffitt, D. R. (2001b). Imagining physically impossible transformations: geometry is more important than gravity. *Cognition*, **81**, 41–61.

Creem, S. H., Downs, T. H., Wraga, M., Harrington, G. S., Proffitt, D. R. and Downs, I. J. H. (2001a). An fMRI study of imagined self-rotation. *Cognitive, Affective, and Behavioral Neuroscience*, **1**, 239–249.

Creem-Regehr, S. H. (2003). Updating space during imagined self- and array-translations. *Memory & Cognition*, **31**, 941–952.

Creem-Regehr, S. H., Neil, J. A. and Yeh, H. J. (2007). Neural correlates of two imagined egocentric spatial transformations. *NeuroImage*, **35**, 916–927.

Creem-Regehr, S. H., Sargent, N. M. and Neil, J. A. (2002). Egocentric spatial transformations of bodies and body-parts. *Poster presented at the 43rd annual meeting of the Psychonomic Society*, Kansas City, MO.

Decety, J., Chaminade, T., Grèzes, J. and Meltzoff, A. N. (2002). A PET exploration of the neural mecahnisms involved in reciprocal imitation. *NeuroImage*, **15**, 265–272.

Decety, J., Perani, D., Jeannerod, M., Bettinardi, V., Tadary, B., Woods, R., Mazziotta, J. C. and Fazio, F. (1994). Mapping motor representations with positron emission tomography. *Nature*, **371**, 600–602.

de'Sperati, C. and Stucchi, N. (1997). Recognizing the motion of a graspable object is guided by handedness. *NeuroReport*, **8** (12), 2761–2765.

Downs, T. H. (2002). Movement and motor imagery of phantom limbs: a study in cortical reorganization. Unpublished dissertation, University of Virginia, Charlottesville, VA.

Farah, M.J. and Hammond, K.M. (1988). Mental rotation and orientation-invariant object recognition: dissociable processes. *Cognition*, **29** (1), 29–46.

Funk, M. and Brugger, P. (2002). Visual recognition of hands by persons born with only one hand. *Cortex*, **38**, 860–863.

Gallagher, S. (1995). Body schema and intentionality. In J. L. Bermudez, A. Marcel and N. Eilan (Eds.), *The body and the self*. Cambridge, MA: MIT Press.

Gerardin, E., Sirigu, A., Lehericy, S., Poline, J., Gaymard, B., Marsault, C., Agid, Y. and Le Bihan, D. (2000). Partially overlapping neural networks for real and imagined hand movements. *Cerebral Cortex*, **10**, 1093–1104.

Grafton, S.T., Arbib, M.A., Fadiga, L. and Rizzolatti, G. (1996). Localization of grasp representation in humans by positron emission tomography. *Experimental Brain Research*, **112**, 103–111.

Grèzes, J. and Decety, J. (2001). Functional anatomy of execution, mental simulation, observation, and verb generation of actions: a meta-analysis. *Human Brain Mapping*, **12**, 1–19.

Grèzes, J., Armony, J. L., Rowe, J. and Passingham, R. E. (2003). Activations related to "mirror" and "canonical" neurones in the human brain: an fMRI study. *NeuroImage*, **18**, 928–937.

Grèzes, J., Folupt, P., Berenthal, B., Delon-Martin, C., Segebarth, C. and Decety, J. (2001). Does perception of biological motion rely on specific brain regions? *NeuroImage*, **13**, 775–785.

Head, H. and Holmes, G. (1911-1912). Sensory disturbances from cerebral lesions. *Brain*, **34**, 102–254.

Hinton, G. E. and Parsons, L. M. (1981). Frames of reference and mental imagery. In J. Long and A. Baddeley (Eds.), *Attention and performance IX* (pp. 261–277). Hillsdale: Erlbaum.

Just, M. A. and Carpenter, P. A. (1985). Cognitive coordinate systems: accounts of mental rotation and individual differences in spatial ability. *Psychological Review*, **92**, 137–172.

Kosslyn, S.M., Digirolamo, G.J., Thompson, W.L. and Alpert, N.M. (1998). Mental rotation of objects versus hands: neural mechanisms revealed by positron emission tomography. *Psychophysiology*, **35**, 151–161.

Kosslyn, S. M., Thompson, W. L., Wraga, M. and Alpert, N. M. (2001). Imagining rotation by endogenous versus exogenous forces: distinct neural mechanisms. *NeuroReport*, **12**, 2519–2525.

Melzack, R. (1990). Phantom limbs and the concept of a neuromatrix. *Trends in Neurosciences*, **13**, 88–92.

Munk, H. (1890). *Uber die Finctionen der gross Hirnrinde*. Aufl. Berlin: Aug. Hirschwald.

Pani, J. R. and Dupree, D. (1994). Spatial reference systems in the comprehension of rotational motion. *Perception*, **23**, 929–946.

Parsons, L. M. (1987a). Imagined spatial transformations of one's hands and feet. *Cognitive Psychology*, **19**, 178–241.

Parsons, L.M. (1987b). Imagined transformations of one's body. *Journal of Experimental Psychology: General*, **116**, 172–191.

Parsons, L.M. (1994). Temporal and kinematic properties of motor behavior reflected in mentally simulated action. *Journal of Experimental Psychology: Human Perception and Performance*, **20**, 709–730.

Parsons, L.M., Fox, P.T., Downs, J.H., Glass, T., Hirsch, T.B., Martin, C.G., Jerabek, P.A. and Lancaster, J.L. (1995). Use of implicit motor imagery for visual shape discrimination as revealed by PET. *Nature*, **375**, 54–58.

Porro, C.A., Francescato, M.P., Cettolo, V., Diamond, M.E., Baraldi, P., Zuiani, C., Bazzocchi, M. and di Prampero, P. E. (1996). Primary motor and sensory cortex activation during motor performance and motor imagery: a functional magnetic resonance imaging study. *Journal of Neuroscience*, **16**, 7688–7698.

Presson, C. C. (1982). Strategies in spatial reasoning. *Journal of Experimental Psychology: Learning, Memory, and Cognition*, **8** (3), 243–251.

Reed, C. (2002). What is the body schema? In W. Printz and A. Meltzoff (Eds.), *The imitative mind: development, evolution, and brain bases* (pp. 233–243). Cambridge: Cambridge University Press.

Reed, C. and Farah, M. J. (1995). The psychological reality of the body schema: a test with normal participants. *Journal of Experimental Psychology: Human Perception & Performance*, **23**, 334–343.

Shepard, R. N. and Metzler, J. (1971). Mental rotation of three-dimensional objects. *Science*, **171** (3972), 701–703.

Shiffrar, M. and Freyd, J. J. (1990). Apparant motion of the human body. *Psychological Science*, **1** (4), 257–264.

Tarr, M. (1995). Rotating objects to recognize them: a case study on the role of viewpoint dependency in the recognition of three-dimensional objects. *Psychonomic Bulletin & Review*, **2** (1), 55–82.

Tarr, M. J. and Pinker, S. (1989). Mental rotation and orientation-dependence in shape recognition. *Cognitive Psychology*, **21** (2), 233–282.

Tversky, B., Kim, J. and Cohen, A. (1999). Mental models of spatial relations and transformations from language. In G. Rickheit and C. Habel (Eds.), *Mental models in discourse processing and reasoning*. Amsterdam: Elsevier.

van Lier, R. (2003). Differential effects of object orientation on imaginary object/viewer transformations. *Psychonomic Bulletin & Review*, **10**, 455–461.

Vingerhoots, G., de Lange, F. P., Vandemaele, P., Deblare, K. and Achten, E. (2002). Motor imagery in mental rotation: an fMRI study. *NeuroImage*, **17**, 1623–1633.

Wexler, M., Kosslyn, S. M. and Berthoz, A. (1998). Motor processes in mental rotation. *Cognition*, **68**, 77–94.

Wilson, M. (2002). Six views of embodied cognition. *Psychonomic Bulletin & Review*, **9**, 625–636.

Wraga, M. (2003). Thinking outside the body: an advantage for spatial updating during imagined versus physical self-rotation. *Journal of Experimental Psychology: Learning, Memory, and Cognition*, **29**, 993–1005.

Wraga, M., Creem, S. H. and Proffitt, D. R. (2000). Updating displays after imagined object and viewer rotations. *Journal of Experimental Psychology: Learning, Memory, and Cognition*, **26** (1), 151–168.

Wraga, M., Thompson, W. L., Alpert, N. M. and Kosslyn, S. M. (2003). Implicit transfer of motor strategies in mental rotation. *Brain and Cognition*, **52**, 135–143.

Zacks, J., Mires, J., Tversky, B. and Hazeltine, E. (2002a). Mental spatial transformations of objects and perspective. *Spatial Cognition and Computation*, **2**, 315–322.

Zacks, J., Ollinger, J. M., Sheridan, M. A. and Tversky, B. (2002b). A parametric study of mental spatial transformations of bodies. *NeuroImage*, **16**, 857–872.

Zacks, J., Rypma, B., Gabrieli, J. D. E., Tversky, B. and Glover, G. H. (1999). Imagined transformations of bodies: an fMRI investigation. *Neuropsychologia*, **37** (9), 1029–1040.

Zacks, J. M., Vettel, J. M. and Michelon, P. (2003). Imagined viewer and object rotations dissociated with event-related fMRI. *Journal of Cognitive Neuroscience*, **15**, 1002–1019.

Zacks, J. M., Michelon, P., Vettel, J. M. and Ojemann, J. G. (2004). Functional reorganization of spatial transformations after a parietal lesion. *Neurology*, **1**, 239–249.

ATSUSHI IRIKI

18

"Understanding" of external space generated by bodily re-mapping: an insight from the neurophysiology of tool-using monkeys

INTRODUCTION

Consider the varieties of bodily "actions" we perform in our daily life. Most of them are executed so as to interact with objects existing in external space. Hence, in order to make the executed action successful, our intrinsic bodily space must be constantly re-mapped automatically to match proper relations with the current status of the extrinsic space. The most fundamental idea regarding such a brain function was first pointed out by Head and Holmes (1911) in a patient with a parietal cortex lesion exhibiting peculiar symptoms, something like the following:

1 Normally, when we pass through a gate of low height, we bend our head or back, almost unconsciously, so that our head will not hit the top of the gate. However, this patient just walked through, striking her head repeatedly.
2 Also, in those days, it was the fashion for women like this patient to often wear a hat with a long feather on it. To pass under the gate as above, we, again almost unconsciously, would bend our head so that the "tip of the feather" would not hit the gate, but this patient could never do this.

Head and Holmes referred to these phenomena as a loss of the "body schema" that would normally reside in the parietal cortex.

The above report would suggest two crucial concepts concerning the "body schema".

1 "Body schema" is formed by integrating the somatosensory (tactile and joint) information about the current "intrinsic"

Spatial Cognition, Spatial Perception: Mapping the Self and Space, ed. Francine L. Dolins and Robert W. Mitchell. Published by Cambridge University Press.

status of one's own bodily structures on one hand, and on the other hand the visual information about the extensive limit, configuration, posture and motion of one's body parts in relation to the external structures (such as obstacles or location of the goal to reach) defined in the environmental space, or "extrinsic" coordinate frames.

2 "Body schema" is not just coding our innate body, but could be extended to an object attached to and protruding from our body surface, moving congruently with the part of the body to which the object is attached.

Psychological studies have repeatedly postulated, on the basis of our experience, that such a "body schema" should be extended to the tip of the "tools" we use in our daily life, or even to the mechanical machinery (such as automobile or crane) we generally operate (Paillard, 1993). However, empirical observations have not yet been demonstrated, and such experience could only be examined through verbal reports of humans, making it practically impossible to study this in infants or nonhuman primates including macaque monkeys. In the first part of this chapter, I will describe the neurophysiological evidence reported from our laboratory earlier to show how somatosensory and visual information is integrated to code the above "body schema" in the intraparietal cortex. Also, in trained monkeys, it will be described how such a body image could be extended along the length of tools and other objects to project the introspective "body" into the external space.

However, are we merely able to calibrate and reform the intrinsic coordinates of ourselves in order to match the environmental extrinsic coordinates? The majority of our daily actions are performed using our arms and hands, the posture and movements should be planned and executed in the intrinsic bodily spatial coordinates; these are usually purposeful, such as reaching, catching, handling and acquiring objects in external space. Another major purpose is to feel, examine or explore the external space. Eventually, if we are able to arbitrarily transform our bodily image by intention or purpose, we may be able to intentionally "understand" the external world by integrating modes of such transformations performed upon various complex actions in space. In the last part of this chapter, I would like to describe how "flexible" our ability is to re-map our intrinsic spaces onto extrinsic spaces in actual time in various hand/arm movements.

Using these abilities preprogrammed in monkeys, our hominid ancestors were most likely able to develop highly organized tools and higher technologies, and thereafter also develop an intellectual understanding of external space. Finally, based on these functions, understanding the external world through these bodily re-mapping mechanisms becomes a distinct possibility.

MODIFICATION OF "BODY SCHEMA" WITH TOOL-USE

Monkeys, unlike apes and humans, generally do not use tools in their natural habitat. Yet, their brain might have the ability to use and manipulate tools even though they cannot construct them. They may simply not use tools because there are no environmental or social demands for their survival and therefore would not benefit potential tool-users. Indeed, we have shown (Iriki *et al.*, 1996) that, when "properly" trained for about two weeks, macaque monkeys could use rake-shaped tools to retrieve food that could not be reached by their own hands and arms (Figure 18.1c). By observing this behavior, one can estimate the monkeys' "introspection," whether they are regarding the rake as a tool that is incorporated or assimilated into their body as its extension, or just holding it as a mere stick that is an object external to their own body. Using the above trained behavior as an experimental paradigm, we inserted electrodes into the monkey's intraparietal cortex to study the nature of bimodal (somatosensory and visual) integrations (Figure 18.1b) to code the "body schema."

In the anterior bank of the intraparietal sulcus, immediately posterior to the forearm representation in the postcentral somatosenory cortex, we found bimodal neurons which had tactile receptive fields (Figure 18.1a,i) on the distal extremity of the hand. Such neurons tend to be activated when a visible object approaches the somatosensory receptive fields (Figure 18.1a,ii; a,iii), and hence, the space encompassing the somatosensory receptive field was defined as the "visual receptive field" of these neurons (Figure 18.1a,iv) (see legend for the method for determination of the visual receptive fields). Figure 18.1d illustrates a representative example of the receptive field properties of such neurons. Since this visual receptive field was formed (Figure 18.1d,ii) and "followed" the hand (where the somatosensory receptive field was located (Figure 18.1d,i)) moving into space, one possible interpretation for the response properties of these neurons could be "coding the image of hand in space." However, an alternative interpretation that these response properties represent the body-centered coordinate frame could also be equally valid.

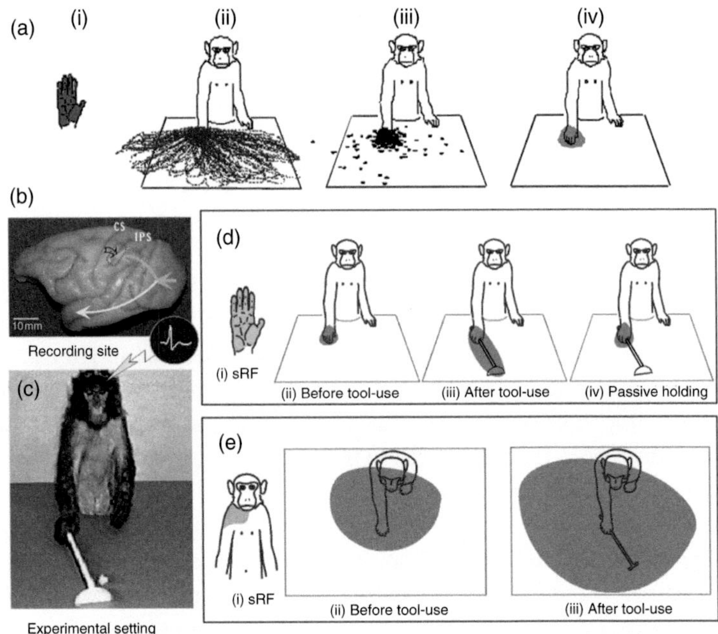

Figure 18.1. (a) the somatosensory receptive field of the neurons was identified as the area of the skin surface where light touches induce spike discharges (i). In other neurons, discharges could be also induced by passive manipulation of joints or active hand-use. The visual receptive field was defined as the area (iv) in which cellular responses ((iii) location of the probe when the neuron discharges) were evoked by visual probes ((ii), trajectories for scanning the whole space around the hand). (b) recording sites of the intraparietal bimodal neurons (red circle) shown in the lateral surface of the left cerebral cortex. CS, central sulcus. IPS, intraparietal sulcus. Blue arrow, somatosensory information processing pathway. Pink arrows, visual information processing pathway. (c) Neural responses are recorded (yellow arrow) while monkeys retrieve bits of food using a hand-held rake. (d) (i) Somatosensory receptive field (blue area) of the "*distal type*" bimodal neurons and their visual receptive fields (pink areas) before (ii), immediately after (iii) tool-use, or when just passively holding the rake (iv). (e) (i) Somatosensory receptive field (blue area) of "*proximal type*" bimodal neurons, and their visual receptive field (pink area) before (ii) and immediately after (iii) tool-use. (Adapted from Maravita and Iriki, 2004.) Parts (b) and (c) are reproduced in the colour plate section.

With tool-use, a peculiar phenomenon was induced (Figure 18.1d, iii). That is, the above visual receptive field extended along the length of the tool when examined immediately after the monkey retrieved food using the rake as an extension of the forearm. Assuming that these neurons are coding an image of the body-parts in space the rake was incorporated into the monkey's image of the hand, most likely in accordance with his interpretation at the time. After this, an even more interesting phenomenon occurred, when physically identical scanning was performed immediately after the monkey retrieved the food by his own hand and held the rake as a merely external object, not as a tool as an extension of his arm. In this situation (Figure 18.1d,iv), the "visual receptive field" (namely his "image of the hand" as was defined above) shrank back to be restricted around his innate hand. The only difference between these physically identical scans should be the monkey's interpretation, that is, whether or not the rake was assimilated into his own body image. Thus, we were able to empirically observe the subjective body image of the monkey, and were able to demonstrate that monkeys can arbitrarily manipulate their body image free from the physical constraints of its structure and appearance.

How these transformations between intrinsic and extrinsic spaces were calculated and represented in the brain is still an open question. However, the following analyses suggest that there might be an intermediate process. This process, which could be illustrateted in the retinotopic space, is also modified by the changes of the experienced body image. As the raw data shown in Figure 18.2a and e, (which are identical with the schematic illustrations shown in Figure 18.1d,ii and d,iii), the location of the visual receptive field was mapped independent of freely moving eye gaze directions during the scanning of the entire space using a visual probe. At the moment when the neuron discharged, we know both the locations of this scanning probe (Figure 18.2b and f, dots) and the gaze directions (Figure 18.2c and g, dots). Hence, we can calculate at which spot on the retina the probe was projected onto (Figure 18.2d and h) corresponding to respective spike discharges. Because the somatosensory receptive field of this neuron is located on the right hand, and the hand was set still in the natural position during the scanning, the location of the images of the probe projected onto the retina falls into the bottom-right part of the retinal coordinate systems before the tool-use. After the tool-use, site of projections extended towards the center of the field, where the image of the tool should be projected. Thus, the analyses indicate that the "receptive field" calculated to be expressed in the conventional retinotopic coordinate system is also elongated along the axis of the hand-held tool upon its usage.

Another aspect of the tool is that it changes the range of the reaching distances. Behaviorally, tool-use trained monkeys have extremely good sense of the reaching distance, so that they can chose the most

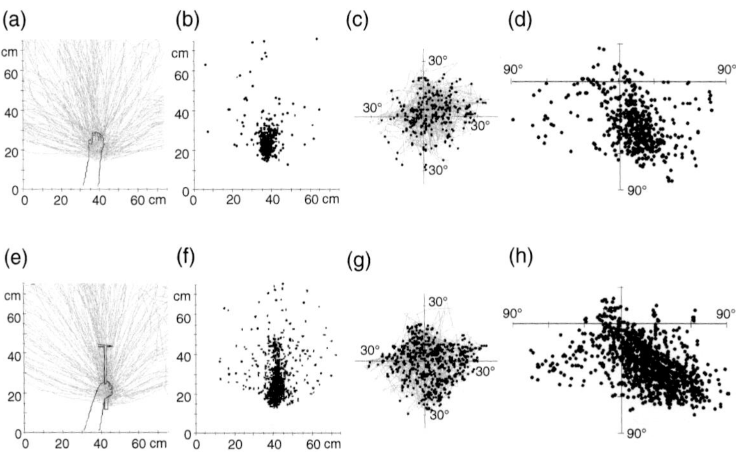

Figure 18.2. Responses of "*proximal type*" bimodal neurons to visual stimuli, of which schematic drawings were depicted in Figure 18.1d. Data were taken during the monkey either not holding (**a–d**) or holding (**e–h**) the rake. (**a,e**) trajectories (shaded lines) of a scanning object projected to a horizontal plane. (**b,f**) Locations of the scanning object in the horizontal plane effective to make neurons fire (each dot represents 1 spike discharged at an instantaneous frequency higher than 3.0 Hz). (**c,g**) Eye movements (shaded line) and gaze positions at which neuronal discharges occurred (dots). (**d,h**) Positions represented in the retinal space at which the scanning probe should be projected at the time the neurons discharged, calculated from the data points shown in b,c for d, and f, g for h, together with their timing data (not shown in the graph).

efficient way to retrieve the food placed at various distances on the table. That is, the monkey uses the rake if it is beyond reach of his hand but reachable with the rake, and he immediately uses his hand to grasp the rake. However, he never uses the tool if the food is placed close enough. If the food is placed further than the reach of the rake, at a glance, he does not even attempt to reach the food. Further, when he is given the rake to hold at its middle portion initially, he tries to extend the rake by re-holding it if the food can be reached by doing so. Figure 18.1e depicts the receptive field properties of neurons with somatosensory receptive fields located at proximal body parts including upper arm, shoulder, neck and face, namely the "*proximal type*" (Figure 18.1c,i). These neurons had visual receptive fields that cover the space accessible with the innate arm (Figure 18.1c,ii). After tool-use, these neurons became responsive to stimuli presented in a wider region, namely, within the space accessible to the handheld rake (Figure 18.1c,iii).

PROJECTION OF "BODY IMAGE" ONTO VIDEO MONITOR

The above described "body schema" is often utilized to organize movement or actions at a subconscious level, and thus in a sense is close to the "internal model" of the bodily status. In contrast, we can vividly realize a visual image of the body in space, and are able to manipulate it consciously. I define, for this chapter, such a conscious visual experience as a "body image." The difference in definition of the concept of the "body schema" and the "body image" has been a matter of considerable debate. In this chapter, I would rather simply regard body schema as the "enactive type" of representation of the body, and body image as a "visual or iconic form" of representation of the body (Bruner *et al.*, 1966).

The next experimental question would be whether monkeys have or can handle those visual images of the body parts. In order to examine this, we trained monkeys with a task similar to a video game, as follows (Iriki *et al.*, 2001). By placing a board in front of the monkeys, we prevented them from directly seeing their hands, training them to retrieve food while viewing the real-time video images of their hands on the screen of a monitor (Figure 18.3a). The whole area of the monitor screen was scanned via a superimposed spot using the chromakey effect technique (Figure 18.3b). This time, the visual receptive fields of the bimodal neurons were formed to appear on the monitor, onto which their hands were projected (Figure 18.3d). Hence, monkeys should have recognized that the hands projected on the monitor were their own hands, namely, an indication of having the iconic representation of the hand image, and those images were coded by parietal bimodal neurons.

The phenomenon demonstrated here would provide clues to developmental as well as evolutionary pictures of higher cognitive functions. During the early childhood of humans, their field of view is restricted to personal and immediately adjacent peripersonal space. An internal representation during this period should be a type of a dexterous sensorimotor intelligence that is acquired unconsciously through experiencing various actions in the environmental space as accustomed patterns of action, and should correspond to the "body schema." While becoming familiar with the surrounding space, children achieve (by the age of nine to ten years old) abilities to handle action-free visual images of their own body, which dissociates an internal schema from existing actions, namely, the "visual or iconic form" of representation of the body (Bruner *et al.*, 1966) they have acquired.

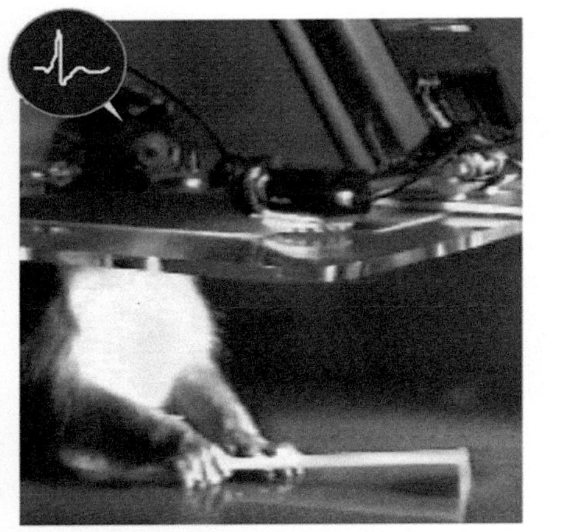

(a) experimental setting

Actual image (hand-rake actions)

Superimposed images (visual probes)

Camera 1

Camera 2 (Blue back)

Effector

Chromakeyer

Monitor

(b) video monitoring and signal filtering system

(c) sRF

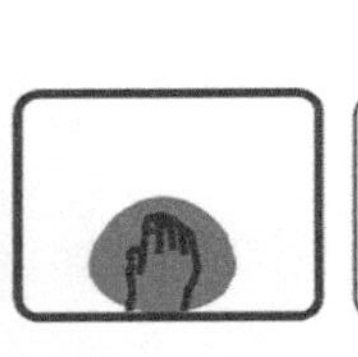

(d) vRF: on monitor

(e) after tool-use

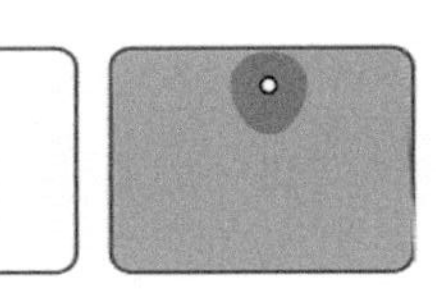

(f) tool-tip only

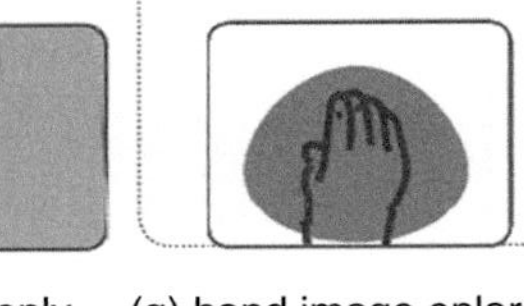

(g) hand image enlarged

(h) compressed

(i) displaced

Figure 18.3. Neural correlates of tool-use under indirect visual control. (a) Neural responses are recorded (white arrow) while monkeys retrieve bits of food and observe their actions on a video monitor, as captured by a video camera (camera 1). (b) The visual scene observed by the animal on the monitor can be modified by adding visual input (via a special device called "*Chromakeyer*") on a neutral background recorded by a second video camera (camera 2), or altering the position and size of visual images by special filters ("*Effector*"). (c) Somatosensory and (d–i) visual neural responses in different viewing conditions. (Adapted from Maravita and Iriki, 2004 and Iriki, 2005.)

Having this mechanism in the brain, we might be able to feel "reality" in the virtual reality or tele-existence apparatus.

Further, by mentally manipulating this visual type of image independent of the actual state of the existing body parts, we can mentally rotate objects and our own body parts independent of supporting angles. Corresponding to this psychological experience, the size and position of visual receptive fields of the presently observed bimodal neurons were modified according to expansion (Figure 18.3g), compression (Figure 18.3h) or change of position (Figure 18.3i) of the visual images in the video monitor. These changes were induced even if the posture or the position of the hand was not actually altered during modification of the screen image. Thus, the presently observed properties of the visual receptive fields would represent neural correlate for this sort of internal representation, indicating that macaque monkeys (along the course of primate evolution, akin to human children during development) attained neural machinery that has the potential, when extensively challenged by the environment, to represent an intentionally controllable visual representation of their own body parts.

The coincidence between the development of human mental representations and response properties of parietal neurons discovered in our decently trained monkeys suggested that one may assume that the evolutionary precursors for introspective manipulation of an abstract sign, or eventually a symbolic representation of one's own body, might be already reserved as neural machinery in the monkey brain. But usually, they are probably not in operation and are able to be recruited only when extensively reinforced. Therefore, we expect that by extending the present experimental paradigms, monkey studies would potentially lead us to a good understanding of the neural mechanisms of our higher cognitive functions such as symbol manipulation, and perhaps eventually language and metaphysical thoughts.

EXTENDING "ABSTRACT BODY IMAGES" INTO EXTERNAL SPACE

As humans mature, we acquire a symbolic mode of internal representations, as, for example, we can describe and explain things verbally, or think logically using symbols (Bruner *et al.*, 1966). At a similar developmental point, of course, the monkey is not able to fully express the same intellectual ability. However, when focusing on a few of the various aspects of the characteristics that the symbol is furnished with, we might be able to observe some precursory components constituting symbolic representations. One such peculiar characteristic may be that symbols are expressed by an arbitrarily defined abstract appearance, and another characteristic may be that symbols have no direct representation of the actual physical objects or events to which they are connected, and are therefore free from physical constraints. Codings of the above two aspects of the abstract and perhaps (pre-)symbolic body images were examined experimentally in the monkey's intraparietal cortex.

The experiments to examine if the monkeys can perceive abstract appearances as their body parts were performed using the same video system as described in the previous section. In this experiment, the monkey's hand image was altered using a video effect called "luminacekey," by which the darker part of the monitor could be erased, leaving only the most brilliant part of the monitor screen visible. Firstly, we confirmed that the visual receptive field was elongated along the image of the hand-held tool in the monitor (Figure 18.3e) in the same manner as it was elongated along the tool when viewing the scanning directly in the space around the actual hand and tool as shown previously. Then, only the spot at the tip of the rake was left visible. This situation is similar to that of using a computer cursor in the monitor, when many of us might occasionally have a "feeling" that the cursor becomes our fingertip. Therefore, the intraparietal bimodal neurons that code the hand image should treat this spot in an identical form as the hand. In other words, the visual receptive field must be formed to include this spot. The results shown in Figure 18.3f indicate that this actually is the case.

One important characteristic of the above data is that the receptive field originally located at the hand "shifted" to the spot at the tip of the tool and eliminated the original hand from the monkey's body image. This is in contrast with the results described in the previous section when the visual receptive field became extended, originating from the hand to include the tool when the whole image of the tool "being held and manipulated" by a hand was visible. The body image is detached from the hand and projected onto the arbitrarily (shape, size, color) defined artificial cursor. In other words, the neurons presented here respond not only to the natural image of the hand on the monitor, but also to a "sign" that functionally substitutes for the actual hand on the monitor. However, the sign with an identical appearance per se was not effective if it appeared in the monitor with no contextual relation to the function of the hand. Thus, monkeys (by training) can handle these more abstract forms of body images, and they are coded by intraparietal bimodal neurons. Further, these abstract body-part

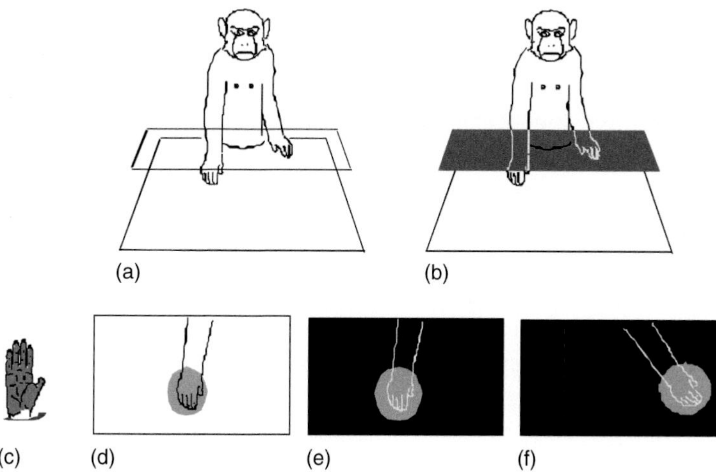

Figure 18.4. Monkeys were trained to retrieve food pellets placed on the table. The front portion of the table was covered by a plate that allowed a 10-cm space between it and the table for working. The plate was made of a liquid crystal device which could be turned either transparent (**a**) to enable the monkey to see his hand retrieving the bait, or opaque (**b**) to prevent him from seeing his hand. (**c**) Somatosensory receptive field. (**d**) Visual receptive field defined around the hand when it was visible through the transparent plate. (**e**) Visual receptive field persisted over an opaque plate right under which the hand stayed immobile from d. (**f**) Visual receptive field followed the arm spontaneously moved under the opaque plate invisibly to the left.

images could be projected onto the monitor screen, placed in a location anywhere in the external space.

The next experiment focused on the second aspect of the symbolic representations, which are not directly connected to the actual physical objects or events, and are therefore free from physical constraints. To represent such body images, the results below (Obayashi *et al.*, 2000) show that, after the hand was hidden under an opaque plate (Figure 18.4b), the visual receptive field, once confirmed to exist around the hand (Figure 18.4d), persisted over the plate above the nonvisible hand (Figure 18.4e). Furthermore, when the hand was moved under the plate so that it was not visible, the visual receptive field moved over the plate to follow the invisible hand (Figure 18.4f). Thus, monkeys can maintain and update subjective body-part images in the mind, and they are coded by intraparietal bimodal neurons.

Those results showed that the visual receptive fields persisted even after the hand was hidden (visual input ceased), and its location trailed the displacement of the occluded hand. Taken together with our previous finding concerning the extension of the visual receptive field to the mentally and functionally incorporated tool in the intraparietal neurons, these lines of evidence suggest that this brain area has a long-lasting representation of the hand in space, which could be updated referring, perhaps, to proprioceptive information arising from the more proximal part of the arm. Indeed, a human parietal lesion can result in neglect or extinction of existing body parts, possibly because the body image stored in this brain area is damaged, and conversely, amputees sometimes perceive a subjective experience of the absent body part (the phantom limb phenomenon) (Berlucchi and Aglioti, 1997), probably due to somesthetic imagery originating from the body image acquired prior to amputation. Thus, subjective representation of the body parts should have perceptual characteristics providing its internal estimate. Now we demonstrated that subjective representation of such body images resides in the intraparietal area of the monkey cerebral cortex. Indeed, phenomena directly relevant to this have been shown to exist in the human parietal cortex (Blakemore and Sirigu, 2003).

The present results represent a demonstration of neural correlates of the body image at the advanced stage similar to that during human postnatal development described earlier in this chapter, indicating that macaque monkeys (along the course of primate evolution, like human children during development) attained an intentionally controllable representation of their own body. When the above-described representation is further advanced, it will become totally free from physical constraints of the actual world to become a symbolic one. The present objective observations of the subjective mind, free from actual visual input, may represent evolutionary precursors of such pre-symbolic abilities, whose neural mechanisms remain to be uncovered.

"UNDERSTANDING" EXTERNAL WORLD THROUGH INTRINSIC ACTIONS

We can mentally calibrate directions of bodily movements into visual coordinate systems to achieve purposeful actions in space. Alternatively, we can apprehend characteristics of the space around us through actions performed by our own body parts. Such interactions between representations of our body motion and extrinsic space should also occur in the intraparietal cortices where the above described bimodal neurons reside. But, instead of "tactile" receptive fields on the skin as have been described above, such an integration might well occur between "joint" displacement information and visual information. In

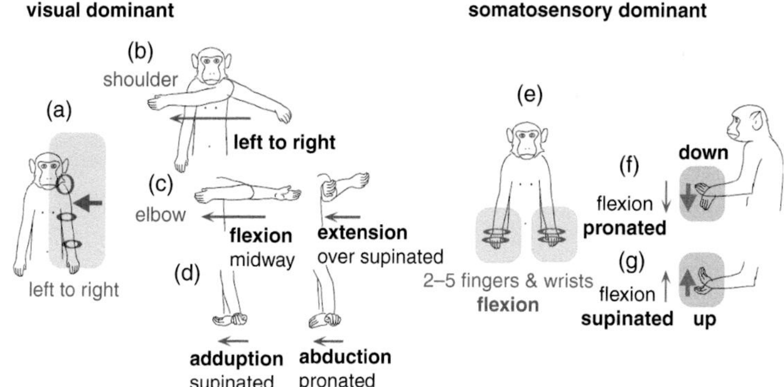

Figure 18.5. Bimodal neurons in which their preferred directions of movements of joint displacements or visual stimuli changed depending on the posture of the forearm. (**a–d**) A representative example of "visual dominant neurons," in which the preferred direction of the visual receptive field was constant (**a**), whereas that for displacements of the shoulder joint stayed constant (**b**) but that for the elbow (**c**) and wrist joints (**d**) inverted depending on the posture of the forearm. (**e–g**) An example of "somatosensory dominant neurons," in which the preferred direction of the joint (wrist) displacement was constant flexion (**e**), whereas that for the visual receptive field inverted depending on the forearm posture (**f, g**). (Adapted from Tanaka *et al.*, 2004.)

this brain area of monkeys, we analyzed the response properties of "bimodal joint neurons," which responded to forearm joint displacements and, at the same time, to visual stimuli moving towards one direction in space (Tanaka *et al.*, 2004). In the majority of these neurons, directions of the hand movement in space as a result of adequate joint displacement were congruent with the preferred directions of the moving visual stimuli in space. When arm posture was rotated, the preferred direction of joint displacement inverted (Figure 18.5a–d), so as to match the induced hand movement in space with the visual preferred direction. Contrariwise, in some neurons, the visual preferred direction in external space inverted, when the arm posture was rotated so that direction of preferred joint displacement (unchanged in intrinsic coordinates) was rotated in extrinsic spatial coordinates. Hence, intraparietal neurons seem not merely to represent mental recalibration of intrinsic movement into extrinsic coordinates, but also render delineation of exterior space through intimate actions (Figure 18.5e–g).

Association between joint displacements and resultant motion in extrinsic spatial coordinates have been shown to exist in neuronal activities recorded from motor related cortices, reserving the question of whether movements are implemented in either extrinsic or intrinsic coordinates in those areas. So far, a widely accepted picture from the viewpoint of motor production would be that motions are executed by the neural commands described in intrinsic coordinates (or muscles and joints) in lower motor-related structures within the CNS, which are gradually "translated" from experienced images of motions planned and coded in extrinsic space. Recent studies by Kakei *et al.* (2001) manifested this picture by demonstrating that there are more neurons in the premotor cortex than in the primary motor cortex, that the motion is coded in extrinsic coordinates, and that intrinsic representation could be recalibrated to hold congruency with the current status of the arm posture.

In addition to the above described and already reported (though in motor related cortices other than the present intraparietal area) properties of bimodal neurons, the most peculiar findings of the present study would be the demonstration of the plasticity that has never been reported before anywhere in the brain. That is, the existence of the "somatosensory dominant" bimodal neurons, in which preferred visual (i.e., extrinsic) direction changed depending on the alternation of the posture defining the mode of intrinsic coordinates. This phenomenon is not a requisite for creation of the goal-directed action in space, but rather vice versa, a rudiment for the perception of the gestalt of space to be implemented by the intrinsic coordinate framework. Therefore, information processing in this brain area seems to map a surrounding environmental space by manipulating intrinsic mental space grounded to our body parts, rather than mere coding of movements in extrinsic space. In other words, this area would handle not just spatial representation of actions, but also understanding of space through actions. The latter function would necessitate more intentional requirements for the brain than just passively calibrating internal states to the external situations to actively delineate structural characteristics of intimately planned and executed motions. Thus, linkage of exterior space and intimate body might emerge through the perception of actions in monkeys.

TOWARD PERCEPTION OF "CAUSAL STRUCTURE" BY BODILY-SPACE REMAPPING

Unlike the simple visual "imagery" which is the retrieval of the actually or previously existing things in the world, "symbols" could be combined, restructurized and modified in the mind. Thus, symbols could be the tool for thinking, i.e., the elements of thought. In the last section of

Sequential tool-use

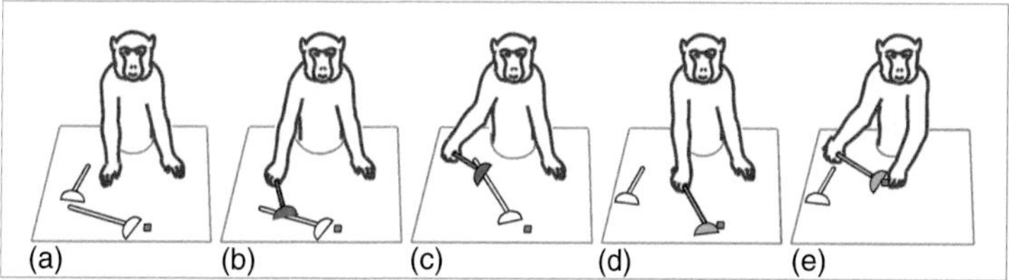

Applied combination

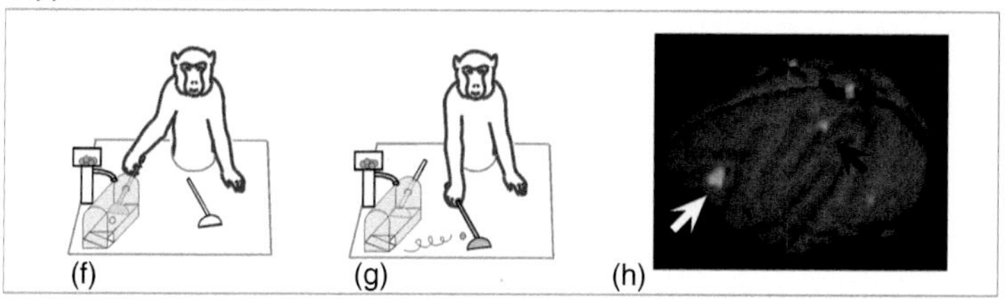

Figure 18.6. Complex tool-use in monkeys. (**a–e**) Experimental setting for the double-rake reaching study in monkeys. (**f–g**) Experimental setting and (**h**) PET imaging brain activation (critical intraparietal and prefrontal activation foci shown by black and white arrows, respectively) for the complex tool-use experiment in monkeys. (Adapted from Maravita and Iriki, 2004.)

this chapter, the idea that monkeys can exhibit such thoughtful abilities perhaps by using the above described (pre-)symbolic body images coded in the intraparietal cortex will be addressed.

Monkeys were exposed to the situation shown in Figure 18.6a–e (Hihara *et al.*, 2003; Maravita and Iriki, 2004). The food was presented out of reach and rakes of two different lengths were placed in front of the monkey in such a way that he could reach the short rake but he could not reach the food using it, whereas he could reach the food with the long rake but couldn't reach the long rake with his hand. Given these spatial arrangements, to our surprise, the monkey who had already been trained to use the rake to retrieve otherwise unattainable food items could almost immediately, after a brief series of trial attempts, perform the following purposeful combinations of actions. That is, as would be mostly evident for us humans, take the short rake first with his hand, obtain the long rake with the short rake, exchange the rakes, and get the food with the long rake.

This indicates that, once the basic skill was learned through extensive training, its application could be accomplished rather easily, perhaps subserved by different brain mechanisms. The latter mechanism was studied by monkey PET imagings using a similar but slightly different tool-using skill (Figure 18.6f and g) (Obayashi *et al.*, 2002; Maravita and Iriki, 2004). In this task, a food pellet was delivered into transparent tubing and the monkey had to initially push it using a rake to roll it out of the tubing, and then retrieve it by pulling it in using another rake. The brain activation pattern obtained by subtraction of single-tool-use from this applied tool-use task showed that prefrontal activity was evident in addition to that of the intraparietal area where the basic learning took place (Figure 18.6h).

The result suggests that prefronto-intraparietal interaction is essential for this applied usage of tools. This mechanism could be further extended to the perception of the causal relationships of the mode of elementary tool usages through intentionally controlled manipulation of the internally created body images in relation to external space. Further, this would lead to the development of an "insight" or perception of the causal structure of events, and intelligent usages of mechanistic tools. Thus, these would represent evolutionary precursors of the seeds of modern technology beyond simple tools, which might well constitute bases for understandings of the causal structures of the external space and world.

CONCLUSION

Let us be reminded of our experience when using a most simple type of tool, such as a stick. By holding a stick in the hand, we can extend our reaching distance beyond the limit of our innate hand. Then, by this "extended" hand, we can feel with the tip, hit with the tip, and retrieve something using the tip. We can perceive these interactions with external space similar to the interaction performed by our innate hand. Hence, we can (1) physically, (2) perceptually, and (3) functionally extend our innate body structure. Further, a great advantage of having a tool at hand is not just extending our body parts in space in a fixed manner. It can be grasped any time we want, and can be abandoned any time we do ***not*** want to use it. Thus, one can select at any time any adequate tool that exactly matches the given situation. This would mean that we can intentionally and purposefully re-map our intrinsic bodily space so as to match the mode of interactions with extrinsic space. If this faculty is further advanced, one may become able to plan to use and even manufacture the tool to exactly fulfill one's purpose in one's mind, such as to explore and "understand" the external world.

In conclusion, the principal advantage of the tool as an extension of our innate body is that it does not constrain various aspects of the image of the body and we can modulate and choose it as we like at any given moment according to the current intention – we acquired a "freedom" to configure our functional body structure according to our free will. Therefore, flexible modification of internal representations free from physical constraints of the actual physical world should be the key element of the expression of the tool-using abilities. These abilities might be extrapolated into higher intellectual cognitive abilities of hominids, such as language and metaphysical thoughts, which might co-evolve with tool-use and technology.

REFERENCES

Berlucchi, G. and Aglioti, S. (1997). The body in the brain: neural bases of corporeal awareness. *Trends in Neuroscience*, **20**(12), 560–564.

Blakemore, S.-J. and Sirigu, A. (2003). Action prediction in the cerebellum and in the parietal lobe. *Experimental Brain Research*, **153**(2), 239–245.

Bruner, J. S., Olver, R. R. and Greenfield, P. M. (1966). *Studies in cognitive growth*. New York: Wiley.

Head, H. and Holmes, G. (1911). Sensory disturbances from cerebral lesions. *Brain*, **34**, 102–254.

Hihara, S., Obayashi, S., Tanaka, M. and Iriki, A. (2003). Rapid learning of sequential tool use by macaque monkeys. *Physiological Behaviour*, **78**(3), 427–434.

Iriki, A., Tanaka, M. and Iwamura, Y. (1996). Coding of modified body schema during tool use by macaque postcentral neurones. *NeuroReport*, **7**(14), 2325–2330.

Iriki, A. (2005). A prototype of *Homo faber*: a silent precursor or human intelligence in the tool-using monkey brain. In S. Dehaene *et al.* (Eds.), *From monkey brain to human brain* (pp. 253–271). Cambridge: The MIT Press.

Iriki, A., Tanaka, M., Obayashi, S. and Iwamura, Y. (2001). Self-images in the video monitor coded by monkey intraparietal neurons. *Neuroscience Research*, **40**(2), 163–173.

Kakei, S., Hoffman, D. S. and Strick, P. L. (2001). Direction of action is represented in the ventral premotor cortex. *Nature Neuroscience*, **4**, 1020–1025.

Maravita, A. and Iriki, A. (2004). Tools for the body (schema). *Trends in Cognitive Science*, **8**, 79–86.

Obayashi, S., Tanaka, M. and Iriki, A. (2000). Subjective image of invisible hand coded by monkey intraparietal neurons. *NeuroReport*, **11**(16), 3499–3505.

Obayashi, S., Suhara, T., Nagai, Y., Maeda, J., Hihara, S. and Iriki, A. (2002). Macaque prefrontal activity associated with extensive tool use. *NeuroReport*, **13**(17), 2349–2354.

Paillard, J. (1993). The hand and the tool: the functional architecture of human technical skills. In A. Berthelet and J. Chavaillon (Eds.), *The use of tools by human and non-human primates* (pp. 36–46). Oxford: Oxford University Press.

Tanaka, M., Obayashi, S., Yokochi, H., Hihara, S., Kumashiro, M., Iwamura, Y. and Iriki, A. (2004). Intraparietal bimodal neurons delineating extrinsic space through intrinsic actions. *Psychologia*, **47**, 63–78.

WILLIAM D. HOPKINS AND CLAUDIO CANTALUPO

19

Left–right spatial discrimination and the evolution of hemispheric specialization: some new thoughts on some old ideas

One aspect of spatial cognition that has been of historical and contemporary interest to psychologists from various sub-divisions is left–right mirror image discrimination. From the standpoint of spatial cognition, left–right discrimination presents an interesting problem because "right" and "left" are relative with respect to the allocentric or egocentric perspective of the participant. Neuropsychological problems in left–right confusion have been reported in several clinical populations such as Gerstmann's syndrome (Suresh and Sebastian, 2000). Moreover, lesions in Brodmann's cortical regions 7 and 40 have been linked to deficits in left–right discrimination (Benton, 1959; Semmes *et al.*, 1963), particularly when damage occurs in the left hemisphere (McFie and Zangwill, 1960). Indeed, left–right confusability has been linked to aspects of reading disability such as orthographic dyslexia (Orton, 1937). However, left–right discrimination problems are not restricted to clinical populations. Physicians, college professors and individuals with IQ scores exceeding 150 have all been reported to exhibit higher than normal self-reported problems in distinguishing left from right (Hannay *et al.*, 1990; Harris and Gitterman, 1978; Storfer, 1995). There is also some evidence that females report higher incidences of left–right discrimination problems than males (Harris and Gitterman, 1978; Stofer, 1995). Because sex differences have been found in spatial cognition, particularly for mental rotation (Kimura, 1999) a cognitive task predicated on left–right mirror-image discrimination, further evaluation of mirror-image discrimination seems highly warranted. Of specific interest to this chapter is the possible relationship between hemispheric specialization and the evolution of left–right discrimination in animals, including humans.

Spatial Cognition, Spatial Perception: Mapping the Self and Space, ed. Francine L. Dolins and Robert W. Mitchell. Published by Cambridge University Press.

Hemispheric specialization refers to perceptual, cognitive and motor specializations of the left and right cerebral hemispheres (Corballis, 2002; Hellige, 1993; McManus, 2002). Historically, hemispheric specialization has been considered a hallmark of human evolution and associated with the evolution of complex cognitive abilities such as language or tool use (Bradshaw and Rogers, 1993; Corballis, 1992, 2002). One longstanding theoretical model of the evolution of hemispheric specialization and its uniqueness to humans is that functional and neuroanatomical asymmetries were selected as means of distinguishing egocentric and allocentric space in relation to the left–right axis. In the specific context of left–right, egocentric space reflects the identification of left and right from the viewer's perspective whereas allocentric space reflects left and right from the opposite observer's perspective.

Specifically, it has been suggested that discriminating left from right would be impossible for bilaterally symmetrical organisms (Corballis and Beale, 1970, 1976). For example, looking into a mirror, one can see one's mirror-image reflection. In the absence of an inherent asymmetrical brain that provided a reference point for left and right, being able to identify the left and right sides of the mirrored image would be impossible from both an ego- and allocentric perspective. Indeed, being able to identify your own right hand and the right hand of the mirrored reflection would require that the subject be able to reference left and right from one's own perspective (egocentric) and from the perspective of the mirror-image (allocentric). In other words, an egocentric knowledge of left and right would be a necessary prerequisite for the ability to perform left–right spatial transformation of visual information. According to Corballis (2000), the top–bottom and back–front spatial axes have functional priority over the left–right one because there are external cues that provide reference points for discrimination on these planes, notably gravity in the case of top–bottom discrimination and asymmetric input of visual information (eyes in front of head but not back of head) that allow for back–front reference points. In contrast, there are no extrinsic cues that provide for discrimination of left and right, making the problem of left–right discrimination and lateralization of specific interest, particularly for terrestrial animals with binocular vision, such as primates. Given the historical view that hemispheric specialization is unique to humans, it follows that left–right discrimination should also be uniquely human.

With respect to findings in human subjects, there is at least some evidence in support of the theory by Corballis and Beale (1976). In two separate studies, less lateralized individuals, as assessed by self-reported handedness, report having greater problems in left–right discrimination than more strongly lateralized individuals (Harris and Gitterman, 1978; Ofte and Hugdahl, 2002; Storfer, 1995). With respect to the comparative data, the basic proposal of Corballis and Beale (1976) has largely been borne out when considering results on mirror-image discrimination in nonhuman animals. Studies on mirror-image discrimination in nonhuman animals, including rats, birds, cats, dogs, monkeys and chimpanzees have shown that they have significantly greater difficulty in learning to discriminate mirror-image stimuli that are rotated on the vertical (left–right) compared to horizontal (up–down) axis (Beale and Corballis, 1968; Corballis and Beale, 1976; Lohmann *et al.*, 1988; Mello, 1964; Nissen and McCulloch, 1937; Riopelle *et al.*, 1964; Todrin and Blough, 1983; Weiss and Hodos, 1986). The exception to this observation is split-brained animals (Beale *et al.*, 1972; Noble, 1966; Noonan and Axelrod, 1991; but see Hamilton *et al.*, 1973) or animals with laterally placed eyes when tested monocularly, which apparently do not have these same difficulties (this is discussed below). As noted above, the difficulty for neurologically intact animals with binocular vision to perform left–right, mirror-image discrimination has historically been attributed to the lack of inherent functional and neuroanatomical asymmetries (Corballis and Beale, 1976). According to this theory, the ability to discriminate mirror images on the left–right plane would depend on the presence of asymmetry between the perceiver's left and right cerebral hemispheres. Thus, like language, the argument follows that left–right discrimination would be unique to humans and virtually absent in other animals, presumably because they lack asymmetries in the brain.

One problem with this theory is the assumption that animals lack inherent functional and neuroanatomical asymmetries. In the past fifteen years, investigators have described a plethora of evidence of population-level behavioral and neuroanatomical asymmetries in various vertebrate species, including lower vertebrates. For example, behavioral asymmetries in perceptual and cognitive processes have been described in rats, frogs, chicks and fish (see Rogers and Andrew, 2002 for review). In primates, population-level behavioral and neuroanatomical asymmetries have been described in a host of species (Bradshaw and Rogers, 1993; Hopkins *et al.*, 2003; Rogers and Andrew, 2002). Thus, there is a paradoxical set of findings in the literature. Behavioral studies suggest nonhuman animals, particularly nonhuman primates, are very poor at mirror-image discrimination, yet recent studies suggest the existence of behavioral and brain asymmetries. Moreover, there is at least some evidence that animals can successfully perform mental rotation tasks, an ability requiring the ability to perform left–right mirror-image discrimination (Burmann *et al.*, 2005; Ettlinger and Elithorn, 1962; Hollard and Delius, 1982; Hopkins *et al.*, 1993; Köhler *et al.*, 2005; Mauck and Dehnhardt, 1999; Stich *et al.*, 2002; Vauclair *et al.*, 1993).

One interpretation of these paradoxical findings might be that mirror-image discrimination has nothing to do with the existence of behavioral and neuroanatomical asymmetries. However, this does not seem likely in light of the fact that individual differences in the ability to discriminate left from right are associated with increasing levels of behavioral lateralization in humans and possibly other animals. Alternatively, there might be limits to the methods and approaches that have been historically used in data collection that has precluded an accurate assessment of mirror-image discrimination abilities in nonhuman animals. In this chapter, we reconsider the theory that left–right discrimination is associated with the evolution of asymmetries in animals. First, we present some new approaches and findings on left–right discrimination in monkeys and chimpanzees. Second, we discuss the potential relationship between left–right discrimination, corpus callosum morphology and the evolution of neuroanatomical asymmetries in primates.

DO NONHUMAN PRIMATES HAVE AN INTRINSIC KNOWLEDGE OF LEFT AND RIGHT?

Most previous studies on mirror-image discrimination in nonhuman primates have focused on visual discrimination or perceptual problems in either a standard two-choice discrimination paradigm or matching-to-sample. Rather than address the issue of mirror-image discrimination entirely as a *perceptual* problem, we have recently tried to consider left–right discrimination from the standpoint of motor processing as well as a perceptual problem. One of the observations that prompted our renewed interest in the relationship between left–right discrimination and brain asymmetry comes from the acquisition of data on an automated joystick testing system developed at the Language Research Center referred to as the LRC-CTS (Language Research Center's Computerized Testing System). The LRC-CTS is an automated test system that presents stimuli and records output responses via the manipulation of a joystick by the subject. Initially, the SIDE program is the task used to train naive animals to reliably manipulate a joystick in relation to the movements of the cursor displayed on the computer monitor. The task begins with a computer-generated 2.5 cm blue border appearing on and completely covering each edge of the computer monitor. The cursor, a white 1-cm circle, is presented in the middle of the screen, surrounded by these four blue target walls. Moving the joystick in any direction causes the cursor to move at a speed of 8 cm/s, eventually resulting in contact between the cursor and one of the four target walls. Such contact produces audio feedback and the delivery of food reinforcement to the subject. Task difficulty is automatically titrated, based on performance, by decreasing first the number and then the size of the target walls. Specifically, hitting the target in a mean latency of 5 s or less, averaged over five consecutive trials, results in a decrease in the number/size of target walls (four walls, three walls, two walls, one full wall, partial wall-1 (target = 3 × 16 cm), partial wall-2 (target = 3 × 6 cm), or partial wall-3 (target = 1.5 × 3 cm, located on any border of the screen). Responses in excess of 20 s, averaged over five consecutive trials, produce an increase in the number/size of target walls. In all conditions, target walls are assigned to positions about the edge of the screen randomly each trial. The position of the target is either up, down, left or right of the cursor.

Although different training paradigms have been used over the years with this software (e.g., Hopkins *et al.*, 1996; Rumbaugh *et al.*, 1989), seven rhesus monkeys and seven chimpanzees were trained on the SIDE task using identical procedures (Hopkins *et al.*, 1996). On the whole, the chimpanzees reached criterion faster than the rhesus monkeys but the relevant finding from this study was the type of errors made by the subjects. Both chimpanzees and rhesus monkeys made significantly more errors when the target stimuli were located on the left–right compared to the top–bottom positions of the screen.

The discrepancy between performance on the left–right compared to the top–bottom axis was particularly evident for the rhesus monkeys compared to the chimpanzees. One interpretation of this finding is that the chimpanzees have an inherent knowledge of left and right compared to the monkeys and that they use this intrinsic knowledge and map it onto the spatial and motor demands of the SIDE task.

To further explore the relationship between target position and movement demands within the LRC-CTS SIDE task, we have recently conducted a preliminary study using a task referred as the motor inversion task. This task is similar to the SIDE task described above.

Briefly, the task involves the presentation of a cursor on the center of a computer screen and a rectangular target centrally placed on one of the four sides of the screen, in random order across trials. The main difference between the SIDE and motor-inversion task is that, in the former, the movement of the cursor is always isomorphic or congruent with that of the joystick, whereas, in the latter, the cursor moves in the opposite direction to that of the joystick, depending on which axis (left–right or top–bottom) the target is placed on. For example, when the target is placed either on the left or right margin of the screen, pushing the joystick to the left makes the cursor move to the right and vice versa, whereas joystick and cursor movements on the top–bottom axis remain congruent. Conversely, when the target is placed either on the top or bottom margin of the screen, pushing the joystick to the top makes the cursor move to the bottom and vice versa (in this case, joystick and cursor movements on the left–right axis remain congruent). In order to bring the cursor to collision with the target, the subject needs to learn to reverse the correspondence between joystick and cursor movements on a given axis. Four common chimpanzees were tested on this task. Figure 19.1 shows the mean time (in msec) required to reach the target as a function of test session. In the first session, the chimpanzees spent a significantly longer time (i.e., had greater difficulty) in reversing the correspondence between joystick and cursor

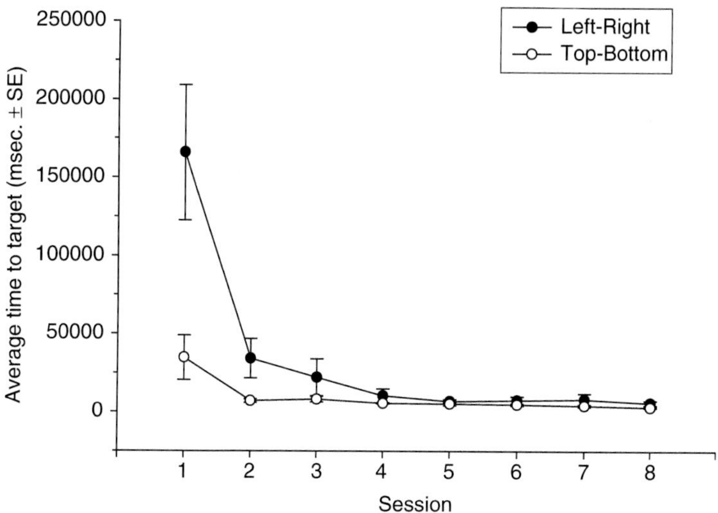

Figure 19.1. Mean latency to contact target when cursor movements were rotated on the left–right compared to top–bottom axis in chimpanzees.

movements on the left–right axis as compared to the top–bottom one ($t(3) = 3.323$, $p < 0.05$). This evidence clearly converges with the results reported above, that chimpanzees experience greater difficulty in discriminating mirror images on the left–right as compared to the top-bottom axis. These studies differ from a recent finding in capuchin monkeys in which different movements of the cursor relative to the joystick had no significant influence on SIDE task acquisition and re-learning (Leighty and Fragaszy, 2003).

MIRROR-IMAGE DISCRIMINATION IN RHESUS MONKEYS AND COMMON CHIMPANZEES USING MTS

With respect to the perception of mirror-image stimuli, rather than focus specifically on mirror-image stimuli, we have recently tested four joystick-trained rhesus monkeys and four common chimpanzees on their sensitivity to the parity or sidedness of arbitrary visual stimuli that were asymmetrically shaped either on the left–right or top–bottom axis. The subjects were tested using a computerized implementation of the MTS paradigm commonly used with joystick-trained nonhuman primates (Washburn and Rumbaugh, 1992). On each trial, one of the stimuli (i.e., the sample), randomly selected, was presented on the top portion of a computer screen. The subject was trained to move a star-shaped cursor by means of a joystick toward the sample. Once collision between the cursor and the sample had occurred, two comparison stimuli were shown on the bottom portion of the computer screen. One stimulus was identical to the sample, whereas the other was the mirror image of the sample. To complete the trial successfully, the subject had to bring the cursor to collision with the comparison stimulus identical to the sample. Each subject was tested up to a minimum of 600 trials. As shown in Figure 19.2, both monkeys and chimpanzees did poorly in discriminating on the basis of parity, although discrimination along the top–bottom axis seemed to be better than along the left–right axis. These results are remarkably congruent with those of other studies in the literature (Riopelle *et al.*, 1964), as well as those of a number of other MTS tests on mirror-image discrimination previously conducted with the monkeys and chimpanzees at the LRC, using a variety of visual stimuli (e.g., geometrical shapes, letters of the alphabet, Arabic numerals and pictures of objects). However, this does not warrant concluding that rhesus monkeys or chimpanzees are unable to (1) perceive left–right asymmetry and

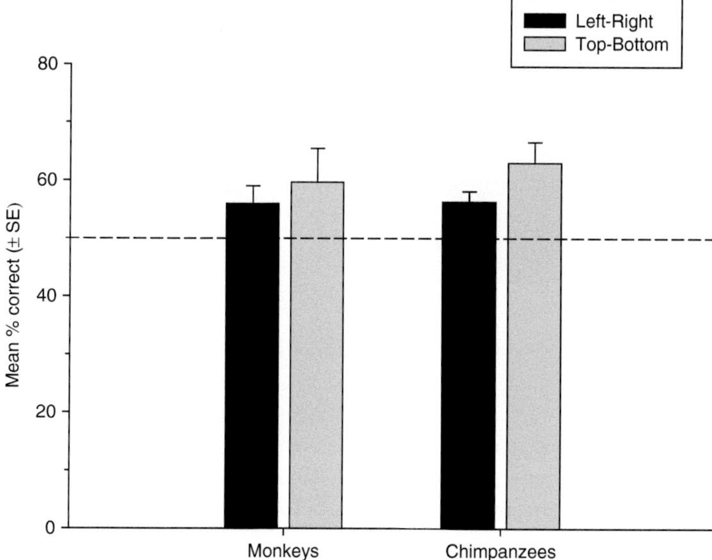

Figure 19.2. Mean percentage correct on discrimination of stimuli differing in parity on the left–right compared to top–bottom axis in rhesus monkeys and chimpanzees.

(2) utilize this information to achieve mirror-image discrimination. Possibly, the difficulty in detecting and measuring such abilities in these species could be in part a reflection of limitations of the specific test paradigms so far implemented.

MIRROR-IMAGE DISCRIMINATION USING CONDITIONAL DISCRIMINATION

In this line of reasoning, we have started data collection on a new computerized task. This task is designed to assess whether joystick-trained rhesus monkeys and chimpanzees can perceive the asymmetry of a visual stimulus on the left–right and top–bottom axes (e.g., an arrow pointing left or right, or up or down, respectively), and utilize such information successfully in learning to produce a lateralized motor response (e.g., move the arrow toward the side it is pointing to). This task involves the presentation of an arrow-shaped cursor on the center of a computer screen, pointing either to the left, right, top or bottom margin of the screen (in random order across trials). Two identical rectangular targets are shown equidistant from the cursor and arranged either horizontally (i.e., one on the left, the other on the

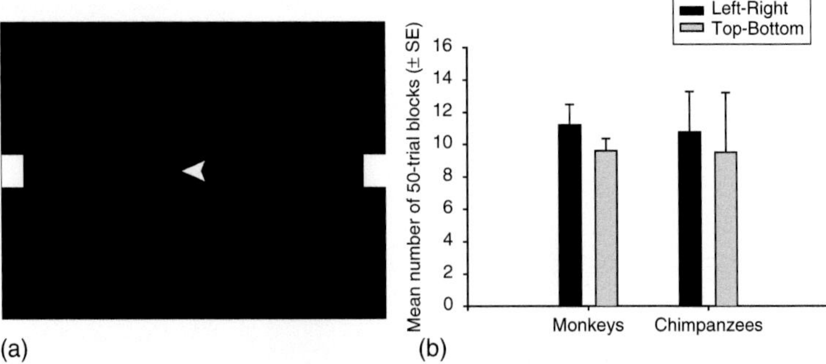

Figure 19.3. (a) Example screen shot appearance of the parity discrimination task with the asymmetry cue positioned in the center of the screen. A correct response would require that the subjects move the center stimulus to the left. (b) Mean number of trials needed to learn conditional motor movement on the left–right compared to top–bottom axis in chimpanzees and monkeys based on the directional visual cues.

right screen margin) or vertically (i.e., one on the top, the other on the bottom screen margin), depending on the orientation of the main axis of the cursor (i.e., horizontal or vertical respectively, see Figure 19.3a). The subject can move the cursor across the screen by means of a digital joystick. In order to perform a trial successfully, the subject needs to bring the arrow-shaped cursor to collision with the target the cursor is pointing toward. Five rhesus monkeys and four common chimpanzees were tested on this task until they reached a criterion of 90 percent correct trials in a fifty-trial block. Figure 19.3b shows the mean number of fifty-trial blocks needed by the subjects to reach criterion. In contrast with the relatively poor performance on the mirror-image discrimination MTS task described above, both monkeys and chimpanzees were in this case clearly able to perceive and successfully utilize the directional information afforded by the asymmetrical stimulus. Interestingly, also in this case, there are indications that processing directional information was more challenging on the left–right axis than on the top–bottom one.

MATCH-TO-SAMPLE (REVISITED)

Following the previous results, we returned to assessing whether the rhesus monkeys (the chimpanzees were not tested on this task) could now perform some type of mirror-image discrimination in a MTS format. The revised MTS task involves the presentation on center screen of

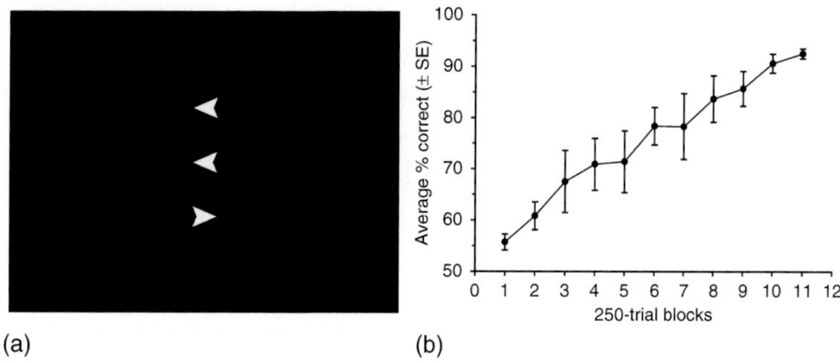

Figure 19.4. (a) Screen shot of the revised mirror-image discrimination MTS task presented to the monkeys. (b) Percentage of correct trials across blocks of trials in revised MTS task in rhesus monkeys.

the same arrow-shaped cursor shown in the previous task, again pointing to the left, right, up or down, in random order. In this case, however, the two comparison stimuli are identical copies of the cursor, one having the same orientation as the cursor, the other rotated by 180 degrees (thus, representing the mirror image of the cursor). The targets were placed equidistant from the cursor and aligned with it along their common axis of asymmetry (see Figure 19.4a). To perform a trial successfully, the subject needed to bring the cursor into collision with the target having the same orientation as the cursor. The target stimuli were left–right mirror-image stimuli. Thus, this task is a variation of the MTS test paradigm described above. The main difference between the two tasks is that in the new task the sample stimulus *is* the cursor, whereas in the previous task the sample and the cursor were separate entities (the possible implications of this difference are discussed below). The preliminary results are shown in Figure 19.4b and suggest that, with the new MTS paradigm, rhesus monkeys can indeed learn to discriminate between the mirror-image stimuli, although asymptotic performance is reached more slowly as compared to the conditional discrimination task, which requires no direct comparison between mirror-image stimuli.

MIRROR-IMAGE DISCRIMINATION AND THE CORPUS CALLOSUM

One of the more interesting observations from a clinical and experimental standpoint is the relationship between mirror-image discrimination and the presence or absence of the corpus callosum. Experimental work with birds, rats and monkeys (Beale *et al.*, 1972;

Noble, 1966; Noonan and Axelrod, 1991) has demonstrated that cutting the corpus callosum and effectively isolating communication between the two hemispheres enhances mirror-image discrimination performance. It has also been argued that severing the anterior commissure is principally responsible for enhanced left–right mirror image discrimination because it connects inferior temporal regions of the cortex, an area of the brain involved in higher-order visual processing (Achim and Corballis, 1977). In other words, split-brained animals have less difficulty discriminating left–right mirror images compared to neurologically intact subjects. It has also been reported that unilateral damage to the inferior temporal cortex spares mirror-image discrimination but results in decreased performance for other forms of visual discrimination (Cowey and Gross, 1970). More recently, it has been demonstrated in neurologically intact baboons that isolating visual information to one hemisphere using tachistoscopic presentations enhances mirror-image discrimination compared to bilateral presentation of visual stimuli (Hopkins *et al.*, 1993). The neurological explanation for these findings is that, in the intact condition, there is a dual representation of visual stimuli in each hemisphere. These two representations are connected via homotopic fibers that run through the corpus callosum and connect homologous regions between the two hemispheres. In the absence of an inherent dominant hemisphere in the intact subject, the two hemispheres compete for representation in discrimination performance and this leads essentially to random responding. In contrast, cutting the corpus callosum and isolating information to one hemisphere eliminates competition between the memory traces of the two hemispheres and allows mirror-image discrimination to take place because the visual stimulus is not represented in each hemisphere.

The experimental evidence of enhanced mirror-image discrimination in split-brain compared to intact animals is mimicked, to some degree, by phylogenetic variation in the size of the corpus callosum relative to brain size and neocortical surface area. Depicted in Figure 19.5 are data published by Rilling and Insel (1999) on the relationship between corpus callosum (CC) size and brain volume in various primate species. In this study, MRI scans were collected in a sample of nonhuman primates representing four different taxonomic families. Rilling and Insel (1999) divided the size of the corpus callosum by both the overall brain volume (CC:VOL) and total neocortical surface area (CC:NEO). As can be seen, species more distantly related to humans (i.e., New World monkeys) have larger ratios than species more closely

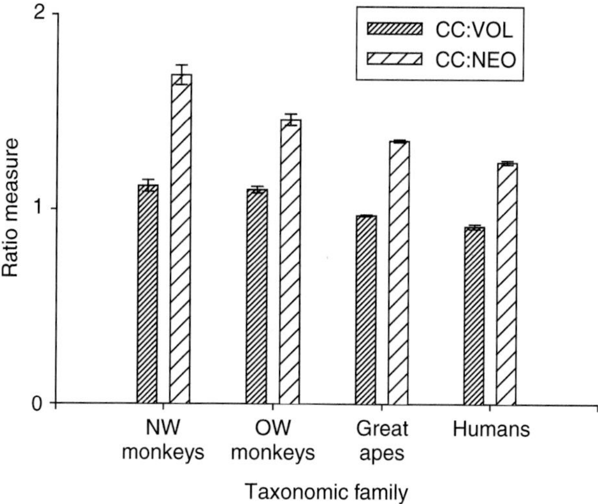

Figure 19.5. Ratio of CC size to brain volume (CC:VOL) and neocortical surface (CC:NEO) area in different primate families. Redrawn from Rilling and Insel (1999).

related (i.e., great apes). Overall, humans have the lowest ratio followed by great apes, lesser apes, Old World monkeys and New World monkeys. The results reported by Rilling and Insel (1999) are consistent with a recent report on the size of the CC relative to brain size in a host of species including rats, rabbits, cats, dogs, horses, cows and humans. Olivares *et al.* (2000) reported a negative association between CC size and brain weight and these effects were particularly evident in the posterior portion of the CC, where visual information would be shared between hemispheres. Olivares *et al.* (2000) attributed these findings to inherent differences in the degree of bilateral overlap of the visual system. One interpretation of the species differences in CC ratio measures is that as the brain increased in size during evolution, each hemisphere became increasingly independent and this led to greater specialization within a hemisphere. According to Rilling and Insel (1999), the increased behavioral and neuroanatomical lateralization in humans is reflected in greater intrahemispheric compared to interhemispheric connectivity.

There is some evidence in support of this view. In a recent study, Hopkins and Rilling (2000) regressed the CC:VOL and CC:NEO measures on measures of brain asymmetry in different primates species. Brain asymmetries were measured by comparing cortical volume at 10 percent of the total length of the left and right frontal and occipital poles.

This is a relatively crude measure of brain asymmetry but one that could be employed with all the species. In the analysis, we assessed whether brain asymmetry was associated with either (1) the ratio of CC size to brain volume (total CC/Volume; CC:VOL) or (2) the ratio of CC size to neocortical area (totalCC/Neocortex; CC:NEO) by regressing the brain asymmetry scores on each ratio score for the entire sample. In the CC:VOL ($R = 0.57$, $F(3,36) = 4.43$, $p < 0.009$) and CC:NEO ($R = 0.58$, $F(3,36) = 4.00$, $p < 0.009$) measures, brain asymmetry accounted for a significant proportion of variance. The asymmetry in occipital pole was the single best and only significant predictor of CC:VOL ($r = -0.56$, $p < 0.01$) and CC:NEO ($r = -0.57$, $p < 0.01$) scores. Thus, individuals with larger left occipital lobe asymmetries had lower brain volume to CC size ratios as well as lower neocortical area to CC size ratios.

These analyses indicated that the assumption that smaller ratios in the size of the CC relative to brain volume are associated with increasing brain asymmetry may be valid. Moreover, combining the CC:VOL, brain asymmetry and mirror-image discrimination data results, the results are consistent with Corballis and Beale's theory that poor mirror-image discrimination is associated with smaller brain asymmetries and larger CC:VOL ratios. Recall that humans are more lateralized and have the smallest CC:VOL values and are presumably better at mirror-image discrimination than other animals. Moreover, more lateralized subjects report having less problems in left–right discrimination than less lateralized individuals (Storfer, 1995). In one of the only studies that directly compared asymmetry and left–right discrimination, Sanford and Ward (1986) found no association between strength of hand preference and mirror-image discrimination in a sample of eight bushbabies. In both the human and animal literature, what is lacking is direct experimental evidence of an association between mirror-image discrimination, relative CC size and neuroanatomical asymmetry. The prediction would be that relative size of the CC to whole brain volume would predict performance on left–right mirror-image discrimination performance. Arguably, the same theoretical prediction could be made for explaining both within and between species performance.

CONCLUSIONS

With respect to the behavioral data, the cumulative results here reported suggest that rhesus monkeys and chimpanzees may indeed (1) perceive left–right asymmetry and (2) utilize such information to achieve mirror-image discrimination. Although preliminary, this evidence nonetheless clearly shows the possibility of devising test paradigms that are sensitive enough to assess and measure systematically the potential ability of great apes and monkeys to discriminate mirror images on the left–right as well as other axes. The data yielded by tasks like these would offer an adequate empirical basis to directly assess the tenability of the hypothesis of Corballis and Beale of a positive relationship between the ability to tell left from right and the degree of hemispheric asymmetry. Further, by comparing the results from different tasks, we would gain valuable insights into the nature of the cognitive mechanisms and demands involved in mirror-image discrimination in nonhuman primates. For example, the fact that the monkeys could learn to discriminate mirror images only when the sample stimulus could be moved closer to the comparison stimuli (as in Task Two, where the cursor itself was the sample) could indicate a crucial role played by (1) the amount of attentional resources as well as (2) the size of the focus of attention required to encode and remember information on the orientation of visual stimuli. Seemingly, rhesus monkeys can successfully encode and compare the different orientation of two otherwise identical visual stimuli only when both stimuli can be brought under a relatively narrow attentional focus. In other words, the difficulty of rhesus monkeys in mirror-image discrimination could be a reflection of a limitation in the amount of attentional resources available for processing directional information in the visual domain, as has been suggested by others as an explanation for mirror-image discrimination difficulties in animals (Hamilton *et al.*, 1973). That this might be the case seems to be supported also by the fact that the monkeys reached asymptotical performance earlier when the task required processing directional information of only one visual stimulus at a time (as in Task One presented above), as compared to when the task involved encoding and comparing the directional information of two stimuli simultaneously.

The methods we have developed may also be of interest in the context of comparative studies of visual imagery. A requisite skill for visual imagery as revealed by studies on mental rotation is mirror-image discrimination. Mental rotation is a well-known documented human ability to recognize objects or stimuli rotated to various degree of angularity from their original orientation. There is a strong linear association in the time needed to recognize the shape and the degree of angulation from 0 degrees to 180 degrees. To eliminate the use of featural cues to recognize rotated stimuli, the foil or response demand requires that the subject indicate whether the rotated object is presented in its "normal" or mirror-image orientation at zero-degree rotation. There have been a number of recent studies in animals including pigeons, sea lions and monkeys reporting evidence of mental rotation but the results are somewhat ambiguous (see Stich *et al.*, 2002 for review). In these reports, the animals can solve the mental rotation problems but do not uniformly show a linear association between the degree of angular rotation and response time (or accuracy) either within or between species. Some have attributed these differences to ecological and sensory differences between birds and aquatic species compared to terrestrial primates (e.g., Hollard and Delius, 1982; Stich *et al.*, 2003). It can certainly be argued that the sensory and perceptual processes underlying left–right mirror-image discrimination is likely quite different between species that vary in the degree of binocular vision. For species with binocular vision, neural processing of visual stimuli occurs in parallel and is bilaterally represented at the sensory level. Higher level processing of mirror-image stimuli would have to occur between hemispheres. In contrast, for species with little or no binocular vision, neural processing of mirror-image stimuli occurs sequentially because the sensory registration is or can be initially unilateral. Alternatively, the reliance on the standard MTS paradigm may limit the extent to which the animals may or may not use a "mental rotation" strategy to solve the task.

In summary, in our view, there is no overwhelming direct evidence in support of the relationship between the evolution of left–right mirror image discrimination and brain asymmetry. Data from rats and birds are certainly suggestive of the role of interhemispheric commissures in the formation of visual representations in each hemisphere but how these generalize to primates remain relatively unstudied but warrants further investigation. Beyond the focus on asymmetry and left–right mirror-image discrimination, continued studies on this topic would be of interest in terms of understanding neurobiological systems involved in spatial cognition and associated handedness and sex effects that have been reported in the literature. Lastly, we have focused on the relationship between hemispheric specialization and the potential evolution of left–right discrimination but clearly other brain systems may be involved and this warrants further investigation. For instance, the hippocampus, particularly the right one, has been implicated in spatial cognition (e.g., Maguire *et al.*, 2000). Interestingly, a study in monkeys with bilateral hippocampus lesions demonstrated that they could learn the spatial location of food items using allocentric but not egocentric cues (Hampton *et al.*, 2004). Chimpanzees have been reported to show a right-hemisphere asymmetry in the size of the hippocampus (Freeman *et al.*, 2004) and at least one study has reported a right-hemisphere asymmetry in visual-spatial discrimination in chimpanzees (Hopkins and Morris, 1989). Collectively, these results imply that spatial cognition associated with left–right discrimination may be associated with increasing isolation of the two hemispheres (see Rilling and Insel, 1999) which had the concomitant effect of increasing the right hemisphere's role in spatial cognition.

ACKNOWLEDGMENT

This work was supported in part by NIH grants RR-00165, NS-29574, NS-36605, NS-42867 and HD-38051.

REFERENCES

Achim, A. and Corballis, M. C. (1977). Mirror-image equivalence and the anterior commissure. *Neuropsychologia*, **15**, 475–479.

Beale, L. L. and Corballis, M. C. (1968). Beak shift: an explanation for interocular mirror-image reversal in pigeons. *Nature*, **220**, 82–83.

Beale, L. L., Williams R. J., Webster, D. M. and Corballis, M. C. (1972). Confusion of mirror-images by pigeons and interhemispheric commissures. *Nature*, **238**, 348–349.

Benton, A. L. (1959). *Right-left discrimination and finger localization.* New York: Hoeber-Harper.

Bradshaw, J. and Rogers, L. J. (1993). *The evolution of lateral asymmetries, language, tool use and intellect.* San Diego: Academic Press, Inc.

Burmann, B., Dehnhardt, G. and Mauck, B. (2005). Visual information processing in the lion-tailed macaque (*Macaca silenus*): mental rotation or rotational invariance? *Brain, Behavior and Evolution*, **65**, 168–176.

Corballis, M. C. (1992). *The lopsided brain: evolution of the generative mind.* New York: Oxford University Press.

Corballis, M. C. (2000). Much ado about mirrors. *Psychonomic Bulletin & Review*, **7**, 163–169.

Corballis, M. C. and Beale, I. L. (1970). Bilateral symmetry and behavior. *Psychological Review*, **77**, 451–464.

Corballis, M. C. and Beale, I. L. (1976). *The psychology of left and right.* Hillsdale, NJ: Erlbaum.

Cowey, A. and Gross, C. G. (1970). Effects of foveal prestriate and inferotemporal cortex lesions on visual discriminations by rhesus monkeys. *Experimental Brain Research*, **11**, 128–144.

Ettlinger, G. and Elithorn, A. (1962). Transfer between the hands of a mirror-image tactile shape discrimination. *Nature*, **194**, 1101–1102.

Freeman, H., Cantalupo, C. and Hopkins, W. D. (2004). Asymmetries in the hippocampus and amygdala of chimpanzees (*Pan troglodytes*). *Behavioral Neuroscience*, **118**, 1460–1465.

Hamilton, C. R., Tieman, S. B. and Brody, B. A. (1973). Interhemispheric comparison of mirror-image stimuli in chiasm-sectioned monkeys. *Brain Research*, **58**, 415–425.

Hampton, R. R., Hampstead, B. M. and Murray, E. A. (2004). Selective hippocampal damage in rhesus monkeys spatial cognition in an open field test. *Hippocampus*, **14**, 808–818.

Hannay, H. J., Ciaccia, P. J., Kerr, J. W. and Barrett, D. (1990). Self-report of left-right confusion in college men and women. *Perceptual and Motor Skills*, **70**, 451–457.

Harris, L. J. and Gitterman, S. R. (1978). University professors' self-descriptions of left-right confusability: sex and handedness differences. *Perceptual and Motor Skills*, **47**, 819–823.

Hellige, J. B. (1993). *Hemispheric asymmetry: what's right and what's left?* Cambridge, MA: Harvard University Press.

Hopkins, W. D. and Morris, R. D. (1989). Laterality for visual-spatial processing in two language-trained chimpanzees (*Pan troglodytes*). *Behavioral Neuroscience*, **103**, 227–234.

Hopkins, W. D. and Rilling, J. K. (2000). A comparative MRI study of the relationship between neuroanatomical asymmetry and interhemispheric connectivity in primates: implication for the evolution of functional asymmetries. *Behavioral Neuroscience*, **114**, 739–748.

Hopkins, W. D., Fagot, J. and Vauclair, J. (1993). Mirror-image matching and mental rotation problem solving by baboons (*Papio papio*): unilateral input enhances performance. *Journal of Experimental Psychology: General*, **122**, 61–72.

Hopkins, W. D., Pilcher, D. and Cantalupo, C. (2003). Brain substrates for communication, cognition, and handedness. In D. Maestripieri (Ed.), *Primate psychology* (pp. 424–450). Cambridge, MA: Harvard University Press.

Hopkins, W. D., Washburn, D. A. and Hyatt, C. (1996). Video-task acquisition in rhesus monkeys (*Macaca mulatta*) and chimpanzees (*Pan troglodytes*): a comparative analysis. *Primates*, **37**, 197–206.

Kimura, D. (1999). *Sex and cognition*. Cambridge, MA: The MIT Press.

Köhler, C., Hoffman, K. P., Dehnhardt, G. and Mauck, B. (2005). Mental rotation and rotational invariance in the rhesus monkey (*Macaca mulatta*)? *Brain, Behavior and Evolution*, **66**, 158–166.

Leighty, K. A. and Fragaszy, D. M. (2003). Joystick acquisition in tufted capuchins (*Cebus apella*). *Animal Cognition*, **6**, 108.

Lohmann, A., Delius, J. D., Hollard, V. D. and Friesel, M. F. (1988). Discrimination of shape reflections and shape orientations by *Columba livia*. *Journal of Comparative Psychology*, **102**, 3–13.

Maguire, E. A., Gadian, D. G., Johnsrude, I. S, Good, D. C., Ashburner, R. Frackowiak, S. J. and Frith, C. D. (2000). Navigation-related structural change in the hippocampi of taxi drivers. *Proceedings of the National Academy of Sciences*, **97**, 4398–4403.

Mauck, B. and Dehnhardt, G. (1999). Mental rotation in a California sea lion (*Zalophus californianus*). *The Journal of Experimental Biology*, **200**, 1309–1316.

McFie, J. and Zangwill, O. L. (1960). Visual-constructive disabilities associated with lesions of the left cerebral hemisphere. *Brain*, **83**, 243–260.

McManus, C. (2002). *Right hand, left hand: the origins of asymmetry in brains, bodies, atoms, and cultures*. Great Britain: Weidenfeld & Nicolson

Mello, N. K. (1964). Interhemispheric reversal of mirror-image oblique lines after monocular training in pigeons. *Science*, **148**, 252–254.

Nissen, H. W. and McCulloch, T. L. (1937). Equated and non-equated stimulus conditions learning by chimpanzees: comparison with unlimited response. *Journal of Comparative Psychology*, **23**, 165–189.

Noble, J. (1966). Mirror images and the forebrain commissures of the monkey. *Nature*, **211**, 1263–1265.

Noonan, M. and Axelrod, S. (1991). Improved acquisition of left-right response differentiation in the rat following section of the corpus callosum. *Behavioural Brain Research*, **46**, 135–142.

Ofte, S. H. and Hugdahl, K. (2002). Right-left discrimination in male and female, young and old subjects. *Journal of Experimental and Clinical Neuropsychology*, **24**, 82–92.

Olivares, R., Michalland, S. and Aboitiz, F. (2000). Cross-species and intraspecies morphometric analysis of the corpus callosum. *Brain, Behavior and Evolution*, **55**, 37–43.

Orton, S. T. (1937). *Reading, writing and speech problems with children*. New York: Norton

Rilling, J. K. and Insel, T. R. (1999). Differential expansions of neural projection systems in primate brain evolution. *NeuroReport*, **10**, 1453–1459.

Riopelle, A. J., Rahm, U., Itoigawa, N. and Draper, W. A. (1964). Discrimination of mirror-image patterns by rhesus monkeys. *Perceptual & Motor Skills*, **19**, 383–389.

Rogers, L. J. and Andrews, R. (2002). *Comparative vertebrate lateralization*. Oxford: Oxford University Press.

Rumbaugh, D. M., Richardson, W. K., Washburn, D. A., Savage-Rumbaugh, E. S. and Hopkins, W. D. (1989). Rhesus monkeys learn complex video tasks: stimulus-response discontinuity revisited. *Journal of Comparative Psychology*, **103**, 67–75.

Sanford, C. G. and Ward, J. P. (1986). Mirror image discrimination and hand preference in the bushbaby. *The Psychological Record*, **36**, 439–449.

Semmes, J., S. Weinstein, S, Ghent, I. and Teuber, H. L. (1963). Correlates of impaired orientation in personal and extra-personal space. *Brain*, **86**, 747–772.

Stich, K. P., Dehnhardt, G. and Mauck, B. (2002). Mental rotation of perspective stimuli in a California sea lion (*Zalophus californianus*). *Brain, Behavior and Evolution*, **61**, 102–112.

Storfer, M. D. (1995). Problems of left-right discrimination in a high-IQ population. *Perceptual and Motor Skills*, **81**, 491–497.

Suresh, P. A. and Sebastian, S. (2000). Developmental Gerstmann's syndrome: a distinct clinical entity of learning disabilities. *Pediatric Neurology*, **22**, 267–278.

Todrin, D. C. and Blough, D. S. (1983). The discrimination of mirror-image forms by pigeons. *Perception and Psychophysics*, **34**, 397–402.

Vauclair, J., Fagot, J. and Hopkins, W. D. (1993). Rotation of mental images in baboons when visual input is direct to the left cerebral hemisphere. *Psychological Science*, **4**, 99–103.

Washburn, D. A. and Rumbaugh, D. A. (1992). Testing primates with joystick-based automated apparatus: lessons from the Language Research Center's Computerized Test System. *Behavior Research Methods, Instruments, & Computers*, **24**, 157–164.

Weiss, S. R. B. and Hodos, W. (1986). Discrimination of mirror-image stimuli after lesions of the visual system in pigeons. *Brain, Behavior and Evolution*, **29**, 207–222.

Part V Comparisons of human and nonhuman primate spatial cognitive abilities

RICHARD SAMBROOK AND DAVID ZURICK

20

The geographical imagination

Space, in a humanistic sense, is composed of places – lived worlds – and not simply the abstract coordinates of Descartes. We live by moving among places, from which we find meaning and, literally, direction. Over the course of our lives, external sources of spatial information are apprehended and harmonized with internal knowledge of orientation and place based upon personal experience. Our sense of place engages our sense of self; emotional as well as intellectual connections to places frame much of our spatial decision-making. Our spatial behavior depends upon how we perceive and conceive places and their geographical relations. The roles of mental maps and environmental perception in spatial decision-making (Downs and Stea, 1977; Golledge and Rushton, 1976) highlight the importance of individual experience, membership in cultural and social systems, and the bridge between the two (Sack, 2003; Tuan, 1977, 1989). What we observe and the sense we make of it are structured by the discursive practices that make up knowledge, and take into account both the visible world and its representation (Foucault, 1970; Schwartz and Ryan, 2003).

Our spatial decisions follow from our culturally and individually acquired beliefs, values and material concerns. Accordingly, there exist multiple perspectives of space that are influenced by such diverse factors as idiosyncrasy, human culture, representation and technology. The existence of so many spatial perspectives has spurred geographers to think of humans as having "geographical imagination" that is dependent not only on the nature of places but also on how we talk and think about them (Gregory, 1994).

Considerable attention has been paid to deconstructing the notion of a common human spatial experience. All environments – natural and social – contribute external sources of spatial information,

Spatial Cognition, Spatial Perception: Mapping the Self and Space, ed. Francine L. Dolins and Robert W. Mitchell. Published by Cambridge University Press.

but our *processing* of such information through cognitive structures is a decidedly individual experience (Blaut, 1991). This individuality highlights a concern of geographers: all spatial behavior requires knowledge of a setting's spatial structure as well as the setting's social/cultural makeup (Hanson, 1999). Social constructions modulate our access to environmental and spatial systems, and thus partially determine our spatial behavior. Spatial knowledge and behavior varies according to age (Pandit, 1997), gender (Jones *et al.*, 1997), race and class (Pratt and Hanson, 1988), and culture (Tuan, 1977), among other factors. For example, men tend to express more confidence with wayfinding and locational abilities than women. Evolutionary psychologists believe this confidence is "due to the traditional role of men as explorers and hunters of game - activities that often took them to distant unfamiliar places and required large-scale environmental knowledge acquisition" (Golledge, 1999, p. 35), although this is also debated by others (see Wynn *et al.*, 1996).

Moreover, spatial knowledge may be manipulated, with highly individual or societal responses. Maps are an obvious case, as they play a fundamental technological role in our organization and navigation of space. Political goals regularly surface among the misconstrued displays of space commonly found in maps (Monmonier, 1996). The creation of such propaganda maps *intends* to affect spatial behavior and geographic distributions rather than simply to reflect them. Spatial behavior and the perception of place and environment are also commonly manipulated for commercial purposes. The tourism industry, notably, is engaged in such efforts, through its marketing practices, with the intention of influencing the spatial behavior of travelers and tourists (Zurick, 1995). Persons are directed through space toward particular destinations (developed explicitly as tourism places) to achieve desired economic results.

The expression and manipulation of spatial behavior by technological means such as maps, or by the industrial advertising of tourism, have decidedly practical goals. They contribute also, however, to the more complex and problematic phenomenon of our geographical imagination, which determines much of our spatial lives. This imagination compels us to travel, both literally, as we move from place to place - in large and small ways - over the course of our lives, and metaphorically, as it defines the shifting contexts and contours of our existence. In both cases, the contribution of the geographical imagination to our intellectual development lies not so much in seeing different places but in seeing places differently. The geographical imagination thus combines with our experience of place and the fundamentals of physical space to influence our spatial cognition and behavior.

What we imagine about the world becomes part of spatial cognition that plays out on larger scales. It allows us to *comprehend* a larger world - *other* worlds - and to make spatial decisions which have impacts beyond our own lives. To the extent that individual citizens are engaged in public and foreign policies, an enlarged geographical imagination allows us to meaningfully participate in the design of spatial structures that operate on national and global platforms (Gregory, 1994). In this way, the geographical imagination extends spatial behavior from the person to the society and from local to global interests. It also extends the impacts of spatial cognition beyond the *consumption* of space (i.e., spatial behavior) to include the *production* of it (i.e., spatial constructions). These productions invariably draw upon not only social theories but also technological innovations. Cartography is a good example, because maps may frame all manner of spatial cognition, and contribute both to its development and expression.

CARTOGRAPHY AND GEOGRAPHICAL IMAGINATIONS

Maps as navigational instruments have become an essential component in the efficacy and outcomes of human spatial behavior. To the extent that mapping is a science it follows the empirical bent of exploration, which is a hallmark of the geographical imagination. This idea applies equally to the collective imagination of large societies and the idiosyncratic imagination of individuals or highly individualized societies living within tightly proscribed spatial boundaries. The early voyages of explorers such as Cook and Darwin, which sought to describe and explicate distant lands, combine with figurative journeys to trace trajectories across entire cultures and societies, as well as their natural environments, thus rendering new and often highly imperialistic geographical imaginations that may be shared by entire civilizations (Mackay, 1985; Said, 1985). Many of these conceptions remain with us today, and influence not only the strategic matters of geopolitics and commerce, but also more personal spatial decisions regarding cultural identity, security and travel, to the point that some commentators remark upon a "clash of civilizations" to describe the geographical elements of the modern world (Huntington, 1998).

A conventional map is a representation of a portion of the Earth's surface. Because of the mathematical complexities involved in representing aspects of a spherical surface on a two-dimensional graphic, all maps in some way distort reality or project alternative realities. The distortions can be intentional or psychological or simply the result of the map projection process. As a rule, people tend to place a high degree of confidence in what they see on maps (Black, 1997). Consequently, the assumptions that underlie the construction of maps are important to know, both in explicating the purpose of the map and in shaping its effect on individual and collective geographical imaginations.

Since spatial behavior of humans results from our needs to navigate our surroundings, maps become vital exploratory tools. The question arises: how do we go about communicating the experience of spatial behavior? In addition to various literary and artistic devices, the experience of spatial behavior is recorded, revised and communicated through an elaborate array of maps, ranging from topological to topographic (or technical), as well as general purpose to thematic. Maps are created and designed to meet the communication needs of the end-user, such as recreation, research or travel (Robinson *et al.*, 1978), and may be as diverse as expressions of spatial behavior itself. In sum, the communication of spatial behavior derives from our comprehension of the world, gained in part from our movement through it.

On the one hand, we travel for the purposes of exploration and discovery and in search of material and psychological needs. This may be a decidedly personal experience and we modify our spatial behavior accordingly, based upon what we learn as we travel and move about. We may also develop our spatial cognition by means of imagery gained from others, often in the form of maps or other spatial graphic devices. The spatial structures that result from both personal experience and the imagery of maps situates locations and places within an evolving mental map that changes with age, education and spatial experience (Gould and White, 1974; Robinson *et al.*, 1978).

All cultures feel compelled to both define and defend territory (Ardrey, 1966), and all make use of topological maps to facilitate navigation within territories. In the tightly proscribed settings of traditional societies, the maps may take quite simple forms – scratches in sand with a stick, sketches on birch bark, or the tracings of palm fronds and shells (Gattis, 2001; Thrower, 1996) (see Figure 20.1). The use of such maps tends to be functional, limited and linked to the practical requirements of navigation and transportation. They do not

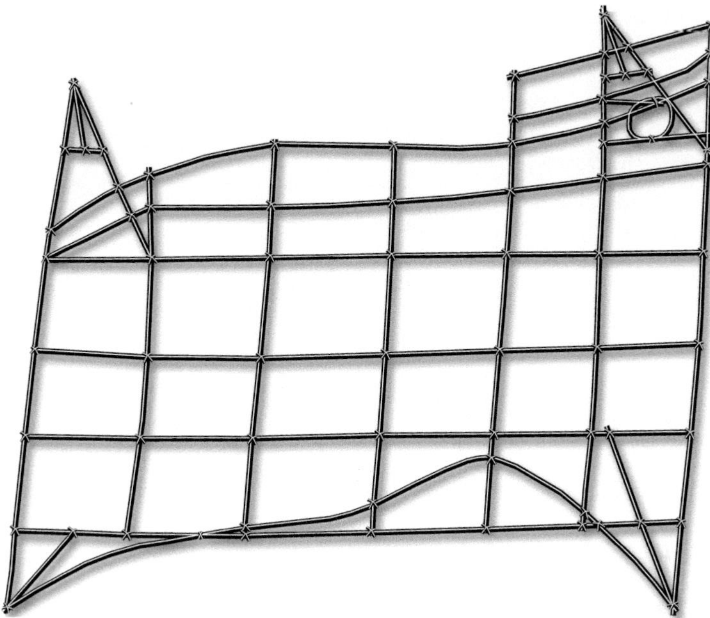

Figure 20.1. "Micronesian 'Rebbelib' or general purpose stick chart from the Marshall Islands." The image sent was approximated from a postcard and/or color photograph among materials collected by Dr David Zurick. Image rendered by Stephanie W. Sambrook

necessarily give order to the world. Among these cultures, the geographical imagination tends to be directed instead toward the mythological world, where the matter of navigation is one of transcendence rather than practicality. The epochal dreamtime of Australian aborigines, the "treasure places" of the Tibetans, and the sacred "heiaus" of Hawaii are all notable examples (see Figure 20.2). Here, maps may be of less consequence than other ways of navigating mythical space, such as stories, trances, art, dance and music.

Among civilizations with a greater spatial reach, maps were developed initially to centrally position a society within an evolving sense of the world (Black, 1997; Brown, 1977; Harley, 1990) (see Figure 20.3). Expanding trade networks, military acquisitions, imperial conquests, all required more sophisticated technologies of mapping. The Astrolabe was developed to provide a means of assessing location north and south of the equator using declinations of the sun. Measurements of the Earth's circumference, initiated with the pioneering studies of Eratosthenes, provided a means of charting the globe and

Figure 20.2. "Sacred landscape map, Himalaya" (painting by Shyam Tamang Lama; used by permission of artist). See also colour plate section.

measuring distance. Ptolemy advanced our navigational abilities by developing a star atlas for purposes of night travel. As the problems of magnetic declination, geometric distortions and longitude measurements were surmounted, the maps of the world gained greater accuracy and currency. They become the tools of global travel across the high seas, and the ultimate weapon of imperial conquest.

It was not simply a matter of technological innovation. The early maps changed the very way persons thought about the world and their place within it. No longer was the world flat, or even round, it was spherical and of immense size and dimension, and exceedingly more complex than anyone had imagined. And no sooner had the broad contours of the world been drawn on the maps, than people sought to immediately position themselves at the center of it all. The Royal Observatory in Greenwich, England, served as the laboratory for the cartographers who designed the global grid of latitude and longitude, and they positioned the Prime Meridian such that it passed directly through their work space, thus denoting not only the beginning point for the measurement of space but also of time. It was an amazingly simple conquest of the Earth through mathematical calibration. Maps continue to be drawn today with Greenwich, and indeed Europe, at the top center of the world, a position of power and prestige, and, of course, centrality.

We have come a long way in our technological abilities to map our world. Electromagnetic signals, energy wavelengths, orbiting

Figure 20.3. "Babylonian clay tablet with cuneiform inscription and 'imagined' map of the world from about 700–500 BC, probably from Sippar, Southern Iraq" © The Trustees of the British Museum.

satellites, global positioning systems, geographic information systems, digital picture elements, all contribute highly sophisticated ways of precisely measuring location, distance, direction and area, and consequently introduce an amazingly high level of precision into our abilities to navigate our worlds. Such technological innovations affect our cognitive development and spatial behavior, and combine with a global media and transportation system to give an overall impression of a "shrinking world." The connections between places have become direct, better known and more immediate.

CONCLUDING OBSERVATIONS

The Gemini astronauts who took the first pictures of Earth from space in the early 1960s provided the world with an amazing viewpoint of home, and a useful metaphor to think about our place in the world. Perhaps no other technological innovation did as much to create a new geographic image of Earth, for in those early pictures we could see, in ways not possible with maps or ground truth, the exquisite beauty and fragility of our world. All sense of order imposed by humans dissolved into the graceful sweeps of color and form that naturally define our planet. It also gave us an opportunity to imagine our world as being small. Our spatial cognition for the most part is an enlarging one, engaging us with an ever expanding reach of spatial experience, asserting our lives in an ever bigger world, compelled as it were by a geographical imagination that reaches outward. The picture of Earth from space required us to look inward. The territorial conquests of history, driven by a geographical imagination that is largely deterministic, are shown ultimately to be unsupportable. Moreover, the notions of "other," embedded in imperialist geographical imaginations, which pit "us" against "them," along with the spatial behaviors that develop as a result of such a binary image of the world – the strategies of conquest and resistance, geopolitical alignments, monopolistic truth, and so forth – dissolve as well in such a holistic view, leaving open the possibilities for a more fluid, comprehensive and, ultimately, more compassionate geographical imagination.

REFERENCES

Ardrey, R. (1966). *The territorial imperative*. New York: Atheneum.

Black, J. (1997). *Maps and history: constructing images of the past*. New Haven and London: Yale University Press.

Blaut, J.M. (1991). Natural mapping. *Transactions of the Institute of British Geographers*, **16**, 55–74.

Brown, L.A. (1977). *The story of maps*. New York: Dover Publications, Inc.

Downs, R.M. and Stea, D. (1977). *Maps in mind: reflections on cognitive mapping*. New York: Harper and Row.

Foucault, M. (1970). *The order of things: an archaeology of the human sciences*. London: Tavistock.

Gattis, M. (Ed.) (2001). *Spatial schemas and abstract thought*. Cambridge, MA: MIT Press.

Golledge, R. G. (Ed.) (1999). *Wayfinding behavior: cognitive mapping and other spatial processes*. Baltimore and London: The Johns Hopkins University Press.

Golledge, R. G. and Rushton, G. (Eds.) (1976). *Spatial choice and spatial behavior*. Columbus: Ohio State University Press.

Gould, P. and White, R. (1974). *Mental maps*. Baltimore: Penguin Books.

Gregory, D. (1994). *Geographical imaginations*. Cambridge: Blackwell Publishers.

Hanson, S. (1999). Isms and schisms: healing the rift between the nature-society and space-society traditions in human geographer. *Annals of the Association of American Geographers*, **89** (1), 133–143.

Harley, J. B. (1990). *Maps and the Columbian encounter*. Milwaukee: University of Wisconsin Press.

Huntington, S. P. (1998). *The clash of civilizations and the remaking of the world order*. New York: Simon and Schuster.

Jones, J. P., Nast, H. J. and Roberts, S. M. (Eds.) (1997). *Thresholds in feminist geography: difference, methodology, and representation*. Lanham, MD: Rowman and Littlefield.

Mackay, D. (1985). *In the wake of Cook: exploration, science and empire, 1780–1801*. London: Croom Helm.

Monmonier, M. (1996). *How to lie with maps*. Chicago: University of Chicago Press.

Pandit, K (1997). Cohort and period effects in U.S. migration: how demographic and economic cycles influence the migration schedule. *Annals of the Association of American Geographers*, **38** (3), 439–450.

Pratt, G. and Hanson, S. (1988). Gender, class, and space. *Environment and Planning D: Society and Space*, **6**, 15–35.

Robinson, A., Sale, R. and Morrison, J. (1978). *Elements of cartography*. New York: John Wiley and Sons.

Sack, R. (2003). *A geographical guide to the real and the good*. New York: Routledge.

Said, E. (1985). *Orientalism*. London: Penguin Books.

Schwartz, J. M. and Ryan, J. R. (Eds.) (2003). *Picturing place: photography and the geographical imagination*. London: I. B. Tauris.

Thrower, N. (1996). *Maps and civilization: cartography in culture and society*. Chicago: University of Chicago Press.

Tuan, Y. F. (1977). *Space and place: the perspective of experience*. Minneapolis: University of Minnesota Press.

Tuan, Y. F. (1989). *Morality and imagination: paradoxes of progress*. Madison: University of Wisconsin Press.

Wynn, T. G., Tierson, F. D. and Palmer, C. T. (1996). Evolution of sex differences in spatial cognition. *American Journal of Physical Anthropology*, **101**, (S23), 11–42.

Zurick, D. (1995). *Errant journeys: adventure travel in a modern age*. Austin: University of Texas Press.

JUDY S. DELOACHE AND MEGAN BLOOM PICKARD

21

Of chimps and children: use of spatial symbols by two species

The symbolic abilities of humans convey an enormous advantage to us over other animals. We enjoy a depth of communication with one another not available to species that lack a formal linguistic code. In addition, we have invented a wealth of symbolic artifacts to expand our cognitive and communicative grasp. Symbols make it possible to preserve information gained from direct experience and communicate it to others, who thus can share our knowledge without direct experience. Symbolization is the bedrock of human culture, as it makes possible the transmission of information from one generation to another.

Because humans have created a variety of symbolic artifacts to support spatial cognition and action, individuals do not have to actually move around in the world to form cognitive maps of space. Instead, we can use real maps to form functional mental representations of space. On a daily basis, citizens of modern societies obtain spatial information from a variety of symbolic artifacts, including maps, models, globes, photos and videos.

As the chapters in this volume make clear, spatial cognition is essential for all animals. They must have the capacity to mentally represent the space through which they move and the paths to follow through that space, as well as the location of significant landmarks in the space. Also necessary is the ability to update mental representations of spatial landmarks as a result of experience. If a favorite stump used to fish for ants has been destroyed, a chimpanzee's mental representation of that location needs to be updated to contain the new information. If a child's mother has moved the cookie jar to a new location, finding a cookie will depend on an updated spatial representation.

In their natural environment, animals like chimpanzees must rely on direct experience to form functional spatial representations.

Spatial Cognition, Spatial Perception: Mapping the Self and Space, ed. Francine L. Dolins and Robert W. Mitchell. Published by Cambridge University Press.

This experience can include seeing other animals acting in space, but it does not include any kind of indirect or symbolic information. What about in a different environment - one rich with creatures who use symbolic artifacts? Could chimpanzees in close contact with humans benefit from symbolic representations of space?

As we will see, the answer is clearly yes. Chimpanzees in a captive environment, accustomed to interacting with humans, are able to exploit information from symbolic representations of space. Here we explore the extent to which chimpanzees' use of one kind of symbolic artifact - scale models - is similar to that of humans. Because both chimpanzees and young children have been tested in similar tasks, we can compare their performance. The specific question we ask is whether chimpanzees and children who successfully exploit the spatial information provided by a model do so on the same basis. That is, we examine in what ways the underlying processes through which the two species succeed are similar and how they are different.

We begin with a brief discussion of the nature of symbolic artifacts and their use. Next we review research on very young children's use of a scale model as a source of information about the location of a hidden object, and we then summarize the performance of chimpanzees in similar tasks. Finally, we compare how the two species go about solving the task.

SYMBOLS AND SYMBOL USE

A symbol is something that someone intends to stand for or represent something other than itself (DeLoache, 2002, 2004). In this definition, *someone* calls attention to the fact that humans are the "symbolic species" (Deacon, 1997) and that symbol use is the "most characteristic mental trait of mankind" (Langer, 1942, p. 72). Nothing distinguishes more sharply between humans and nonhuman primates than our extraordinarily flexible and creative use of symbols. This does not, of course, mean that humans are the only species capable of using symbols. Rather, it refers to the fact that only humans create symbols and have symbol systems. The very indefinite term *something* signifies that almost anything can serve as a symbol for almost anything else. Another defining characteristic of a symbol is that it *represents*; it stands for something that is different from itself. Last, but certainly not least, is the role of *intention*: one entity stands for another only if some person *intends* for it to do so. Nothing is inherently a symbol, but anything that someone elects to use as a symbol can function in that capacity. Thus, intention is both necessary *and* sufficient to establish a symbolic relation.

One unique aspect of symbolic objects is that they have an inherently dual or double nature (Gregory, 1970; Kennedy, 1974; Potter, 1979; Werner and Kaplan, 1963). An artifact, such as a map or model that is created and/or used to serve a symbolic function, is an object in itself and at the same time a representation of something entirely different. A map is perceived to be a flat surface covered with an arrangement of markings and fields of color, but the marks and colors are intended by the map-maker and interpreted by the map user to stand for features of the environment. A scale model is perceived as an arrangement of miniature objects, but interpreted as standing for a set of larger entities. The physical relation between the elements on the map or in the model and what they represent in the world can vary greatly, but the spatial relations among them do not. A model or map that is useful for finding one's way preserves the spatial relations among the elements in the represented space.

Because of the dual nature of symbolic objects, both aspects of their reality must be mentally represented to use them symbolically. *Dual representation* must be achieved; one must think about both the concrete object itself and its abstract relation to what it stands for (DeLoache, 1987, 2002, 2004). As will be clear in the next section, the necessity for achieving dual representation constitutes a substantial challenge to young children's understanding and use of symbolic objects. The younger the child, the more likely he or she is to focus on the symbolic object itself, missing the higher-level relation between the model or map and what it represents.

THE UNDERSTANDING AND USE OF SPATIAL SYMBOLS BY YOUNG CHILDREN

Research on very young children's understanding and use of symbolic artifacts has revealed dramatic age-related changes in symbolic functioning and particularly in the achievement of dual representation. In the scale-model task used to study very young children's use of symbolic artifacts (DeLoache, 1987), children watch as a small toy is hidden in a scale model, roughly one-seventh the size of a room furnished with standard furniture. They are then asked to find a larger version of the toy that has been hidden in the corresponding location in the room. The model and the room each contain the same pieces of furniture used as hiding places in the same spatial arrangement. Surface similarity between the model and the room is very high – the items of furniture and the surroundings in the two spaces look very much alike.

The standard model task begins with an extensive orientation to the room, the model, and the correspondence between the two. The child is first familiarized with the room and its furniture, then with the model; the experimenter explains that they are "just the same except one is little." All of the furniture from the little room is then brought into the room and the experimenter draws attention to the corresponding pieces, explicitly pointing out that they are just the same except for size (e.g., "Little Snoopy's couch is just like Big Snoopy's couch only smaller"). After returning the furniture to the model, the experimenter explains the game in detail, stressing the two toys will always be hidden in corresponding places (e.g., "Wherever Little Snoopy hides in his little room, Big Snoopy hides in the same place in his big room.").

The experimental trials directly follow. The child watches as the experimenter hides the little toy in a location in the model. The experimenter explicitly tells the child that the large toy will be hidden in the same place in the room without explicitly naming the hiding location (e.g., "Remember, Big Snoopy is hiding in the same place as Little Snoopy"). Finally, the child goes to the room to find the hidden big toy.

Successful searching requires that children map the furniture from the model to the room and that they infer that the hiding event they witnessed in the model corresponds to an unseen event that took place in the large room. A correct search is scored only when the child finds the large toy in the *first* location he or she searches. After finding the large toy, the child returns to the model to retrieve the small toy.

This second retrieval serves as an important control. Success is evidence that the child *remembers* the location of the hidden toy and is *motivated* to search for it. Thus, the second retrieval confirms that neither memory nor motivation problems could account for poor performance on the first retrieval task.

The model task has several advantages for investigating early symbolic understanding (Marzolf and DeLoache, 1997). An important one has to do with its relatively low verbal demands. Children – and chimpanzees – do have to understand the simple instructions they are given, but they are not required to respond verbally. Also, performance is unambiguous: subjects either find the toy in the first location they search, or they do not. Finally, young children's natural enjoyment of searching for hidden objects produces consistent data from this often uncooperative age group.

The results from testing very young children in many model studies reveal a remarkable developmental change between the ages of 2.5 and 3. In the original scale model study (DeLoache, 1987), a striking performance difference was found between the two age groups. Whereas three-year-olds were highly successful on both retrievals, two-and-a-half-year-olds showed a very different pattern. They had little trouble finding the miniature toy they had observed the experimenter hiding in the model. However, they almost never found the large toy in the room. Thus, older children used their knowledge of where they had seen the small toy being hidden in the model to infer where to find the large toy in the room. The younger children failed to use their memory for the hiding event in the model to figure out where to search in the room.

Multiple replications of the original study confirmed these age differences. Two-and-a-half-year-old children typically retrieve the large toy in the room less than 20 percent of the time, whereas they retrieve the small toy in the model from 80 to more than 90 percent of the time. On the other hand, children just six months older retrieve the toy in the room and in the model over 80 percent of the time (see DeLoache, 2002, for a summary of results). Sex differences have almost never occurred in the model task.

POSSIBLE ACCOUNTS OF YOUNG CHILDREN'S DIFFICULTY WITH THE MODEL TASK

What accounts for such striking age differences? In particular, why do the younger children fail to use their memory for the location of the miniature toy in the model to infer the location of the large toy? One possibility is that two-and-a-half-year-old children are simply unaware of the correspondence between the small and large items of furniture. Clearly, being able to detect the object correspondences is necessary for success; if a child does not recognize the relation between the miniature and full-sized objects, it would be impossible to succeed on the task.

Accordingly, Troseth *et al.* (2007) investigated the relation between matching the corresponding objects in the two spaces and success in the retrieval task. The two-and-a-half-year-old children first experienced the same extensive orientation phase used in the standard model task. Then, they were given a simple matching task in which the experimenter pointed out an object in the model and asked the child to find "the one that looks like this one" in the room. The children had no trouble with this task; they were successful 79 percent of the time. After watching the experimenter point to the model chair, they went into the room and designated the large chair. However, when these children were given the standard retrieval task a few minutes later, they performed at the same low level as two-and-a-half-year olds in other studies, retrieving the toy only 21 percent of the time. They failed to use the knowledge they clearly possessed of the correspondence between the objects in the two spaces to solve the retrieval task.

The results of the study make two important points. First, although associating objects across the two spaces is *necessary* for success in the model task, it is not *sufficient*. Second, one could solve the model task simply by mapping from the miniature object in the model to the corresponding larger object in the room. However, that is apparently not how young children succeed in the task. Some additional factor is responsible for children's success.

Several attempts to improve two-and-a-half-year olds' performance testify to the difficulty that young children have appreciating the model-room relation. In one attempt, the hiding places were overtly labeled as the toy was hidden in the model: "Look, I'm hiding Little Snoopy behind the couch" (DeLoache, 1989). Even with such explicit labeling, two-and-a-half-year olds still failed to retrieve the hidden toy, doing so on only 16 percent of trials. In another study, Troseth and DeLoache (2003) asked two-and-a-half-year-olds to find a highly motivating object – their own mother – in one of several large hiding places in a room. The experimenter pointed to one of the locations in the model and told each child, "Your mom is hiding right behind/under here." These children performed poorly, finding their mother only 38 percent of the time, indicating that the difficulty of the task does not stem from lack of motivation to find the hidden object.

Of the mistakes that young children make in the model task, the most common is to search in the previously correct hiding location, that is, the place where they found the toy before. This well-known maladaptive search tendency does not fully explain their poor performance on the model task (Kuhlmeier, 2005; O'Sullivan *et al.*, 2001; Sharon and DeLoache, 2003). Sharon and DeLoache (2003) pooled and analyzed data from several published and unpublished scale model studies. The pattern of performance across studies revealed that children tend to go back to previously correct locations when they don't know where to find the toy. Thus, re-searching the previously correct location reflects poor knowledge of the correct location, not just an inability to inhibit repeating a previously correct response.

Two studies have shown that making it impossible for children to search repetitively does not improve performance. Sharon and DeLoache (2003) physically altered the correct hiding place at the end of each trial to make it clear that the toy could not be hidden there on the next trial. For example, after the toy had been hidden in the basket, the basket was left lying on its side, making it obvious that nothing was concealed in it. Employing the same basic logic, O'Sullivan *et al.* (2001) removed the piece of furniture that had just served as the hiding place after each trial, thereby eliminating the possibility of the child returning to that location. In both studies, the two-and-a-half-year-old children retrieved the toy only about 30 percent of the time. Their performance thus reflects a lack of understanding of the symbol–referent relation between the model and the room rather than a maladaptive search strategy.

Recent research suggests that a tendency toward repetitive searching may play a small role in performance in the model task that shows up only in combination with other factors. Kuhlmeier (2005) improved two-and-a-half-year-olds' performance in the model task by making two modifications. First, to increase children's motivation to find the toy in the first place they searched, they were rewarded if their *first* search was in the correct location. (This procedure was modeled after one used with chimpanzees in the scale model task, discussed later.) In addition, instead of hiding a toy in the model, the experimenter simply pointed to the correct location, a modification that Kuhlmeier argued should be more readily understood by the young children. With these two modifications (but not with either of them alone), the children's performance was substantially better (60 percent) than that of a group in the standard task (20 percent). Thus, poor inhibitory control is not the sole cause of two-and-a-half-year-olds' difficulty with the model task, but its contribution may be apparent only in combination with other variables.

In summary, a number of studies have ruled out several factors as causes of two-and-a-half-year-old children's failure on the standard model task, indicating that two-and-a-half-year-olds are relatively insensitive to the relation between the model and the room it represents. By three years of age, children are highly successful in the standard task, in which explicit instructions about the relation are provided and there is a high level of surface similarity between the two spaces.

However, several modifications of the standard task result in marked changes in the level of success of both age groups.

WHAT IS NECESSARY FOR CHILDREN'S SUCCESS IN THE MODEL TASK?

The three-year-olds' success can be dramatically diminished by seemingly minor changes to the standard task. For example, there must be a high level of physical similarity between the articles of furniture in the two spaces. In studies in which the objects in the two spaces were of different colors and different materials, three-year olds succeeded only about 25 percent of the time (as opposed to over 80 percent with a high level of physical similarity between model and room).

Further evidence for the importance of perceptual similarity is the fact that two-and-a-half-year-olds' performance improves when the two spaces are made more similar by decreasing their size discrepancy. When the larger space is only two times the size of the model, children are quite successful at the task, retrieving the object 70 percent of the time (DeLoache *et al.*, 1991). Thus, physical similarity plays a key role in children's performance in the model task. It is not, however, sufficient for success.

Three-year-olds also have trouble if the experimenter provides relatively little information about the relation between the model and the room up front. As described earlier, the model task typically begins with the experimenter explicitly explaining and describing all the relations between the individual items in the two spaces. When the direct correspondence between the two rooms and the objects in the rooms is not explicitly pointed out, three-year-olds' performance is relatively poor (25 percent) (DeLoache, 1989). The correspondence between the model and real space must be explicitly and exhaustively communicated for these children to understand the model/room relationship.

Interestingly, if the spatial relations among the items of furniture in the model and the room are different, three-year-old children perform well if the correspondence between model and room is explicitly pointed out to them (Marzolf *et al.*, 1999). However, if the object correspondences are not established for them, three-year-olds perform poorly. Thus, children need *both* a high level of physical similarity (similarity both of objects and their spatial arrangement) *and* full instructions to succeed in the standard task. Neither is sufficient on its own.

THE ROLE OF DUAL REPRESENTATION IN CHILDREN'S PERFORMANCE IN THE MODEL TASK

What is the source of the surprising degree of difficulty that very young children have in the apparently simple scale model task? As mentioned earlier, a challenge for young children in using symbolic artifacts is achieving *dual representation*, representing the symbolic object itself as well as its relation to something else. The younger the children, the more they are drawn to the model itself, making it difficult to notice its abstract relation to something other than itself (DeLoache, 1987, 2002, 2004).

The first test of dual representation was based on the counter-intuitive prediction that young children would find a picture of the room easier to use as a source of information about the room than a scale model. The reasoning was that pictures are *less salient* as objects, so it should be *easier* to achieve dual representation, that is, easier to appreciate the relation between the picture and what it stands for. This prediction was confirmed and has been replicated several times. In these studies, two-and-a-half-year-old children were shown a picture (color photograph, colored drawing, line drawing) of the room, and the experimenter pointed to the relevant item of furniture in the picture, telling the child that it was the hiding place of the toy without naming it (e.g., "Here's where Snoopy's hiding in the room"). In every picture study, the performance of the two-and-a-half-year-old children has averaged around 80 percent, significantly better than their performance (or that of same-age children) in the model task (DeLoache, 1987, 1991; Marzolf and DeLoache, 1994). The same superiority with two-dimensional stimuli is found when two-and-a-half-year-old children view a hiding event via video (Troseth and DeLoache, 1998).

Dual representation predicts that the more engaging an object is on its own, the harder it is to regard it as a symbol for something else. In two other tests of the hypothesis, altering the physical salience of the model to the children affected their performance on the task (DeLoache, 2000). For instance, two-and-a-half-year-olds were successful in a study in which the model was placed behind a window, which served to decrease its physical salience. The children found the toy 54 percent of the time, a success rate much higher than their normal 25 percent on the standard model room task. On the other hand, three-year-olds were allowed to play with the model before the model task, thus increasing the model's physical salience to them. Subsequently, they retrieved the toy only 44 percent of the time (markedly lower than three-year-olds' usual performance of about 80 percent).

The most stringent test of dual representation involved removing the need for it altogether (DeLoache *et al.*, 1997). This was accomplished by the experimenter convincing two-and-a-half-year-old children that she had a machine that could shrink things, including toys and a room. First, a troll doll was placed in front of the shrinking machine (which looked suspiciously like an oscilloscope), the machine was "turned on," and the child and experimenter waited in the room next door, listening to the sounds the "machine makes while it's working." They returned to find a miniature version of the troll. A demonstration of the power of the machine to shrink a room (a large tent) followed; while the child and experimenter were out of the room, a small model was substituted for the tent. Then, for the actual study, the child watched as the large troll was hidden somewhere in the tent, and the child and experimenter retired to wait for toy and tent to be miniaturized. When they returned, the child was asked to retrieve the toy. Just as in the standard model task, the children had to use their memory for the hiding event observed in one space to know where to search for the toy in a different space. However, if they believe that the room and model are one and the same, there is no symbolic relation between them; the model *is* the room. As predicted on the basis of dual representation, the children in the shrinking room study performed much better than children of the same age in the standard model task, retrieving the toy on 76 percent of trials.

The same pattern of results was found in a video analogue of the shrinking room study (Troseth and DeLoache, 1998). In the previously mentioned video study, two-and-a-half-year-olds easily used the information provided by video to find the toy hidden in the room, but two-year-olds did not. However, the result was different for another group of two-year-olds who were led to believe they were watching through a window as an experimenter hid the toy in the room next door. In fact, they were watching a previously recorded video of the event. (The video monitor was located directly behind the window with everything but the monitor screen hidden from view.) After observing the hiding event, the children were invited to go into the room and find the toy.

The point of this study was that if the children believed they were watching a live hiding event in the room, dual representation would not be required – the children would not have to link a video representation of the hiding event with the real event. As predicted on the basis of dual representation, performance was better in this task than in the standard model task; children retrieved the object 63 percent of the time.

The shrinking room and deceptive video studies provide very strong support for the concept of dual representation. When children think the model actually is the room, or when they think they are watching a real rather than a video event, they use the information provided to them. Thus, it is the presence of a symbol–referent relation that is the source of difficulty for very young children.

COMPARISON OF CHILDREN'S AND CHIMPANZEES' PERFORMANCE IN THE MODEL TASK

Sally Boysen, Valerie Kuhlmeier, and their colleagues, have tested chimpanzees in scale model tasks similar to those used with children. Comparison of the patterns of success and failure shown by chimpanzees and children sheds light on how both species interpret and solve the task.

Seven chimpanzees, all of whom understand spoken language, have been tested in scale model tasks very similar to those used with children (Boysen and Kuhlmeier, 2002; Kuhlmeier *et al.*, 1999; Kuhlmeier and Boysen, 2001, 2002). Some studies employed a scale model of a familiar room, and others involved a model of an outdoor enclosure. Just as in the research with children, the chimpanzees watched as a desirable miniature object (a juice or soda container) was hidden in a scale model, and they were then encouraged to find the real drink in the larger space.

These studies established that adult chimpanzees are capable of solving the scale model task. However, the nature of their performance differs in several interesting ways from that of children. For example, the early studies revealed striking gender differences in the animals' use of a model as a source of information for searching (Kuhlmeier *et al.*, 1999), which is not the case with children. After watching the hiding event in the model, females generally went directly to the corresponding place in the larger space. Their first searches were at the correct location at rates above chance. In contrast, the majority of male chimpanzees adopted rigid search patterns – for example, starting with the first location in the space and then searching the other locations in a rigid pattern until finding the soda. Consequently, their first search scores were typically not above chance.

The nature of the male chimpanzees' searching pattern is notably different from young children's behavior in these tasks. Children typically adopt the goal of finding the hidden object on their first search (even when they perform very poorly). The male chimpanzees apparently adopt a goal of finding the reward, but whether it is found on the first or last search does not matter. In addition, children almost never show the chimpanzees' search strategy of starting at the same place each time and following a set pattern of searching the other locations. Although children's errors often involve perseverative searching, those errors typically involve searching the location that had been correct on the immediately preceding trial (O'Sullivan *et al.*, 2001; Sharon and DeLoache, 2003), not following a routinized pattern of searches.

A subsequent series of studies was conducted to examine the nature of the male chimpanzees' dominant search strategy. In this case, the search task was modified so that the animals received the reward only if their *initial* search was in the correct location (Kuhlmeier and Boysen, 2001). First, the animals were trained to exchange an empty juice bottle for a full one. Then they participated in the modified model task in which the hidden object in the larger space was an empty container that they could exchange for a full one – but *only if they found it on their first search.* The effect on the retrieval performance of the male chimpanzees was remarkable. Now, the males were quite successful, all of them performing at rates above chance. All but one female performed above chance as well. Recall that children were not immediately successful with a procedure involving contingencies (Kuhlmeier, 2005).

A second study explored whether the chimps would transfer their now successful search strategies to the standard task, without the response contingencies. They were generally successful, with all but one male and one female performing above chance in the standard model task. Thus, once the male chimpanzees were prevented from following their preferred search strategy, their behavior revealed an ability to exploit the information from the model.

Such an improvement indicates that successful searching for male chimpanzees was inhibited by their perseverative search tendencies, unlike the perseverative searching in children. For children such behavior is a fall-back strategy employed only when they have no idea what else to do.

A final comparison can be made between chimpanzees and children when it comes to variations in perceptual similarity between the model and large space. Such studies suggest that perceptual similarity is an important cue for both chimpanzees and children with respect to solving the model task (DeLoache *et al.*, 1991, 2004; Marzolf and DeLoache, 1994). For chimpanzees, as long as at least two out of three perceptual cues (color, shape and position) between the model and the larger space were available, most (five out of seven) retrieved the hidden objects at rates above chance. However, performance deteriorated substantially when only position remained as a perceptual cue. When four identically colored and shaped objects were placed in different locations within the spaces, only two out of seven chimpanzees retrieved the hidden objects above chance (Kuhlmeier and Boysen, 2002). As discussed previously, young children also benefit from high levels of physical and relational similarity between the two spaces.

Another set of studies (using the original, non-contingency procedure) addressed whether chimpanzees would, like children, be more successful if photographs and video provided information about the location of the hidden object. The fact that two-and-a-half-year-old children perform substantially better with two-dimensional stimuli than with a three-dimensional model provides crucial evidence for the role of dual representation in their use of a scale model.

However, the performance of the two chimpanzees tested in this study did not improve with the two-dimensional stimuli; one was already quite successful in the model task (and hence could not improve), and the other performed poorly in both tasks. With the new contingency procedure, all the chimpanzees perform fairly well in the model task, making it impossible to compare picture and model task performance. It is possible that a picture superiority result would emerge in a developmental study of chimpanzees, that is, a study comparing young chimpanzees of different ages on both tasks. As it is, there is no evidence regarding the role of dual representation, which is central to children's success in the model task, in chimpanzees' performance in the task.

THE BASIS FOR THE SUCCESS OF CHILDREN AND CHIMPANZEES IN THE MODEL TASK

Given the convincing evidence that chimpanzees are capable of solving the scale model task an important question to ask is whether the basis for their success is the same as it is for children. Do adult chimps, like three-year-olds, have a symbolic interpretation of the task? Does an appreciation of the higher-level, stands-for relation between the model and larger space underlie their ability to use their knowledge of the location of the object in the model to find the analogous object in the larger space? Or do chimpanzees and children solve the model task in different ways?

There does not seem to be any evidence that chimpanzees solve the task symbolically. As noted earlier, the model task can be solved simply on the basis of mapping between the corresponding objects in the two spaces. Having observed the hiding event with one particular hiding place in the model, all that is necessary is picking out the similar hiding place in the larger space. There is no evidence that anything further is involved in chimpanzees' success in the task.

In contrast, there is both direct and indirect evidence that children do *not* solve the model task simply by mapping from the corresponding objects in one space to those in the other. The direct evidence is the fact that, as described earlier, children who can match the miniature and large objects in the two spaces nevertheless fail the model task. Detecting object correspondences between the two spaces is not sufficient for children.

Indirect, but even stronger evidence that more than object matching is involved in children's successful performance in the model task comes from the explicit tests of and evidence for the important role of dual representation. Manipulations to increase, decrease and eliminate the need for dual representation led to predicted effects. Children perform better with pictures, which reduce the difficulty of dual representation, than with models. As noted before, chimpanzees showed no difference in performance with pictures versus models.

Further, increasing the salience of a model by letting children play with it makes the task harder, whereas decreasing its salience by putting it behind a window makes it easier (DeLoache, 2000). Finally, convincing children that a machine can shrink a room into a model-sized space – thereby eliminating the need for dual representation – leads to improved performance in the task (DeLoache *et al.*, 1997). All of these counterintuitive results were predicted on the basis of dual representation, and none can be explained in any other way.

There is no evidence that dual representation plays a role in adult chimpanzees' performance in the model task. However, the absence of evidence that chimpanzees interpret the model task symbolically does not necessarily mean that they cannot do so. The best argument for the possibility that chimpanzees would be capable of accomplishing the task symbolically is the fact that they have succeeded on a number of other tasks involving numerical and linguistic symbols (e.g., Boysen and Hallberg, 2000; Rumbaugh and Savage-Rumbaugh, 1996). One problem in determining whether chimpanzees solve the task symbolically is that, for the most part, these adult chimpanzees are quite successful. A number of the most informative manipulations with young children regarding their interpretation of the scale model problem involve making the task easier, a strategy that is essentially impossible with the highly successful chimpanzees. It would be very interesting to have developmental data for chimpanzees in the model task.

In conclusion, adult chimpanzees and young children can successfully use a scale model as a source of information for solving a search problem. The results of a large number of studies provide strong evidence that young children solve the task by mentally representing the higher-level, representational relation between the model and the larger space it represents. The underlying basis for chimpanzees' success in the task is not presently clear. It may be that only research with young chimpanzees of varying ages could provide a clear answer.

REFERENCES

Boysen, S. T. and Hallberg, K. I. (2000). Primate numerical competence: contributions toward understanding nonhuman cognition. *Cognitive Science*, **24**, 423–443.

Boysen, S. and Kuhlmeier, V. (2002). Representational capacities for pretense with scale models and photographs in chimpanzees. In R. W. Mitchell (Ed.), *Pretending and imagination in animals and children* (pp. 210–228). New York: Cambridge University Press.

Deacon, T. W. (1997). *The symbolic species: the co-evolution of language and the brain.* New York: W. W. Norton.

DeLoache, J. S. (1987). Rapid change in the symbolic functioning of very young children. *Science*, **238**, 1556–1557.

DeLoache, J. S. (1989). Young children's understanding of the correspondence between a scale model and a larger space. *Cognitive Development*, **4**, 121–139.

DeLoache, J. S. (1991). Symbolic functioning in very young children: understanding of pictures and models. *Child Development*, **62**, 736–752.

DeLoache, J. S. (2000). Dual representation and young children's use of scale models. *Child Development*, **71**, 329–338.

DeLoache, J. S. (2002). Early development of the understanding and use of symbolic artifacts. In U. Goswami (Ed.), *Blackwell handbook of childhood cognitive development* (pp. 206–226). Malden, MA: Blackwell.

DeLoache, J. S. (2004). Becoming symbol-minded. *Trends in Cognitive Sciences*, **8**, 66–70.

DeLoache, J. S., Kolstad, V. and Anderson, K. N. (1991). Physical similarity and young children's understanding of scale models. *Child Development*, **62**, 111–126.

DeLoache, J. S., Miller, K. F. and Rosengren, K. S. (1997). The credible shrinking room: very young children's performance with symbolic and nonsymbolic relations. *Psychological Science*, **8**, 308–313.

DeLoache, J. S., Simcock, G. and Marzolf, D. (2004). Transfer by very young children in the symbolic retrieval task. *Child Development*, **75**, 1708–1718.

Gregory, R. L. (1970). *The intelligent eye*. New York: McGraw-Hill.

Kennedy, J. M. (1974). *A psychology of picture perception*. Oxford: Jossey-Bass.

Kuhlmeier, V. A. (2005). Symbolic insight and inhibitory control: two problems facing young children on symbolic retrieval tasks. *Journal of Cognition and Development*, **6**, 365–380.

Kuhlmeier, V. A. and Boysen, S. T. (2001). The effect of response contingencies on scale model task performance by chimpanzees (*Pan troglodytes*). *Journal of Comparative Psychology*, **115**, 300–306.

Kuhlmeier, V. A. and Boysen, S. T. (2002). Chimpanzees (*Pan troglodytes*) recognize spatial and object correspondences between a scale model and its referent. *Psychological Science*, **13**, 60–63.

Kuhlmeier, V. A., Boysen, S. T. and Mukobi, K. (1999). Comprehension of scale models by chimpanzees (*Pan troglodytes*). *Journal of Comparative Psychology*, **113**, 396–402.

Langer, S. K. (1942). *Philosophy in a new key*. Cambridge: Harvard University Press.

Marzolf, D. P. and DeLoache, J. S. (1994). Transfer in young children's understanding of spatial representations. *Child Development*, **65**, 1–15.

Marzolf, D. P. and DeLoache, J. S. (1997). Search tasks as measures of cognitive development. In N. Foreman and R. Gillet (Eds.), *A handbook of spatial research paradigms and methodologies*, (Vol. 1, pp. 131–152). Hove, England: Psychology Press.

Marzolf, D. P., DeLoache, J. S. and Kolstad, V. (1999). The role of relational similarity in young children's use of a scale model. *Developmental Science*, **2**, 296–305.

O'Sullivan, L. P., Mitchell, L. L. and Daehler, M. W. (2001). Representation and perseveration: influences on young children's representational insight. *Journal of Cognition & Development*, **2**, 339–365.

Potter, M. C. (1979). Mundane symbolism: the relations among objects, names, and ideas. In N. R. Smith and M. B. Franklin (Eds.), *Symbolic functioning in childhood* (pp. 41–65). Hillsdale, NJ: Erlbaum.

Rumbaugh, D. M. and Savage-Rumbaugh, E. S. (1996). Biobehavioral roots of language: words, apes, and a child. In B. M. Velichkovsky and D. M. Rumbaugh (Eds.), *Communicating meaning: the evolution and development of language* (pp. 257–274). Hillsdale, NJ: Erlbaum.

Sharon, T. and DeLoache, J. S. (2003). The role of perseveration in children's symbolic understanding and skill. *Developmental Science*, **6**, 289–296.

Troseth, G. L. and DeLoache, J. S. (1998). The medium can obscure the message: young chidren's understanding of video. *Child Development*, **69**, 950–965.

Troseth, G. and DeLoache, J. (2003). *Young children's use of scale models: testing alternatives to representational insight*. Unpublished manuscript.

Troseth, G. L., Bloom, M. E. and DeLoache, J. S. (2007). Young children's use of scale models: testing an alternative to representational insight. *Developmental Science*, **10**, 763–769.

Werner, H. and Kaplan, B. (1963). *Symbol formation*. Oxford: Wiley.

SARAH T. BOYSEN AND KIM A. BARD

22

Chimpanzee spatial skills: a model for human performance on scale model tasks?

For the past several years, Boysen and colleagues have explored the spatial capacities of the chimpanzee (*Pan troglodytes*), using scale models based on the creative series of experiments by DeLoache and her colleagues (e.g., 1987, 1991, Chapter 21, this volume). We have been interested in examining the flexibility and breadth of representational capacities used by chimpanzees, and have assessed performance with a variety of "model" formats, including scale models, miniature objects and color photographs, compared with their real-world, full-scale referents. We hypothesized that an understanding of the scale model task designed by DeLoache for testing young children could be accomplished by our test-sophisticated group of enculturated chimpanzees. However, initially male chimpanzees and female chimpanzees used different strategies to solve scale model tasks. In a series of studies (Kuhlmeier and Boysen, 2001, 2002; Kuhlmeier *et al.*, 1999), special emphasis was placed on how the chimpanzees could learn to understand the spatial and perceptual correspondence of representations and their referential context (for male chimpanzees, in particular). Some chimpanzees (females, it seems) readily and immediately recognized and acted upon the topographic relationships between models, photos and the actual full-sized space without special training Here we explore the implications of these experiments for specifying the cognitive basis and the evolutionary foundation for solving spatial tasks. Equivalent performance by chimpanzees and older children on the scale model tasks would suggest equivalent cognitive bases for solving scale model tasks.

DeLoache and her colleagues argue that a "dual representation" must be available to a child in order for them to recognize the relationship between a model and the real room. If so, what would the requisite

Spatial Cognition, Spatial Perception: Mapping the Self and Space, ed. Francine L. Dolins and Robert W. Mitchell. Published by Cambridge University Press.

perceptual and spatial characteristics be that contribute to forming such representations? DeLoache's model posits that similarities existing between the perceptual features of a scale model and its real-world counterpart facilitate the ability of subjects to use these to compare the two (e.g., DeLoache *et al.*, 1991). Thus, close similarity between model and reality should facilitate recognition of the representational nature of the model, and dissimilarities should diminish such recognition. Our experiments examine the potential interrelated perceptual characteristics replicated when a representational scale model is constructed (e.g., size and spatial topography of the physical spaces, and colors of objects within them) that are used to make the comparison between model and reality. Consequently, by varying the explicit features or perceptual relationships necessary for constructing this comparison, the essential requirements can be detailed.

In the present chapter, we describe our findings on numerous variations of the original DeLoache scale model task as presented to our adult chimpanzees. First, we report the findings when the scale model task was introduced to two adult chimpanzee subjects, Sheba and Bobby. Because of Bobby's extreme difficulties with the task, and the relative ease with which Sheba demonstrated her understanding of the relation between objects in the model and those in the real environment, subsequent variations on the task were completed. For Experiment 1, these conditions included the following: presenting the information to the animals in the form of color photographs of the hiding sites instead of the actual scale model; presenting a panoramic photograph of the actual space while highlighting the specific hiding site; and finally, presenting a highly demonstrative videotape of an experimenter going into the real room, which included identifying each possible hiding site and then hiding the reward. Differences in the animals' performance persisted, necessitating additional experiments that included the entire group of seven adult chimpanzees available in our facility. These experiments were designed to examine whether individual differences in the cognitive skill level of the two chimpanzees previously tested (Sheba, Bobby) could account for the performance disparity, or if there were possible sex differences with the task. Each of these experiments is presented in further detail in the chapter.

The second experiment was devised to test all the adult animals in our group by constructing a second model that was a 1:7 scale model of the animals' outdoor playground. Once again, differences in performance between the male and female chimpanzees were apparent. To address more specific parameters of the testing context and stimuli, in Experiment 3 we examined several variations in the task by manipulating the perceptual schema in the actual space, while continuing to use the scale model for demonstrating the hiding event. Numerous questions were addressed in this series of studies, including whether the chimpanzees could locate the correct hiding site if the objects representing the sites within the model and those in the real space had a high degree of perceptual similarity (both sets of objects were the same shape and color). Second, we addressed whether the animals could correctly identify the hiding site if the objects within the model were the identical shape (to scale) to those in the real space, but were different in color. Next, the color of the objects in the scale model were changed, as well as providing four novel colors for the hiding sites in the playground.

The fourth and final experiment in this series that we report here was specifically designed to focus on the male chimpanzees' attention on the actual hiding event, as we hypothesized that they were not paying attention to that aspect of the task. Consequently, even when they moved immediately to the playground sites, they appeared to have little or no real memory of the correct site. A significant change in the response contingencies in this final study required that the animals respond correctly on their first attempt, where they would discover an empty juice bottle that could be immediately exchanged for a full bottle of juice from the experimenter. If the subject had not paid close attention to the actual location where the "reward" was hidden, they were required to exit the playground area, and repeat the entire trial. In devising the modified task, we hypothesized that the newly imposed response contingencies would motivate the male chimpanzees, still lagging in performance, to pay more attention to the experimenter's demonstration within the scale model. The explicit procedures and results for all experiments addressing the understanding of scale models by chimpanzees follows.

THE CHIMPANZEE SCALE MODEL TASK

We have completed numerous experiments exploring the ability of adult chimpanzees to recognize the topographic relationship between a scale model and the full-size area it represents. We begin here by introducing our initial study, which used a scale model of an indoor room that was very familiar to the two chimpanzee subjects tested. Both were adult animals, and had been involved in cognitive and behavioral studies for most of their lives. Although they were quite large and fully grown, both subjects were still able to interact safely outside their home cage area. Modifications were made in a second scale model task, representing one of the animals' outdoor play enclosures, so that in subsequent experiments, all seven adult chimpanzees in the group could be tested.

EXPERIMENT 1: INITIAL INDOOR SCALE MODEL STUDY

Sheba, a fifteen-year-old adult female, and Bobby, a ten-year-old adult male, were tested individually in a room familiar to both. Throughout testing, the room contained four distinct furnishings: a large blue metal cabinet, an artificial tree, a large blue plastic tub, and a multicolored fabric chair. A scale model that was one-seventh the size of the room was placed directly outside in an adjacent hallway. The model contained miniature versions of the furniture, carpet, and other permanent features of the room, including small versions of the four furnishings found in the full-size space. An aluminum can of soda and a miniature version of the can were used as the items to be hidden in the model and the real room, although the chimpanzees were only permitted to watch when the miniature was "hidden" in the model.

Both chimpanzees completed eight trials that included three phases, the first being the Hiding Event, when the chimpanzee watched as the experimenter placed the miniature soda can behind one of the four furnishings that served as hiding sites in the model. Next, the experimenter showed the chimpanzee a real can of soda, and left to hide it in the full-size room, out of view of the animals. After hiding the can, the experimenter returned to the hallway where the chimpanzee was waiting by the model. The subject was next allowed access to the room to find the real soda can (Retrieval 1). After searching the room and locating the can of soda (that the subject was allowed to drink), the chimpanzee returned to the model. They were encouraged to indicate the location where the miniature object had originally been hidden (Retrieval 2). This second retrieval served as a memory test for the chimps' ability to remember the hiding location in the model, and assured that if they failed to find the hidden object in the room, it was not due to memory failure. Given both chimps' extensive experience on a wide range of cognitive tasks and their success with two initial orientation trials prior to formal test trials, we were surprised to find that only Sheba was able to find the soda at a level that was significantly above chance (88 percent correct response, $p<.05$; all data analyzed with the binomial distribution). She entered the room and typically moved directly to the hiding site. Consequently, she found the soda on seven out of eight trials (two trials per site). Sheba's responses suggested that she was able to use the model as a source of information to determine the analogous location in the actual room (Kuhlmeier *et al.*, 1999).

The second chimp, Bobby, approached the task quite differently. Although he was usually successful during Retrieval 2 at the model (63 percent correct response, $p<.05$), indicating he remembered where the miniature item had been hidden, he was not using that information effectively to find the soda in the room. With the exception of one trial, Bobby entered the room and immediately searched at the location that had been assigned randomly as the correct hiding site on the first test trial. If, by chance, he was unsuccessful (usually the case), he would search the other sites in the room following a highly stereotyped, counter-clockwise route until he found the soda can.

We hypothesized that presenting the information in a photographic format might help him locate the object hidden in the room. This result would provide evidence for the "picture superiority" effect reported by DeLoache (1987) for her younger subjects. The same actual room and hiding sites used with the scale model task were used in the new task using photographs. Individual photographs of the four hiding locations were outside the room, and the experimenter showed the soda can to the subject and pointed to the photograph of the hiding location. Once again, Sheba was successful (63 percent correct response, $p<.05$), and Bobby, a failure.

The procedures for the photograph version of the task were modified to see if Bobby's performance would improve. One change minimized any interference of the presence of all four photos being visually available at the same time. Eight more trials were conducted with just the photo of the specific hiding site used for that trial. As before, Sheba was the only subject who found the soda can in the room at a level that was significantly above chance (63 percent correct response, $p<.05$). Bobby, in sharp contrast, responded in the same perseverative manner as he did with the model, although his initial preference for rotating among the sites changed (he chose the blue cabinet on six of the eight trials). In a final attempt for Bobby to succeed with photographs, we presented a panoramic photograph of the actual room, and indicated on the picture (by pointing) the location of the hidden food. His performance was not facilitated by changing the stimuli, while, once again, Sheba had little difficulty using the panoramic image (75 percent correct response, $p<.05$).

A final modification to our approach was to present a pre-recorded video filmed as an experimenter identified each correct location in the real room. We hypothesized that this format, while having the two-dimensionality and some perceptual features of photographs such as color, might convey more information. With the added elements of sound and action, we hoped this would facilitate Bobby's performance. The animals were presented with a brief (10 s) videotaped scenario of the experimenter hiding the soda can in the room. After the video presentation, each subject was allowed to search the room. Sheba, as usual, was highly successful with this shift in modality; and continued to locate the hidden item at levels well above chance (five out of eight correct response, $p < .05$). Despite our best efforts, however, Bobby continued to search sub-optimally, and demonstrated perseverative choices or, more often, clockwise, location-by-location, inefficient attempts.

The indoor scale model, diverse photographs, and video tasks clearly demonstrated that it was within the cognitive capacity of the chimpanzee to recognize the relationship between the various representational modes of a scale model and the larger space they represented. These findings also suggested that one subject, Sheba, was able to understand that the room and model were related, and that the elements of one were analogous to the elements in the other. Consequently, events that occurred in one site represented events that would also occur in the other. However, the reason for Bobby's poor performance was unclear – was it due to his inability to form a "dual orientation" for the model, as DeLoache's model proposed was necessary for children's success, or was it more simply an inability to inhibit a rigid, perseverative search pattern?

COMPARISON OF PERFORMANCE

DeLoache's original work with young children (1987, 1991) found that children showed a definitive age difference in understanding how scale models compared with their real-world counterparts. Children, aged two-and-a-half or three years, were asked to find a full-size toy hidden in a room after witnessing the hiding of a miniature toy placed in the analogous hiding site within a 1:7 scale model. The older group of children (age three) readily retrieved the toy, while the two-and-a-half-year-olds had greater difficulty finding the toy. The younger children's poorer performance was not due to failure to remember the original hiding site in the model, since they were able to find the miniature toy in the model when asked to return to the model after searching unsuccessfully in the full-size room.

DeLoache (1987, 1991, 2000) proposed that younger children's failures were based on an inability to form what she termed a "dual representation" of the model. She argued that to solve the scale model task, children must: (1) represent the model as a real, tangible object, with its toy-like characteristics and perceptual features; and (2) they need to see the model as a symbolic representation for the actual room. Support for the dual representation hypothesis was derived from results with trials during which the same younger children were shown panorama photographs of the room, rather than the scale model. With this procedural change, both age groups succeeded at the task. Thus, DeLoache argued that using the scale model required greater representational skills than using photographs, and two-year-olds performed poorly in situations where they needed to form a dual representation. According to DeLoache, however, photographs did not require a dual representation, and were readily interpreted representationally. This difference in representational format thus allowed the younger subjects to succeed with the task.

We had hypothesized that Bobby's difficulties with the task may have been on his inability to represent the model as both object and symbol, since he was younger and less experienced than Sheba. We reasoned that his performance might improve if we presented color photographs and videotaped scenarios, similar to the procedural changes that DeLoache made for her younger subjects. However, unlike the successful performance revealing a "picture superiority" effect by two-year-old children, Bobby continued to use ineffective search strategies, typically exploring the room in a highly stereotyped, clockwise pattern when searching the four possible sites.

SPATIAL COGNITION, NONHUMAN ANIMALS, SEX DIFFERENCES AND MODELS

Our studies using scale models have demonstrated that the ability to recognize the topographic, perceptual and spatial relationship between a scale model and its real-world referent is within the capacity of enculturated chimpanzees (Kuhlmeier and Boysen, 2001, 2002; Kuhlmeier *et al.*, 1999). We were initially surprised when the scale model experiments revealed sex differences in spatial abilities in our chimpanzees, with females' performance superior to males. While sex differences have been shown previously in humans (e.g., Halpern,

1986; Kimura, 1999; Linn and Peterson, 1985) and laboratory rodents (e.g., Einon, 1980; Gaulin 1995; Gaulin and Fitzgerald, 1986, 1989; Williams *et al.*, 1990; Williams and Meck, 1991), we are not aware of any experimental studies of chimpanzees that have revealed similar sex differences.

Throughout the series of experiments recognition of a scale model as a representation for its full-sized analogue space revealed that female chimpanzees showed a decided advantage in performance compared with males (Kuhlmeier *et al.*, 1999). While these results may seem counterintuitive when compared to sex differences in spatial abilities reported for humans (e.g., male bias for mental rotation) and polygynous rodents, a closer examination of the findings suggests otherwise. Remarkably, the sub-optimal performance of male subjects persisted over numerous experiments during which the perceptual and spatial features of the models were experimentally manipulated. After several experiments were completed and the male chimpanzees continued to have difficulties, a specific intervention was devised to facilitate their performance. It is important to note that this additional training regimen was unnecessary for the females, as their performance was consistently strong, regardless of experimental manipulation of the stimuli or testing context.

In the following, a description of the specific results for each of the scale model experiments will be discussed in further detail. There is little question that the task demands of the scale model experiments resulted in poor performance by some subjects, specifically the four adult males. Moreover, while our initial interest was the chimps' capacity for manipulating different symbol types, the limitations and constraints on their performance based on sex that emerged from the initial studies indicated potentially interesting interference effects. Hypothesized explanations for the differences observed between males and females included individual differences among the subjects, attentional differences between male and female chimpanzees, or potential species-typical, behavioral predispositions toward differing spatial strategies used by both sexes. Indeed, the task demands of the scale model task revealed spatial strategies that differed significantly between males and females.

In general, the observed differences in the use of space may be related to foraging and territoriality patrol in male wild chimpanzees (Goodall, 1986; Hunt, personal communication), which may have had a powerful impact on the male chimpanzees' spatial performance under the specific conditions of the scale model task. It has been well documented that male chimpanzees have a much more extensive range and engage in territorial border patrols. Females travel less and have a finite and constrained range (Goodall, 1986). It has also been effectively argued (Gaulin, 1992; Gaulin and Hoffman, 1986) that for polygynous rodent species, males and females would differ in their spatial abilities, notably because males of polygynous species have larger home ranges or territories than females, and monogamous species pairs share similar-sized ranges. These hypotheses have been further supported by laboratory studies of two vole species that differed in their social and sexual structure, and the polygynous males outperformed females of the same species on spatial tasks (Galea *et al.*, 1995). However, there were no differences in performance between males and females of the monogamous species (Gaulin and Fitzgerald, 1986). Williams *et al.* (1990) suggested that preferential use of geometric cues would be advantageous for acquiring, defending or using a large territory in which foliage and other types of landmarks changed seasonally.

EXPERIMENT 2: OUTDOOR SCALE MODEL STUDY

In order to test all seven of our adult chimpanzees housed at the center, a 1:7 scale model of one section of their outdoor enclosure was constructed. All but two of the animals (Digger and Abby) had extensive experience on many other cognitive tasks (e.g., Brown and Boysen, 2000; Limongelli *et al.*, 1995). The outdoor scale model task was procedurally the same as the indoor study, with both the model and outdoor enclosure containing four hiding sites marked by miniature and full-sized toys familiar to all the chimps. The toys were arranged in four positions that roughly approximated an elongated rectangle and remained stationary for all trials in the first test, and were moved to new locations on each trial during the second test. Slender plastic bottles filled with fruit juice, along with a miniature replica of the bottle, were used as hiding items. During Test 1 of the outdoor version of Experiment 2, three of the chimpanzees performed at levels significantly above chance (Sheba: eleven out of twenty correct responses, $p<.05$; Sarah: thirteen out of twenty correct responses, $p<.05$; Darrell: eleven out of twenty correct responses, $p<.05$). These results replicated Sheba's original performance with the indoor model (Experiment 1), and confirmed that other chimpanzees could understand the relationship between a model and its referent. In addition, analyses of the response patterns of the unsuccessful chimpanzees were also of great interest. All four animals showed a significant preference to visit a particular site first (usually Position 1) on each trial, regardless of whether it was the correct site. Second choice of sites were also analyzed, and one female, Abby, chose the correct site in her second search attempt at a level that was significantly above chance (twelve out of fourteen correct responses, $p < .05$). She would first search at the site in Position 1, and then more directly to the correct site to retrieve the juice bottle. The other three subjects, all males, did not use the same strategy. Instead, if their initial choice of Position 1 was incorrect (usually the case), they typically moved to the adjacent site at Position 2. If a third choice was necessary, it was usually the site located in the same clockwise direction (Positions 3 or 4). Thus, all the males used a search strategy that involved checking each site, moving in a clockwise pattern around the enclosure until locating the hidden juice bottle, precisely the same pattern we saw when Bobby was tested originally with the model room. Interestingly, none of the animals ever had the opportunity to see another chimpanzee participating in the task, so any response strategies that were seen could not have been copied or emulated in any way.

To tease apart what factors might be contributing to these difficulties, we attempted to disrupt their search pattern by moving the hiding sites to different positions in the model and the real enclosure on each trial. With this change, the animals encountered a different object in all positions for all trials. The results of these test conditions revealed that two females, Sarah and Sheba, were again highly successful, and as well as the third adult female, Abby, who also responded significantly above chance (for all: five out of eight correct responses, $p < .05$). Abby's success was striking given her performance on the first version of the outdoor task, when the sites were fixed on every trial. Moving the sites with the new test conditions (moving all sites on each trial) appeared to break up her routine of always checking Position 1 first before going to the correct site. Interestingly, Abby had no previous cognitive training whatsoever, had serious retinal damage from diabetes, and was a former ex-pet who had lived in a human home for the first twenty years of her life in species isolation.

The other four chimpanzees tested, all males, did not perform optimally. Darrell, whose performance had been comparable to Sheba and Sarah's in the first test, showed deterioration with the procedural change in Test 2. His score fell below statistical significance, although both older adult males (Darrell and Kermit, both aged nineteen) chose the correct site on 50 percent of trials. The two young adult males, Digger and Bobby, were both unsuccessful and had the poorest performance of all subjects. When we examined the response patterns of the males, they showed a significant preference for initially searching at Position 1, regardless of the item located there, precisely as they had in the first test, where the objects remained in the same location on each trial. The unsuccessful chimpanzees also showed a strong tendency to visit the adjacent site as their second choice, and continued the search in the same direction. Moving the sites between each trial did not attenuate or eliminate the males' highly stereotyped search strategy, though it benefited one female's performance (Abby).

Following on from Experiment 2, in the third experiment we were interested in clarifying the relative contributions of the varied perceptual features inherent in the scale model task: color, shape and spatial configuration. Thus, the next experiment addressed each of these variables separately. First, the spatial arrangement of the objects in the model and their corresponding referent was modified. Individual color and shape cues of the objects were still available, but the spatial-relational information was no longer informative. The chimps would have to rely upon the color and shape of the miniature objects and the full-sized items in the enclosure. Though we had expected the animals to show some decrement in performance, five of the seven chimpanzees tested located the hidden bottle at levels that were statistically significant. Under this experimental condition, five out of eight correct trials, with chance at .25, was reliable, above-chance performance using the binomial test ($p < .05$). Nonetheless, with spatial cues eliminated, all successful subjects were correct on at least five of the eight trials (Sarah six out of eight, Sheba seven out eight, Bobby seven out of eight, Darrell six out of eight, and Kermit five out of eight). In addition, no sex differences were revealed, nor did the animals perform differently between the first half and last half of the trials, indicating that the chimps had an easy transition in using the available cues from color and shape.

EXPERIMENT 3: DISCORDANT CUES BETWEEN MODEL AND REFERENT

Immediately following Experiment 2, we created discordant color cues between the hiding sites in the model and those in the enclosure. This discrepancy was accomplished in two ways. First, the color of the items serving as hiding sites in the model was confounded with those in the testing area. For example, the tire in the model was red, but the tire in the testing area was black, and so on, for all four hiding sites. That is, the same colors were represented in the model and the real space, but color cues between them were not correlated. We predicted that performance would deteriorate under conditions where the one-to-one perceptual relationship between the model and the referent was decreased. Consequently, we did not expect that the formerly unsuccessful chimpanzees would perform well under these conditions, and predicted that the female chimpanzees, all of whom had done well in Experiment 2 in particular, would also show some deterioration. Thus, we predicted that all subjects would show poorer performance, since we had hypothesized that any manipulation that would undermine the perceptual comparability would affect performance negatively. Since color was a particularly potent cue for stimulus generalization and matching-to-sample strategies, and primates are principally visual animals, the discordant color manipulation would likely diminish performance. Under these conditions, all but two animals were able to find the hidden bottle with reliable, above-chance performance. Performance was considered successful if the chimps were correct on five out of eight trials, and with no reliable object color available, all of the chimps that correctly had five or more correct retrievals. As a group, their performance was also statistically significant at $p<.05$. No differences were detected between the first and second half of the trials, as well, indicating that the chimps had performed consistently through the eight trials of testing.

EXPERIMENT 4: PERCEPTUAL IDENTITY OF HIDING SITES

In the fourth experiment, we examined the effects of increasing the perceptual similarity of the four hiding sites. In previous studies, each hiding site and its comparable miniature site were marked by different items, each a different color, offering the possibility that the successful subjects were solving the task by simply matching the color of the correct site in the model and the same color object in the outdoor enclosure. Completing the task correctly would not necessarily be dependent on understanding the relationship or referential nature of the scale model, but simply through a matching-to-sample strategy, based on the perceptual features of objects in the model and the enclosure. To address this, we conducted test trials using four large, red plastic tubs (.75 m × .75 m) at all the hiding sites. This manipulation controlled for the contribution that color cues were making towards the animals' choices. We predicted that there would be a decrement in overall performance, and such disruption was evident in the poorer performance by three of the chimps (Sarah, Sheba and Bobby), relative to their performance in Experiment 2. However, two subjects (Darrell and Kermit) found the hidden bottle on their first search attempt, and continued to have successful retrievals that resulted in statistically significant performance (six out of eight trials, 75 percent correct responses). Thus, when color cues were not available, the task became more difficult for all the chimps, except Darrell and Kermit, indicating that this test condition was more difficult, overall. As a group, however, the chimps' performance was statistically significant (sign test, $p < .05$), with no evidence of sex differences. These results demonstrated that the chimpanzees could, indeed, rely upon just the spatial-relational correspondence between hiding sites and the real enclosure. In general, the results are similar to those reported for three-year-olds. Apparently, for chimpanzees and young children, recognition of the model/real referent correspondence was challenging when object cues such as color and shape were not available.

EXPERIMENT 5: CHANGING RESPONSE CONTINGENCIES

After considerable experience with different versions of the scale model task over several years, the male chimpanzees persisted in having difficulty performing optimally. These four subjects did not appear to be using a consistent, model-to-referent mapping strategy possibly used by the females. Instead, they relied on the same rigid search strategy, searching the sites clockwise. This response pattern resulted in ultimate discovery of the hidden food reward, but their behavior made it unclear whether they recognized the perceptual or spatial, much less the representational, features of the scale model. We hypothesized that a procedural change that emphasized the importance of attending to the initial hiding event in the model might influence the males' response. To that end, a procedural change was devised so that only a strategy based on mapping the model-enclosure relationship would result in reward (Kuhlmeier and Boysen, 2001). The new approach modified the task so that attainment of the reward was contingent on locating an empty juice bottle on the first search attempt. This first attempt was highlighted – if an incorrect hiding site was selected first, the chimpanzee was required to exit the enclosure, return to the transfer chute, and repeat the entire trial. If the empty juice bottle was located on the first search attempt, the chimpanzee could exchange the bottle through the cage wire for a full bottle of juice immediately. We essentially instituted a training regimen to teach the males to pay attention specifically to the hiding event. If they had a capacity to recognize the relationship between the model and the enclosure, this new procedure would favor a mapping strategy over their previous search strategy.

All subjects first participated in training trials with the new procedure, and were tested for generalization of the new contingency task using novel hiding sites. After thirty-two training trials, each chimpanzee was performing above chance (correct response >15: binomial test, $p<.01$). When the novel hiding sites were introduced, as a group, the chimpanzees performed above chance (six out of seven were above chance, $p<.05$), with no difference in performance from the first half of the trials to the last half. There was also no difference between the performance of the male and female chimpanzees.

A second test compared the animals' performance with the new contingency procedures and the original methods used with the standard scale model task. Under both of these test conditions, their group performance was significant. When the standard procedures (with full juice bottle hidden), five of the seven chimps were significant at the individual level, while under the contingency procedure, six out of seven subjects' performance was significant. Additional analyses (Wilcoxon test) compared the order of the two different procedures, as well as overall performance on both procedures, and there was no significant effect for either variable. The chimpanzees were able to solve the scale model task under conditions in which a full juice bottle was not present during the search phase (contingency procedure), and now under the original test conditions during which the four males had shown consistently poor performance (Kuhlmeier and Boysen, 2001; Kuhlmeier *et al.*, 1999). Overall, five of the seven chimps were able to solve the problem when either a full juice bottle was present (standard procedure) and when an empty bottle had to be exchanged (contingency procedure). It is interesting that during the majority of test trials with the standard procedure, most subjects did not revert to the original search strategy which could have resulted in them finding the full bottle through trial-and-error. Instead, the chimps continued to use the mapping strategy that was encouraged by the contingency procedure. However, even though the animals demonstrated immediate transfer of the mapping strategy during trials when the contingency procedure was necessary, three of the four animals that had been unsuccessful during the original study, the former search strategy began to emerge during the second half of the eight-test trials. This suggested that, had additional trials been run, the original search strategy could possibly have re-emerged. It would be interesting to test the standard and contingency procedures with young children (less than three years old) to determine whether experience with the contingency approach might improve performance.

PRINCIPLES OF GENERALIZATION FROM MODEL TO REALITY

Consistent with the innovative paradigms and results reported by DeLoache (1987, 1991, 2000) and our preliminary studies of chimpanzees' understanding of scale models, we think it is important to have a closer examination of the features of scale models that allow for recognition of their usefulness in predicting reality. Clearly, some key perceptual features contribute toward increasing or decreasing the facile transposition of the key characteristics of a scale model, permitting mapping of these elements to those of the real-world referent of the model. In our own studies, reducing the usefulness of color cues diminished performance by previously successful chimpanzees. For example, deviations from the correspondence of color characteristics between landmarks in the model and those in the real space detracted from performance. As suggested by DeLoache's results, when younger children's performance was facilitated after experience with a larger model that was more similar in scale to the actual room, increasing the size of the model (originally on a scale of 1:7) would likely increase the chimpanzees' successful performance (and vice versa). That is, as the perceptual similarity between the model and the real space diminished, correct choice of the analogous hiding sites in the real space would deteriorate.

SUMMARY

The overall results from the scale model studies with our chimpanzees demonstrate that they are capable of understanding a scale model and photographs of a corresponding real space. The results replicate and extend results from studies of similar skills in children by DeLoache and her colleagues (e.g., DeLoache, 1987, 1991, 2000; Marzolf and DeLoache, 1994; Marzolt *et al.*, 1999). However, in this case, we are assessing the capabilities of a nonverbal species with whom previous attempts to demonstrate an understanding of models had failed (see Premack and Premack, 1983). Initially, recognition of these relationships was not fully demonstrable in all chimpanzee subjects, as evidenced by the failure of most male subjects to respond optimally early in the series of experiments. The second experiment that used a model of the animals' outdoor enclosure allowed us to examine error patterns exhibited by all our chimpanzees, including all four of our adult males, who had performed more poorly. Their search strategies differed from those reported for young children. While two-and-a-half-year-old children had difficulties with DeLoache's scale model task, they did not show similar perseverative search patterns that we saw in our male chimps. However, once the male chimpanzees' reward attainment was contingent upon using the model to find the object in the real space, they showed no difficulty in finding the object by using the model.

Given these findings, one question that arises is whether using the scale model approach in general is of any interest ecologically, and indeed, cognitive relevant for testing with chimpanzees. First, it is of significance that they were able to perform well on the task at all, and thus had we not made an effort to examine their responses, with respect to DeLoache's innovative studies (e.g., 1987; see Chapter 21, this volume), we would not have been able to answer the question. Second, the set of experiments has shown us that chimpanzees can use spatial information when that is all the information available to them, just as they would have to rely significantly on their spatial-relational environment in the wild. With a preference for fruit, wild chimps would have to encode the spatial configuration of their foraging territory, so as to re-visit trees when the fruit was more likely to be ripe. From one site within the territory, it would be most advantageous to expend the minimal caloric resources to move on to the next foraging site, and thus the ability to represent a spatial configuration within the territory, relative to the current foraging site. In addition, the results of the scale model experiments during which the spatial arrangement, color and shape of the items within both the model and the actual enclosure indicated that chimpanzees can represent color and shape cues, if available, but are capable of using whatever perceptual features are available. This skill set would be highly adaptive in the wild, as it reflects greater flexibility in chimpanzees' finding their way within the home territory, as well as maximizing opportunities for using optimal foraging strategies.

REFERENCES

Brown, D. A. and Boysen, S. T. (2000). Spontaneous discrimination of natural stimuli by chimpanzees (*Pan troglodytes*). *Journal of Comparative Psychology*, **114**, 392–400.

DeLoache, J. S. (1987). Rapid change in the symbolic functioning of very young children. *Science*, **238**, 1556–1557.

DeLoache, J. S. (1991). Symbolic functioning in very young children: understanding of pictures and models. *Child Development*, **62**, 736–752.

DeLoache, J. S. (2000). Cognition and language: dual representation and young children's use of scale models. *Child Development*, **71**, 329–338.

DeLoache, J. S., Kolstad, D. V. and Anderson, K. N. (1991). Physical similarity and young children's understanding of scale models. *Child Development*, **62**, 111–126.

Einon, D. (1980). Spatial memory and response strategies in the rat: age, sex, and rearing differences in performance. *Quarterly Journal of Experimental Psychology*, **32**, 473–489.

Galea, L. A. M., Kavaliers, M., Ossenkopp, K.-P. and Hampson, E. (1995). Sexually dimorphic spatial learning in meadow voles, *Microtus pennsylvanicus* and deer mice, *Peromyscus maniculatus*. *Journal of Experimental Biology*, **199**, 195–200.

Gaulin, S. J. C. (1992). Evolution of sex differences in spatial ability. *Yearbook of Physical Anthropology*, **35**, 125–151.

Gaulin, S. J. C. (1995). Does evolutionary theory predict sex differences in the brain? In M. S. Gazzaniga (Ed.), *The cognitive neurosciences* (pp. 1211–1224). Cambridge, MA: MIT Press.

Gaulin, S. J. C. and Fitzgerald, R. W. (1986). Sex differences in spatial ability: hypothesis and test. *American Naturalist*, **127**, 74–88.

Gaulin, S. J. C. and Fitzgerald, R. W. (1989). Sexual selection for spatial learning ability. *Animal Behaviour*, **37**, 322–331.

Gaulin, S. J. C. and Hoffman, H. A. (1986). Evolution and development of sex differences in spatial ability. In L. Betzig, M. Borgerhoff Mulder and P. Turke (Eds.), *Human reproductive behaviour*. Cambridge: Cambridge University Press.

Goodall, J. (1986). *The chimpanzees of Gombe: patterns of behavior*. Cambridge, MA: Belknap, Harvard University Press.

Halpern, D. (1986). *Sex differences in cognitive abilities*. Hillsdale, NJ: Erlbaum.

Kimura, D. (1999). *Sex and cognition*. Cambridge, MA: MIT Press.

Kuhlmeier, V. A. and Boysen, S. T. (2001). The effect of response contingencies on scale model task performance by chimpanzees (*Pan troglodytes*). *Journal of Comparative Psychology*, **115**, 300–306.

Kuhlmeier, V. A. and Boysen, S. T. (2002). Chimpanzees (*Pan troglodytes*) recognize spatial and object correspondences between a scale model and its referent. *Psychological Science*, **13**, 60–63.

Kuhlmeier, V. A., Boysen, S. T. and Mukobi, K. (1999). Comprehension of scale models by chimpanzees (*Pan troglodytes*). *Journal of Comparative Psychology*, **113**, 396–402.

Limongelli, L., Boysen, S. T. and Visalberghi, E. (1995). Comprehension of cause-effect relations in a tool-using task by chimpanzees (*Pan troglodytes*). *Journal of Comparative Psychology*, **109**, 18–26.

Linn, M. C. and Peterson, A. C. (1985). Emergence and characterization of sex differences in spatial ability: a meta-analysis. *Child Development*, **56**, 1479–1498.

Marzolf, D. P. and DeLoache, J. S. (1994). Transfer in young children's understanding of spatial representations. *Child Development*, **65**, 1–15.
Marzolf, D. P., DeLoache, J. S. and Kolstad, D. V. (1999). The role of relational similarity in young children's use of a scale model. *Develop. Sci.*, **2**, 296–305
Premack, D. and Premack, A. (1983). *The mind of an ape*. New York: Norton.
Williams, C. L. and Meck, W. H. (1991). Organizational effects of gonadal steroids on sexually dimorphic spatial ability. *Psychoneuroendocrinology*, **16**, 155–176.
Williams, C. L., Barnett, A. M. and Meck, W. H. (1990). Organizational effects of early gonadal secretions on sexual differentiation in spatial memory. *Behavioral Neuroscience*, **104**, 84–97.

AMY E. LEARMONTH AND NORA S. NEWCOMBE

23

The development of place learning in comparative perspective

Upon exiting the shopping mall, it is common to look out over a sea of cars extending for great distances. How can people locate their own vehicle? One important and useful method of performing this task is to use landmarks around the edges of the parking lot. For example, a well-equipped parking lot may have a series of poles with colors and numbers on them. An individual who has been reasonably attentive in the morning while parking may remember that the car is slightly closer to the doors of the mall than the pole with the green three, and that it is two rows over in the direction of the pole with the purple three. In the sea of glinting metal it is then suddenly not hard to walk unerringly to the parking place. This kind of event is part of everyday navigation, and is an example of the powerful spatial location system called place learning.

This chapter is a discussion of the development of place learning in rats and humans. We concentrate on these two species because, unfortunately, less is known about development of this spatial learning system in other species, including nonhuman primates (a focus of this book). Indeed, although there is a good deal of research on spatial cognition in non-human primates (Gomez, 2005; Haun and Call, 2009; Tomasello and Call, 1997; see also Penn *et al.*, 2008; Dolins, 2009 and Chapter 8, this volume), place learning has apparently not been assessed even in adults in these species. However, this gap in the literature may be less crucial than it first seems, because we shall see that there are substantial parallels in development in rats and humans, suggesting that common mammalian ancestry is at work here.

We begin with a discussion of what place learning is and why it is an important component of spatial cognition. We proceed to focus on describing the development of place learning in rat pups and what

Spatial Cognition, Spatial Perception: Mapping the Self and Space, ed. Francine L. Dolins and Robert W. Mitchell. Published by Cambridge University Press.

mechanisms of development may account for acquisition of place learning in the rat. Following that section, we summarize knowledge concerning the development of place learning in humans and then examine again proposed mechanisms for development. Finally, we directly compare the developmental facts concerning rats and humans, and draw some conclusions about the similarities and differences in developmental trajectories.

PLACE LEARNING

Any discussion of the development of place learning must begin by differentiating the place learning system from other systems of spatial competence. Place learning is one of the externally referenced spatial systems because it is reliant on the distance and direction of landmarks without reference to the navigating animal. Place learning is achieved by specifying the distance and direction of the desired object or location from landmarks within an environment. The distance and direction from two landmarks is usually sufficient to clearly specify a location in space, but more landmarks are sometimes used to more exactly specify a location. In the place learning system landmarks can be a number of distinct features of the environment including objects, such as trees, mountains and buildings, boundaries, such as the fence surrounding the playing field, or regions, such as the little park west of the psychology building. Information from all of these landmarks can be used to locate either the observer in space or a desired object.

In any discussion of place learning as a system of spatial cognition it is important to distinguish place learning from other systems of navigation. Newcombe and Huttenlocher (2000) suggested that there are four systems of navigation, two simple and two complex. Place learning is one of the complex systems. The simpler externally referenced spatial system is cue learning. Like place learning, cue learning is based on the location of the to-be-located object in relation to a landmark, but in cue learning the landmark and the object are coincident, meaning the landmark is on top of or within close proximity to the desired object. This system uses the landmark as beacon or anchor point for a search for the desired object and has no recourse if the desired object is not under, on top of or otherwise adjacent to the landmark. This system is useful much of the time, but can only be used when there is a coincident landmark. This system is useful for things like remembering that the keys are on the hall table. However, it can no longer be used as soon as the to be located object is not coincident with the landmark.

The other two systems besides place learning and cue learning are response learning and dead reckoning (also called inertial navigation). Both of these systems are viewer referenced, meaning that they use spatial information that is reliant on an individual's awareness of his location in space. A continued awareness of one's location in space allows one to locate objects in reference to the self. The simpler system, response learning, involves the use of a learned motor routine that runs off regardless of the location of the individual or that individual's movement in space. This system is the one in use when the writer reaches for a tissue without looking away from the computer screen. The problem is that if someone has moved the tissue box the reach is a failed attempt and does not yield a tissue. The more complex viewer centered system, dead reckoning, takes into account the movements of the viewer in the environment using vestibular and proprioceptive information. This system keeps track of the movements of the viewer over time and updates the locations of objects relative to the viewer's new position. This system is quite powerful and can be used in situations, such as total darkness, where visually guided navigation is not possible. It is the system that allows you to find your way around the house during a blackout. The weakness of this system is that it is subject to drift, when the estimations of distance traveled or rotations made are not exact the errors carry into the next movement and over time compound into large inaccuracies.

In normal navigation the place learning system is used in concert with the other systems mentioned above. In adults, the four systems are combined such that the weaknesses of each are compensated for with one of the other systems. Each system can be experimentally distinguished from the others and its features and development examined.

DEVELOPMENT OF PLACE LEARNING IN RATS

Research on the development of place learning in rats is impressive. A coherent description of the timing and sequence of acquisition has been worked out in quite a detailed fashion, and has influenced research with human children. A very powerful and often used tool for examining place learning in rats is the Morris Water Maze (MWM). In the MWM, the rat is placed in a circular pool of murky water and must swim to a platform hidden just under the surface of the water in order to escape the water, which functions as an aversive stimulus. The task usually takes place in a lab, many features of which are visible from within the pool. Because there is nothing in the pool itself to guide their search, the rats can only find the platform if they use landmarks outside of the pool, thus demonstrating place learning. This simple but clever task has been modified for use with humans – who have never been asked to swim! Instead, they have searched for targets in unfeatured areas where they must use distal landmarks to succeed, including a round enclosure filled with Styrofoam chips (Overman *et al.*, 1996), a baby pool with pillows in it (Bushnell *et al.*, 1995) and a computer-generated space (Laurance *et al.*, 2003).

Initial research with the MWM task showed that preweanling rats become able to solve the Morris water maze and escape the water at twenty days (Rudy and Paylor, 1988; Rudy *et al.*, 1987). However, there was some disagreement about the exact age at which rats can first place learn; Schenk (1985) claimed the ability did not appear until twenty-six days. Nevertheless, at between the twenty and twenty-six day mark, it seemed clear that rats find the platform more reliably than chance and over trials their latency to the platform decreases, indicating they have learned to use the distal landmarks to guide their search for the platform (Brown and Whishaw, 2000; Rudy and Paylor, 1988; Rudy *et al.*, 1987; Schenk, 1985).

One may see earlier competence if one makes allowances for some facts about infant rats. Rats are born with their eyes closed. Thus, because place learning relies on the ability to use distal visual landmarks to locate an object, it cannot be present prior to the time that rat pups open their eyes, in the end of the second week of life. From that point they have the possibility of using visual information to solve a place learning task. The other difficult developmental issue with the MWM task is that, although rats instinctively swim, young rats may have difficulties that are related to swimming endurance and body temperature regulation that affect their performance in the task (Brown and Whishaw, 2000; Caeman and Mactutus, 2001). Additionally,when training for the task takes more than one day it may not lead to an accurate assessment of the first emergence of place leanring ability because the ability is assessed on the testing day (Akers and Hamilton, 2007). Therefore, studies seeking to determine the earliest point at which rat pups can solve the MWM task make procedural allowances for the special difficulties faced by young rat pups.

One allowance involves using an MWM task that takes place in a pool one-third the size of the one normally used. In the smaller pool,

Caeman and Mactutus (2001) found that seventeen-day-old rats demonstrated a preference for the quadrant of the pool with the platform and reduced latencies to find the platform across training. In these special conditions, place learning appears earlier than in any other study. Brown and Whishaw (2000) found place learning at nineteen days. In their study the pool was the same size as the classic conditions, but the rats were carefully maintained such that their body temperatures remained close to normal. The rat pups were dried and put in a warming cage between trials, preventing any issue with the young pups' inability to maintain their body temperature.

Another issue that obscures an exact day for the emergence of place learning in preweanling rats is that there are procedural differences that lead to the claim that in some studies the platform can be located by directional responding instead of true place learning (the rats learn to swim in a specific direction instead of to a specific place) (Akers *et al.*, 2009). Akers *et al.* (2009) found that preweanling rats learn to respond directionally in the MWM task at twenty to twenty-one days and do not show unambiguous place learning until twenty-six or twenty-seven days. These findings may explain some of the different ages at which different researchers find success in the MWM, but the task used may also be more difficult in general than the procedures of the previous researchers (the MWM itself was moved between trials to clearly differentiate between directional responding and place learning).

Although rats become able to place learn as preweanlings, they continue to improve in their ability to place learn. In a detailed study, Rudy *et al.* (1987) found a developmental trend in the ability of young rats to solve the MWM. Independent groups of rats between nineteen and twenty-three days show steady reduction of the errors in heading and increases in time spent in the correct quadrant of the pool (Rudy *et al.*, 1987). Each day older afforded the rats an improvement in their ability to find the hidden platform, but still the twenty-three-day-old rats did not perform like the adults in the original Morris (1981) study.

There is a steady increase in both the speed and accuracy with which young rats solve the MWM task (Schenk, 1985). According to Schenk, performance does not reach adult levels until forty days. There was steady improvement as the rats got older, with each day bringing further improvement in the frequency with which rat pups found the platform. By thirty-five days, the rat pups acquired the task like the adults and were only different in their search behavior during the probe trial when the platform that had allowed them to escape the water was no longer present. This inefficient search seemed to reflect a lack of persistence once the rat had gone to the location where the platform was supposed to be and found it missing; the adult rats continued to search the correct area, while the thirty-five-day-old pups wandered farther from the correct location. This lack of persistence may be more due to fatigue or some other factor, such as fear, than place learning ability (Schenk, 1985).

Independent groups of preweanlings (twenty days), periadolescents (thirty-four days) and adults (sixty to seventy days) learn in the MWM, but rats at the three ages show differences in how they remembered and were reminded of the task (Brown and Kraemer, 1996). The rats became progressively better at acquiring the task with age and adult rats showed the least forgetting of the location of the platform over time. A second experiment demonstrated that a single reminder trial just before the probe trial alleviated forgetting in periadolescent rats. The differences in the rats' forgetting of the location of the platform mirrored their differences in acquiring the task (Brown and Kraemer, 1996).

These studies show that, although place learning can be seen as early as seventeen days, early competence is not equal to the place learning demonstrated by adults. Over the next twenty days the rat place learning system develops slowly, not reaching clear adult competence until forty days. The early competence seen in the seventeen-day-olds transitions into adult competence as the rats solve the MWM task more and more quickly and efficiently.

MECHANISMS OF DEVELOPMENT IN RATS

Maturation of the hippocampus in rats is seen as one of the principal mechanisms of development for the place learning system. The involvement of the hippocampus in the ability of rats to solve the MWM task is demonstrated by studies that show adult rats who have lesions of the hippocampal formation do not solve the MWM unless a landmark is in the pool coincident with the platform (as opposed to the situation where the landmarks are in the room, external to the pool), making the task one of cue learning instead of place learning (Morris, 1981). In rats the hippocampus shows extensive development in the first week after birth and continues to develop for at least a few weeks after that. The cell density in the dentate gyrus of the hippocampus reaches adult levels at about twenty-five days and the hippocampus shows metabolic activity from around thirty days (Schenk,

1985). The development of the hippocampus in the rat has been suggested as an explanation for the developmental trajectory of place learning. That the developmental milestones seen with the behavioral place learning tasks correspond roughly to the developmental course of the hippocampus is evidence that the development of the hippocampus is involved with the development of place learning (Nadel and Zola-Morgan, 1984; Schenk & Morris, 1985). Akers *et al.* (2009) suggest that subtle differences in the rates of development within different parts of the hippocampus within the first few months of a rat's life could lead to the uneven findings between seventeen and twenty-seven days.

The other means that could be involved in the development of place learning in the rat is that of experience. Tees *et al.* (1990) show that rats reared in a dark room do not perform as well as rats reared in a standard light/dark schedule on several versions of the Morris Water Maze. Also, the earliest ability to solve the MWM comes not long after the rats open their eyes for the first time, and there is marked improvement in their ability to solve the task over the next few days. That the young rats are gaining visual experience over the four weeks of improvement after their first ability to show place learning could be another means by which place learning develops in rats.

Maturation of the brain and experience with the world occur simultaneously and synergistically, providing a powerful source for the development of place learning in the rat. The maturation of the brain could allow the infant rats that have experience with place learning to gain more from the experience and the experience of place learning could shape the maturation that occurs in the brain. Caeman and Mactutus (2001) suggest that the rats in their study improve not only because of maturation, but because over the course of training experience with the maze itself allows for some of the maturation in the hippocampus of the young rats. They also argue that the immature performance seen in the young rat pups could be seen as an accomplishment in itself and the slow progress to adult performance reflects the changing needs of the young rat as it begins to explore its world to a greater and greater extent.

Place learning emerges in the rat sometime between seventeen and twenty-six days with what may be an early directional responding, and clearly is place learning by twenty-six days. The rats continue to improve their place learning until about forty days. The mechanisms involved in the emergence and improvement of their navigation in general and place learning in particular include an interaction of the maturation of the neural systems through experience. The maturation factors may also lead to increased endurance and better temperature regulation in the MWM task, leading to better performance with age.

THE DEVELOPMENT OF PLACE LEARNING IN HUMANS

Spatial competence in humans begins long before infants take their first step. Infants demonstrate that they are aware of the locations of objects and expect objects to behave in ways that follow basic physical laws (Newcombe *et al.*, 1999, 2005; Quinn *et al.*, 2002). But this is not yet place learning. As soon as they can move around an environment on their own, human infants begin their exploration of the world. That exploration requires that they be able to locate items and return to places they have been.

Studies that have required infants to use non-coincident, but very close-at-hand landmarks have found that at least by sixteen months infants can successfully locate a hidden object. In these studies the infants were asked to find a toy hidden in a large rectangular sandbox. In this task the edges of the sandbox itself functioned as the landmark as well as an enclosing geometric space. The infants in this study were quite accurate in their search and their pattern of search, with the most accurate searched at the ends and center of the box, which indicates that they were using the edges of the sandbox to guide their search. The sixteen- to twenty-four-month-old infants could use the edges of the sandbox to code the location of the desired object (Huttenlocher *et al.*, 1994). Another study reported a similar result with twelve-month-old infants and a baby pool. The infants searched for objects hidden under pillows in the round pool. The infants searched quite accurately, indicating they were coding the distance and direction of the hidden object (Bushnell *et al.*, 1995). This may not be the first instance of place learning, however; because the these infants did not move from the location from which they had seen the object hidden before being encouraged to search, we do not know if they are coding location relative to themselves or the landmarks. Infants as young as eighteen months old can also use the geometric configuration of the room as a landmark (Hermer and Spelke, 1994; Learmonth *et al.*, 2001; Smith *et al.*, 2007). Experimenters disoriented the children before allowing search, thereby disabling the viewer-centered spatial coding systems.[1] These studies demonstrate that at a very young age children can use surrounding frames to guide their search behavior, be that frame the edges of the sandbox, the enclosing walls of the room, or the surrounding natural environment. However, it is unclear that this is a demonstration of place learning. The surrounding landmark could be used as a landmark in the place learning system, but it could also be a coincident landmark, therefore only requiring the cue learning system.

The pattern found in these studies implies that infants place learn using very proximal landmarks or geometric enclosures. It is an impressive accomplishment that these infants use the edges of the array to locate their search, but other researchers have found that children as old as twenty-six months are not able to find an object they have seen hidden in one of four identical containers placed within a space containing multiple distal landmarks (DeLoache and Brown, 1983). It is possible that the use of several identical containers was confusing to the children who might have had difficulty with the inclusion of a number of identical containers. An interpretation of this finding offered by Huttenlocher and Newcombe (1984) is that the children in these studies were not coding the location of the hidden toy with respect to the distance and direction of the available distal landmarks.

To examine the ability to use distal landmarks, which are required for true place learning Newcombe *et al.* (1998) used a sandbox similar to the one used in the Huttenlocher *et al.* (1994) study. Children between sixteen and thirty-six months watched as a toy was hidden in the large rectangular sandbox and the sand was smoothed to remove any solution based on the look of the sand, then the child was distracted to break fixation on the correct location and moved around to the other side of the sandbox before being encouraged to search for the object. The movement around the sandbox eliminates the ability to find the object successfully by remembering its distance and direction from the infant as they observed the object being hidden. In order to differentiate place learning from dead reckoning, the researchers manipulated the available landmarks to see if distal landmarks improved the ability of the infants to find the hidden object. For half the infants, the sandbox area was surrounded by a white curtain that eliminated visual access to distal landmarks, for the other half the curtain was removed and the room contained a number of distal landmarks, such as a poster on each of the four walls. This study found a particular age at which infants become able to use distal landmark information to guide their search. Children between sixteen and twenty-one months did not show any effect of condition, indicating that they did not use the distance and direction of the landmarks to refine their search. However, the searches of children twenty-two months old and older were significantly more accurate in the condition without the curtain than the one with the curtain, indicating that they were using the distal landmarks to refine their search and using distance and direction from those landmarks in their search for the hidden object.

The results of the Newcombe *et al.* (1998) studies delineate the emergence of place learning between eighteen and twenty-four months. Therefore, the first emergence of the place learning system is part of an abrupt change. However, even though children over the age of twenty-two months could use the distal landmark information to improve the accuracy of their search, they still are a long way from adult competence.

In a task designed to mimic the MWM where children searched under panels on the floor for a puzzle, Balcomb *et al.* (2009) found an increase in spatial searching between sixteen and seventeen months as well as replicating the sharp developmental change around twenty months found by the previous research, indicating that the place learning system is undergoing changes within a few months of the children learning to walk. Even though the twenty-four-month-olds were more successful than the younger children, their success rate, though far above chance, was around 50 percent, which is a long way from adult competence (Balcomb *et al.*, 2009).

Competence in place learning comparable to that in adults is not seen until children are between seven and ten years old (Aadland *et al.*, 1985; Jansen-Osmann and Wiedenbaur, 2004; Lehnung *et al.*, 1998; Overman *et al.*, 1996). How children get from their first ability to use distance and direction information from distal landmarks to adult competence is explored in a study using children between three and ten years of age, and adults (Laurence *et al.*, 2003). Children were asked to find a colored square on the floor of a computer-generated space. The computer-generated arena is a space created on a desktop computer that allows navigation of an observer (the computer-generated observer was not visible on the screen, but as it moved the view of the arena changed as it would if the observer were moving through the computer-generated space) in a circular area that is surrounded by a larger square room with pictures on the walls. The pictures act as distal landmarks and can be used to locate the colored square. The search pattern and latency to the target location were recorded for all the children. All of the children succeeded in finding the square when it was visible, demonstrating that they understood the task and were following the instructions of the experimenter. In the next phase, the square was in a different location from the visible trials and invisible until the child moved to the correct location. The invisible square was always in the same location, allowing the children to demonstrate learning across trials. The data from this study shows a gradual developmental trend in the ability of these children to locate the colored square using the landmarks on the walls of the room.

The three- and four-year-olds did not successfully locate the colored square once it was invisible. On some of the search trials they did come across the correct location, but they were not able to locate it again on the next trial. At each age between three and nine there was improvement in the ability of the children to locate the invisible target. (Performance of nine- and ten-year-old children did not differ from the adults.) Not only did the children find the invisible square more frequently with each passing year, but also the amount of time it took them to find the square decreased with each year and how that time was spent changed. The youngest children moved around the space with no particular discernable pattern, but the pattern emerged and changed as the age of the group increased. The first pattern to emerge was a tendency to search the correct distance from the edge of the circular space in which they were navigating. This was followed by a tendency to move directly into the correct quadrant and then search that area, followed finally by ballistic search in which the child moved directly to the correct location. This was also the search pattern seen in the adults. This study shows steady improvement over the early and middle childhood in place learning until adult competence is achieved by age nine (Laurence *et al.*, 2003).

In addition to the changing search pattern over developmental time, there was also a change in the self-reported use of the landmarks on the walls of the computer-generated space. The youngest children did not report noticing the landmarks at all. The four- and five-year-olds started to report the use of one of the available cues to locate the target. However, in this situation, one cue was often not enough to clearly specify the location of the target because it was not directly in front of one of the landmarks. The six-year-olds started to report the use of multiple landmarks, with about half of the group using this technique. The older the children, the more likely they were to report using more than one cue to find the target (Laurence *et al.*, 2003). The use of more than one landmark is important to effective place learning. The children who used just one of the landmarks had difficulty when their approach to that landmark was not angled exactly right and many of them ended up moving along the same path away from and back toward the landmark repeatedly, missing the target because their initial approach had been from an unfamiliar location in the space. The use of more than one landmark makes place learning much more effective because the location of the target can be triangulated using integrated information from both of the landmarks.

Jansen-Osmann and Wiedenbaur (2004), using a different virtual maze, found that six- to eight-year-old children recalled fewer landmarks after navigating a virtual maze than ten- to twelve-year-olds, who were not different from adults. A second study found an increase in performance across all three age groups indicating some improvement in place learning between sixth grade and adulthood (Jansen-Osmann *et al.*, 2007).

In a real-space task similar to the virtual task used in the Laurence *et al.* (2003) study, Overman *et al.* (1996) found a similar pattern of steady improvement in the ability of the children aged between three and seven to locate the desired object. They used a round enclosure filled with Styrofoam chips as an apparatus and found that the distance traveled to find a hidden box of candy steadily decreased in children from ages three to seven. They could not look at latency to find the box because as the children got bigger, moving through the Styrofoam was easier due to the change in size.

All of these studies indicate that place learning can be seen in its most rudimentary form in twenty-two-month-old human infants in their use of distance and directional information from distal landmarks. From that point children place learn, but they require the next seven years to develop adult competence.

MECHANISMS OF DEVELOPMENT IN HUMANS

In humans place learning matures slowly over the course of the first eight or ten years of life. The mechanisms responsible for this development are several and possibly different at different times. The hippocampus, which is important to place learning, develops late in humans; it may be that development within the brain is what leads to that first leap in ability that appears abruptly at twenty-two months. The behavioral evidence of this leap at twenty-two months corresponds roughly to evidence that hippocampal maturation continues until about twenty-one or twenty-two months of age (Kretschmann *et al.*, 1986; Seress, 1992).

In a study to examine the change in spatial ability at about twenty-one months, seen in the Newcombe *et al.* (1998) study,

Sluzenski *et al.* (2004) used other tasks that rely on the hippocampus with young children between eighteen and forty-two months. This study offers some convergent evidence that the dramatic change in the ability of toddlers to use the distal landmarks in the Newcombe *et al.* (1998) study is a result of maturation in the hippocampus. In the first task, infants were asked to retrieve two identical hidden objects from the same sandbox used in the 1998 study. They saw both objects hidden sequentially by an experimenter. Then, after a brief head turn to break eye contact with the objects' location, they were asked to find both objects. There were no reliable age differences in the ability to find the first object, indicating that the children at all ages understood the task and were motivated to find the hidden object, however there was a difference between the eighteen- and twenty-four-month-old children in the accuracy, if not the enthusiasm, of their search for the second object. Not only did the twenty-four-month-old children search more accurately for the second object, they also showed less perseveration between the first and second search. There were no differences between the groups that were twenty-four months and older. In another task a delay of two minutes was introduced between the child's observation of the hiding of an object in the sandbox and search. Again, the eighteen-month-olds were less able than the older children to find the hidden object after a delay (Sluzenski *et al.*, 2004).

The gradual maturation between three and nine may be due to experience. The Laurence *et al.* (2003) study indicates that one of the changes that occurs over the early and middle childhood years is a change in the search strategy used. The older children reported that they were using the landmarks to find the square and tended to move directly toward the target location, while the younger children were less likely to report the use of the landmarks and had more meandering search patterns. It is possible that the older children, who had more experience navigating independently, learned to use the landmark strategy from that experience. An interesting direction to take this research would be to see if the younger children could be trained to use the more successful strategy of the older children.

Another possible explanation for the slow maturation could be that the hippocampal formation may continue to mature and change shape long after it has reached mature size. Gogtay *et al.* (2006) found that the posterior third of the hippocampus, with the possible exception of the posterior pole, increases between the ages of four and twenty-five, but a corresponding decrease in the anterior portion leaves the total volume constant. They suggest that the posterior hippocampus is important for spatial learning. This continued structural change could be an important clue to explain the slow maturation of place learning.

Experience and brain maturation work and develop together. The children under twenty-two months who did not use the distal landmarks to guide their search have both less mature hippocampi and less experience with upright navigation than the children one month their senior who did use the distal landmarks. Furthermore, maturation of brain systems and experience may happen together and influence each other. The experience of upright walking, which occurs at around twelve months and allows for much more visual exploration of distal landmarks than crawling, may instigate the maturation of the hippocampus which is necessary for using distance and direction of distal landmarks to determine location.

COMPARISON OF RATS AND HUMANS

The importance of a comparison of the developmental trajectories of place learning in rats and humans is illustrated by the fact that many studies of spatial learning in rats have led to studies in humans (for example, Cheng's 1986 study of reorientation in rats has led to a series of recent studies of reorientation in humans by Hermer and Spelke, 1994 and Learmonth *et al.*, 2002). In most cases the rat experiments are conducted with adult rats, which may be a less useful model of spatial behavior in children. The comparative analysis of the development of place learning leads to conclusions about the potential use of young rats as a model of place learning in young children.

The Laurance *et al.* (2003) study used a computer-generated space designed to be analogous to the Morris Water Maze. The attempts to make the human task as similar as possible to the task used with rats makes comparison of the two species easier. Although the demands of the tasks are different, the similarity of the developmental trends are suggestive. Rats showed the same tendency to localize their search that the children did, indicating that they were using the same progression from the disorganized search pattern through several noticeably similar intermediary steps to the search pattern that allowed them to move directly from the starting point, at any place in the pool, to the target location. It may be that the similarities express a necessary progression to mature place learning: in order to achieve place learning any organism must follow the course of steps from the first ability to use distance and direction from a landmark to find an object to adult competence in place learning.

The mechanisms of development in place learning in rats may be very similar to those in humans. In both cases maturation in the brain and experience, most likely in combination, can be cited as involved in the development of place learning. That rat pups do not succeed in finding the platform in the Morris Water Maze until after seventeen days could be analogous to the inability of human infants to use the distal landmark information to guide their search for a desired toy in the sandbox until twenty-two months. The timeframe for maturation in the two species are very different when we just look at the absolute time, however, when we look at developmental time relative to the lifespan of the animal we see some similarities emerge and some more differences. The rat shows mature place learning at forty days, roughly adolescence in the rat. The human develops adult like place learning by about nine years, several years before the start of adolescence.

Differences between rats and humans are also present. Unlike rats, humans take a very long time to demonstrate adult-like place learning. This slow development, while going through the same progression of steps as the rats, is quite different in overall time spent at each step. Nonetheless, humans have mature place learning long before they are mature individuals. These two ideas, that place learning takes quite a long time in the process of emerging and that place learning emerges fully in its mature form long before the human is mature, may be connected. The human development of place learning takes longer because all of human development takes longer (Gould, 1985). The human infants have more time and, thus, the luxury of exploring components of the place learning system before the next develops. This expanded time for development may allow humans to retain some flexibility into adulthood that is lost in the shorter developmental course of the rat. The system that develops over the longer time course retains some of the features of the developing system, which allows for more change and reconstruction of the final spatial system, that is to say that it allows for neoteny (Gould, 1985; Parker and McKinney, 1999).

CONCLUSION

The development of navigational ability, and particularly place learning, is important to the survival of any mobile organism. It allows for the location of objects to be remembered even if the location of the organism changes and the available landmarks are not near the object. Although the other systems of navigation and object localization are also important, place learning is the most powerful of those systems and the one that allows the organism to find an object under the least favorable conditions. Place learning is what we use to find our favorite camping spot in the mountains, but it could also be what we use to find food, shelter or some other things necessary to survival.

All mobile organisms would benefit from the use of the place learning system, and thus as a mechanism for survival it may be an old system that has been retained by numerous branches of the evolutionary tree. From this perspective the similarities in the developmental trajectories of different species may be related to the need of all mobile species to place learn (Jacobs and Schenk, 2003). Rats and humans share some striking similarities in their spatial development while other forms of cognition are pretty dissimilar. In fact, comparative cognition is exciting for space exactly because there are some similar adaptive challenges faced by both species and some of the systems may well be conserved across evolution.

The developmental progression in the human infants and children took longer than that of rat pups in absolute time, but followed the same path. The development in both species seemed to follow the same progression of search patterns in the same order, indicating either that the steps to adult place learning seen in both human and rats are necessary to the place learning system or that the steps trace an evolutionarily old path followed by ancestors we have in common.

NOTE

1. The research investigating the use of the geometric properties of an enclosing space are discussed in Chapter 5, this volume.

REFERENCES

Aadland, J., Beatty, W. W. and Maki, R. H. (1985). Spatial memory of children and adults assessed in the radial maze. *Developmental Psychobiology*, **18**, 163–172.

Akers, K. G. and Hamilton, D. A. (2007). Comparison of developmental trajectories for place and cued navigation in the Morris water task. *Developmental Psychobiology*, **49**, 553–564.

Akers, K. G., Candelaria-Cook, F. T., Rice, J. P., Johnson, T. E. and Hamilton, D. A. (2009). Delayed development of place navigation campared to directional responding in young rats. *Behavioral Neuroscience*,**123** (2), 267–275.

Balcomb, F., Newcombe, N. S. and Ferrara, K. (2009). Convergence and divergence in representational systems: place learning and language in toddlers.

In N. Taatgen and H. Van Rijn (Eds.), *Proceedings of the 31st Annual Conference of the Cognitive Science Society* (pp. 596–601). Austin, TX: Cognitive Science Society.

Brown, R.W. and Kraemer, P.J. (1997). Ontogenetic differences in retention of spatial learning tested with the Morris water maze. *Developmental Psychobiology*, **30**, 329–341.

Brown, R. W. and Whishaw, I. Q. (2000). Similarities in the development of place and cue navigation by rats in a swimming pool. *Developmental Psychobiology*, **37**, 238–245.

Bushnell, E.W., McKenzie, B.E., Lawrence, D.A. and Connell, S. (1995). The spatial coding strategies of one-year-old infants in a locomotor search task. *Child Development*, **66**, 937–958.

Caeman, H. M. and Mactutus, C. F. (2001). Ontogeny of spatial navigation in rats: a role for response requirement? *Behavioral Neuroscience*, **115** (4), 870–879.

Cheng, K. (1986). A purely geometric module in the rat's spatial representation. *Cognition*, **23**, 149–178.

DeLoache, J.D. and Brown, A.L. (1983). Very young children's memory for the location of objects in a large scale environment. *Child Development*, **54**, 888–897.

Dolins, F. L. (2009). Captive cotton-top tamarins' (*Saguinus oedipus oedipus*) use of landmarks to localize hidden food items. *American Journal of Primatology*, **71**, 316–323.

Gogtay, N., Nugent, T. F. III, Herman, D. H., *et al.* (2006). Dynamic mapping of normal human hippocampal development. *Hippocampus*, **16**, 664–672.

Gomez, J. (2005). Species comparative studies and cognitive development. *Trends in Cognitive Science*, **9** (3), 118–125.

Gould, J.J. (1985). *Ontogeny and phylogeny*. Cambridge, MA: Harvard University Press.

Haun, D.B.M. and Call, J. (2009). Great apes' capacities to recall relational similarity. *Cognition*, **110**, 147–159.

Hermer, L. and Spelke, E. S. (1994). A geometric process for spatial reorientation in young children. *Nature*, **370**, 57–59.

Huttenlocher, J. and Newcombe, N. (1984). The child's representation of information about location. In C. Sophian (Ed.), *Origins of cognitive skills* (pp. 81–111). Hillsdale, NJ: LEA.

Huttenlocher, J., Newcombe, N. and Sandberg, E.H. (1994). The coding of spatial location in young children. *Cognitive Psychology*, **27**, 115–147.

Jacobs, L.F. and Schenk, F. (2003). Unpacking the cognitive map: the parallel map theory of hippocampal function. *Psychological Review*, **110** (2), 285–315.

Jansen-Osmann, P. and Wiedenbaur, G. (2004). The representation of landmarks and routes in children and adults: a study in a virtual environment. *Journal of Environmental Psychology*, **24**, 347–357.

Jansen-Osmann, P., Schmid, J. and Heil, M. (2007). Spatial knowledge of children and adults in a virtual environment: the role of environmental structure. *European Journal of Developmental Psychology*, **4** (3), 251–272.

Kraemer, P.J. and Randall, C.K. (1995). Spatial learning in preweanling rats trained in a Morris water maze. *Psychobiology*, **23** (2), 144–152.

Kretschmann, H.J., Kammradt, G., Krauthausen, I., Sauer, B. and Wingert, F. (1986). Growth of the hippocampal formation in man. *Bibliotheca Anatomica*, **28**, 27–52.

Laurence, H., Learmonth, A.E., Nadel, L. and Jacobs, W.J. (2003). Maturation of spatial navigation strategies: convergent findings from computerized spatial environments and self-report. *Journal of Cognition and Development*, **4** (2), 211–238.

Learmonth, A. E., Newcombe, N. and Huttenlocher, J. (2001). Toddlers' use of metric and landmark information to reorient. *Journal of Experimental Child Psychology*, **80**, 225–244.

Learmonth, A. E. Newcombe, N. and Nadel, L. (2002). Children's use of landmarks: Implications for modularity theory. *Psychological Science*, **13** (4), 337–341.

Lehnung, M., Leplow, B., Friege, L., Herzog, A. and Ferstl, R. (1998). Development of spatial memory and spatial orientation in preschoolers and primary school children. *British Journal of Psychology*, **89**, 463–480.

Morris, R. G. M. (1981). Spatial localization does not require the presence of local cues. *Learning and Motivation*, **12**, 239–260.

Morris, R. G. M., Garrud, P., Rawlins, J. N. P. and O'Keefe, J. (1982). Place navigation in rats with hippocampal lesions. *Nature*, **297**, 681–683.

Nadel, L. and Zola-Morgan, S. (1994). Infantile amnesia: a neurobiological perspective. In M. Moscovich (Ed.), *Infant memory* (pp. 145–172) New York: Plenum Press.

Newcombe, N. and Huttenlocher, J. (2000). *Making space: the development of spatial reasoning*. Cambridge, MA: The MIT Press.

Newcombe, N., Huttenlocher, J. and Learmonth, A. (1999). Infants' coding of location in continuous space. *Infant Behavior and Development*, **22**, 483–510.

Newcombe, N. S., Sluzenski, J. and Huttenlocher, J. (2005). Preexisting knowledge versus on-line learning. *Psychological Science*, **16** (3), 222–227.

Newcombe, N., Huttenlocher, J., Drummey, A. B. and Wiley, J. G. (1998). The development of spatial location coding: place learning and dead reckoning in the second and their years. *Cognitive Development*, **13**, 185–200.

Overman, W. H., Pate, B. J., Moore, K. and Peleuster, A. (1996). Ontogeny of place learning in children as measured in the radial arm maze, Morris search task, and open field task. *Behavioral Neuroscience*, **110** (6), 1205–1228.

Parker, S. T. and McKinney, M. L. (1999). *Origins of intelligence*. Baltimore: The Johns Hopkins University Press.

Penn, D. C., Holyoak, K. J. and Povinelli, D. J. (2008). Darwin's mistake: explaining the discontinuity between human and nonhuman minds. *Behavioral and Brain Sciences*, **31** (2), 109–178.

Quinn, P. C., Polly, J. L., Furer, M. J., Dobson, V. and Narter, D. B. (2002). Young infants' performance in the object-variation version of the above-below categorization task: a result of perceptual distraction or conceptual limitation? *Infancy*, **3**, 323–347.

Rudy, J. W. and Paylor, R. (1988). Reducing the temporal demands of the Morris place-learning task fails to ameliorate the place-learning impairment of preweanling rats. *Psychobiology*, **16**, 152–156.

Rudy, J. W., Stadler-Morris, S. and Albert, P. (1987). Ontogeny of spatial navigation behaviors in the rat: dissociation of "proximal"- and"distal"-cue-based behaviors. *Behavioral Neuroscience*, **101** (1), 62–73.

Schenk, F. (1985). Development of place navigation in rats from weanling to puberty. *Behavioral and Neural Biology*, **43**, 69–85.

Schenk, F. and Morris, R. G. M. (1985). Dissociation between components of spatial memory in rats after recovery from the effects of retrohippocampal lesions. *Experimental Brain Research*, **58**, 11–28.

Seress, L. (1992). Morphological variability and developmental aspects of monkey and human granule cells: differences between the rodent and primate dentate gyrus. In C. E. Ribak, C. M. Gall and I. Mody (Eds.), *The dentate gyrus and its role in seizures* (pp. 3–28). Elsevier Science Publishers.

Sluzenski, J., Newcombe, N. and Satlow, E. (2004). Knowing where things are in the second year of life: implications for hippocampal development. *Journal of Cognitive Neuroscience*, **16** (8), 1–9.

Smith, A. D., Gilchrist, I. D., Cater, K., Ikram, N., Nott, K. and Hood, B. M. (2007). Reorientation in the real world: the development of landmark use and integration in a natural environment. *Cognition*, **107**, 1102–1111.

Tees, R. C., Buhrmann, K. and Hanley, J. (1990). The effects of early experience on water maze spatial learning and memory in rats. *Developmental Psychobiology*, **23** (5), 427–439.

Tomaselle, M. and Call, J. (1997). *Primate cognition*. London: Oxford University Press.

CHARLES MENZEL

24

Spatial cognition and memory in symbol-using chimpanzees

The knowledge base of great ape navigation is poorly known compared to what is known about other species. In this chapter, I present some of the questions that arise from a comparative and naturalistic perspective on chimpanzee navigation and memory. I then describe studies of spatial memory in symbol-competent apes (genus *Pan*), focusing on experimental studies of apes in tasks that require locomotion. These studies address two main issues in spatial learning. The first issue is how flexibly apes can remember and move to different goals from different starting locations. The ability to find a straight-line route between any two points, such that either point can serve as the starting point and the other point serve as the goal location, is traditionally viewed as one of the three definitive types of evidence for generalized mapping skills in animals (Muller *et al.*, 1996). The second issue is the extent to which chimpanzee memory of events and places is independent of stimulation from the environment in question. The ability to recall events, environmental features and locations in some detail, from situations that are outside the situations in which the objects and locations were originally encountered, traditionally has been viewed as a central factor in the emergence of human thinking (Lorenz, 1971), but until recently there have been almost no data on recall capabilities in nonverbal animals.

Throughout this chapter, emphasis will be on the types of information that chimpanzees acquire and use regarding spaces that are middle- or large- scale in size, for example, some 100 to 20,000 square meters in studies at the Language Research Center. These areas typically include trees, trails, visual barriers, and other three-dimensional environmental features. For an ape to perform efficiently in the tasks presented, it might need to remember events and environmental

Spatial Cognition, Spatial Perception: Mapping the Self and Space, ed. Francine L. Dolins and Robert W. Mitchell. Published by Cambridge University Press.

features over time spans of minutes to weeks. I do not review studies of behavior in tasks that primarily entail manual reaching to targets, studies of tool use, or studies of simulated, computer-presented spatial tasks (e.g., Fragaszy *et al.*, 2003; Menzel and Menzel, 2007). Nor am I concerned with linguistic or quasi-linguistic questions per se.

NATURALISTIC BACKGROUND

Chimpanzees show a combination of traits especially rare among animals: they are large-bodied animals specializing on a type of food, ripe fruit, that is rare and ephemeral (Ghiglieri, 1984). Chimpanzees at Kibale National Park, Uganda, devote nearly three times as much of their feeding time to ripe fruit as sympatric monkeys do, and they go for species of fruit that are rare (Wrangham *et al.*, 1998). By using ripe fruits, chimpanzees maintain a much higher quality diet than would be predicted for a mammal of comparable body size (Conklin-Brittain *et al.*, 1998).

Chimpanzees often travel rapidly and continuously in straight lines over long distances to ripe fruit trees. Field observers frequently state that the chimpanzees appear to know where they are going and the types of food they are likely to find there (Wrangham, 1975, 1977; Ghiglieri, 1984; Goodall, 1986). Although experimental data are lacking from the field, the spatial memory capabilities of chimpanzees and their ability to organize an efficient itinerary among food sources can be expected to improve their foraging efficiency substantially (Wrangham, 1977). Ghiglieri (1984) suggests that chimpanzees' spatiotemporal memory capabilities are highly developed and play a crucial role in helping them obtain adequate energy from rare, scattered, ephemeral fruit sources in the face of feeding competition at Kibale forest, where for each individual chimpanzee there are approximately 200 monkeys, plus numerous squirrels, birds and other chimpanzees going for the same fruit. Based on paleo-environmental data and the behavior of living great apes, Potts (2004) hypothesizes that specialization on rare, ephemeral ripe fruit in tropical forests was a central organizing factor in the emergence and subsequent evolutionary development of chimpanzee cognitive capabilities, including their ability to assign attributes to distant locations.

Boesch and Boesch-Achermann (2000) regard the low visibility of the tropical rainforest as a factor of utmost importance in understanding how chimpanzee spatial memory operates in the present day at the Taï forest, Ivory Coast. The findings from this field site are of special interest because the dense tropical forest is the type in which most wild chimpanzees live. The home range size of a chimpanzee at Taï forest can be up to 25 square kilometers, and the terrain is relatively flat and heavily forested. From ground level where the chimpanzees travel, visibility is typically less than 20 meters. The chimpanzees appear to remember the locations of food sources, tools and other chimpanzees, while still at a distance and out of sight of these targets. For example, in a 30-hectare section of rainforest that Boesch and Boesch-Achermann (2000) selected for intensive study, chimpanzees picked up stones from previous nut-cracking sites, carried the stones to other tree-root anvils over an average transport distance of 120 m, and used them to crack Panda nuts. The chimpanzees selected stones that were closer to the nut trees they were preparing to use, over stones that were farther away. They seemed to be influenced simultaneously by both weight and distance, but to give preference to a "least distance" strategy (Boesch and Boesch, 1984). Because the stones and Panda-nut trees were dispersed in space, at varying angles, and completely out of sight of one another, and because the positions of the stones changed from one occasion to the next, the authors (Boesch and Boesch, 1984; Boesch and Boesch-Achermann, 2000) suggested that the following cognitive operations were required to explain their data: (1) measurement and conservation of distance; (2) comparison of distances oriented in different directions; (3) permutation or mental rearrangement of objects (stones) within this "map;" and (4) permutation or mental variation of reference points (nut-bearing trees), such that the organism can measure the distance from varied starting points to any goal location.

At Gombe National Park, Tanzania, some female chimpanzees visit the same termite mounds daily during the termite season, often approaching the heaps from a different angle or in a different order. They may select a grass stem, carry it to a mound that is still 100 m or more away and out of view, and subsequently use the stem as a tool for termite fishing (Goodall, 1986; see also Sanz *et al.*, 2004). Captive bonobos and orangutans will select and carry a tool that will be needed an hour or more later (Mulcahy and Call, 2006).

MEMORY IN CAPTIVE CHIMPANZEES

Experimental studies of captive chimpanzees indicate that chimpanzees not only can remember the locations of a large number of food items (Tinklepaugh, 1932) but also that they retain information about the relative positions of those food items (E. Menzel, 1978, 1984). E. Menzel conducted an extensive series of studies of chimpanzee memory and communication in a 1-acre field cage (see Chapter 4, this volume). In one study (E. Menzel, 1973a), one chimpanzee from a group of six juveniles was carried around the experimental field and shown a piece of fruit in each of eighteen randomly selected sectors of the field. Unlike Tinklepaugh's (1932) classic study of multiple delayed reactions in chimpanzees, each trial used a different set of locations. During the cue giving, the chimpanzee was not allowed to do anything other than cling to his carrier and watch as the food was hidden. This eliminated locomotor practice and primary reinforcement from the cue-giving phase of a trial. The field included trees, areas of relatively tall grass, small hills, and other visual barriers. Not all food locations were visible to the chimpanzee from a single vantage point. After the experimenters showed the chimpanzee the food, they returned him or her to the group. The experimenters then climbed an observation tower and within two minutes released the group into the field. The question was how the chimpanzee organized its route while collecting the eighteen hidden food items. Four different chimpanzees were used as test animals.

Results were that the animal that had been shown the food obtained most of the hidden items. In sixteen trials, the test animal found 200 of 288 hidden foods, and the five control animals that had not been shown the food found a total of just seventeen. Usually, the test animal ran in a direct line to the exact clump of grass, tree stump, or hole in the ground where hidden food lay, grabbed and ate the food, and then ran directly to the next location, no matter how far away or obscured by visual barriers that location was. The control animals obtained food mainly by searching near the test animal or by begging from him. The test animal used a very efficient route during the collection of food, and its route bore no relation to the route along which it had been carried during cue giving. The test animals clearly took into account more than one or two food locations at a time, because on any given choice they not only generally went to one of the closer locations, of all those that still contained food, but also they usually went from the outset of a trial toward larger rather than smaller "clusters." Thus, in a further experiment, they typically went toward whichever half of the field contained three food locations versus two food locations. In other tests, the experimenters hid a piece of preferred food (fruit) in nine locations and a piece of non-preferred food (vegetable) in nine other locations. The test chimpanzee went first to the fruit locations and rarely re-inspected a location that he had already emptied.

Other experiments in the series (E. Menzel, 1973b) showed that a chimpanzee could retain and transmit information to group members about object quality (fruit versus vegetable), the relative goodness or badness of object (food versus snake), quantity of food, and location. The velocity at which a chimpanzee approached a distant, hidden goal provided sufficient information for discriminating whether the item was food or a novel object and who knew best its location (Menzel and Halperin, 1975). The group of chimpanzees searched very differently after an operationally designated "leader" had been shown a real snake than they did after the leader had been shown a piece of food which was later removed. If the leader had been shown a snake, group members showed piloerection as they emerged from the release cage, and they moved together slowly toward the location. Not only would the group members use a stick to probe the last location where the snake had been seen, but they also would broaden the area searched, as if taking into account the potential of the snake for movement. If the leader had been shown food, the group typically raced for the location, and they restricted their search to the vicinity of the original location.

These studies showed not only that chimpanzees could remember many food locations in a given trial, but that they also remembered the spatial relationships among the locations (E. Menzel, 1973a; Gallistel, 1990). The studies also demonstrated transmission of information about the nature and location of hidden objects from one chimpanzee to another at an unexpectedly high level of efficiency (E. Menzel, 1973b).

STUDIES OF RECALL MEMORY IN *PAN*

The studies reviewed above, and those on macaque learning (Chapter 9, this volume), assess an animal's knowledge of resource locations primarily from its path of travel and from the locations that it inspects. Such analyses potentially underestimate knowledge (Marler and Terrace, 1984, p. 12). In principle, an animal might recall the presence, type, relative quantity and location of a resource but not move to it for several days. A current limitation in studies of how well nonverbal animals know their habitat is that there are almost no data on recall memory, as contrasted with recognition (Shettleworth, 1998). To what extent can a nonverbal animal recall and compare distant locations that lie well beyond its immediate range of sight and hearing (Byrne, 2000)? Which various attributes of past events are recalled? Can the animal perform recall in different situations and in different ways

(Tulving, 1983)? What are all the types of information that a primate can retrieve and convey regarding past events and distant locations, from a single location in space?

For example, a hamadryas baboon sitting on its sleeping cliff in the morning can see only a fraction of the 30 or so square kilometer area that constitutes the home range of its band. From its sleeping cliff, can the baboon recall the varying water levels at distant drinking sites, recall whether each pool is now overgrown with algae, and recall the locations of dwarf carrots, of beetle larvae, of ripe acacia pods, and of hyenas' dens that are hidden behind ridges? Can it recall its own recent encounters with leopards in some detail? The ability to retrieve detailed information about areas of the home range that are out of view, and the ability to recall past encounters with predators might, in principle, be important in selecting a safe and efficient foraging route (Kummer, 1995). The problem is how to prove the existence of these types of recall and representational memory capabilities in nonverbal animals, and how to work with them systematically and analyze them in further detail. Until recently, there have been almost no data available on recall capabilities in nonhuman animals to take the issue further (Ruggiero and Flagg, 1976; Suddendorf and Corballis, 1997; Shettleworth, 1998; C. Menzel, 1999; 2005; Roberts, 2002; Hampton and Schwartz, 2004; Schwartz, 2005).

An extension of classic delayed-response methods for studying spatial cognition and memory in nonhumans is to assess which features of events an animal conveys to other beings through behavior or a symbol system (E.W. Menzel, 1973b; Herman and Forestell, 1985; Tulving 1984, p. 224; Savage-Rumbaugh *et al.*, 1986). Symbol-competent chimpanzees provide a powerful method for investigating the "time in which the chimpanzee lives" (Köhler, 1925, p. 272) and the nature of the information they possess and convey about the structure of their environment and their past experiences. Symbol-competent apes at Georgia State University's Language Research Center have learned lexigrams, which are arbitrary geometrical forms that refer to types of food and other objects. These animals have been immersed in the visual symbol system and exposed to spoken English since an early age. As a result of their rearing history, the apes can comprehend lexigrams and can use lexigrams in a productive manner, and some of them can understand spoken English at approximately the level of a very young child (Savage-Rumbaugh *et al.*, 1993). The symbol system allows one to ask questions regarding memory that would be difficult to pose otherwise; it is possible to obtain more information from the apes with the symbol system than without it. The apes potentially can convey specific events and environmental features they have seen, through use of the lexigram system. They can be tested outside the spatial and temporal contexts in which they encountered the events. Some of the Language Research Center apes can use lexigrams outside the context in which the objects were originally encountered (Savage-Rumbaugh *et al.*, 1983, 1986, 1993).

CHIMPANZEE LABELING OF NON-VISIBLE OBJECTS AFTER BRIEF DELAYS

Savage-Rumbaugh *et al.* (1983) permitted each of two male common chimpanzees, Sherman and Austin, to inspect a group of five objects on a tray. The objects were drawn at random from a set of twenty photographs of foods and ten tools, the lexigram "names" for which were already very familiar to the chimpanzees. The ape then had to move to a lexigram keyboard that was out of sight of the objects and to select a lexigram corresponding to one of the objects that he had seen. The remainder of the ape's task was to return to the objects, to collect the same type of object that he had indicated on the keyboard, and to carry that object around a corner to an experimenter, who had not been visible to the chimpanzee when he made any of his choices. The chimpanzee was praised and rewarded when he brought the experimenter the same type of object as he had indicated on the keyboard. Both Sherman and Austin learned this task. The findings showed that chimpanzees could indicate by lexigrams which type of object they had seen several seconds earlier in a now out-of-view location.

In the study described above, testing was mainly limited to indoor rooms and used a small number of locations, and the study was geared for specialists in language learning. Of greater interest to field biologists are studies involving longer timespans and larger spatial scales. The research reviewed next typically used a larger number of locations, and locations that were more dispersed. Test objects were at greater distances from the keyboard. On some tests, the number of objects was increased, and in some the timespans ranged up to hours or days.

A BONOBO'S ANNOUNCEMENT OF ITS INTENDED DESTINATION IN A 20-HECTARE FOREST

Savage-Rumbaugh *et al.* (1986) tested a three-year-old male pygmy chimpanzee, or bonobo, Kanzi, in a heavily wooded area of 20 hectares (55 acres). The forest included 3.5 miles of trails. Kanzi had traveled with human companions in the forest on a daily basis, and he had learned many lexigrams and the locations of all seventeen of the food stations in the forest. Each station in the forest had a different type of food reliably present. Kanzi's rearing history, and the manner in which he acquired facility in the use of lexigrams and became exposed to the forest, are presented in Savage-Rumbaugh *et al.* (1985, 1986 and 1993). The principles by which the lexigrams were originally designed are described in Rumbaugh (1977) and Savage-Rumbaugh (1986).

Four months after food was first placed in the forest, blind trials were conducted. A person who had never entered the wooded area accompanied Kanzi and allowed him to lead the way from place to place. The person did not know the location of any of the foods or the locations of any of the trails. At each stopping point the person presented Kanzi with an array of six to ten alternatives. These alternatives were lexigrams and photographs representing food and locations. Kanzi made a selection from this array. The person recorded which lexigram or photograph Kanzi touched at the stopping point, and also recorded any lexigrams or photographs that Kanzi touched or carried along the way. The results were that when Kanzi was at a stopping point, he touched photographs on five occasions, touched lexigrams on seven occasions, and touched both photographs and lexigrams on three occasions, corresponding to his next destination, with 100 percent accuracy. After he had touched a photograph or lexigram, Kanzi proceeded to guide the person to the location of the food he had selected, sometimes traveling over thirty minutes to reach the food site. Distances between the successive food sites ranged up to 600 meters. On all but one occasion Kanzi moved to the indicated destination in the most direct route possible within the trail system. To reach his destination, Kanzi had to choose a direction at up to seven different trail crossings. He used the smallest possible number of turns and the shortest possible route along the trail, on all but one occasion. Along the way, Kanzi often pointed to the photograph or to the lexigram that he had selected (Savage-Rumbaugh *et al.*, 1986).

A BONOBO'S USE OF ROAD SIGNS IN A 20-HECTARE FOREST

A current limitation in our understanding of great ape navigation is that we know little of the signs that apes use to discriminate the condition of distant areas. Can apes use arbitrary signs to discriminate the presence and location of objects that are great distances away? It is

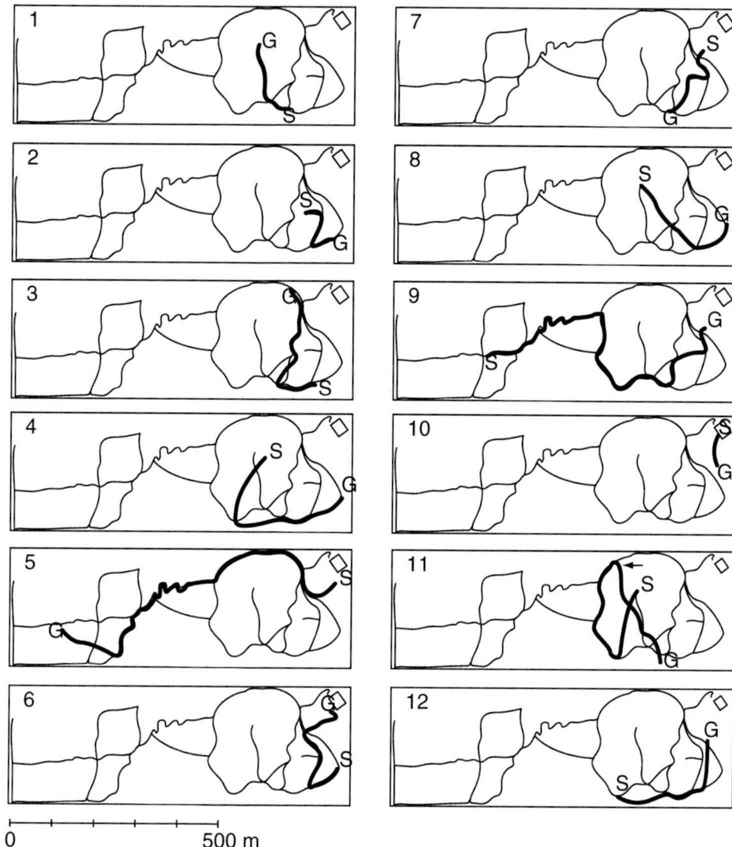

Figure 24.1. Kanzi's travel routes in the 20-hectare forest during twelve trials. S = start point at which he was given cue. G = goal location. Thick line = his path of travel. Thin lines = human-made trails. The rectangle in the upper right is the laboratory building. The arrow in Trial 11 shows the location at which Kanzi was given the cue a second time. Reprinted from Menzel *et al.* (2002).

a safe bet that they can match to sample as tamarins and macaques do (Chapter 9, this volume) and also recall where various classes of foods were once found, but this is not the same as using an arbitrary sign to learn about parts of the environment that are completely out of view.

Menzel *et al.* (2002) used Kanzi's lexigram competence as a tool to study his spatial memory organization when he was four years old. The focus of the study was on Kanzi as a receiver rather than a sender of information. In the first experiment, we showed Kanzi a "road sign" just outside his indoor sleeping area. The sign indicated, by lexigrams, the location where food was hidden. Only two of the fifteen locations were visible from the sign. Distances ranged up to 170 meters from the sign. In 99 of 127 test trials Kanzi went to the designated location on his first move. In a second experiment, we presented Kanzi with the road sign at varied points in the forest rather than at the original fixed place. In these trials the goal was a preferred toy. Kanzi's human companions were never informed about the location of the goal, and distances ranged up to 650 m. In all twelve trials Kanzi led his companions to the designated location, using an efficient path (Figure 24.1). In sum, Kanzi appeared to be able to navigate, based on the information provided by a lexigram, from almost any arbitrarily designated starting location in his 20-hectare environment to any one of the numerous goal locations.

Travel and spatial memory performances with arbitrary cues so widely separated in space from rewards have not previously been demonstrated experimentally for nonhuman primates. The data suggest that Kanzi had acquired a detailed knowledge of the human-made trail system, if not the more general layout and structure of his sizable environment. He typically seemed to know where he was at any given time and what led to what, because he could go on request from any given place to almost anywhere else, at least if the destination was a place for which he and his human companions had a name. The tasks required Kanzi to assess at the very outset of a trial, from a mere lexigram discriminandum, what route to take and which distant, invisible location to head for. The distances were huge by the standards of discrimination learning (cf. Rumbaugh and Washburn, 2003), the directions were seldom a simple straight line, and there was nothing iconic or "natural" about a lexigram. Kanzi's performances are of further interest because he was a juvenile. At the same age wild bonobos would still be some years away from being independent travelers.

BONOBO MEMORY OF LOCATION AND OBJECT NAMES DURING ENGLISH COMPREHENSION TASKS

Savage-Rumbaugh *et al.* (1993) studied language comprehension by Kanzi when he was eight years old. An experimenter made a verbal statement in English to Kanzi from behind a one-way mirror. One form of sentence was "take object X to location Y." There were usually seven to eighteen objects in an array facing Kanzi and fourteen different locations available (e.g., bedroom, colony room, outdoor cage). Kanzi chose both the correct object and the correct location on forty-five of fifty-eight trials, and he was correct on the location part of the task on 91 percent of trials. Another form of sentence was "go get object X that's in location Y" (e.g., go get the carrot that's in the bedroom). Six to nine different words indicating locations were used in each session. There were at least three different types of objects in each distal location, and another exemplar of object X was in the array of objects that faced Kanzi. The results of this task were that Kanzi went to the correct location and selected the correct object there on twenty-seven of thirty-five trials. He did not simply take the nearby, visible exemplar of object X. Kanzi also responded appropriately to statements of fact (e.g., "the surprise is hiding in the bucket") that were not requests for specific actions.

UNPROMPTED RECALL AND REPORTING OF HIDDEN OBJECTS BY A COMMON CHIMPANZEE

I will next describe in some detail my first study on memory in a lexigram-competent female common chimpanzee named Panzee. The aim of this study (C. Menzel, 1999) was to identify the types of information that Panzee could recall and transmit about hidden objects. Panzee was eleven years old at the outset of the experiment, and she had already learned more than 120 lexigrams (Beran *et al.*, 1998). Savage-Rumbaugh, Brakke, and associates had reared Panzee in an exceptionally enriched environment that included daily exposure to spoken English and to lexigrams that referred to foods, tools, actions and locations in the 20-hectare forest at the Language Research Center. Panzee was exposed to the lexigrams in the everyday contexts of food preparation, travel and play, rather than taught lexigrams through discrete trial training (Brakke and Savage-Rumbaugh, 1995, 1996; also Savage-Rumbaugh *et al.*, 1993). There had been no previous systematic study of Panzee's retention time beyond about forty seconds for visual stimuli or locations. Panzee had not been presented with objects outside her outdoor enclosure that she could later name or request indoors. Recruitment of people to the outdoor enclosure and inducing them to search the terrain was never a part of her experimental history. This is not to claim, however, that she brought no relevant experience to the experimental situation. For example, Panzee had used gestures and lexigrams in routine interactions with caregivers to request movement between cages, activities (e.g., chasing, grooming, drawing, television), foods, and other objects (e.g., blanket, clay). As a juvenile,

Panzee had played the game of "hide and seek" with caregivers, and it was in the context of searching for a hidden person that she originally used the "hiding" gesture described later.

During this experiment (C. Menzel, 1999), Panzee was housed in the Lanson building, which was an indoor–outdoor facility. When Panzee was indoors she could not see the outdoor cage or the forest. The outdoor cage was bordered by forest, and an area of forest adjacent to the cage, measuring about 160 square meters, served as a test area for the introduction of objects. Panzee had not entered any forest for at least six years prior to this study. The indoor and the outdoor cage each contained a keyboard, and each keyboard displayed 256 lexigrams. To obtain something from outside the outdoor cage in this study, Panzee had to recruit the assistance of a person and in effect tell them where to go. Panzee was not required also to tell people what type of object to look for. If she did provide this information, it would be as a result of her own initiative or due to past experience.

Design and procedure

This initial experiment consisted of thirty-four trial-unique tests, each consisting of a cue-giving phase, a delay phase, and a response phase. The test objects were twenty-six food types and seven types of nonfood objects. During cue giving, the experimenter stood outside Panzee's outdoor enclosure, held up an object while Panzee was watching, then walked to a predetermined hiding place in the forest and placed the object on the ground. Each trial used a different location, and distances from the enclosure to the object ranged up to 8 m. During cue giving, all locations were visible to Panzee. On thirty-two of the thirty-four trials, the experimenter placed the object under natural cover, so that the object was concealed from view, and then smoothed over the area. On these thirty-two trials, the object was buried under leaf litter and completely concealed from view, not simply placed behind a bush. On the remaining two trials the object was not concealed, and Panzee could see it from her outdoor cage. After placing the object, the experimenter left the area, and this began the delay phase of the trial. On ten trials (Trials 9, 17, 18, 19, 20, 21, 24, 25, 26 and 34) the experimenter imposed an overnight delay between the cue-giving phase and the response phase by showing Panzee the object outdoors after all caregivers had left for the day. Panzee did not have a chance to interact with a person until the next morning. On the other twenty-four trials, Panzee had an opportunity to interact with a person indoors on the same day, typically within a few minutes after seeing the object. On any given opportunity, Panzee might or might not choose to recruit the assistance of the person to recover the hidden object. The delay phase lasted until Panzee initiated recruitment of a person (details below). Thus, Panzee determined the exact time of recruitment and the total length of delay.

In the response phase of a trial, Panzee interacted with a person inside the Lanson building. This person did not know what had gone on outside the Lanson building and did not know what the object was nor where it was located. Three different people served as "uninformed persons" during the study. On at least twenty-four of the thirty-four trials, the person did not even know in advance that a trial had been set up. Once a person was involved in the study, he or she knew that potentially a trial could happen but did not know when. Experimental trials were interspersed with non-test days on which no object was presented, over a total study period of 268 days.

The uninformed person went about his or her routine caregiving activities and let Panzee initiate any interactions regarding what had gone on outside the Lanson building. During the study period, persons made detailed notes regarding any occasion on which Panzee recruited them outdoors, regardless of whether they thought a test actually was being conducted. The person recorded on each such occasion all lexigrams that Panzee touched on the indoor and outdoor keyboards, Panzee's gestures, and her vocalizations. Gestures recorded included pointing outdoors with an extended arm, beckoning with a sweeping movement of her hand, and covering her eyes with the palm of her hand (the gesture previously associated with the game of "hide and seek" with caregivers). Spatial data recorded during each trial included the location where Panzee sat when she pointed manually out of the outdoor enclosure and each location in the woods that the uninformed person inspected manually. If Panzee guided the uninformed person to a hidden food item, then the person carried the food indoors and offered it to Panzee, regardless of whether Panzee had touched the lexigram corresponding to the food.

Results

The uninformed persons found all thirty-four objects that were used in these tests as the result of Panzee's behavior. "False positives" occurred on only three of the 268 days of the experiment. That is, Panzee almost never recruited the uninformed person and caused him or her to search when there was no test object in the woods. When a test object was present outdoors, Panzee never pointed to a location that had contained an object on a previous trial. In fact, in thirty-four recruitment sequences with an object present in the woods, Panzee never pointed in an incorrect direction. In total, there were fifty-seven days on which an object was present. Panzee recruited a person and pointed toward a single, specific location on thirty-four of these fifty-seven days and did not recruit a person on the remaining twenty-three days. (Panzee did not fail to report the objects that were hidden on these twenty-three remaining days; she simply did not always recruit a person to the object at her first opportunity. She might wait a day or more.) In contrast, there were 211 non-test days on which no object was present in the woods, and Panzee recruited a person and pointed toward a specific location on just three of these 211 days.

From Trial 1, Panzee was highly effective in attracting the uninformed person's attention, in conveying the type of item hidden, and in directing the person to the location of the hidden item. A summary account of how the uninformed persons described Panzee's behavior is as follows. Panzee gained the person's attention by vocalization, by gesturing toward the indoor keyboard, or by moving to the keyboard when the person happened to come near her cage. She held her index finger on a lexigram until the person came over and acknowledged her lexigram use verbally. Panzee then covered her eyes with her hand, held her arm extended in the direction of the tunnel leading outdoors, moved to the tunnel, and went outdoors. If the person did not follow, then Panzee came back inside, again gestured "hide," beckoned, and pointed outdoors.

Once both Panzee and the person were outdoors, Panzee beckoned manually and moved to the edge of the enclosure, across from the hidden object. She sat facing the object, extended her index finger through the cage wire, and jabbed it in the direction of the object. She prompted the person to search the terrain by gesturing "hide," by pointing manually toward the object location with her index finger extended, by giving low vocalizations, and by staring toward the location, with interspersed looks toward the person. She might leave her position, walk to the outdoor keyboard several meters away, touch a lexigram, and then return to her original position and resume pointing. During interactions outdoors, the person was outside the enclosure; Panzee was inside the enclosure and did not have physical access to the object. Because of the orientation of Panzee's body and gaze and the persistence of her pointing in a given direction, the person restricted most of his or her searching to within 1 m of a straight line between Panzee and the object location and found the object within a few minutes, typically within 1 min. The person determined the distance to the object, in part, simply by following the direction of Panzee's gaze and responding to her relative degree of excitement. Panzee kept pointing, showed intensified vocalization, shook her arm, and bobbed her head or body, as the person got closer to the site. Once the person found the object, Panzee stopped pointing outside the enclosure and stopped gesturing "hide."

Lexigrams used during recruitment

The lexigrams that Panzee touched when she recruited a person typically corresponded to the type of object she had seen (Table 24.1). Considering just the thirty-three lexigrams corresponding to the thirty-three types of objects used in the experiment to be the relevant available choices, then 84 percent of Panzee's seventy-six lexigram touches corresponded to the specific type of experimental object presented, rather than to one of the other thirty-two types of experimental objects. It is obvious that the observed frequency of correct lexigram touches far exceeded the frequency expected by chance. Furthermore, when Panzee recruited a person, she touched lexigrams that she rarely touched under routine conditions. On twenty-five of thirty-four trials, the lexigram corresponding to the object hidden was one that accounted for 1 percent or less of Panzee's routine lexigram use. Panzee used the correct lexigram on twenty-one of these twenty-five trials, and on the four remaining trials Panzee relied effectively on gesture to direct the person to the hidden object.

Additional findings of interest were as follows. First, Panzee indicated specific types of nonfood objects (snake, balloon, paper) as well as food types. Second, the method did not involve any deliberate training. Even on Trial 1, Panzee used lexigrams, the gesture "hide," arm pointing and manual pointing toward the object, to recruit and direct the person. Third, Panzee indicated object types after overnight delays, including seven of the ten trials with experimentally imposed overnight delays. For example, on Trial 34 she touched the lexigram corresponding to the type of object without having seen the outdoor area in fifteen hours. I have replicated this specific finding in more than thirty additional trials. Finally, on some occasions Panzee did not recruit a person until two hours or more had passed, even when the

Table 24.1. *Lexigrams Panzee touched in the context of recruiting a person as a function of which object Panzee had been shown outdoors*

			Lexigrams Panzee used in context of gesturing to recruit uninformed person	
Trial	Object shown	Delay (in hours)	Indoor keyboard	Outdoor keyboard
1	Kiwi	0.17	*grape, kiwi*	–
2	Monster mask[b]	0.18	*monster mask*	–
3	Banana	3.77	*stick, banana, stick*	*banana, banana*
4	Grapes[a]	0.15	*grape*	*grape, grape, grape, grape*
5	Apple	3.33	*hide, apple*	*apple*
6	Coke[b]	1.17	*Coke*	*Coke*
7	Pineapple	0.85	*pineapple*	–
8	Blanket[a]	2.52	–	–
9	Orange drink (can)[b]	16.20	*Coke*	*orange drink, peach, Coke*
10	Strawberries[b]	0.88	*strawberry*	–
11	Snake[b]	1.62	*snake*	*snake*
12	String rope[b,c]	0.22	*surprise, surprise*	*Coke, Coke, string rope, Coke, stick, string rope*
13	Blueberries[b]	0.07	*blueberries*	*blueberries*
14	Melon[b,d]	0.22	*orange*	*melon, melon*
15	Raisins[a]	0.03	*raisin*	–
16	Peanuts and balloon[b]	0.17	*peanut*	*peanut, peanut, hide, balloon*
17	M & M candies[a]	15.62	*M & M, M & M*	*M & M, M & M*
18	Popsicle[b]	16.35	*surprise, hide, hide, hide, hide, surprise, hide*	*hide, hide, hide, stick, stick*
19	Pear	19.12	*stick, hide, hide, stick, stick, stick*	*stick, hide, stick, stick*
20	Orange	19.07	*orange* (×4); *melon, orange*	*orange*
21	Paper[a] and drawing pen[b]	90.17	–	*hide* (×4); *stick, hide, stick, hide, hide, hide, paper*
22	Kiwi	98.85	–	*hide, stick, hide, stick, stick, stick, hide, kiwi, kiwi*
23	Cherries[b]	0.15	*stick, cherries, cherries*	*cherries*

Table 24.1. (*cont.*)

			Lexigrams Panzee used in context of gesturing to recruit uninformed person	
Trial	Object shown	Delay (in hours)	Indoor keyboard	Outdoor keyboard
24	Peaches[a]	90.80	*hide*	*peach, hide*
25	Jelly[b]	15.17	*jelly, stick*	*jelly*
26	Cereal[b]	304.40	*stick, stick*	*peach, peanut, pineapple*
27	Yogurt[a]	22.50	*stick*	*yogurt*
28	Egg[a]	0.46	*stick, egg*	*stick, stick*
29	Jell-O gelatin[b]	22.25	–	*Jell-O*
30	Bread	0.88	*bread*	*bread*
31	Sugar[b]	19.97	*stick, stick, stick*	*stick, stick*
32	Juice[a]	0.05	*juice, juice*	*juice, juice*
33	Raisins[a]	0.35	*stick, stick*	*raisin*
34	Pear	15.75	*stick, stick, pear*	*pear, pear, pear, pear*

Note: Lexigrams are those that Panzee touched prior to the person finding the object. Dashes indicate that Panzee did not touch any lexigrams on the keyboard in question. Delay = time elapsed from Panzee seeing the object until she began to recruit the uninformed person.

[a] Lexigram accounted for 1 percent or less of Panzee's total lexigram touches in a twelve-day baseline sample.

[b] Lexigram rarely used by Panzee under routine conditions and never touched in a twelve-day baseline sample. Panzee had not seen strawberries, string rope, blueberries, melon, balloon or cherries in more than four months.

[c] When Panzee was offered the string rope, she ignored it and again touched *Coke* on the keyboard; the person interpreted this as a request for Coke.

[d] Panzee traditionally uses *orange, melon* to refer to melon.

object was a highly preferred type of food; the longest delay with correct reporting of object type exceeded ninety hours (Table 24.1).

Several aspects of Panzee's performance suggested that her memory involved a fairly rich information retrieval process. First, Panzee, rather than the experimenters, determined the exact time of reporting, and she reported items after extended delays. Second, Panzee selected lexigrams from the whole keyboard of 256 lexigrams, rather than from a small set of alternatives as in traditional primate matching-to-sample tasks. Third, she selected the lexigrams indoors, without an immediate view of the area in which the object was hidden. Fourth, she did not simply touch lexigrams; she pointed toward the outdoor area from the vicinity of the indoor keyboard, and she persisted in the interaction using a variety of response outputs until the person found the object outdoors. The findings strongly suggest that Panzee could recall which one of several dozen types of objects she had been shown from a distance (and had not been allowed to navigate to or to touch) for overnight periods of at least sixteen hours.

FURTHER TESTS OF RECALL WITH PANZEE

I have replicated these findings in additional experiments including variations in the quantity of items hidden, the types of items hidden, the modality of cue giving (direct view versus video representation), the modality of recruitment (direct pointing versus reporting on a video representation), and the outcome of recruitment (whether or not the item was removed and given to Panzee). I will briefly describe several of the main variations.

Use of indirect cues of object type

Panzee retained indirect signs of the type of object hidden. First, if the experimenter told Panzee in English the type of food that was inside an opaque container, Panzee touched the lexigram corresponding to the type of item hidden after delays ranging from several minutes to approximately one hour. (Panzee understands approximately 150 spoken English words; Beran *et al.* 1998.) Second, if Panzee saw an experimenter hide an opaque container in the forest with a lexigram on it corresponding to the type of food inside the container, she reliably retained the specific lexigram for one or more days.

Use of televised cues

Panzee used a video representation of the forest as a guide to locating objects in the forest. In these tests, an experimenter showed Panzee a brief videotape of a food-hiding event that had been filmed in the forest behind her enclosure. Each trial used a different location. On some of the trials, Panzee was confined indoors overnight immediately after she viewed the cue tape and could not see the forest for more than fifteen hours. The angle of the camera during the filming of the food-hiding event differed by up to 120 degrees from Panzee's familiar angle of view of the forest. On some trials, the video cue tape presented her with a perspective of the forest that she could not possibly have had from her outdoor enclosure. Some of the video cue tapes were in black and white rather than color. Results were that Panzee reliably recruited an uninformed person and directed him or her to the hidden food on the next day. Panzee performed accurately across a series of more than forty trials, including Trial 1, and including the black-and-white video trials. She also typically touched the lexigram corresponding to the type of item she had seen on the cue tape. Panzee also performed quite accurately when presented with a black-and-white still image that showed the experimenter touching a location in the woods.

Reporting of hidden objects on a video representation

Panzee could transfer in the reverse direction, from real life to television. If she watched the experimenter hide an object in the forest in real life, and the experimenter then locked her indoors overnight, she was able to report the exact location of the hidden object to an uninformed person the next day by touching the corresponding location on a televised image of the forest, using a stick or a joystick-controlled laser pointer. Panzee did this accurately and reliably in a series of more than twenty trials, including on Trial 1. Panzee also can use a laser pointer outdoors on an overcast day to pinpoint the location of hidden items in the woods (Menzel, 2005).

NEW ASPECTS OF THE STUDY

The key empirical issue of this and many other recent studies of animal memory is how information about the environment and about the animal's own past experiences is organized. The data strongly suggest that, for Panzee, information is to a significant extent organized independently of stimulation from the environment in question and constitutes recall as opposed to recognition. The "time in which chimpanzees live" remains an open question, but Panzee could certainly recall which one of at least thirty different types of objects she had seen outdoors, after delays of more than forty-eight hours.

This study presents several new features compared to other work on animal memory. The most obvious aspect is that Panzee used lexigrams to report hidden objects, and that she did so from locations visually separated from the place where the event occurred, after substantial time delays. Panzee transmitted multiple features of events, not just a single feature. In particular, she used lexigrams in conjunction with a video representation of the woods to convey both the type and the location of objects she had seen hidden in the woods on the previous day. The caregivers' inferences from Panzee's behavior about when, where and what new objects were hidden were highly accurate.

Secondly, evidence for generalized spatial skills came from the fact that Panzee used a video representation of the woods as a guide to locating objects in the woods, after overnight delays. Her use of televised spatial information exceeded that reported for other nonhuman organisms, including juvenile chimpanzees (E. Menzel *et al.*, 1978), baboons (Vauclair, 1990) and dolphins (Herman *et al.*, 1993), in regard to extracting spatial information from a video representation and using that information at a much later time in an outdoor area. What is new is the long timespan over which Panzee retained televised spatial information, the large number of different locations used in the test, her ability to withstand changes in visual perspective, her use of a video representation to report locations, and her labeling of the item type by lexigram. For a review of findings on chimpanzees' comprehension of scale models, see Chapters 21 and 22, this volume.

Another relatively novel aspect of this study compared to other research on animal memory concerns the physical separation between Panzee and the physical properties of the objects. During the information-gathering phase of each trial, Panzee was not allowed to handle the objects herself. She had to watch the object-hiding events from a distance. She did not always even see the hiding event directly; sometimes she saw only a video representation of the event. Later, she had to solicit the assistance of an uninformed person and then had to guide the person to the hidden object by pointing from a distance, or by using a stick to touch the corresponding location on a video representation of the forest. She was not allowed simply to take the person by the arm to the locations. Furthermore, she had to watch the object-removal events from a distance; and she did not always even see the removal event directly. As stated earlier, Panzee had not entered any forested area within six years prior to the onset of this study. Thus, her memory of the presence, type and location of test objects in the forest was based on observations of events from a distance. These aspects of her performance seem very different from memory studies using food-storing animals. In such studies, animals usually hide the items themselves and receive proprioceptive and close visual feedback from the location during the information-gathering part of the trial, both of which can improve memory. To paraphrase Nissen (1951, pp. 379–380) and Pavlov (1904), the more remote and indirect the basis of association between cues, objects and activities, the more likely we are to assume that some form of symbolic or sign process contributes to the problem solving.

Finally, an aspect of the study that is relatively unique, compared to most laboratory studies of human memory, as well as compared to animal studies, is that Panzee's recall and reporting of hidden objects were unprompted. In studies of human recall memory, the experimenter typically tells the participant when to provide information. Even in so-called "free-recall" tasks, the experimenter typically tells the person when to report. In this study, Panzee, rather than the human caregivers, initiated the exchanges of information about hidden objects, and she persisted in reporting a given location for days, if the search was interrupted by outside disturbances or if the person failed to find the hidden item. Putney (2007) characterizes Panzee's memory retrieval and her directing behavior in theoretical terms as aspects of future-oriented control of action at two levels. One of these levels is the direct control of fine-tuned, spatially oriented motor performance in the execution of plans. The other is the supervisory control of planning of upcoming actions within the working memory system. If we envision a continuum between recall and recognition, then Panzee's performance is farther in the direction of recall than any other animal performance of which I am aware. I do not claim that she is a genius, and it does seem quite conceivable that other animals might be capable of doing the same or analogous things, at least if they have been reared from infancy in appropriate environments. What I do claim is that if we return to the general issue of spatial representation and attempt to incorporate into it an improved psychological concept of recall memory in animals, here are some of the details we must include, assuming that Panzee is representative.

1. Memory can, to a significant extent, be independent from sensory feedback from the original environment in question.
2. Recall can be performed in different ways and in different situations.
3. Specific types of objects can be recalled after hour or days, not merely after a few seconds or minutes.
4. Recall and reporting can be unprompted. A chimpanzee can initiate exchanges of information about locations and the types of objects likely to be found there.

5. A variety of indirect signs can serve as inputs for long-term memory.
6. A wide range of distracting events can occur during the retention interval without causing an animal to forget hidden items.
7. A wide range of response outputs can be used to convey a given message.
8. The use of retrieved information can be strategic. A chimpanzee can adjust the time and place of recruitment to the ongoing activities of humans. A chimpanzee can modify the manner of recruiting according to outcomes, including whether the object is found or is not found.
9. Memory of object types as well as locations can be accurate even when a sizeable number of trials occur in succession (e.g., twelve trials in ninety minutes using 5-minute retention intervals, C. Menzel, 2005).
10. Memory of an initial object-hiding event can remain accurate after seeing multiple intervening object-hiding events. For example, Panzee showed accurate reporting of object types when presented successively with six different types of items, one at a time, in six different locations per trial.
11. A chimpanzee can sustain high accuracy of memory for object types and locations across multiple years of study in an outdoor area, despite hundreds of trials and the possibility of proactive interference.

CONCLUSIONS

A primate's knowledge of distant, remembered portions of its environment is no doubt much less acute than its visual and tactile perception of nearby objects (Dominy *et al.*, 2001). Nevertheless, the past few decades have furnished compelling evidence against the ancient belief that nonhuman animals are totally lacking in true symbolic capabilities and thereby are limited entirely to the "here and now." It seems clear that tamarins, baboons, mangabeys, macaques and apes sometimes discriminate locations and objects that are beyond seeing or hearing. The extent to which memory is independent of feedback from the immediate environment in question can be studied in a variety of ways, but the use of apes trained in an artificial language provides an especially powerful approach. Some apes at the Language

Research Center behave as if distant areas are real, worth communicating about, and structured in terms of objects. The distal, remembered environment, as we currently conceive it for symbol-competent chimpanzees (genus *Pan*), includes the types and locations of food- and non-food objects, signs, trails, landmarks and goals, and temporal properties of objects.

An important question is, of course, how representative such apes are of their species generally. It would be remarkable if apes in their natural habitat did not employ well developed and impressive recall memory capabilities in their daily foraging. A pressing problem is to characterize the operation of recall capabilities experimentally in situations that are large enough and that contain enough key structures to engage a chimpanzee's memory adequately, and to check these findings against everyday foraging patterns in the natural habitat while opportunities still exist.

ACKNOWLEDGMENTS

I thank John Kelley, Michael Beran, Mary Beran, Stephanie Berger, Betty Chan, Christopher Elder, Cameron Hastie, John Mustaleski, Isabel Sanchez and Shelly Williams for their help conducting the experiments with Panzee. Research on apes was supported by the L. S. B. Leakey Foundation, the Wenner-Gren Foundation, NSF grant SBR-9729485, and NIH grants HD-38051, HD-056352, MH-58855 and HD-06016.

REFERENCES

Beran, M. J., Savage-Rumbaugh, E. S., Brakke, K. E., Kelley, J. W. and Rumbaugh, D. M. (1998). Symbol comprehension and learning: a "vocabulary" test of three chimpanzees (*Pan troglodytes*). *Evolution of Communication*, **2**, 171–188.

Boesch, C. and Boesch, H. (1984). Mental map in wild chimpanzees: an analysis of hammer transports for nut cracking. *Primates*, **25**, 160–170.

Boesch, C. and Boesch-Achermann, H. (2000). *The chimpanzees of the Taï Forest*. New York: Oxford University Press.

Brakke, K. E. and Savage-Rumbaugh, E. S. (1995). The development of language skills in bonobo and chimpanzee: I. Comprehension. *Language and Communication*, **15**, 121–148.

Brakke, K. E. and Savage-Rumbaugh, E. S. (1996). The development of language skills in *Pan*: II. Production. *Language and Communication*, **16**, 361–380.

Byrne, R. W. (2000). How monkeys find their way: leadership, coordination, and cognitive maps of African baboons. In S. Boinski and P. Garber (Eds.), *On the move: how and why animals travel in groups* (pp. 491–518). Chicago: University of Chicago Press.

Conklin-Brittain, N. L., Wrangham, R. W. and Hunt, K. D. (1998). Dietary response of chimpanzees and cercopithecines to seasonal variation in fruit abundance. II. Macronutrients. *International Journal of Primatology*, **19**, 971–998.

Dominy, N. J., Lucas, P. W., Osorio, D. and Yamashita, N. (2001). The sensory ecology of primate food perception. *Evolutionary Anthropology*, **10**, 171–186.

Fragaszy, D., Johnson-Pynn, J., Hirsh, E. and Brakke, K. (2003). Strategic navigation of two-dimensional alley mazes: comparing capuchin monkeys and chimpanzees. *Animal Cognition*, **6**, 149–160.

Gallistel, C. R. (1990). *The organization of learning*. Cambridge, MA: MIT Press.

Ghiglieri, M. P. (1984). *The chimpanzees of Kibale Forest: a field study of ecology and social structure*. New York: Columbia University Press.

Goodall, J. (1986). *The chimpanzees of Gombe*. Cambridge, MA: Belknap Press of Harvard University Press.

Hampton, R. R. and Schwartz, B. L. (2004). Episodic memory in nonhumans: what, and where, is when? *Current Opinion in Neurobiology*, **14**, 1–6.

Herman, L. M. and Forestell, P. H. (1985). Reporting presence or absence of named objects by a language-trained dolphin. *Neuroscience and Biobehavioral Reviews*, **9**, 667–681.

Herman, L. M., Pack, A. A. and Palmer, M. (1993). Representational and conceptual skills of dolphins. In H. L. Roitblat, L. M. Herman and P. E. Nachtigall (Eds.), *Language and communication: comparative perspectives* (pp. 403–442). Hillsdale, NJ: Lawrence Erlbaum.

Köhler, W. (1925). *The mentality of apes*. New York: Liveright.

Kummer, H. (1995). *In quest of the sacred baboon: a scientist's journey*. Princeton, NJ: Princeton University Press.

Lorenz, K. (1971). *Studies in animal and human behaviour*, vol. 2. Cambridge, MA: Harvard University Press.

Marler, P. and Terrace, H. S. (1984). Introduction. In P. Marler and H. S. Terrace (Eds.), *The biology of learning* (pp. 1–13). New York: Springer-Verlag.

Menzel, C. R. (1999). Unprompted recall and reporting of hidden objects by a chimpanzee (*Pan troglodytes*) after extended delays. *Journal of Comparative Psychology*, **113**, 426–434.

Menzel, C. R. (2005). Progress in the study of chimpanzee recall and episodic memory. In H. Terrace and J. Metcalfe (Eds.), *The missing link in cognition: origins of self-reflective consciousness* (pp. 188–224). New York: Oxford University Press.

Menzel, C. R., Savage-Rumbaugh, E. S. and Menzel, E. W., jr. (2002). Bonobo (*Pan paniscus*) spatial memory and communication in a 20-hectare forest. *International Journal of Primatology*, **23**, 601–619.

Menzel, E. W., jr. (1973a). Chimpanzee spatial memory organization. *Science*, **182**, 943–945.

Menzel, E. W., jr. (1973b). Leadership and communication in young chimpanzees. In E. Menzel, jr. (Ed.), *Symposia of the fourth International Congress of Primatology, vol 1: precultural primate behavior* (pp. 192–225). Basel: Karger.

Menzel, E. W., jr. (1978). Cognitive mapping in chimpanzees. In S. H. Hulse, H. Fowler and W. K. Honig (Eds.), *Cognitive processes in animal behavior* (pp. 375–422). Hillsdale, NJ: Lawrence Erlbaum.

Menzel, E. W., jr. (1984). Spatial cognition and memory in captive chimpanzees. In P. Marler and H. S. Terrace (Eds.), *The biology of learning* (pp. 509–531). New York: Springer-Verlag.

Menzel, E.W., jr. and Halperin, S. (1975). Purposive behavior as a basis for objective communication between chimpanzees. *Science*, **189**, 652–654.
Menzel, E.W., jr. and Menzel, C.R. (2007). Do primates plan routes? Simple detour problems reconsidered. In D.A. Washburn (Ed.), *Primate perspectives on behavior and cognition* (pp. 175–206). Washington, DC: American Psychological Association.
Menzel, E.W., jr., Premack, D. and Woodruff, G. (1978). Map reading by chimpanzees. *Folia primatologica*, **29**, 241–249.
Mulcahy, N.J. and Call, J. (2006). Apes save tools for future use. *Science*, **312**, 1038–1040.
Muller, R.U., Stead, M. and Pach, J. (1996). The hippocampus as a cognitive graph. *Journal of General Physiology*, **107**, 663–694.
Nissen, H. (1951). Phylogenetic comparison. In S.S. Stevens (Ed.), *Handbook of experimental psychology* (pp. 347–386). New York: Wiley.
Pavlov I. (1904). Physiology of digestion. Nobel lecture. December 12, Stockholm.
Potts, R. (2004). Paleoenvironmental basis of cognitive evolution in great apes. *American Journal of Primatology*, **62**, 209–228.
Putney, R.T. (2007). Willful apes revisited: the concept of prospective control. In D.A. Washburn (Ed.), *Primate perspectives on behavior and cognition* (pp. 207–219). Washington, DC: American Psychological Association.
Roberts, W.A. (2002). Are animals stuck in time? *Psychological Bulletin*, **128**, 473–489.
Ruggiero, F.T. and Flagg, S.F. (1976). Do animals have memory? In D.L. Medin, W.A. Roberts, and R.T. Davis (Eds.), *Processes of animal memory* (pp. 1–19). Hillsdale, NJ: Lawrence Erlbaum.
Rumbaugh, D.M. (1977). *Language learning by a chimpanzee: the LANA project*. New York: Academic Press.
Rumbaugh, D.M. and Washburn, D.A. (2003). *Intelligence of apes and other rational beings*. New Haven: Yale University Press.
Sanz, C., Morgan, D. and Gulick, S. (2004). New insights into chimpanzees, tools, and termites from the Congo basin. *American Naturalist*, **164**, 567–581.
Savage-Rumbaugh, E.S. (1986). *Ape language: from conditioned response to symbol*. New York: Columbia University Press.
Savage-Rumbaugh, E.S., McDonald, K., Sevcik, R.A., Hopkins, W.D. and Rubert, E. (1986) Spontaneous symbol acquisition and communicative use by pygmy chimpanzees (*Pan paniscus*). *Journal of Experimental Psychology: General*, **115**, 211–235.
Savage-Rumbaugh, E.S., Murphy, J., Sevcik, R., Brakke, K.E., Williams, S.L. and Rumbaugh, D.M. (1993). Language comprehension in ape and child. *Monographs of the Society for Research on Child Development*, **58** (3–4, Serial No. 233).
Savage-Rumbaugh, E.S., Pate, J.L., Lawson, J., Smith, S.T. and Rosenbaum, S. (1983). Can a chimpanzee make a statement? *Journal of Experimental Psychology: General*, **112**, 457–492.
Savage-Rumbaugh, E.S., Rumbaugh, D.M. and McDonald, K. (1985). Language learning in two species of apes. *Neuroscience and Biobehavioral Reviews*, **9**, 653–665.
Schwartz, B.L. (2005). Do nonhuman primates have episodic memory? In H. Terrace and J. Metcalfe (Eds.), *The missing link in cognition: origins of self-reflective consciousness* (pp. 225–241). New York: Oxford University Press.
Shettleworth, S.J. (1998). *Cognition, evolution and behavior*. New York: Oxford University Press.

Suddendorf, T. and Corballis, M. C. (1997). Mental time travel and the evolution of the human mind. *Genetic, Social, and General Psychology Monographs*, **123**, 133–167.

Tinklepaugh, O. L. (1932). Multiple delayed reaction with chimpanzees and monkeys. *Journal of Comparative Psychology*, **13**, 207–243.

Tulving, E. (1983). *Elements of episodic memory*. Oxford: Clarendon Press.

Tulving, E. (1984). Precis of Tulving's *Elements of episodic memory*. *Behavioral and Brain Sciences*, **7**, 223–268.

Vauclair, J. (1990). Complex cognitive processes: study of mental performances in baboon. In J.-J. Roeder and J. R. Anderson (Eds.), *Primates: recherches actuelles* (pp. 170–180). Paris: Masson.

Wrangham, R. W. (1975). *The behavioural ecology of chimpanzees in Gombe National Park, Tanzania*. PhD dissertation, Cambridge University.

Wrangham, R. W. (1977). Feeding behaviour of chimpanzees in Gombe National Park, Tanzania. In T. H. Clutton-Brock (Ed.), *Primate Ecology* (pp. 503–538). London: Academic Press.

Wrangham, R. W., Conklin-Brittain, N. L. and Hunt, K. D. (1998). Dietary response of chimpanzees and cercopithecines to seasonal variation in fruit abundance. I. Antifeedants. *International Journal of Primatology*, **19**, 949–970.

Index

Page numbers in *italics* refer to figures and tables, those with suffix *n* refer to footnotes.

Printed in the United States
By Bookmasters